Global Issues in Water Policy

Volume 17

More information about this series at http://www.springer.com/series/8877

Steven Renzetti • Diane P. Dupont
Editors

Water Policy and Governance in Canada

Springer

Editors
Steven Renzetti
Department of Economics and Environmental Sustainability Research Centre
Brock University
St. Catharines, ON, Canada

Diane P. Dupont
Department of Economics and Environmental Sustainability Research Centre
Brock University
St. Catharines, ON, Canada

ISSN 2211-0631 ISSN 2211-0658 (electronic)
Global Issues in Water Policy
ISBN 978-3-319-42805-5 ISBN 978-3-319-42806-2 (eBook)
DOI 10.1007/978-3-319-42806-2

Library of Congress Control Number: 2016952625

Printed on acid-free paper

This Springer imprint is published by Springer Nature
The registered company is Springer International Publishing AG Switzerland

To our children, Alexandra and Nicholas,
who have provided much inspiration.

—Diane P. Dupont and Steven Renzetti

Foreword

In March 2000, the World Commission on Water in the twenty-first century chaired by Dr. Ismail Serageldin, vice president of environment of the World Bank, issued its report *A Water Secure World: Vision for Water, Life, and the Environment* (World Water Council 2000). The commission's vision was that every human being should have access to safe water for drinking, appropriate sanitation, and enough food and energy at reasonable cost. Providing adequate water to meet these basic needs would be done in a manner that works in harmony with nature.

However, the commission insisted:

> The future of hope that we sketched out in that Vision can come about only if there is wide adoption of the best existing technologies and practices, and rapid innovation and adoption of both institutional arrangements and new technologies. These institutional and technological innovations and changes will not come about unless water's economic, social, environmental, and political dimensions are adequately taken into account.

The commission report was speaking of the global challenge. Today nearly two billion people suffer from water insecurity defined as not having enough water to fulfill this vision. A recent study by the World Bank projects that, unless action is taken soon, water will become scarce in regions where it is currently abundant (such as Central Africa and East Asia) and scarcity will greatly worsen in regions where water is already in short supply (such as the Middle East and the Sahel in Africa). These regions could see their growth rates decline by as much as 6 % of GDP by 2050 due to water-related impacts on agriculture, health, and incomes (World Bank 2016).

Canadians attending meetings seeking to raise awareness to this issue and help find solutions are often confronted with a reference to the fact that we are "so lucky" to be a country with the highest level of water resource availability per capita. However, as the first chapter in this book points out, given the distance between where the water flows or is held and where most of the population is concentrated, water of sufficient quantity and quality is not always available for all Canadians. Furthermore, even many Canadians believe in this "myth of abundance," making it challenging, for example, to get them to reduce their per capita water use to levels comparable to other developed nations. Water quality also is an issue. In remote areas, especially ones serving indigenous peoples, drinking water supply systems are often similar to those found in developing nations, and ongoing boil water advisories exist (Office of the Auditor General of Canada 2005). Lake Winnipeg, the sixth largest freshwater lake in Canada, is reputed to be the largest eutrophic lake in the world (Lake Winnipeg Foundation 2014). Floods are frequent and droughts create major challenges, especially in the southern Prairie Provinces and the interior of British Columbia.

This book reviews where we stand in Canada in taking into account the economic, social, environmental, and political dimensions of technologies and practices of water management. The valuable Introduction to the book provides an overview of the contents, an excellent summary of the highlights of each chapter, and a review of the major policy issues addressed in the book. The first of seven parts describes the challenges facing managers of Canada's water. The second part places these in an international and transboundary context.

The part on economics which follows is important to me because it treats an issue which many consider the key to making change happen toward better water management: valuing and pricing water. As stated by Dupont and Adamowicz in *Water Valuation*:

> Canadian policies and regulations related to water at different levels have mostly failed to incorporate water value information aimed at promoting more efficient decision-making. In particular, the absence of information on the value of water in its alternative uses means that we are not maximizing the wellbeing of Canadian society from its valuable water resources. Ultimately, this can lead to misallocation of water resources and inefficient estimates in water quality.

The fourth part of the book treats issues of governance and management of water. It includes a description of the exemplary management of the Fraser River Basin in British Columbia. Included too is an excellent review of how indigenous sociocultural relations to water and traditional strategies of water resource use, monitoring, management, and protection could change our way of thinking about water governance. An example of such an approach in action could be *Northern Voices, Northern Waters: NWT Water Stewardship Strategy* (the Water Strategy), released in 2010 following a collaborative effort by the government of the Northwest Territories (GNWT), Aboriginal Affairs and Northern Development Canada (AANDC), and representatives from Aboriginal governments. Regulatory boards, agencies, environmental organizations, industry, academic institutions, and the general public were also involved in developing this water stewardship strategy (NWT Water Stewardship n.d.).

As the Introduction points out, given our recent history, drinking water safety may be Canada's most crucial water management issue. The need for coherent policies to address this across Canada is thoroughly explored in the four chapters of the fifth part. Part six presents case studies in the fields of energy and agriculture to illustrate how evolving technologies require policy interventions. Hydraulic fracturing for oil and gas production and management of farm-level decisions in the interest of the common good are the subjects treated.

Part seven concludes the book with two chapters quite different in approach. The first invites us to consider resilience, the tendency of a social-ecological system subject to change to remain within a stability domain, continually changing and adapting yet remaining within critical thresholds, as essential when planning and taking decisions. The second reminds us that the relatively new field of geomatics, including remote sensing and geographic information systems, provides us with a tool to acquire, store, manage, and visualize water data which can be used in water models and other decision-support tools.

Canadians should be concerned about management of our waters. An adequate reliable quantity of safe water is essential to our lives, and water of appropriate quality is essential to our livelihoods. Canada supported the adoption of the Sustainable Development Goals (Agenda 2030) at the United Nations General Assembly in September 2015 (UN 2015). Goal 6 is to *ensure the availability and sustainability of water for all.* The reality is that most of the other 15 Sustainable Development Goals cannot be achieved unless the water goal is achieved. This is as true in Canada as in the rest of the world. Thus, as de Loë points out in *Coordinating Water Policies: Necessary, but not Sufficient*, the most viable way to establish a more coordinated, strategic approach to water policy in Canada may be to shift attention to the ways in which water matters to other actors and sectors whose support is needed to achieve water policy objectives.

If Canadian policy makers take into account the economic, social, environmental, and political dimensions to which this book provides insight, we will move in the right direction in managing Canada's water. This is not just for ourselves. There are hundreds of millions who do not have enough water to meet their needs, no matter how well they manage what they have. We can help them to meet their needs by producing for them and at the same time reap economic benefits for Canadians. A Canada covered with healthy ecosystems will help to maintain a healthy planet.

Marseille, France William J. Cosgrove

References

Lake Winnipeg Foundation. (2014). *Harmful algae blooms on Lake Winnipeg*. http://www.lakewinnipegfoundation.org/harmful-algae-blooms-lake-winnipeg

NWT Water Stewardship. (n.d.). *Government of the Northwest Territories*. http://www.nwtwaterstewardship.ca/

Office of the Auditor General of Canada. (2005). *2005 September report of the commissioner of the environment and sustainable development*. http://www.oag-bvg.gc.ca/internet/English/parl_cesd_200509_05_e_14952.html

UN. (2015). *Sustainable development goals*. United Nations. http://www.un.org/sustainabledevelopment/sustainable-development-goals/

World Bank. (2016). *High and dry: Climate change, water, and the economy*. Water Global Practice. http://www.worldbank.org/en/topic/water/publication/high-and-dry-climate-change-water-and-the-economy

World Water Council. (2000). *A water secure world, vision for water, life, and the environment*. World Water vision commission report. http://www.worldwatercouncil.org/fileadmin/wwc/Library/Publications_and_reports/Visions/CommissionReport.pdf

Contents

Contributors

Wiktor L. Adamowicz Department of Resource Economics and Environmental Sociology, University of Alberta, Edmonton, AB, Canada

Zafar Adeel Pacific Water Research Centre, Simon Fraser University, Burnaby, BC, Canada

Julia Baird Research Associate, Environmental Sustainability Research Centre (ESRC), Brock University, Catharines, ON, Canada

Karen Bakker Department of Geography, Program on Water Governance, University of British Columbia, Vancouver, BC, Canada

Lalita A. Bharadwaj School of Public Health, University of Saskatchewan, Saskatoon, SK, Canada

Lori E.A. Bradford School of Public Health, University of Saskatchewan, Saskatoon, SK, Canada

Oliver M. Brandes POLIS Project on Ecological Governance and Centre for Global Studies, University of Victoria, Victoria, BC, Canada

Martin J. Bunch Faculty of Environmental Studies, York University, Toronto, ON, Canada

Heather Castleden Health, Environment, and Communities Research Lab, Department of Geography and Planning, Queen's University, University Ave, Kingston, ON, Canada

Ashlee Cunsolo Departments of Nursing and Indigenous Studies, Cape Breton University, Sydney, NS, Canada

Deborah Curran Faculty of Law and School of Environmental Studies, University of Victoria, Victoria, BC, Canada

Marius Cutlac Alberta Innovates Technology Futures, Edmonton, AB, Canada

Rob de Loë School of Environment, Resources and Sustainability, University of Waterloo, Waterloo, ON, Canada

Gemma Dunn Program on Water Governance, University of British Columbia, Vancouver, BC, Canada

Diane P. Dupont Department of Economics and Environmental Sustainability Research Centre, Brock University, St. Catharines, ON, Canada

Theresa Fresco Fraser Basin Council, Vancouver, BC, Canada

Jennifer Fresque-Baxter Department of Environment & Natural Resources, Government of the Northwest Territories, Yellowknife, NT, Canada

Graham A. Gagnon Centre for Water Resources Studies, Dalhousie University, Halifax, Canada

Dustin Evan Garrick Smith School of Enterprise and the Environment, University of Oxford, School of Geography and the Environment, Oxford, UK

Lars Hallström Resource Economics and Environmental Sociology, University of Alberta, Edmonton, AB, Canada

Sherilee Harper Department of Population Medicine, University of Guelph, Guelph, ON, Canada

Leila Harris Institute for Resources, Environment, and Sustainability (IRES), Center for Women's and Gender Studies (CWAGS), University of British Columbia, Vancouver, BC, Canada

Catherine Hart Health, Environment, and Communities Research Lab, Queen's University, Kingston, ON, Canada

B. Timothy Heinmiller Department of Political Science, Environmental Sustainability Research Centre, Brock University, St. Catharines, ON, Canada

Ted Horbulyk Principal Researcher – Economics, International Water Management Institute, Colombo, Sri Lanka

Jack Imhof Director of Conservation Ecology, Trout Unlimited Canada, Calgary, AB, Canada

Carolyn M. Johns Department of Politics and Public Administration, Ryerson University, Toronto, ON, Canada

Megan Kot Centre for Water Resource Studies, Dalhousie University, Halifax, NS, Canada

Katrina Krievins Masters Student, Sustainability Science and Society Program, Brock University, Catharines, ON, Canada

Steve Litke Fraser Basin Council, Vancouver, BC, Canada

David Marshall Fraser Basin Council, Vancouver, BC, Canada

Debbie Martin School of Health and Human Performance, Dalhousie University, Halifax, NS, Canada

Edward A. McBean School of Engineering, University of Guelph, Guelph, ON, Canada

Simon J. Mitchell St John River Advisor, WWF, Waterlution and the Canadian Rivers Institute, St. John River, NB, Canada

Bruce Mitchell Department of Geography and Environmental Management, University of Waterloo, Waterloo, ON, Canada

Madjid Mohseni Department of Chemical & Biological Engineering, University of British Columbia, Vancouver, BC, Canada

Michele-Lee Moore Department of Geography, and the Water, Innovation, and Global Governance Lab, University of Victoria, Victoria, BC, Canada

Karen Morrison Faculty of Environmental Studies, York University, Toronto, ON, Canada

Emma S. Norman Department of Native Environmental Science, Northwest Indian College, Bellingham, WA, USA

Nicholas Ovsenek College of Medicine, Department of Anatomy and Cell Biology, University of Saskatchewan, Saskatoon, SK, Canada

Ryan Plummer Environmental Sustainability Research Centre (ESRC), Brock University, St. Catharines, ON, Canada

Joanna Reid Independent Scholar, Vancouver, BC, Canada

Steven Renzetti Department of Economics and Environmental Sustainability Research Centre, Brock University, St. Catharines, ON, Canada

Manuel J. Rodriguez École supérieure d'aménagement du territoire et de développement régional (ÉSAD), Laval University, Ville de Québec, QC, Canada

Karena Shaw School of Environmental Studies, University of Victoria, Victoria, BC, Canada

Marian Weber Alberta Innovates Technology Futures, Edmonton, AB, Canada

Wanhong Yang Department of Geography, University of Guelph, Guelph, ON, Canada

Part I
Introduction and Background

Chapter 1
Introduction

Steven Renzetti and Diane P. Dupont

Abstract Canada is widely seen as a water-rich country as it possesses 7 % of the world's reliable flow of freshwater and has many of the world's largest rivers. Nonetheless, Canada's water resources face a number of significant challenges arising from population growth, natural resource –based developments, the looming implications of climate change, a growing reliance on large-scale irrigation, and a legacy of past laws and regulations unable to address these challenges adequately. This chapter sets out the objectives of this volume and summarizes the contributions made by each of its chapters.

1.1 Background and Objectives

Canada is widely seen as a water-rich country. It possesses 7 % of the world's reliable flow of freshwater and has many of the world's largest rivers. These features, however, are to a certain degree misleading as Canada faces a number of significant water-related challenges. These stem in part from supply-demand imbalances. Approximately 60 % of Canada's freshwater flows end in the Arctic Ocean while most of the population lives within 100 km of Canada's southern border. Furthermore, population growth, natural resource–based developments, the looming implications of climate change, and growing reliance on large-scale irrigation have created growing pressures on Canada's water resources – often in relatively arid areas such as Canada's prairie provinces. Additional challenges are related to water quality and stem from significant levels of nutrients leaving farming operations, ageing sewage treatment facilities and the toxic legacies of past industrial developments.

S. Renzetti (✉) • D.P. Dupont
Department of Economics and Environmental Sustainability Research Centre, Brock University, 1812 Sir Isaac Brock Way, L2S 3A1 St. Catharines, ON, Canada
e-mail: steven.renzetti@brocku.ca; diane.dupont@brocku.ca

S. Renzetti, D.P. Dupont (eds.), *Water Policy and Governance in Canada*,
Global Issues in Water Policy 17, DOI 10.1007/978-3-319-42806-2_1

The changing supply-demand water balance in Canada has played out against a backdrop of past water policies and governance models that have often been criticized as being inadequate, fragmented, and inefficient (Bakker 2007). The Canadian water policy landscape, however, has changed significantly in recent years. Some of these changes include: the decline of federal government involvement, major new policy initiatives by several provinces, the growth in decentralized governance with greater transparency and public participation, the growing recognition of the stark challenges associated with water in First Nations communities, the adoption of integrated water resources and basin-wide management as a guiding management principle, the growth in interest in economic instruments, and evolving challenges associated with transboundary management.

The objective of this volume is to provide readers with an insightful and critical assessment of the state of Canadian water governance and policy. Its scope is Canadian but its chapters adopt a variety of disciplinary and multidisciplinary perspectives, as well as having a focus upon different scales (local, basin, provincial and national). It seeks to examine water-related issues that are not only important for the future of Canadian water management but also provide insights into other topics in water management that are vital to many jurisdictions globally. These include transboundary management, non-market valuation of water, decentralized governance methods, and the growing importance of the role of Indigenous peoples.

1.2 Contents of Volume

The volume has 23 chapters and they are organized in seven parts. The first part 'sets the stage' by highlighting a number of challenges that are key to understanding the Canadian water policy experience. Mitchell (2016) identifies a number of the important hydrological and policy contexts that inform many of the subsequent chapters. In particular, Mitchell addresses the seemingly apparent abundance of water in Canada and argues:

> water of sufficient quantity and quality is not always available for all Canadians. Furthermore, a 'myth of abundance' has made it challenging to get all Canadians to reduce their per capita water use to levels comparable to other developed nations. (The Hydrological and Policy Contexts for Water in Canada).

Horbulyk reviews and critically assesses the central policy issue of water allocation throughout Canada. Based on his assessment of the challenges facing the current patchwork of allocation rules across Canada, Horbulyk (2016) comes to the important conclusion that:

> for most parts of the country it appears that the status quo will no longer apply, and that water shortages, and the effects of these shortages on water quality, will present a need to address policies for water allocation more carefully and comprehensively than has historically been the case. (Water Policy in Canada).

Brandes and Curran provide an overview of the Canadian legal system as it relates to water and then examine changes that have been taking place in water law in the province of British Columbia. These changes are aimed at recognizing explicitly the role of water in ecosystem functioning. Castleden and her coauthors (2016) then explore:

> current methods and models that are being used to integrate/implement Indigenous ontologies, epistemologies, and methodologies with Western ones to inform water resource management in First Nations, Inuit, and Métis communities, and to outline how such approaches would actually benefit all Canadians. (Reconciliation and Relationality in Water Research and Management in Canada: Implementing Indigenous Ontologies, Epistemologies, and Methodologies).

The complex set of relationships between Canadian water policy and the fate and future of Canada's Aboriginal peoples makes up an important theme in the volume that is revisited later in Parts II, IV, V and VI.

The second part places Canadian water policies and management approaches into an international perspective. Adeel's chapter compares Canada to other countries with similar hydrologic conditions and levels of economic development. An important point of comparison is the role played by national water agencies in other countries. Because of the absence of a national water agency in Canada, the chapter concludes by examining the potential role for such an agency in Canada. Garrick considers the experiences of Mexico, the United States and Canada with management of transboundary waters. Using the Rio Grande, Colorado and Columbia Rivers as cases, the author considers institutional reforms needed to provide enhanced water security in the face of the looming hydrologic impacts of global climate change. Norman and Bakker consider the historical evolution of Canada and the United States' efforts to manage transboundary water. They then examine how this relationship may be reshaped in the future by the growing importance of Aboriginal/Indigenous approaches to water management. Johns also examines transboundary management and focuses on the critically important issue of shared Canadian-American management of the Great Lakes that support 33 million people (including 9 million Canadians, about 25 % of the entire population of Canada). Johns (2016) concludes:

> clear transboundary and domestic water policy goals and transboundary and domestic policies at every level, policy implementation is critical and requires significant political will and resources. (The Great Lakes, Water Quality and Water Policy in Canada).

The third and fourth parts provide an appraisal of Canada's experiences with water management from an economic and institutional/political focus, respectively. The third part contains two papers that address economic dimensions of Canadian water policies. Dupont and Adamowicz consider the application of economic nonmarket valuation methods to Canada's water resources. Despite the potential for these methods to assist in a variety of types of decision-making, they find a history of rather uneven application of these methods to water resources management. They conclude by pointing to steps needed to enhance the use of these methods. Renzetti's findings regarding the state of water pricing mirror those of Dupont and Adamowicz.

Despite some recent advances, water prices at the municipal and provincial levels frequently fail to reflect the full social costs associated with water withdrawals (and resultant changes in water quality). Taken together, these two chapters illustrate the relatively limited role economics has played in Canadian water resources management in the past.

The chapters in the fourth part consider a number of issues related to the institutional and political dimensions of water governance. Heinmiller examines a central but often unappreciated aspect of water policy making and management. Namely, management is often the result of decisions taken in the political sphere. Heinmiller illustrates how the diversity of water management across Canada can be explained, in part, by the interplay between regional hydrologic conditions and local interest groups. In his chapter, de Loë espouses the view – long-held by many water researchers in Canada – that a central problem with Canadian water governance arises from a lack of policy coordination across jurisdictions. While acknowledging this has been the case in the past, the author cautions against seeing improved cooperation and coordination as a panacea for current management problems. de Loë argues that, while the principle of coordination may remain valid, its focus may need to be broadened so that water practitioners reach out to local agencies and actors not directly involved with water issues in order to ensure the efficacy of their efforts. Marshall and coauthors provide a very valuable overview of the evolution and current workings of the Fraser River Council. The Council is widely seen as one of Canada's 'success stories' as it has pioneered collaborative watershed governance over a large and complex river system. Bharadwaj and her coauthors continue the analysis by comparing models of water governance. The first type, favoured by governments, is infused with Western values and notions of rationality. The second type, promoted by First Nations scholars and communities, heavily relies on spirituality and interconnectedness. The authors explore opportunities for moving forward with approaches that bridge the two seemingly disparate models.

Part V includes several papers all addressing what may be considered the most crucial management issue: that of providing potable drinking water. The motivation for these chapters is Canada's distinctly mixed record regarding the provision of safe drinking water. As highlighted by Mohseni and coauthors (2016), there have been numerous long-standing boil-water orders in remote and First Nations communities, as well as sporadic instances of bacterial outbreaks. Indeed, recent research (Murphy et al. 2015) suggests that bacterial infections arising from drinking tap water in Canada exceed 100,000 cases annually. Furthermore, a number of authors point to the lack of legally enforceable national drinking water standards as an important contributing factor to the diversity of approaches to drinking water protection observed across provinces. Dunn, Harris and Bakker examine the uneven uptake of Canada's Drinking Water Quality guidelines across jurisdictions in Canada and conclude that Canada's highly decentralized approach has resulted in data gaps and urban-rural disparities in the provision of safe drinking water. Kot, Castleden and Gagnon focus on the province of Alberta which, in 2011, became the first North American government to require proactive, source-to-tap 'Drinking Water Safety Plans' (DWSPs) for all of its public water suppliers. In particular, the authors consider the challenges faced by small and rural water agencies to implement the DWSP and identify factors that could facilitate adoption in other jurisdictions in the future. Morrison and her coauthors (2016) explore the potential for enhanced coordination between two seemingly related policy issues: the provision of safe drinking water and the broader concerns of public health. The authors argue that enhanced coordination and a re-focusing of public health issues at the watershed level may hold promise to "reinvigorate policies and programs designed to protect the environment and promote human health and wellbeing" (Public Health at the Watershed Scale). The final chapter in this part (Chlorination of Drinking Water – Scientific Evidence and Policy Implications) grounds the discussion of providing potable drinking water in the realities of water treatment processes and the trade-offs inherent in chlorinating drinking water. The authors point to the growing challenge of acceptance of chlorination especially amongst First Nations communities and the policies and procedures needed to find working alternatives.

The next part has two chapters that focus on two high profile cases of pressing water quality issues. These are the challenges posed by hydraulic fracturing ('fracking') and large-scale agricultural operations. Moore and her coauthors (2016) survey the rapidly changing policy landscape that exists at the intersection of First Nations' water rights and the burgeoning industry seeking to exploit Canada's substantial unconventional energy deposits. These authors point out that the lack of clearly defined regulatory regimes for either First Nations water rights or fracking create a serious impediment to the development of sound water governance. They argue that, in addition to the need for water governance to be made "more transparent and accountable, and adequately resourced" (Patchy resources for the governance of Canada's resource patches: How hydraulic fracturing is illuminating the need to improve water governance in Canada), there is a strong need to ensure Indigenous peoples are fully engaged and that their rights are protected. Weber and her coauthor point out that agricultural operations impact water quality in every province of Canada and are most pointedly evidenced by the massive algal blooms observed in Lakes Winnipeg and Erie. There is evidence also that past programs and policies – often voluntary and fragmented – have not been particularly effective. As a result the authors conclude by identifying avenues for future policy development, which appear to be more promising.

The last part is forward looking as it deals with emerging perspectives and approaches that may lead to better water management policies. Plummer and his coauthors employ resilience analysis to assess governance at the watershed level. This type of analysis, adapted from ecological researchers, provides valuable insights into the ways in which an understanding of the 'co-evolution' of human society and complex ecosystems can inform improved watershed governance. Yang provides a novel perspective on water governance issues by highlighting the growing importance of geomatics in policy development and analysis. A number of trends including a growing emphasis on multidisciplinary approaches to water policy research and the availability of very large data sets all indicate that this will be an increasingly important approach to water policy work. Yang identifies a number of factors that will support this development in the future.

1.3 Major Policy Issues Addressed in Volume

The previous section provided thumbnail sketches of the chapters in this volume. Another way to consider these chapters is to highlight the themes that run through them. Doing so helps to identify what these leading scholars believe to be the major issues facing Canadian water management today and in the future.

1.3.1 Institutional Fragmentation

This is perhaps the most commonly cited issue across chapters. Many authors have pointed to the lack of coordination across federal, provincial and local level of governments. The direction of the fragmentation is both 'vertical' and 'horizontal'. The former is evidenced by the lack of coordination between the national and provincial governments. The latter is seen in the lack of coordination amongst provincial governments. These two types of fragmentation are seen in a wide range of specific policy areas including the establishment and enforcement of drinking water regulations, policies to address non-point source pollutants and the pricing of potable water at the municipal level. The prevailing consensus amongst the authors is that this feature of the Canadian water policy landscape has important and largely negative consequences for promoting sustainable water use. The implications of this lack of coordination are manifested in differing provincial rules coverage for water allocation frameworks, drinking water and water quality regulations that vary by province, and regulations for cost accounting and pricing that vary municipality to municipality.

It appears that much of the explanation for the lack of coordination and fragmentation of Canada's water policies lies in the withdrawal of successive federal governments from active engagement on water related issues. While the federal government remains active in areas clearly within its constitutional mandate (transboundary waters, waters on federal lands), a combination of budget cuts and shifts in priorities has meant that the promise found in the Federal Water Policy of 1987 has never been realized. As de Loë points out, since that time the provinces and territories have tried on their own to coordinate water policy nationally through the Council of the Federation with limited success. In addition, non-government organizations such as the Forum for Leadership on Water and the Canadian Water Resources Association have tried to motivate governments to take action with limited results.

The tragic events of Walkerton, Ontario in 2000 sharply focused attention on the need for better integration and flow of information across the many institutions that play a role in providing good quality and safe drinking water. E coli contamination in that community's drinking water ultimately led to 7 deaths and many hospitalizations (O'Connor 2002). While there is still an absence of Federal coordination, individual provinces have made efforts to improve decision making by local municipalities. Since that time, for example, the province of Ontario passed the Clean Water Act (2006). A direct outcome of this act was the creation of local Source (Water) Protection Committees that drafted source water protection plans that currently form the basis for decisions about good water management.

1.3.2 First Nations and Water

To the international reader, the number of chapters that examine First Nations issues related to water might be surprising. There are several reasons for this. First, the state of water quality in many remote and First Nations communities presents an immediate policy challenge. For example, many communities face frequent and continuous boil water orders associated with a lack of good quality drinking water (Dupont et al. 2014). Recently, the focus has shifted to potential damages from fracking developments (Expert Panel on Harnessing Science and Technology to Understand the Environmental Impacts of Shale Gas Extraction 2014). Second, as set out in the chapters by Bharadwaj and by Dunn, recent court decisions have clarified a number of First Nations' rights regarding land tenure. This development comes at a time of heightened pressure for natural resource development in many contested lands (with attendant concerns over impacts on water quality). This combination of events has accelerated efforts to clarify First Nations' rights to water and also to integrate First Nations communities into planning and policy processes in a number of jurisdictions. Third, from a scholarly perspective, there is an increased awareness amongst researchers of the limitations of modes of analyses that exclude insights derived from First Nations' knowledges.

1.3.3 Economic Dimensions of Water

One of the most dramatic developments in water resources management in the last two decades is the recognition of the economic dimensions of water use and the rapidly growing acceptance of the role for incentive-based policy approaches in promoting efficient and sustainable water use. These approaches include reforms to municipal water pricing, increased efforts to value water's ecological goods and services and the introduction of markets to address water quality challenges.

Despite the widespread acceptance of these approaches by transnational, national and local agencies and jurisdictions, these developments have largely bypassed Canada. With isolated exceptions, fees levied to directly access water are remarkably low, polluters typically are not compelled to pay charges that reflect the social costs associated with decreased water quality and municipal water prices are not yet sufficient to encourage efficient use while supporting sustainable infrastructure.

As the chapters by Dupont and Adamowicz and Renzetti demonstrate, there have been some improvements in municipal water prices and some efforts to employ nonmarket valuation methods in support of water management but these have been isolated initiatives rather than being part of a concerted national effort, in part due to the absence of the Federal Government in the water arena. Canadian efforts in this regard are very much in their infancy. However, it is unclear to what extent this is shaping or influencing actual water policies and government programs. Almost a three-quarters majority of people state that they believe that resources like the Great Lakes need to be protected for reasons associated with ecosystem services (e.g., protection of human health or habitat for wildlife and fish species) (Great Lakes Water Quality Board Public Engagement Work Group 2015). However, only 39 % state that they are willing to see their taxes increase in order to pay for improvements to drinking water systems (RBC Blue Water Project 2015) while 34 % would be willing to support higher prices that would lead to wastewater/stormwater infrastructure renewal.

1.3.4 Transboundary Issues and Canadian Water Management in International Context

It can be argued that one of Canada's 'success stories' related to water governance lies in the long history of cooperation and largely successful co-management with the United States of its boundary waters. Working together through the International Joint Commission, the two countries have brought the Great Lakes back from substantial ecological damage (although wetlands loss may be irreversible) and are meeting new challenges such as non-point source agricultural pollution. The countries are also showing a willingness to redirect institutions, e.g. there has been a fundamental shift at the IJC towards a focus on adapting to hydrologic impacts of climate change in Great Lakes and away from controlling and managing lake levels. Current directions undertaken for renegotiation of Columbia River Treaty show promise by including First Nations communities and placing much greater weight on ecological health of river than in the past. While these successes can be viewed in isolation, several chapters in the volume demonstrate that Canada can learn from other countries' experiences and approaches. This is particularly true if we look to other federal states which have achieved a greater degree of involvement by the national government as well as and greater coordination across jurisdictions.

1.3.5 Emerging Issues and Policy Responses

The sheer scale and rapidity of development of Canada's oil sands poses daunting challenges to Canadian water resource managers. Indeed, some analysts have argued that the combined efforts of provincial and federal regulators have thus far have failed to create a comprehensive monitoring and enforcement program. The equally rapid advent of hydraulic fracturing as a means of extracting hydrocarbons threatens to create at least as many challenges for Canadian governments (Expert Panel 2014). There are several reasons for this. The technology associated with 'fracking' is evolving rapidly. The impacts of fracking on groundwater resources are difficult to monitor and vary significantly with the methods used and the local hydro-geologic formations. Fracking is taking place in several regions of Canada and many of these are in areas where First Nations groups are contesting ownership.

References

Bakker, K. (2007). Introduction. In K. Bakker (Ed.), *Eau Canada: The future of Canada's water* (pp. 1–22). Vancouver: UBC Press.

Castleden, H., Cunsolo-Willox, A., Harper, S., Martin, D., & Hart, C. (2016). Reconciliation and relationality in water research and management in Canada: Implementing indigenous ontologies, epistemologies, and methodologies. In S. Renzetti & D. Dupont (Eds.), *Water policy and governance in Canada*. Springer.

Dupont, D., Waldner, C., Bharadwaj, L., Plummer, R., Carter, B., Cave, K., & Zagozewski, R. (2014). Drinking water management: Health risk perceptions and choices in first nation and non-first nations communities in Canada. *International Journal of Environmental Research and Public Health, 11*(6), 5889–5903.

Environment Canada. (1987). *Federal water policy*. Ottawa: Environment Canada.

Expert Panel on Harnessing Science and Technology to Understand the Environmental Impacts of Shale Gas Extraction. (2014). *Environmental impacts of shale gas extraction in Canada*. Ottawa: The Council of Canadian Academies. http://www.scienceadvice.ca/uploads/eng/assessments%20and%20publications%20and%20news%20releases/shale%20gas/shalegas_fullreporten.pdf. Accessed 18 Apr 2016.

Great Lakes Water Quality Board Public Engagement Work Group. (2015). *2015 Binational great lakes Basin poll*. Submitted to the International Joint Commission. http://ijc.org/files/tinymce/uploaded/WQB/WQB_GreatLakesPollReport_March2016.pdf. Accessed 19 Apr 2016.

Horbulyk, T. (2016). Water policy in Canada. In S. Renzetti & D. Dupont (Eds.), *Water policy & governance in Canada*. Springer.

Johns, C. (2016). The great lakes, water quality and water policy in Canada. In S. Renzetti & D. Dupont (Eds.), *Water policy and governance in Canada*. Springer.

Mitchell, B. (2016). The hydrological and policy contexts for water in Canada. In S. Renzetti & D. Dupont (Eds.), *Water policy & governance in Canada*. Springer.

Mohseni, M., McBean, E., & Rodriquez, M. (2016). Chlorination of drinking water – Scientific evidence and policy implications. In S. Renzetti & D. Dupont (Eds.), *Water policy and governance in Canada*. Springer.

Moore, M., Shaw, K., & Castleden, H. (2016). Patchy resources for the governance of Canada's resource patches: How hydraulic fracturing is illuminating the need to improve water governance in Canada. In S. Renzetti & D. Dupont (Eds.), *Water policy and governance in Canada*. Springer.

Morrison, K., Bunch, M. J., & Hallström, L. (2016). Public health at the watershed scale. In S. Renzetti & D. Dupont (Eds.), *Water policy and governance in Canada*. Springer.

Murphy, H. M., Thomas, M. K., Schmidt, P. J., Medeiros, D. T., Mcfadyen, S., & Pintar, K. D. M. (2015). Estimating the burden of acute gastrointestinal illness due to Giardia, Cryptosporidium, Campylobacter, E. coli O157 and norovirus associated with private wells and small water systems in Canada. *Epidemiology and Infection, 144*, 1355–1370. doi:10.1017/S0950268815002071.

O'Connor, D. (2002). Part One of the report of the Walkerton commission of inquiry. Queen's printer for Ontario.

RBC Blue Water Project. (2015). *RBC Canadian water attitudes study 2015*. http://www.rbc.com/community-sustainability/_assets-custom/pdf/CWAS-2015-report.pdf. Accessed 19 Apr 2016.

Chapter 2
The Hydrological and Policy Contexts for Water in Canada

Bruce Mitchell

Abstract The chapter focuses on the hydrological and policy contexts for water in Canada. Regarding the hydrological context, attention focuses upon available water quantity and quality related to the needs of humans and other living species; and, situations related to floods, droughts, wetlands and permafrost, especially in the context of anticipated climate change. Specific examples highlight the complexity and uncertainty involved. With regard to the policy context, consideration first is given to arrangements for the authority and responsibilities of the federal, provincial, territorial and municipal governments, as well as Indigenous peoples. Other aspects considered are the steady reduction in the federal commitment to, and engagement in, water illustrated by the federal water policy, flood damage reduction program, experimental lakes research area, and the 'war on science'. A final matter addressed is the concept of water as a basic human right, with attention to evolution of thinking both internationally and within Canada. It is essential to have data, knowledge and appreciation for the hydrological reality across the nation, as well as of actions at various spatial scales to facilitate or limit such understanding. Context for policy is also very important. In particular, it is important to appreciate policy choices for related aspects, such as climate change.

2.1 Introduction

This chapter has two purposes. First, it explores the hydrological context for Canada, with particular attention to water quantity and quality related to the needs of humans and other living species; and related to floods and droughts as well as to wetlands and permafrost. Second, attention turns to the policy context, especially the significance of roles by various levels of government, the steady disengagement of the federal government regarding water, and the issue of water as a basic human right.

B. Mitchell (✉)
Department of Geography and Environmental Management, University of Waterloo, Waterloo, ON N2L 3G1, Canada
e-mail: mitchell@uwaterloo.ca

S. Renzetti, D.P. Dupont (eds.), *Water Policy and Governance in Canada*, Global Issues in Water Policy 17, DOI 10.1007/978-3-319-42806-2_2

2.2 Hydrological Context

2.2.1 Water Quantity

Writing over 35 years ago, Foster and Sewell (1981) suggested that Canadians generally believed that their country enjoys a 'superabundance' of water, and observed that such a belief was based more on 'myth' than reality. Indeed, given the size of Canada, and its number of major and minor rivers and lakes, it is not surprising that many Canadians believe there is abundant water in their country, notwithstanding that Foster and Sewell identified six regions (Okanagan, Milk, North Saskatchewan, South Saskatchewan, Red-Assiniboine, and southern Ontario) for which current and anticipated future demands either exceeded or were within 20 % of total available monthly flow. As Sprague (2007, p 32) later summarized:

> Although Canada has a relatively large supply of water per capita, it does not necessarily have a large supply per region. Most Canadians live in the south of the country, far from some of the larger sources. Certain regions already have water shortages, and climate change could well exacerbate this situation.

Sprague (2007) explained how such a 'myth of superabundance' could persist. The main reason, he argued, is that many commentators, and members of the public, do not differentiate between the water in Canadian lakes, approximately 20 % of the amount of all lakes in the world, and the portion of that water which is renewable. He noted that the water in lakes is similar to the sum of all deposits in a bank account, whereas for sustainable use the account holder should only be withdrawing the interest, and leaving the principal intact. In this analogy, if only the interest, or the renewable supply is considered, then Canada has a bit over 6 % of the world's water supply, placing it in fourth place, after Brazil (1st with 12.4 %) and Russia (2nd with 10 %), and being tied with the same renewable volume (6.4–6.5 %) as Indonesia, the United States, and China (Sprague 2007, p 24).

2.2.2 Water Use

Shrubsole and Draper (2007, p 40) reported that annual per capita residential water use in Canada for 2001 was 335 l per day. For 2009, Environment Canada (2011) reported that the average residential water use had lowered to 274 l per person per day, and by 2011, had fallen to an average of 250 l per person per day. Notwithstanding the steady decline in annual average per capita daily residential water use, Canadians continue to be the second highest per capita users in the world, below only the United States. The steady decline has been due to improved technology, and water conservation programs, including a steadily increasing charge for water provided from municipalities to their users with the purpose of the charges being to move towards full cost recovery rather than achieve a profit from water sales.

Regarding variation in per capita residential use across the country, Environment Canada's (2011) data for 2009 showed that Prince Edward Island had the lowest per capita residential use at an average of 189 l, followed by Manitoba at 199 l and Alberta at 209 l. In contrast, the highest per capita residential use was in Newfoundland and Labrador (395 l/person/day), followed by New Brunswick (394), the Territories (390), Quebec (386), British Columbia (353), Nova Scotia (292), Saskatchewan (238), and Ontario (225). In a broader context, in environmental indicators using 1999 data, Canada was ranked 28th out of 29 OECD countries in terms of per capita daily water consumption. The lowest nations at that time were Luxembourg, followed by Denmark and the United Kingdom each tied for second place (Environmental Indicators 2015).

Environment Canada (2013, p 9) also reported that just over 30 % of Canadians depend on groundwater for municipal, domestic and rural use. Based on 1996 data, percentages ranged from 23.1 % in Alberta to 100 % in Prince Edward Island, with two thirds of groundwater users across the country being in rural areas. In terms of municipalities, the Kitchener-Waterloo-Cambridge area in southern Ontario and Fredericton in New Brunswick receive most of their water from aquifers. The users dependent on groundwater vary across the country, with municipalities being the largest users in Ontario, Prince Edward Island, New Brunswick and the Yukon; agriculture for livestock watering in Alberta, Saskatchewan and Manitoba; industry in British Columbia, Quebec, and Northwest Territories; and rural domestic use in Newfoundland and Labrador (Environment Canada 2013).

2.2.3 *Water Quality*

With regard to water use, it should be highlighted that at any given year for the past 5 years there have usually been about 100 communities under a 'boil water' advisory. Such advisories normally occur in relatively small and remote communities, many on First Nations reserves, for which serious challenges exist to provide, operate and maintain safe potable water supply systems. This matter is discussed in more general terms in Sect. 2.3.3 which focuses on 'water as a human right'. However, water quality problems are not confined to remote communities. A dramatic example was the boil water advisory declared for a short time for the 700,000 residents of Winnipeg in late January 2015, due to tests which indicated presence of *E. coli* in the public water system. Other earlier high profile experiences occurred at Walkerton, Ontario in 2000 and in North Battleford, Saskatchewan in 2001. In Walkerton, a community of 5000 people, 2300 became ill and 7 died due to the deadly bacteria *Escherichia* coli 0157:H7, or *E. coli.* in the water supply system (Perkel 2002), and in North Battleford thousands of residents became ill due to exposure to the parasite *cryptosporidium*, both due to malfunctioning water treatment systems. An inquiry following the Walkerton incident led to a recommendation to use a multi-barrier approach to ensure water quality, starting with initiatives to ensure quality of the aquatic ecosystems which become the source of domestic drinking water (O'Connor 2002a, b). In addition, it was recognized that ongoing provincial government cutbacks in Ontario had reduced monitoring and testing capacity to an extent that users of water systems had become at risk.

Another concern related to water quality is the potential for pollution of ground water due to LUST (leaking underground storage tanks), as well as from other sources such as landfill sites and industrial waste disposal sites. Regarding the issue of LUST, prior to the 1980s most underground storage containers were made from steel, which commonly begin to leak after 15–20 years. Furthermore, in many cases there are not accurate records of where LUSTs are located, and therefore problems do not become apparent until after water pollution is discovered. A pervasive and significant concern is pollution of aquatic systems from non-point sources. Such sources include road salt, manure, chemical pesticides and accidental spills. While point sources (e.g., sewage treatment plants, industrial plants) can and have been addressed systematically, non-point sources of pollution are much more challenging to resolve since so many individual land owners have to become engaged if there is to be progress.

An example highlights the challenge, and potential costs. In early 2015, the Supreme Court of Canada awarded a farmer near Brooke-Alvinston in Lambton County, Ontario, compensation of $100,000 for damage to his crops due to salt spread on nearby county roads. The farmer had argued that salt spread onto roads adjacent to his farm between 1998 and 2013 had caused significant crop losses and had reduced the overall value of the farm property. Provincial regulations stipulate how quickly and systematically municipalities must remove snow from roads, and municipalities can become liable if a motorist establishes that snow was not removed fast or thoroughly enough, which could lead to an automobile accident. However, regulations do not specify the amount of salt that should be used, nor what are acceptable amounts carried off onto nearby properties or environments. As a result of the Supreme Court decision, municipalities in Ontario have become more vulnerable. This challenge was noted by the president of the Good Roads Association in Ontario, who was quoted as saying the decision by the Supreme Court "sets a dangerous precedent for Ontario municipalities" (Herhalt 2015).

Transboundary waters provide an especially difficult challenge which requires collaboration. For example, the water quality of Lake Winnipeg has been seriously degrading with a primary cause being non-point source pollutants, such as nitrogen and phosphorus from agricultural land, triggering algal blooms. The Lake Winnipeg basin covers over 1 million square kilometers, and includes parts of Alberta, Saskatchewan, Manitoba and Ontario as well as North and South Dakota, Minnesota and Montana. A main contributor of such pollutants to the lake is the Red River which flows through Minnesota and North Dakota before crossing the international boundary into Manitoba. For a solution to be realized, the upstream riparian states need to work collaboratively with Manitoba as the downstream partner. And, of course, Manitoba will receive most of the direct benefits of improved quality in the Lake.

The idea of a Lake Friendly Accord was introduced in June 2013. The Accord's purpose is to facilitate coordination of action for a shared goal to improve water quality in Lake Winnipeg through reducing nutrient flows into the lake. The Manitoba provincial government began by partnering with local level governments to create a shared goal for reducing nutrient loadings into the south basin of the lake into the waterways, to develop commitments for specific initiatives, and to produce annual reports about the status of plans, actions and accomplishments. The Accord is intended to complement actions underway by the International Joint Commission, the Red River Basin Commission (a non-government organization), and the Prairie Provinces Water Board. Representatives of the Canadian Government, the Government of Manitoba and South Basin Mayors and Reeves signed the Accord on 21 March 2014, and the Red River Basin Commission on 4 September 2014. And, on 20 January 2015, the Commissioner of the Minnesota Pollution Control Agency signed the Accord in Winnipeg, indicating steady progress.

The Great Lakes system also has been under stress related to water quality, with highly visible pollution in Lake Erie during the 1960s leading to commentary that Lake Erie was 'dying'. Through the International Joint Commission, 43 Areas of Concern were identified in 1987 across the Great Lakes, leading to preparation of Remedial Action Plans for each one. Seventeen of the AOCs are in Canada, with 5 of them bi-national situations. By early 2015, 3 of the AOCs in Canada (Collingwood Harbour 1994; Severn Sound 2003; Wheatley Harbour 2010) and 2 in the USA (Oswego River, New York 2006; Presque Isle Bay, Pennsylvania 2013) had been officially delisted. Furthermore, two other AOCs in Canada (Spanish Harbour and Jackfish Bay) were designated in 1999 and 2011, respectively, as Areas of Concern in Recovery, signalling all remedial actions were judged to have been completed successfully, with time needed for those two AOCs to complete the recovery naturally.

2.2.4 Flooding

Up to 200 communities in Canada have experienced significant flooding since records have been kept. Below, examples are provided about major flood events. Caution should be used regarding the damage estimates, as it is not always clear in reports whether 'damages' mean total damages, insured damages and/or indirect impacts. The numbers provided below are understood to represent total damages, unless noted otherwise. Furthermore, the amounts reflect the costs at the time of the flooding events, not present values.

Major floods have occurred in the lower Fraser River valley of British Columbia (1948, $22 million in compensation paid, 16,000 people evacuated from homes), southern Manitoba (1950, nearly $125 million compensation, 107,000 people evacuated; 1997, about $300 million, 28,000 people evacuated), related to Hurricane Hazel in Toronto (1954, $100 million, 4000 left homeless, 81 deaths), Fredericton (1974, no data found), Montreal region (1976, $30 million in compensation), Winisk, Ontario (1986, entire community destroyed), Saguenay River Valley in Quebec (1996, 10 deaths and $800 million), Alberta (2005, 'Flood of Floods', $400 million), St. John River in New Brunswick, 2008 ($50 million), Richelieu River, 2011 ($40 million), Manitoba and Saskatchewan (2011, $1 billion and $360 million, respectively; 2014, no data available, but estimates 'greater than' damages for 2011 and over 920,000 acres unseeded due to inundation), and, Calgary and southern Alberta (2013, $6 billion, over 100,00 people evacuated) (Dearden and Mitchell 2016, p 394–395).

The traditional way to deal with flooding is to strive to control water flows using dams, dykes and diversion channels (Shrubsole 2016, p 158–159). While such a 'structural approach' does provide protection, it protects against a flood event of a specified return period, such as a 1 in 100 year flood. By definition, there will eventually be a 1 in 200 or 300 year flood, whose waters will overtop structural measures designed for a 1 in 100 year event. Experience has shown that once structural measures are built, the flood-prone land is often perceived to be 'safe', resulting in pressure for development on it. However, eventually a flood event of greater magnitude than the design capacity occurs, causing higher levels of damage because of the new and vulnerable development.

Subsequently, non-structural approaches began to be used in place of or to complement structural approaches. While structural measures are designed to control behaviour of the natural system, non-structural measures intend to modify human behaviour. In that context, non-structural measures include flood plain mapping, flood forecasting and warning systems, land use zoning regulations to restrict development in flood-prone areas, and public information.

To advance the application of a mix of non-structural measures, the federal and provincial governments began a national flood damage reduction program in 1975, with a primary goal to prepare flood-risk maps to identify hazardous areas in communities. The maps were complemented with land use regulations to discourage building in areas subject to flood risk. Over 300 locations were mapped under this program, and the information was available to agencies responsible for floodplain management, as well as to municipalities, developers and home owners. Of course, conditions in a river basin change, meaning that floodplain maps need to be updated to ensure they reflect the flood hazard.

The national flood damage reduction program was ended in 1998, as a cost-saving measure by governments, and by then many of the floodplain maps were seriously outdated. In some provinces, such as Ontario where Conservation Authorities have a lead role related to flood-prone land, work continued to maintain and update the flood-risk maps, but in many other provinces the maps simply became outdated.

2.2.5 Wetlands

Wetlands are defined by Environment Canada (2010, p 1) as areas:

> submerged or permeated by water – either permanently or temporarily – and are characterized by plants adapted to saturated soil conditions.

Wetlands are usually identified as five kinds: bogs, fens, marshes, shallow water, and swamps. They are particularly valuable because they absorb and reduce the impact of large waves or floods; filter sediments and toxic materials; provide nutrition and habitat for many fish, shellfish, shorebirds, waterfowl and furbearing mammals; provide food products (e.g., wild rice, cranberries, fish, wildfowl), energy (e.g., peat, wood), and building materials (e.g., wood); and, often become valuable recreational destinations for hunting, fishing and birdwatching (Environment Canada 2010, p 1). Indeed, because of these numerous positive characteristics, some wetlands have been designated for protection under international convention.

Canada has about 1.5 million km^2 of wetlands, about 14 % of the land mass in the country, and also about 25 % of the wetlands in the world. Of the total wetlands in Canada, 37 of them covering about 131,000 km^2 have been designated as of international importance (Biodivcanada 2014).

However, when European settlers arrived in what was to become Canada, the normal view was that wetlands were a nuisance and should be drained or infilled to facilitate agriculture or other settlement activity. While a systematic national inventory of wetlands in Canada does not exist, it has been estimated that between the early 1800s and 1991 more than 200,000 km^2 of wetlands had been lost (Biodivcanada 2014, p 2), primarily through drainage to create more agricultural land. The most complete data for wetlands are for the Prairie Provinces and southern Ontario.

While some increases in wetlands have occurred, losses continue primarily due to their conversion to other uses, construction of water level control infrastructure (including hydroelectric dams and reservoirs), and climate change, as well as due to their fragmentation resulting from various types of development, as well from pollution, invasive species, grazing and recreational activity. Biodivcanada (2014, p 2) has noted that wetlands are particularly vulnerable when they are in or adjacent to large urban areas, and it estimated 80–98 % of wetlands once in or adjacent to major urban areas no longer exist. In contrast, at a global scale it estimated wetlands now cover 5–10 % of the Earth's land surface, even though more than 50 % of wetlands have been lost globally.

Climate change is an obvious threat to wetlands. Especially in arid or semi-arid areas, if and as summers become drier and warmer, combined with more extraction of water for irrigation, the probability increases that the basic supply of water for wetlands, whether surface or ground water, will be reduced. As water volumes decline, then concentrations of pollutants, such as naturally occurring salts or human-induced agricultural chemicals and atmospheric pollutants, are likely to rise. Modest changes in temperatures and/or water supply usually significantly affect biota in wetland systems.

In terms of protecting wetlands, various strategies are used. Specific initiatives for coastal wetlands include adding sediment to counterbalance rising sea levels, planting grasses to protect coastal sand dunes from erosion, and building dykes or protective offshore barrier islands. For inland wetlands, planting grasses also is used, as well as artificial control of water levels. For both coastal and inland wetlands, developing and applying protection policies is also needed.

The North American Waterfowl Management Plan (NAWMP) was signed by Canada and the United States in 1986, with the purpose to protect wetlands in North America. The stimulus was a rapid decline in waterfowl populations as a result of removal of or damage to wetlands. Indeed, it has been calculated that the many small wetlands on the prairies of Canada and the US are the most productive habitat for waterfowl at a global scale, supporting between 50 and 88 % of the breeding populations of several species in North America (Biodivcanada 2014). Mexico became a partner in NAWMP during 1993.

The North American Waterfowl Management Plan (2012) was revised in 2012. The new plan noted that NAWMP had conserved and restored 63,000 km^2 of wetlands, grasslands and other key habitats for ducks, geese and swans in the three countries since 1986. The revised plan has three key goals, one of which is that to achieve waterfowl conservation it is essential to have "Wetlands and restored habitats sufficient to sustain waterfowl populations at desired levels, while providing places to recreate and ecosystem services that benefit society" (p 2), and a key strategy is to achieve "healthy wetland ecosystems that sustain natural functions (e.g., water quality, flood control, carbon storage) that benefits people and wildlife" (p 6).

2.2.6 *Drought*

Droughts are challenging to define, as unlike floods, it is not always obvious when a drought begins or ends. However, there is general agreement that a drought occurs when lower than normal precipitation leads to sharply reduced water supply, which in turn leads both to reduced moisture in soil and replenishment of surface and groundwater sources. In Canada, regions most susceptible to drought are the three Prairie provinces, as well as the interior of British Columbia. Southern Ontario and Quebec also experience droughts, but generally their droughts are less severe, cover a smaller area, and are shorter in duration than droughts on the Prairies. In contrast, both Atlantic Canada and the Canadian North experience relatively few drought conditions.

Records show significant droughts occurred on the Prairies in the 1890s, 1910s, and from 1929 to 1937. Other droughts occurred there in the late 1950s and early 1960s, 1980s and from 1999 to 2005, with the latter being the worst in 100 years for the Prairies (Chipanski et al. 2006). The drought period during 2001 and 2002 covered almost the entire southern portion of Canada, from British Columbia to Atlantic Canada.

The drought in the 1930s led to establishment of the Prairie Farm Rehabilitation Administration (PFRA) by an Act of Parliament, passed in 1935. The PFRA was created to facilitate technical and financial assistance to farmers in order to allow them to construct small water holding areas, such as dugouts and small dams. In addition, PFRA acquired sub-marginal agricultural land, and, after seeding the land, turned it into community pastures. Initially, PFRA established 16 community pastures, ranging in size from 2430 to 10,120 ha. It presently operates over 80 community pastures, with 75 % being in Saskatchewan. The landowners of the submarginal land were assisted in obtaining other property.

In 1959, PFRA initiated the South Saskatchewan River Project, which resulted in the Gardiner and Qu'Appelle Dams, as well as the 225 km long Diefenbaker Lake. Water from this project supplies drinking water to about 50 % of the population in Saskatchewan, and also a significant amount of water for irrigation.

In 2013, as part of its cost-cutting, the federal government terminated funding for the shelterbelt program that had been operated for decades by PFRA, and through which farmers had obtained free seedlings of trees and shrubs to create shelterbelts.

In addition to PFRA, each of the three Prairie provincial governments developed their own programs, and collaborates with the federal government. Generally, the focus has been on adaptation strategies, including soil and water conservation, irrigation, and infrastructure (e.g., wells, reservoirs, pipelines).

2.2.7 Permafrost

Permafrost involves ground which remains at or below 0° Celsius for at least 2 years. Permafrost areas can be continuous, discontinuous, in isolated patches, or in alpine systems. Permafrost is often covered by a thin layer of soil that regularly melts and freezes.

It has been estimated that permafrost underlies 50 % of Canada's land mass. Turner (2013) has reported that, in the North, permafrost is mainly continuous and may be several 100 m thick, with temperatures averaging −5 ° C. In contrast, in the South, the permafrost areas are often discontinuous and often only a few metres deep.

Permafrost, similar to other natural systems, is susceptible to modification due to climate change. In that context, the International Panel on Climate Change 5th Assessment Synthesis Report (2014, p 6) stated that:

> There is high confidence that permafrost temperatures have increased in most regions since the early 1980s in response to increased surface temperatures and changing snow cover.

Such vulnerability is important, because of inter-connections among permafrost, hydrological processes, soil conditions and vegetation (Smith 2011, p 3).

To elaborate, changes in permafrost environments have implications for both terrestrial and aquatic systems. Related to hydrology, Smith (2011) reminds us that permafrost influences infiltration, surface runoff and aquifer recharge and storage. Furthermore, both frozen ground and the thickness of an active layer of soil influence the hydrologic cycle by affecting evapotranspiration, as well as vegetation succession and growth because of impact on rooting zone depth and soil moisture capacity. In addition, infrastructure on the surface may be damaged when permafrost thaws, as a result of 'thermokarst development'. Such damage can include cracking or breaking of pipelines carrying water, oil or natural gas, and structures built to contain mine tailings and other wastes, as well as buildings, schools and factories.

Furthermore, frozen peatlands retain significant quantities of carbon, and thawing of permafrost can have consequences for capacity to store carbon from greenhouse gas emissions. Finally, changes to permafrost may lead to drier terrestrial conditions, in which shrubs and tundra replace wetlands, with implications for the number and mix of plants, birds and animals.

While relatively small human populations live in the areas underlain by continuous permafrost, the consequences for the hydrological cycle of changing permafrost conditions are significant, especially given the potential for mineral and energy development projects in Canada's North.

2.3 Policy Context

2.3.1 Role of Different Levels of Government

As a federal state, in Canada power and authority are shared between the federal and provincial governments. In turn, provincial governments allocate authority to municipal governments. Important for the policy context is the Canada Act, 1982, which distinguishes between proprietary rights and legislative authority regarding natural resources. Proprietary rights, or ownership, are held by provincial governments for all Crown lands and natural resources not specifically held under private ownership, within their jurisdictions. North of 60° latitude, the federal government has proprietary rights to land and resources until territories gain the same status as a province. As a result of devolution of authority in the past decade and a half, however, both the Yukon and Northwest Territories now have more 'provincial-type' powers and responsibilities related to water.

Legislative authority is divided between the federal and provincial governments. Statutory jurisdiction for trade and commerce is held by the federal government, allocating to it significant authority related to interprovincial and export trade of resources, including water. The federal government's legal authority for navigation and shipping, and fisheries, has been used as the basis for federal water pollution regulations, even though water is a provincial responsibility under the Canada Act. Such arrangements inhibit *national approaches* (federal, provincial, and territorial collaboration) to address water issues, given provincial sensitivities about their authority and responsibility under the Canada Act being infringed upon. At the municipal level, responsibility focuses on providing potable water for residents as well as removing and treating wastewater.

Aboriginal people, not all of whom are First Nations, increasingly have argued that they had, and have retained, traditional rights to natural resources, even if treaties had been signed with the federal government. During the 1980s, increasing recognition began by the federal government regarding traditional rights of Aboriginal peoples. Existing and treaty rights were formally recognized in the Canada Act of 1982, which has led to Aboriginal peoples becoming more involved in decisions related to water development projects located on their traditional areas.

The implications of the above arrangements are non-trivial (Pentland and Wood 2013). It often is difficult for one level of government to take a leadership role, or to act unilaterally, as other levels can claim they have shared jurisdiction. Consequently, some governments or leaders avoid taking action, or initiatives, arguing that they do not have sufficient responsibility or authority, and thus others should be acting. And, finally, an obvious need exists for collaborative and cooperative approaches, but often, it seems, nothing or little happens, or it takes an exceedingly long time for initiatives to get started, because of the often difficult negotiations that extend over a lengthy period.

2.3.2 *Disengagement by the Federal Government*

Engagement in water policy and management by the federal government reached a high point in 1987, when a federal water policy was created (Pearse et al. 1985; Environment Canada 1987). The intent was to clarify goals and actions by the federal government to facilitate efficient and equitable development and use of water through cooperative programs, provision of information, development of expertise, and technology development and transfer. However, by the mid-1990s, the federal government had reduced its engagement related to water (Bruce and Mitchell 1995). The Inland Waters Directorate in Environment Canada was eliminated, and its staff was re-assigned to other divisions. An Interdepartmental Committee on Water also became inactive. As a result, the federal water policy "lost much of its momentum" (Bruce and Mitchell 1995, p 2). Furthermore, funding for the Canada Water Act, 1970 was significantly reduced.

Previously, it was mentioned that the National Flood Damage Reduction Program received less and less funding from the federal government, and stopped being funded in 1998. The federal government has also shown lack of commitment to water in other ways, since that time. For example, in 2012 the federal Conservative government announced it would stop funding the Experimental Lakes Research Area in northwestern Ontario, an area of 58 lakes in which pioneering research had been completed since 1968 on issues such as eutrophication of freshwater lakes due to phosphorus overloading, and the impact of acidification of freshwater systems from acid rain. These two research projects created new and internationally recognized insight and solutions to address the negative impacts from eutrophication and acid rain. Regarding eutrophication, Clancy (2014, p 87) stated that the research:

> …led to recommendations that made Canada the first country to ban phosphates from laundry detergents (which accounted for more than half the phosphorus supply in many lakes) and the compulsory removal of phosphates from municipal sewage discharges into lakes.

The experimental lakes continue to function as a research site only because in 2013 the Ontario provincial government committed funding, and the International Institute of Sustainable Development in Winnipeg took over management of the facility.

In the summer of 2012, the Conservative government introduced omnibus Bill C-38, in which it proposed major changes to the Fisheries Act in order to focus attention on a small number of 'valuable' freshwater fish species; repealed the Environmental Assessment Act; and, significantly amended the Species at Risk Act as well as the Navigable Waters Protection Act, with the latter having the number of water bodies under the authority of the Act reduced from over 3 million to 169 (Turner 2013, p 27).

The above actions by the federal conservative government, and its insistence that federal researchers should not present research findings at conferences unless approved by the federal government, nor respond to questions from the media about research without a federal government spokesperson present to determine which questions could be answered, led Turner (2013) to conclude that the federal government had engaged in a 'war on science'. In Turner's (2013, p 112) view, the role of science was viewed by the federal government as:

> …to create economic opportunities for industry, and the purpose of government is to assist in that process in whatever way it can.

Furthermore, he added, at the core of the federal government's approach was to be 'willfully blind', and also to have an:

> abiding mistrust of expertise and … contempt for any kind of science not being applied directly to an economic activity of immediate benefit to Canadian industry and self-evident appeal to Conservative voters.

Thus, the approach of the federal conservative government was to reduce funding for many kinds of environmental research, and to be selective regarding what kind of research is funded and reported publically. In that regard, water per se had not been the target for less funding. However, the federal election in October 2015 resulted in the Liberal party forming the government. Prime Minister Trudeau indicated that science should inform decisions by the federal government, and also that government scientists should have the right to speak out about their research. Thus, a shift in approach at the federal level may occur.

2.3.3 Water as a Human Right

By the end of the twentieth century, the global population had grown by a factor of 3 and water consumption by a factor of 6. And, by the middle of the second decade of the twenty-first century, almost 800 million people did not have access to safe potable water, and about 2.6 billion did not have access to adequate sanitation facilities for human waste. Furthermore, about 1.5 million children under 5 years of age died annually from water-related illnesses. Given such reality, an important policy question is whether water should be a basic human right, and, if so, what role should countries such as Canada play in realizing such a goal both domestically and internationally (Matthews et al. 2007; Clancy 2014, p 193–194). In Sect. 2.2.3, it already was noted that there are communities in Canada that do not have safe and reliable drinking water. Related policy questions are what rights should other living species beyond humans have related to water, and also what rights should future generations have?

In reflecting on the above policy matters, the approach taken by South Africa is instructive. The Constitution of the Republic of South Africa, 1996, Section 27(1) (a) stipulates the right to basic access to food and water, and also the right of the environment to enough water to meets its requirements. More specifically, it is stated that,

> The quantity, quality and reliability of water required to maintain the ecological functions on which humans depend shall be reserved so that the human use of water does not individually or cumulatively compromise the long term sustainability of aquatic and associated ecosystems.

Thus, South Africa has explicitly recognized the right to water for both humans and other living species.

At a global scale, the United Nations Human Rights Council considered a proposal for water and sanitation to be recognized as a basic human right three times in the 6 year period 2002–2008. Canada opposed this proposal, along with Russia and the United Kingdom. It appears as if at least four issues influenced Canada's position at the Human Rights Council. First, concern existed related to NAFTA (North American Free Trade Agreement). If Canada supported water as a basic human right, how could it oppose requests from the United States or other countries to have water exported from Canada by pipeline or ship? Second, what would be liability implications related to boil water advisories in Canada? The federal government was not keen to assume liability for inadequate water supply systems in remote northern communities, many of which are Aboriginal communities for which the federal government has responsibility. Third, if water became a basic human right, what other matters might be proposed to become a right? And fourth, related to the first point, what would be implications for national sovereignty, and Canada's right to choose how and when its resources were developed and extracted?

The Universal Declaration of Human Rights, adapted by the UN General Assembly in December 1948, was the first global expression of rights for all human beings. The Declaration in the UN Charter, is binding on all member states, and contains 30 articles. Article 25 states that:

> Everyone has the right to standard of living adequate for the health and well-being of himself and of his family, including food, clothing, housing and medical care and necessary social services,…

Striking by omission in Article 25 is the word, 'water'. How could food be included and water be omitted, when conventional wisdom is that a healthy adult in normal weather conditions would live for 3–5 days without water, but for about 3 weeks without food?

At a meeting of the UN General Assembly on 28 July 2010, the following motion was proposed:

> ….states and international organizations to provide financial resources, build capacity and transfer technology, especially to developing countries, in scaling up efforts to provide safe, clean, accessible and affordable drinking water and sanitation for all.

The motion was passed by a vote of 122 countries in favour, 0 against, and 41 abstentions. Canada was among the countries which abstained. Others abstaining included Australia, Austria, Denmark, Ireland, Israel, Japan, Netherlands, New Zealand, Sweden, Turkey, United Kingdom, and the United States along with Botswana, Ethiopia, Guyana, Kenya, Poland, Republic of Korea, Romania, Slovakia, Trinidad and Tobago, Tanzania, and Zambia. For many abstaining nations, including Canada, the rationale was that the motion was premature, given the UN Human Rights Council was still examining this matter.

Two years later, at the Rio+20 Summit in June 2012, in the final report from the Summit, entitled *The Future We Want: Outcome of the Conference*, the following statement appears in Section 121:

> We reaffirm our commitment regarding the human right to safe drinking water and sanitation, to be progressively realized for our population with full respect for national sovereignty.

This statement, along with the vote at the UN General Assembly in 2010, are often cited as evidence that the international community has recognized water as a basic human right. However, careful reading indicates that at the UN General Assembly the nations supported provision of assistance to help, especially for developing countries, to achieve safe and accessible drinking water and sanitation for all. Furthermore, the statement after the Rio+20 Summit was explicit that national sovereignty of nations must be respected, reflecting concern that water-bountiful countries had concerns about other countries pointing to the UN resolution as giving them the right to access to water in another country, a concern identified above as one shared by Canada.

From a policy perspective, an obvious question arises as to what actions the federal and provincial governments will take to respect the 2010 motion and 2012 declaration, given they are binding on all members. Action should be expected at two levels by Canadian governments: (1) to ensure all Canadians have access to safe water and sanitation, and (2) to contribute to support developing nations which require external assistance if the intent of the UN resolution is to be achieved. To date, little has been said by political leaders at federal and provincial levels in Canada as to what will be done.

2.4 Implications

In terms of the hydrological context, there is no doubt that Canada is well endowed with fresh water. However, given where the water flows or is held and where most of the population is concentrated, water of sufficient quantity and quality is not always available for all Canadians. Furthermore, a 'myth of abundance' has made it challenging to get all Canadians to reduce their per capita water use to levels comparable to other developed nations. Water quality also is an issue. In remote areas, especially ones serving Aboriginal peoples, water supply systems are often similar to those found in developing nations, and ongoing boil water advisories exist.

Flooding damages continue, often due to historical development on flood-prone lands. Wetlands continue to be removed, even though Canada has one quarter of all the wetlands in the world. Droughts create major challenges, especially in the southern Prairie Provinces, and the interior of British Columbia. And permafrost areas, which underlie 50 % of the Canadian land area, are susceptible to climate change.

From a policy context, the division of responsibilities and authority among federal, provincial, territorial and municipal governments, and increasingly including First Nations, creates a significant challenge to achieve an integrated, coordinated and collaborative national approach to many water problems. This challenge has been exacerbated as the federal government had steadily reduced its involvement in water matters since the early 1990s. And, the issue of water as a human right should be a real issue for Canada because of inadequate water supply and sanitation systems for some of its communities, and the UN commitment to help developing countries ensure all residents have access to safe potable water for drinking and have capacity to remove human wastes safely. On the issue of water as a human right, both federal and provincial governments have been conspicuously silent.

References

Biodivcanada. (2014). *Wetlands*. http://www.biodivcanada.ca/default.asp?lang=En&n_F07D520A-1. Accessed 25 Jan 2015.

Bruce, J., & Mitchell, B. (1995). *Broadening perspectives on water issues* (Incidental report no IR96-1). Ottawa: Royal Society of Canada.

Chipanski, A. C., Findater, K. M., Hadwen, T., & O'Brien, E. G. (2006). Analysis of consecutive droughts on the Canadian Prairies. *Climate Research, 30*(3), 175–187.

Clancy, P. (2014). *Freshwater politics in Canada*. Toronto: University of Toronto Press.

Dearden, P., & Mitchell, B. (2016). *Environmental change and challenge: A Canadian perspective*. Don Mills: Oxford University Press.

Environment Canada. (1987). *Federal water policy*. Ottawa: Environment Canada.

Environment Canada. (2010). *Wetlands*. https://www.ec.gc.ca/eau-water/default.asp?lang=En&n-24147C37-1. Accessed 25 Jan 2015.

Environment Canada. (2011). *2011 Municipal water use report – Municipal use 2009 and atatistics*. https://www.ec.gc.ca/doc/publications/eau-water/COM1454/survey2-eng.html. Accessed 19 Jan 2015.

Environment Canada. (2013). *Groundwater*. https://www.ec.gc.ca/eau-water/default.asp?lang=En&n=300688DC-1. Accessed 25 Jan 2015.

Environmental Indicators. (2015). *Canada vs. The OECD: An environmental comparison – Water: Water consumption*. www.environmentalindicators.com/hldocs/indicators/6wate.html. Accessed 19 Jan 2015.

Foster, H. D., & Sewell, W. R. D. (1981). *Water: The emerging crisis in Canada*. Ottawa: Canadian Institute for Economic Policy.
Herhalt, C. (2015). Region considers the implications of successful lawsuit over salt runoff. *The Record*, 23 January, B1.
International Panel on Climate Change. (2014). *Fifth assessment synthesis report, Summary for policy makers*. http://ipcc.ch/pdf/assessment-report/ar5/syr/SYR-AR5-LONGERREPORT-Corr2.pdf. Accessed 26 Jan 2015.
Matthews, C., Gibson, R. B., & Mitchell, B. (2007). Rising waves, old charts, nervous passengers: Navigating toward a new water ethic. In K. Bakker (Ed.), *Eau Canada* (pp. 335–358). Vancouver: UBC Press.
North American Waterfowl Management Plan. (2012). *People conserving waterfowl and wetlands*. http://nawmprevison.org/sites/default/files/NAWMP-Plan-En-may23-pdf. Accessed 26 Jan 2015.
O'Connor, D. R. (2002a). *Report of the Walkerton inquiry: Part one, The events of May 2000 and related issues*. Toronto: Ontario Ministry of the Attorney General.
O'Connor, D. R. (2002b). *Report of the Walkerton inquiry: Part two, a strategy for safe drinking water*. Toronto: Ontario Ministry of the Attorney General.
Pearse, P. H., Bertrand, F., & MacLaren, J. W. (1985). *Currents of change: Final report, inquiry on federal water policy*. Ottawa: Environment Canada.
Pentland, R., & Wood, C. (2013). *Down the drain: How we are failing to protect our water resources*. Vancouver: Greystone Books.
Perkel, C. N. (2002). *Well of lies: The Walkerton water tragedy*. Toronto: McClelland and Stewart.
Shrubsole, D. (2016). Planning challenges related to flood management in Canada. In P. Dearden & B. Mitchell (Eds.), *Environmental change and challenge: A Canadian perspective* (pp. 158–159). Don Mills: Oxford University Press.
Shrubsole, D., & Draper, D. (2007). On guard for thee? Water abuses and management in Canada. In K. Bakker (Ed.), *Eau Canada* (pp. 37–54). Vancouver: UBC Press.
Smith, S. (2011). *Trends in permafrost conditions and ecology in Northern Canada. Canadian biodiversity: Ecosystem status and trends 2010* (Technical thematic report no. 9). Ottawa: Canadian Councils of Resource Ministers.
Sprague, J. B. (2007). Great wet North? Canada's myth of water abundance. In K. Bakker (Ed.), *Eau Canada* (pp. 23–35). Vancouver: UBC Press.
Turner, C. (2013). *The war on science*. Vancouver: Greystone Books.

Chapter 3
Water Policy in Canada

Ted Horbulyk

Abstract Canada's provincial, territorial and federal governments face key water policy decisions about the process of water allocation and the outcomes associated with alternative approaches. Water allocation refers to the systems that are used to decide which users are allowed to draw water from which sources, for which purposes, and when. Especially when water resources are under pressure from competing demands, or when water quality is threatened, it is logical to question whether the available water resources are being used in the best way possible. Water allocation policy is a multifaceted and far reaching tool with which governments can influence the structure of the economy and quality of residents' lives. Water allocation decisions influence numerous uses of surface and groundwater resources, including such non-extractive uses as providing environmental, recreational and aesthetic benefits when some of the water is left *in situ*. This chapter illustrates a range of specific water policy issues and considerations that relate to water allocation, and provides an overview of current practices in each jurisdiction. A number of challenges are highlighted that will have to be addressed if society is to benefit fully from the potential gains that effective water allocation can deliver.

3.1 Introduction

A key water policy decision for Canada's provincial, territorial and federal governments concerns the process of water allocation and the expected outcomes associated with alternative approaches. Simply put, this is the issue of who gets to use the available water and how much should be left untapped in service of the environment. Especially when water resources are under pressure from competing demands, or when water quality is threatened, it may be constructive to question whether the available water resources are being used in the best way possible.

Alternative remedies to problems of water shortage could include investments in infrastructure to enhance water supply, storage or distribution, yet reallocating the

---

T. Horbulyk (✉)
Principal Researcher – Economics, International Water Management Institute,
PO Box 2075, Colombo, Sri Lanka
e-mail: t.horbulyk@cgiar.org

S. Renzetti, D.P. Dupont (eds.), *Water Policy and Governance in Canada*,
Global Issues in Water Policy 17, DOI 10.1007/978-3-319-42806-2_3

existing supply, temporarily or permanently, might prove to be more effective and less costly. Even where available water supplies are not scarce, an ineffective system of water allocation could constrain the rate and patterns of economic growth and development. As Canadian jurisdictions continue to adapt to climate change and associated increases in the variability of water events, many will benefit from revisiting important aspects of water allocation.

Broadly defined, water allocation refers to the systems that are used to decide which users are allowed to use water from which sources for which purposes, and when. In Canada, there is currently a mosaic of regulatory approaches for the allocation of groundwater, surface water, and—in the special case of Newfoundland and Labrador—the allocation of water from icebergs. In some cases, water is generally allocated in the way it always has been, with some jurisdictions' water allocation essentially continuing processes that pre-date those jurisdictions' entry into nationhood. In other jurisdictions, water allocation approaches are being actively revised as a policy response to emerging threats and opportunities.

This chapter examines water allocation conceptually, by highlighting some key dimensions of allocation policy, and by identifying important attributes of the systems of water allocation that are currently in use or that might be adopted. Previous studies have documented in considerable detail the history and legal foundations of Canadian water allocation (Nowlan 2005; Guelph Water Management Group 2007; Hill et al. 2007; Adamowicz et al. 2010; Jeerakathil 2014; Living Water Policy Project 2015). The chapter offers a brief review of water allocation practices currently in use and identifies some important challenges that will confront policy makers in the next decades.

3.2 Scope and Dimensions of the Water Allocation Decision

To organize the choices that need to be made about water allocation, the presentation in this chapter sequentially examines (i) the scope of the water allocation challenge, (ii) the key dimensions of a water allocation; and (iii) various attributes that characterize existing and alternative policies. Scope is about what is or is not included in examining water allocation. Three key dimensions considered here are space, time and water allocations across sectors of an economy. Attributes are descriptive characteristics that distinguish one jurisdiction's policy from another's. Comparison of attributes provides a method of understanding differences across jurisdictions and may provide a guide for policy reform.

The examination of water allocation in this chapter should be interpreted in a broad and holistic fashion. Decisions about some uses of water can affect decisions about *all* uses of water. The scope of the allocation process includes not only industrial or municipal uses of water, but must include ecological services derived from water, water as a cultural asset, and so on. Considerable attention is paid here to the potential role of provincial, territorial and federal governments as enacted in their legislation and practices. First Nations also exercise responsibility for some water use decisions and processes, and, in some cases have challenged or contested the governments' allocation approaches (Walkem 2007; Horbulyk 2014).

Multiple policy dimensions of water allocation are in play. Allocation rules can determine geographic or spatial patterns of water use, can influence which economic sectors have access, and can affect numerous aspects of the timing of water use.

Since water is expensive to transport, the spatial dimension of water allocation concerns whether water is being used in the places where it is needed most, and how broad patterns of economic growth and development might change if geographic patterns of water allocation are changed. It is well understood that the historical siting of most of Canada's cities and large towns was influenced by proximity to specific rivers or lakes. In a "post-industrial" Canada, it is less clear whether the future growth and prominence of these centres will continue to depend upon access to the same (or growing) volumes of water resources, especially in per capita terms.

On a large scale, spatial reallocation raises challenges and opportunities associated with inter-basin, inter-provincial, and international water sharing and transfers, some of which were not always fully anticipated under current law. Whereas most water use decisions are made by the provinces or territories directly, the federal government, constitutionally, has a key role to play in overseeing the use of boundary and transboundary surface and groundwater resources. In large surface water basins, such as the Great Lakes Basin, the Mackenzie River Basin and the Saskatchewan River Basin, multiple provinces and territories participate in bilateral and multilateral agreements that, in turn, place limits on each of those provinces' or territories' internal uses of those waters (Partners FOR the Saskatchewan River Basin 2009; Percy 2012; Saunders 2014; Miltenberger and Phare 2015).

There is an important inter-sectoral dimension to water allocation policy, especially when growing urban uses or industries are also motivated by concerns about the security of future supplies. New users sometimes attempt to develop or to acquire access to water already in use by others, such as the water that has historically supported rural or non-consumptive uses. Some controversial water uses, such as water bottling for resale and hydrologic fracturing (fracking) represent new or growing industrial uses that compete with historical uses, and that might have the potential to harm water quality as well. Recent experiences with low water levels on the Great Lakes and the associated costs imposed on the shipping industry and the environment, suggest solutions that include water allocation reforms (de Villiers 2015). In some provinces or territories, the ability to attract and offer an "open door" to new industries that would locate there may rely on an ability to offer assured access to water supplies of sufficient quality and at sufficiently low cost. If perpetual rights to all available flows have already been assigned (whether or not these are actually fully utilized), what water supply options can be created for new entrants?

The inter-temporal dimension of water allocation decisions is apparent when considering appropriate incentives to regulate the inter-seasonal or inter-annual use of water. Regulators can be place limits on when water can be withdrawn or can change the timing of water supplies, such as through investment in storage and conveyance infrastructure or by promotion of such processes as aquifer storage and recovery. The onset of a severe drought that might last for several years in a row presents a special challenge, since it may not be possible to develop any alternate supplies. The promise or threat of new emergency rules for allocating the remaining supply can introduce considerable uncertainty and alarm among new and existing users, especially those, such as orchardists and industrialists, who have already made multiyear investments and commitments to ongoing water uses. Advance publication of a clear set of specific rules and processes to govern such emergencies can create public confidence that a crisis will be avoided. This could allow all users to make appropriate prior investments or other contingency plans that relate to their proposed water use.

Groundwater allocation is increasingly challenged by situations where ongoing withdrawal rates exceed rates of recharge. Allocation decision-making is often complicated by lack of reliable data about groundwater recharge rates, about the available groundwater inventory, and about the precise interactions of surface water flows and groundwater. Some stocks of groundwater do not recharge at all, and effective regulation of these stocks would follow considerably different principles than used with other aquifers. In some settings, it is cheaper and more effective to allow groundwater to move from point to point under the ground than to build an aqueduct on the surface. Similarly, it may be better to store water, even year-to-year, under the ground than in a reservoir on the surface, since there is less water loss by evaporation and the capital cost is much lower. The process of conjunctive water management can allow the selective substitution of surface water for groundwater (and vice versa) at times of the year and at sites where use of each source is most beneficial. Comprehensive water allocation policies can enable and empower such conjunctive management processes.

Aquifer storage and recovery technologies (also known as 'managed aquifer recharge') allow some aquifers to be artificially replenished in order to allow a greater rate of water withdrawal later. In some cases, these aquifers are recharged with surface water that happens to be seasonally abundant, while in other cases, the recharge is a convenient means of re-using or recycling other available water flows. In Canada, most jurisdictions have regulated the withdrawal of groundwater as a unidirectional, extraction process. Some jurisdictions do not have clear rules about the status and priority of users who propose to store water underground with the expectation that it will remain accessible to them for subsequent withdrawal on a priority basis.

Storing surface water behind hydroelectric dams can significantly alter the quality and seasonal patterns of water flow and availability to other users, including instream flows and water for diverse environmental services. In some cases, low winter flows that might occur naturally are augmented by enhanced water releases from reservoirs, and this can be beneficial for diluting treated wastewater and for improving fish habitat. In other cases, the seasonal re-filling of hydroelectric reservoirs presents direct competition to other uses, such as municipal uses and irrigation, even though hydro production is generally classified as a 'non-consumptive' use of the water. Effective water allocation policies can provide clear guidance about how such competition among users for scarce seasonal flows will be resolved.

3.3 Attributes of Water Allocation Policies

A number of public policy objectives could be served by the design or reform of water allocation policies, such as promoting the achievement of economic efficiency, and/or meeting specific goals of fairness, including guarantees of access to water as a human right. Working from a ranking of specific public policy objectives, water allocation policies could then be created with specific design features that promote the achievement of these larger goals. Once it is clear what the policy should achieve, one can focus on which policy or program attributes will reach those goals most effectively.

For instance, in 2015, the Ministers of the Environment of Canada's provinces and territories, working through the CCME, published some options that could guide a jurisdiction that was setting prices for water withdrawals (CCME 2015). These options include some that focus on influencing the behaviour of water users, and others that focus on raising public revenues. The first group of options is especially relevant to water allocation since water pricing is one method to allocate water—i.e., influencing the behaviour of water users. However, it is not an approach that has been relied upon historically in Canada. The ministers' options to influence user behaviour could include pricing water to address selective water challenges; to promote economic efficiency; to promote water conservation in general; or to promote industrial policy, among others. In a parallel fashion, other water policy reforms (that do not involve water pricing) could seek to achieve one or more of these goals. In general, careful choices will be required, since an allocation of water that is precisely targeted to promote economic efficiency will be different than one that promotes industrial policy by favouring specific industries.

Hypothetically, a Canadian province or territory could choose as its objective to promote economic efficiency, or to promote economic efficiency only provided that all citizens were first assured access to some specified quantities of water as a basic right. Such an allocation approach could be geared towards providing households with a so-called "lifeline" quantity of water (of sufficient quality) each month, beyond which the remaining water would be allocated, under the policy, to its highest and best use. Meeting this "highest-and-best-use" goal would necessarily involve comparison of the costs and benefits of providing water to different regions and sectors, and to providing access to water at different times of the years (and across years).

Whatever the water allocation objective that is set, there are then a set of attributes or design characteristics that can be part of the allocation policy to be followed. Examples of these are common in the literature on public policy design, and are adapted here to address water allocation specifically.

- *Administrative ease*: This is promoted where there is use of automated metering and othe monitoring approaches that promote low-cost enforcement and dispute resolution.
- *Low regulatory burden*: This is promoted, for example, with easy-to-access application and reporting processes for users, such as those that use the internet or are enabled for smart phones.

- *Simplicity*: The allocation system is easy to explain and to understand. Such a system might avoid a program structure with multiple layers of users, uses, volume thresholds, sectoral exemptions and regional variations.
- *Transparency*: The systems that are used can be seen to be implemented fairly and consistently for all users. Water use data are public, as is information about enforcement actions.
- *Predictability*: Users have enough information to form accurate expectations about how much water they can access now and in the future, including under extreme events, such as drought.
- *Revenue stability*: If part of the allocation system includes raising funds to operate the system or for other public purposes, then, all else being equal, it would be better to have stable revenue streams. Alternatively, if pricing is used to ration water under scarcity, some prices might be much higher in some years than others, creating revenue windfalls.
- *Procedural fairness*: This is promoted by the inclusion of informal appeal mechanisms, perhaps backstopped by a government ombudsperson and the civil court system. Some principles may be included for compensation when one user is harmed by the actions of others.
- *Horizontal equity and vertical equity*: These attributes usually relate to how fairly the policy affects users with similar versus different levels of income, wealth or ability to pay.
- *Flexibility*: An allocation approach can enable any future recommended changes to be implemented in a timely manner and with low transition costs. This approach can respond to greater hydrologic variability under climate change.
- *Avoids irreversible harm*: The allocation system can include environmental safeguards, for example, so that in times of shortage or flood, there are built-in protections for critical species, habitat, and infrastructure.

3.4 Current Water Allocation Practices in Canada

In Canada, there is currently a mosaic of policy approaches for the allocation of groundwater and surface water. As this discussion will illustrate, in some Canadian jurisdictions the allocation policies and practices are supported by explicit principles that guide the policymakers' actions. In other places, such principles are noticeably absent. In many cases, water is allocated in the same way it has been historically, with some regions' water allocations essentially continuing processes that pre-date those regions' entry into nationhood. In other regions, water allocation approaches are now being revised as a policy response to emerging threats and opportunities.

Current jurisdictional water allocation approaches surveyed here include (i) historical approaches (such as prior allocation licensing systems that assign a right to use water, but not to own it); (ii) a new reliance on pricing approaches for bulk water abstraction (even where these have little apparent effect on water allocations or use); (iii) market-based approaches for short-term or long-term re-assignment of water licences or uses; and (iv) ongoing legislative and regulatory reforms. In some jurisdictions, a water user's allocation of water has the characteristics of a property right, whereas in others it does not. In some jurisdictions, the right to use water is tied permanently to the land title of the land parcel where the water is used, yet in others that right can be transferred, either through administrative means or in the marketplace.

No Canadian jurisdiction is yet offering the full range of water allocation practices observed in Western US states, such as California. Noticeably absent from Canada are systems of freely-transferable annual or multiyear leases of water from one rights holder to another, often enabled via options contracts that signal the terms under which such a lease might become exercised (Howitt 2014). At the individual water-user level in Canada, most water approvals are defined for a specified volume of total withdrawal of water, as is common in the US, and do not assign fractional shares of the current annual flow, which approach is used in Australia. [At the basin scale in Canada, inter-provincial sharing agreements do employ approaches that assign fractional shares of the current annual flow.] Unlike allocation systems used in Australia, Canadian jurisdictions do not explicitly address or prescribe the volumes or quality of any return flows. In Alberta, for example, the percentage of total irrigation water withdrawals that is returned to surface or groundwater has declined in response to improved technical efficiency of irrigation methods and in response to the higher opportunity cost of that water in a system that now allows some limited trading of water entitlements (Horbulyk 2005).

Table 3.1 lists the current allocation approaches employed by all provinces and territories, presented in geographical order from west to east and from south to north. A number of jurisdictions (mainly in the west) assign priority to some users based on seniority (prior allocation), whereas other jurisdictions base priority on the purpose for which the water will be used (or they use a system of priorities that relies upon both use and seniority). Some jurisdictions allocate water to users, but do not specify any system of priority or seniority that would apply in times of shortage.

Some jurisdictions historically allocated water under a riparian rule that assured access to water to landowners whose property adjoined a river, lake or other surface water body. More recently, in a number of cases these riparian rights or allocations have been severely restricted (in volumetric terms) or have been extinguished altogether.

In British Columbia, groundwater was historically an open-access resource with no need for an approval or licence, but the 2016 legislative reforms have introduced a licensing requirement using a "grandfathering" process that seeks to attach the same seniority date to new licences "as if" the licence had been granted in the year the water was first used (British Columbia 2015).

Some of the jurisdictions have legislation that guides the manner in which moderate degrees of water shortage in a specific location would be accommodated, such as by following an established system of priorities. In the case of Ontario's Low Water Response Plan, all users could be expected to make cuts of a specific magnitude once a critical threshold is reached (Kreutzwiser et al. 2004; Roth and Murray

Table 3.1 Jurisdictional approaches to water allocation

Jurisdiction	How is water allocated initially, and reallocated later?	How is water allocated for environmental purposes and instream flow needs?
British Columbia	*Water Sustainability Act*	
	Surface and groundwater water use require a (transferable) water licence or approval. Licence seniority is based on a system of prior allocation. Licences are issued without expiry dates, except for new power generation uses which have 40-year terms. Domestic uses of groundwater are exempt. Newly issued groundwater licences may bear the priority date of first use. In times of shortage, low-priority domestic users of surface and groundwater may continue to divert water for essential purposes (0.25 m^3/household/day).	New applications will be assessed for effects on other users and instream flows. In times of shortage, a temporary Critical Environmental Flow Protection Order will take priority over senior licences to leave water *in situ*. Under the Fish Protection Act, water use can be limited by fish population protection orders.
Alberta	*Water Act*	
	Surface water and groundwater use require a water licence or approval. Licence seniority is based on a system of prior allocation. Household users are exempt from the need to licence (up to 1250 m^3/household/year) and maintain first priority in times of shortage. Traditional agricultural users may use up to 6250 m^3/household/year without a licence, and if registered, have the priority date associated with their first use. Legislation allows for the province to issue itself a licence to hold water in its natural state. Permanent transfers of licences (market based) are subject to an approval process. Short-term transfers are allowed among licensees, but only during times of rationing, and only to make up any shortfall.	Significant areas of the province are closed to new licence applications. New applications must take into consideration approved water management plans (if any), effects on the aquatic environment and on hydrology. Up to 10 % of the volume of water transferred between licenced users may be held back by the Crown in order to protect the aquatic environment or to meet a conservation objective. Some licences have limitations curtailing their use if necessary to meet instream objectives. Government owns some major storage reservoirs that could be managed to meet water conservation objectives in times of shortage.

(continued)

Table 3.1 (continued)

Jurisdiction	How is water allocated initially, and reallocated later?	How is water allocated for environmental purposes and instream flow needs?
Saskatchewan	*Water Security Agency Act*	
	Surface water and groundwater use require a water licence or approval. Riparian users and domestic well users are exempt from the need to licence (up to 5000 m^3/household/year). The historical systems of seniority, based on either prior allocation or hierarchical classes of water use were abolished. Older licences were issued without fixed terms. The Crown may choose to cancel or not to renew existing licences issued after 1984 (in some cases with limited compensation), and the water may be reassigned to other users.	New applications are assessed for effects on other users and instream flows. The Saskatchewan Water Security Agency may reallocate water for such purposes as meeting instream flow needs by canceling or not renewing licences granted after 1984.
Manitoba	*Water Rights Act*	
	Surface water and groundwater use require a water licence or approval. Domestic users are exempt from the need to licence (up to 2.5 m^3/household/day). There is a system of priority by class of use, which is: domestic, municipal, agricultural, industrial, irrigation and other purposes. Within classes, there is a system of precedence by seniority date. Before 1987, licences were issued without fixed terms. The Crown may choose to cancel or not to renew existing licences, such as to protect aquatic ecosystems or to move the water to a higher priority class of use in response to new applications. In the latter case, the user acquiring the water rights must pay compensation to the former rights holder.	New applications are assessed for effects on other users and on aquatic ecosystems. The minister may reallocate water for such purposes as meeting instream flow needs by cancelling or not renewing licences.
Ontario	*Ontario Water Resources Act*	
	Surface water and groundwater use require a permit or other approval to take water. Domestic use and use for livestock and fire-fighting are exempt from the permit requirement (up to 379 m^3/day). Other users are generally exempt on amounts up to 50 m^3/day, but, by exception, can be required to seek a permit, or their use might be prohibited. Users and uses in place before 1962 are generally exempt from permit requirements. Permits issued since 1962 have expiry dates but earlier ones do not. Permits to take water do not represent a water right and do not assign priority or precedence of use to individual users.	New applications are assessed for effects on other users and on the natural environment. Under the Low Water Response Policy, the occurrence of a Level III drought can require a 20 % reduction in water taking by permit holders.

(continued)

Table 3.1 (continued)

Jurisdiction	How is water allocated initially, and reallocated later?	How is water allocated for environmental purposes and instream flow needs?
Quebec	*Environment Quality Act*	
	Surface water and groundwater use require a water withdrawal authorization. Some uses (up to 75 m³/day) are exempt from this requirement, but not those relating to water bottling, food manufacturing or inter-basin transfers from the St. Lawrence. Authorizations typically expire in 10 years. Permits to take water do not represent a water right and do not assign priority or precedence of use to individual users.	New applications are assessed for effects on other users and on the natural environment. The government is to foster the protection of water resources, particularly by fostering sustainable, equitable and efficient management of resources in light of the precautionary principle and the effects of climate change.
		If the Minister is of the opinion that a water withdrawal presents a serious risk for public health or aquatic ecosystems, the Minister may order the temporary cessation or limitation of the water withdrawal. In the case of a water withdrawal authorized by the Minister, the order may direct that the withdrawal cease permanently and without compensation.
New Brunswick	*Clean Environment Act*	
	Surface water and groundwater use require a water withdrawal approval. Uses up to 50 m³/day are exempt from this requirement, as are domestic wells not connected to a distribution system. An approval does not represent a water right and does not assign priority or precedence of use to individual users.	Applications are assessed to evaluate the sustainability of the water supply, to assess the water quality, and to evaluate potential impacts to existing water users.
Nova Scotia	*Environment Act*	
	Surface water and groundwater use require a water withdrawal approval. Uses up to 23 m³/day are exempt from this requirement. An approval does not represent a water right. Groundwater allocations are based on a "first-come, first-served basis" with priority given to drinking water applications. Priority is also given to existing withdrawal approvals over new applications.	Applications are assessed to evaluate the sustainability of the water supply, to assess any adverse effect on the environment and to evaluate potential impacts to existing water users.

(continued)

Table 3.1 (continued)

Jurisdiction	How is water allocated initially, and reallocated later?	How is water allocated for environmental purposes and instream flow needs?
Prince Edward Island	*Environmental Protection Act*	
	Surface water and groundwater use require a permit or approval. Surface water uses up to 50 imperial gallons/minute and total daily withdrawal up to 10,000 imperial gallons are exempt from this requirement, as are groundwater users (up to 4 l/s) if not connected to a water supply system.	The Minister may issue a groundwater extraction permit if the Minister is satisfied that the operation of a well will not have an unacceptable impact on the environment.
	In any watershed where the demand exceeds the supply of surface water resources, available water will be shared in allotments of 400 imperial gallons/minute.	
Newfoundland and Labrador	*Water Resources Act*	
	Surface water and groundwater use require a water licence or approval, as does harvesting water from icebergs. Riparian users are exempt from the need to licence withdrawals used for domestic purposes. There is a system of priority by class of use, which is: domestic; municipal; agricultural; commercial, institutional and industrial; water and thermal power generation; and other purposes. The Crown may choose to cancel or not to renew existing licences to move the water to a higher priority class of use in response to new applications. In such cases, the user acquiring the water rights must pay compensation to the former rights holder. The holder of a licence with higher priority is entitled to receive the whole of the supply of water to which the licence entitles that person before a licence holder using water for a lower priority purpose has any claim to that supply.	New applications are assessed for effects on other users and on the natural environment.
Yukon	*Waters Act*	
	Surface water and groundwater use require a water licence or approval. Licence seniority is based on a system of prior allocation. Domestic users are exempt. Most other users are exempt from the need to licence (up to 300 m^3/day), but oil and gas, and municipal use are exempt up to 100 m^3/day.	New applications are assessed for effects on other users and on the natural environment.

(continued)

Table 3.1 (continued)

Jurisdiction	How is water allocated initially, and reallocated later?	How is water allocated for environmental purposes and instream flow needs?
Northwest Territories	*Waters Act*	
	Surface water and ground-water use require a water licence or approval. Licence seniority is based on a system of prior allocation. Domestic users are exempt from the need to licence (without limit). Water uses for municipal purposes are exempt up to 50 m^3/day, and selected other users are exempt up to 100 m^3/day. A person may use water without a licence if the use has no potential for significant adverse environmental effects, and would not interfere with existing rights of other water users.	New applications are assessed for effects on other users and on the natural environment.
Nunavut	*Nunavut Waters and Nunavut Surface Rights Tribunal Act*	
	Surface water and groundwater use require a water licence or approval. Licence seniority is based on a system of prior allocation, except that water use on 'Inuit-owned land' takes priority over licenced uses. Domestic users are exempt from the need to licence (without limit) as are most other users (up to 50 m^3/day).	New applications are assessed for effects on other users and on the natural environment.

2014). While most jurisdictions grant special powers to the Crown to take steps in a more extreme situation—such as when the allocation decided by priorities alone would be physically or socially unworkable—none of them has spelled out precisely what those steps might be, or what principles would guide the Crown's action in forming a disaster response plan.

3.5 Challenges for Effective Water Allocation

Important challenges to be faced in allocating Canada's water include addressing the effects of climate change; securing the value of water that provides ecological goods and services; developing sufficient metering and monitoring capacities; acquiring better estimates of water resource inventories; and ensuring that sufficient water can be protected to meet obligations to all future generations, including indigenous people. These issues are discussed in turn.

There is wide variation in the meteorological and hydrological forecasts for Canadian provinces and territories that would be associated with climate change and increasing climate variability. Even so, for most parts of the country it appears that the status quo will no longer apply, and that water shortages, and the effects of these shortages on water quality, will present a need to address policies for water allocation more carefully and comprehensively than has historically been the case. Most jurisdictions have put in place fairly comprehensive systems under which the initial allocations of water were made. Few jurisdictions have systems in place that can effectively *re-allocate* water, especially in times of severe or prolonged shortage or drought.

A relatively recent challenge comes from increasing public attention to, and valuation of, instream flows and environmental uses of water. In dry parts of the Canadian west, the population is becoming increasing urban, and urban residents are expressing their interest in maintaining the health of aquatic ecosystems, and the state of their natural environment more generally. Acting as citizens, as voters and as taxpayers, this population appears to be increasingly concerned about such issues as species at risk; the preservation of critical habitat, and safeguarding minimum flows in waterways, especially in times of drought (Droitsch and Robinson 2009).

Even in locations where water shortages are commonplace or have been experienced historically, there are no comprehensive systems for metering or monitoring all surface and groundwater use. With numerous legislated exemptions and exclusions from provincial and territorial licensing and approval processes (such as for low-volume domestic and other uses) the *cumulative* effects of actual water withdrawals and return flows cannot be established by water managers with any certainty. Managers have not historically had to perform the "water mastering" function that is common in many US states, whereby users' actions are monitored, and permissions to extract water according to established allocation rules might have to be granted site-by-site on a daily or weekly basis. Preparation for the day when such actions might be required will require prior public investments in metering and monitoring equipment and systems, including staffing and training.

In some jurisdictions, the properties and magnitudes of water resources are not well established, for groundwater in particular. The interactions of groundwater and hydraulically connected surface water bodies may not be well described by available and up-to-date studies, and so allocations of groundwater that have been made independently from other allocations of surface water may not be sustainable. Effective water allocation will require a clear commitment to estimating sustainable rates of flow from specific aquifers, combined with monitoring systems to assess groundwater resources *in situ*.

In some jurisdictions, the policy challenge for governments is that they have not clearly defined or adjudicated allocations of water to aboriginal (Indian, Inuit and Métis) water users. In some cases, there are treaties or agreements that cover water allocation issues, but without comprehensive treatment of how the terms of those accords would mesh with other statutory allocations of water, especially under low-flow conditions or other water supply emergencies. In other cases, there are no treaties or agreements that explicitly address aboriginal water use, even where such use is expected to grow in future (Walkem 2007). The ability to manage future systems of water allocation to all users will require clear allowance for any unmet obligations to First Nations' users, especially in systems where withdrawal volumes have reached or are approaching the limits of sustainable supply.

To conclude, water allocation policy is a multifaceted and far reaching tool with which governments can influence decisions about how, when and where water is used. Water allocation decisions can give shape to economies as they make numerous uses of surface and groundwater resources, including such non-extractive uses as providing environmental, recreational and aesthetic benefits when some of the water is left *in situ*. A number of challenges are highlighted here that will have to be addressed if society is to benefit fully from the potential gains that effective water allocation can deliver.

References

Adamowicz, W. L., Percy, D. R., & Weber, M. (2010). *Alberta's water resource allocation and management system: A review of the current water resource allocation system in Alberta*. Alberta Innovates – Energy and Environment Solutions. www.seawa.ca/wp-content/uploads/2014/08/Albertas_Water_Resource_Allocation_and_Management_System.pdf

Province of British Columbia. (2015). *Licensing groundwater use under British Columbia's Water Sustainability Act*. Victoria: British Columbia. https://engage.gov.bc.ca/watersustainabilityact/files/2016/02/LicensingGroundwaterUse-Web-Copy.pdf

Canadian Council of Ministers of the Environment (CCME). (2015). *Water pricing options*. Winnipeg: CCME. http://www.ccme.ca/files/Resources/water/water_valuation/Principles%20for%20Water%20Pricing%201.1_e%20PN%201536.pdf

de Villiers, M. (2015). *Back to the well: Rethinking the future of water*. Fredericton: Goose Lane Editions.

Droitsch, D., & Robinson, B. (2009). *Share the water: Building a secure water future for Alberta*. Canmore: Water Matters Society of Alberta and Vancouver: Ecojustice. http://www.water-matters.org/docs/share-the-water.pdf

Guelph Water Management Group. (2007). *Characterization of water allocation systems in Canada*. Technical Report 1. Prepared for the Walter and Duncan Gordon Foundation. Guelph: Guelph Water Management Group, University of Guelph. http://waterwiki.net/images/a/a8/Technical_Report_1_Oct2007.pdf

Hill, C., Furlong, K., Bakker, K., & Cohen, A. (2007). A survey of water governance legislation and policies in the provinces and territories, Appendix 1. In K. Bakker (Ed.), *Eau Canada: The future of Canada's water* (pp. 369–392). Vancouver: UBC Press.

Horbulyk, T. M. (2005). Markets, policy and the allocation of water resources among sectors: Constraints and opportunities. Canadian Water Resources Journal, 30(1), 55–63. doi:10.4296/cwrj300155

Horbulyk, T. M. (2014). Resilience of river basin governance institutions in the Saskatchewan River Basin of Western Canada. In D. E. Garrick, G. R. M. Anderson, D. Connell, & J. Pittock (Eds.), *Federal rivers: Managing water in multi-layered political systems* (pp. 90–104). Cheltenham: Edward Elgar. doi:10.4337/9781781955055.00015

Howitt, R. E. (2014). Are lease water markets still emerging in California? In K. W. Easter & Q. Huang (Eds.), *Water markets for the 21st century: What have we learned?* Global issues in water policy (Vol. 11, pp. 83–102). Dordrecht: Springer. doi:10.1007/978-94-017-9081-9_5

Jeerakathil, R. (2014). *Water rights: Different provincial regimes and what may make sense for Saskatchewan*. Presented to the Saskatchewan Mining Association Environmental Forum, Saskatoon, 29 October 2014. http://www.saskmining.ca/uploads/news_files/112/r-jeerakathil-paper.pdf

Kreutzwiser, R. D., de Loë, R. C., Durley, J., & Priddle, C. (2004). Water allocation and the permit to take water program in Ontario: Challenges and opportunities. *Canadian Water Resources Journal,* 29(2), 135–146. doi:10.4296/cwrj135

Living Water Policy Project. (2015). *Explore Canadian water policy* (by province). http://www.waterpolicy.ca/

Miltenberger, J. M., & Phare, M.-A. (2015). *Evolution in transboundary watershed governance: Lessons from the Mackenzie Basin*. Presented to Creating a Blue Dialogue webinar series, 16 September 2015. http://www.poliswaterproject.org/webinar/851

Nowlan, L. (2005). *Buried treasure: Groundwater permitting and pricing in Canada*. Prepared for the Walter and Duncan Gordon Foundation, with case studies by Geological Survey of Canada, West Coast Environmental Law and Sierra Legal Defence Fund. Toronto: Walter and Duncan Gordon Foundation. http://www.wheatleyriver.ca/wp-content/uploads/2011/02/Buried_Treasure.pdf

Partners FOR the Saskatchewan River Basin. (2009). *From mountains to the sea: The state of the Saskatchewan River Basin*. Saskatoon: Partners FOR the Saskatchewan River Basin. http://www.saskriverbasin.ca/pages/state_of_the_basin_report.html

Percy, D. R. (2012). *Resolving water-use conflicts: Insights from the prairie experience for the Mackenzie River Basin* (C.D. Howe Institute Commentary, no. 341). Toronto: C.D. Howe Institute. https://www.cdhowe.org/pdf/Commentary_341.pdf

Roth, A., & Murray, D. (2014). *Ontario's low water response planning: Making decisions for areas most vulnerable to drought and low water conditions*. Prepared for Conservation Ontario and Ontario Ministry of Agriculture and Food, Water Policy and Governance Group, Waterloo. www.wpgg.ca/sites/default/files/Roth%20and%20Murray%202014.pdf

Saunders, J. O. (2014). Managing water in a federal state: The Canadian experience. In D. E. Garrick, G. R. M. Anderson, D. Connell, & J. Pittock (Eds.), *Federal rivers: Managing water in multi-layered political systems* (pp. 76–89). Cheltenham: Edward Elgar. doi:10.4337/9781781955055.00015

Walkem, A. (2007). The land is dry: Indigenous peoples, water, and environmental justice. In K. Bakker (Ed.), *Eau Canada: The future of Canada's water* (pp. 303–319). Vancouver: UBC Press.

Chapter 4
Changing Currents: A Case Study in the Evolution of Water Law in Western Canada

Oliver M. Brandes and Deborah Curran

Abstract New social, economic and environmental priorities are challengeing the Canadian water law regime. Water law in western Canada, a direct product of the colonial legal system and European settlement, illustrates many of the emerging tensions associated with a modern water management regime in flux. In an age of increasing hydrologic uncertainty with drier summers followed by more extreme storm events, lawmakers are seeking to increase resilience both for the environment and also for the institutions and the laws that govern freshwater resources. In Canada evidence of an evolving water law and management regime is already apparent—from developments in Aboriginal law that are changing how and who governs water, retreat by the federal government as an active participant in water resource management, to increased provincial efforts to fill that void.

This chapter explains the structure and foundations of Canada's approach to water law, in particular in western Canada; and explores how water law is changing, and what this reveals about the potential of a twenty-first century approach to water management and governance. It will explicitly review the primary allocation regimes that exist across Canada: modified common law riparian rights in the Maritime provinces and Ontario; Quebec's civil law tradition; the authority management approach in the North; and the prior allocation system that underpins the prairie provinces and British Columbia. Through this discussion the chapter will set out the foundational principles that characterize the current approach to western water law. Investigation into the recent law reform in British Columbia provides the focus to better understand Canadian western water law and to identify characteristics of an emerging regime based on partnership and with an explicit emphasis on

O.M. Brandes (✉)
POLIS Project on Ecological Governance and Centre for Global Studies, University of Victoria, 1700, STN CSC, Victoria, BC V8W 2Y2, Canada
e-mail: omb@uvic.ca

D. Curran
Faculty of Law and School of Environmental Studies, University of Victoria, 1700, STN CSC, Victoria, BC V8W 2Y2, Canada
e-mail: dlc@uvic.ca

S. Renzetti, D.P. Dupont (eds.), *Water Policy and Governance in Canada*,
Global Issues in Water Policy 17, DOI 10.1007/978-3-319-42806-2_4

protecting water for nature. This case study explores how modern water governance requires a more collaborative approach where all governments, rights holders, and stakeholders have roles and responsibilities, with creative integration of top-down and bottom-up planning and decision-making. The example of British Columbia demonstrates how this water law regime is "changing the current"—evolving gradually toward a more collaborative and adaptable system with the promise of its new *Water Sustainability Act*.

4.1 Introduction

Water law and management in Canada, as in much of the world, is changing. The existing approach to water allocation and the supporting legal entitlement regimes are unable to address the modern challenges of water management. The issues facing freshwater management are numerous and are introduced and explored throughout this volume. Some of the most crucial drivers of change include conflicts associated with increasing water use and intensified resource extraction, under resourced governments, the assertion of rights to water by Indigenous peoples and others seeking a more direct role in decision-making, and changing hydrology due to climate instability.

With the social, economic and environmental priorities of the twenty-first century, changing settlement patterns, and a growing population concentrated along the southern border, it is increasingly clear that Canada's water law regimes are no longer able to address the requirements of modern water management. A changing climate has significant implications for water management and manifests in more extreme extremes such as frequent and increasingly severe drought and floods coupled with a longer-term trend toward persistent and prolonged water scarcity. Top-down management by senior governments is not sufficient in the face of the nuanced challenges facing water managers and watersheds across Canada, and indeed the world. Partnership, stewardship, and dealing with the uncertainty of perpetual change will be the hallmarks of successful water law and management regimes in the future.

In Canada, water law and management regimes are one of the last frontiers of environmental regulation where decisions are based on insufficient real-time science about ecosystem needs. The knowledge gaps include basic information such as how much water is available, and how much water is actually being used (Curran and Brandes 2012). Fundamentally, these regimes still reflect their outdated colonial origins that emphasized settlement of the land for development, agriculture, and mining.

Water law in western Canada, a direct product of the colonial system and European settlement, illustrates many of the emerging tensions and growing pains associated with modern challenges and a water management regime in flux. As Jim Mattison, the former comptroller of water rights for British Columbia recently remarked:

> the water allocation system is—and always has been—about the orderly distribution of water for economic development (Mattison 2016).

This comment from the provincial government's former most senior water decision-maker reflects the fact that water users and governments are still primarily concerned with driving economic development. Access to water is a critical element of growing cities, local breweries, and food production; it also continues to support the existing economic foundation of mining, manufacturing, and energy development.

Early evidence of a new paradigm for water law is emerging in Canada (Brandes et al 2014; Curran 2015). Developments in Aboriginal law are shifting how and who governs water. This shift is accelerated by a significant retreat, since the 1990s, by the federal government as an active participant in water resource management. As a consequence of this dynamic, provincial governments and an increasingly active civil society are attempting to fill the water governance void (Curran 2015).

This chapter has two primary purposes: to explain the structure and foundations of Canada's approach to water law, in particular in western Canada; and, to explore how water law is changing, and what this reveals about the potential of a twenty-first century approach to water management and governance. The chapter first outlines the legal foundation of how Canada governs its freshwater resource and describes the existing jurisdictional arrangement governing water and the primary allocation regimes that exist across Canada. This discussion sets out the foundational principles that characterize the current approach to water law across the country and highlights emerging trends in law reform.

Second, the chapter delves into the British Columbia example to investigate the primary principles of western water law and how they manifest in the existing legal structures. In an age of increasing hydrologic uncertainty, typified by drier summers followed by more extreme and less predictable storm events, lawmakers are seeking to increase resilience both for the environment and for the institutions and laws that govern freshwater resources. Using the recent law reforms in British Columbia, this discussion explores how water law is evolving to address the limits of the existing system. It also explores the potential for a new paradigm based on partnership and with an explicit emphasis on protecting water for nature.

The British Columbia case study provides an example with which to examine the underlying principles related to water entitlements and allocation (quantity), and to plumb more deeply into western water law generally. Recent reforms, including the introduction of the new *Water Sustainability Act* and a more explicit consideration of environmental flows, represent emerging trends and new directions in water management.

There is a growing recognition that water law needs to change—and, indeed, is already changing—in the coming decades more than it has over the past century. The British Columbia example highlights how an updated legal system offers the potential to better balance social, ecological, and economic needs going forward. Although the case study focuses on the West, the implications are relevant across Canada—perhaps even globally—as they reveal emerging trends and the potential for new structures for water laws in Canada and beyond (Curran and Mascher 2016; Brandes et al 2014).

4.2 Water Jurisdiction and Allocation Systems in Canada

4.2.1 Jurisdiction Over Water

Canada is a constitutional federation. The Constitution of Canada distributes powers between the federal and provincial governments to make laws related to ownership and management of resources, including fresh water (Constitution Act 1867; Muldoon et al 2015; Hogg 2007). This structure has an explicit colonial origin, yet hundreds of indigenous nations with rich and varied cultures, governance regimes, and sophisticated indigenous legal systems preceded European contact (Borrows 2002; Napoleon 2007). Most of these nations still exist today and have constitutionally protected Aboriginal rights.

Fundamentally, fresh water is a simple element; however, its flowing nature can make management and governance complicated and challenging. The word "water" does not appear in the Canadian Constitution. It is described as a "fugitive resource" because it defies tidy division into federal and provincial jurisdictions (Pearse and Quinn 1996). The laws governing freshwater management in Canada involve a complex swirl of overlapping jurisdictions, including numerous agencies and departments, and a range of actors including federal, provincial, Aboriginal, and local governments. In essence the Constitution sets out an approach of shared responsibility for water management, but does not specifically articulate overarching responsibility to any one level of government. This shared and multijurisdictional approach is also used in other sectors such as agriculture and health. Constitutional sharing of responsibility creates overlap, and sometimes contradictions in jurisdiction, and usually is only partially able to address the social and ecological interactions associated with water.

Canada's approach to water law in general, and water allocations in particular, varies significantly from province to province. In Canada, the thirteen provinces and territories have primary responsibility for the regulation of ground and surface water, with water generally owned and managed by these sub-national governments due to their ownership and management of public or Crown land (Percy 1988).[1] Clear federal interests also exist in defining Aboriginal water rights, trans-boundary (including interprovincial) waters, waters on federal lands, and issues concerning navigation and fisheries. The most common approach to dealing with conflicts or uncertainties over jurisdiction is for governments to create negotiated agreements with one another, rather than to test the constitutional or legal authority to act unilaterally (Bailey 2008).

The provincial governments are generally understood to hold the primary role in water management with constitutional powers that include:

---

[1]For example, section 5(1) of British Columbia's *Water Sustainability Act* (2014) contains the Crown ownership provision stating "The property in and the right to the use and flow of all the water at any time in a stream in British Columbia are for all purposes vested in the government, except only insofar as private rights have been established under authorizations."

- power to make laws concerning property and civil rights, including regulation of the use of property and land use (e.g., drainage);
- jurisdiction to regulate "local works" and undertakings;
- power over Crown land, with limited provincial ownership of all public lands (including water);
- ownership and management of natural resources;
- regulatory authority over all municipalities/local governments, including the power to authorize and regulate municipal water (e.g., water quality standards and the qualifications of municipal employees engaged in water quality management);
- matters of a local or private nature; and,
- natural resources, forestry and hydroelectric energy.

The three Canadian territories (all in the far North) do not have independent constitutional status but have powers similar to the provinces with the federal government holding the ultimate authority to legislate with respect to any territory in Canada.

The federal government also has constitutional powers over a number of areas that directly or indirectly relate to fresh water. Examples include:

- sea coast and inland fisheries;
- navigation and shipping;
- international or interprovincial works and undertakings, which the courts have interpreted cover pipelines;
- federal works and undertakings;
- canals, harbours, rivers, and lake improvement; and
- Indians and lands reserved for Indians.

The federal government also has the power to implement treaties, such as the 1909 *International Boundary Waters Treaty* with the United States.

Two highly relevant but rarely used broad federal powers related to the environment—therefore affecting water—are the "peace, order and good government" (POGG) and federal criminal law power. "Peace order and good government" provides the power to uphold federal laws, which regulate matters of national importance.[2]In one critical water law case, the federal government was found to have the power under POGG to make laws concerning ocean dumping even when dumping occurred in water under provincial jurisdiction (*R v. Crown Zellerbach Canada Ltd.* 1988). The federal criminal law authority is also important for environmental protection because the federal government can enact a law for the protection of public health and safety that prohibits an activity, or that carries a penal sanction—powers that serve as an important control over pollution.[3]

---

[2] For example, *Interprovincial Co-operatives v. Manitoba* (1975) stands for the principle that interprovincial pollution of fisheries is a matter falling under the federal power over POGG.

[3] In *R. v. Hydro-Québec* (1997), the S.C.C. decided that the federal government has the authority to pass legislation that criminalizes harm to the environment, which can often relate to water or water

4.2.2 *Water Allocation Systems and How They Work in Canada*

Water allocation refers to the system of rules and procedures that give access to water by granting licences or approvals to use water. As de Loë et al (2007) note, by regulating the availability and priority of access to water for consumptive and non-consumptive uses, water allocation systems influence economic productivity, social and cultural wellbeing, and ecosystem quality.

Many provinces and territories have recently passed water policies or new water laws and begun a concerted process of modernizing water allocation systems as a strategic priority. Some examples include:

- Alberta's *Water for Life* reforms, initiated in 2003 with renewal in 2008 and ongoing implementation (Alberta Environment 2003; AEP 2016);
- Ontario's *Clean Water Act* and its comprehensive source water protection program, started in 2006 (OMECC 2016);
- British Columbia's *Living Water Smart Plan* and *Water Act Modernization* process started in 2008 (BC MoE 2008; BC MoE 2016);
- Quebec's updated water laws in 2009, confirming the collective nature of water (Quebec Government 2009);
- Northwest Territories' *Northern Voices Northern Peoples* initiative with a completed territorial strategy released in 2010 (GNWT 2010; GNWT 2016); and
- Saskatchewan's recent launch of its *25 Year Water Security Plan* in 2012 (SWSA 2012).

Included in these broad strategies are explicit reforms to modernize allocation systems such as: the modified water markets in Alberta; a hybridized prior allocation system in British Columbia; updated permitting and licensing in Ontario and Quebec; and initiatives in Saskatchewan to better define priorities to optimize water use in the public interest. All such reforms have at their core a focus on streamlined approvals and licensing, attention to monitoring and compliance, protection of ecosystems and environmental flows, and addressing localized crises of drinking water-quality, droughts, and floods, or persistent water scarcity. Other drivers of reform include longer-term concerns about equitable sharing of the resource, preservation of social and cultural values such as rural economic development, and improving water efficiency and conservation (Woods and Pentland 2013).

In most Canadian law, water rights or entitlements are not a property right. Instead, they are a right of use where the Crown asserts ownership of the water and permits licence holders to use water under terms and conditions that may change (Brandes and Nowlan 2009; Curran 2014b; Percy 1988). This is fundamentally different from the situation for Canada's neighbours in the United States, where once vested these "usefructory" rights (rights to use) can become constitutionally protected property interests. In the United States, water can be sold, leased or trans-

---

bodies. In other words, the protection of the environment may constitute a criminal law purpose under Part 3 of the criminal law test.

ferred in other ways and can be protected as a property right (*Thayer v. California Dev. Co.* 1912; Norman and DuMars 1989, p 351).

In practice, however, although not legally treated as property rights in Canadian law, licensed water entitlements (and even common law rights associated with groundwater) exhibit many of the characteristics of a property rights regime, and are treated by the public and political processes as property rights. This is due, in part, to their perpetual nature—many water licences in Canada have no expiry—and the fact that governments have extremely limited ability to amend licences without concerns regarding the threat of compensation. The ongoing effort in British Columbia to transform groundwater use, historically governed under the common law "rule of capture," into licensed entitlements under the legislative regime demonstrates this tension. Government officials view this historical use as a quasi-property right that affects their ability to deny or substantially limit water use through the licensing process with an active interest in avoiding claims of compensation by existing groundwater users. From a policy perspective, this represents a kind of de facto property rights regime for water (Furubotn and Pejovich 1972; Schlager and Ostrom 1992).

Surface water rights in Canada were initially based on the English common law rule of riparian rights. The riparian system evolved into distinct systems that address the differences in climate, geography, and development priorities across the nation, (Lucas 1990; de Loë et al 2007), including:

1. ***Regulated riparian model***—Licensing and permitting, overlaying the traditional court-made riparian rights doctrine. Under this system, direct water users who use more than a set volume must have a permit from an administrative agency to use a specific volume of water. Ontario and some of the Atlantic provinces use this system.
2. ***Civil law tradition***—A hybrid system based on riparian rights and adapted from a civil law tradition and augmented by legislation. Quebec uses this system.
3. ***Prior allocation***—A system that enshrines the first-in-time, first-in-right (FITFIR) principle, where government grants the right to use water by licence while requiring the "beneficial use" of the diverted water from its source. British Columbia, Alberta, and Manitoba use this system.
4. ***Authority management approach***—A system where government delegates responsibility for allocation decisions to regional or resource boards, or to other administrative bodies. Although this system is primarily used in the North by the three Canadian territories—Nunavut, the Yukon, and the Northwest Territories—Saskatchewan also uses a modified version of this approach with responsibility delegated to a government agency.

Groundwater rights evolved differently from surface water rights. In the English common law tradition, groundwater was treated more often as an exclusive right.[4] English judges applying common law principles to water conflicts extended riparian rights to groundwater flowing in defined channels, while the "rule of absolute

[4] Providing liberty to extract, but not a right to prevent others from doing the same. See, for example, *Acton v. Blundell* (1843), and *Chasemore v. Richards* (1859) where groundwater law was based on the rule of capture (fundamentally a no-liability rule).

capture" applied to all other sources of groundwater: landowners could use the water under their soil regardless of any injury to their neighbours. The law treated these two sources of water differently, despite their interconnectivity as part of the same hydrologic system.

Evidence of the historical distinction between surface and ground water is still apparent, as many water allocation systems in Canada do not adequately protect groundwater (Council of Canadian Academies 2009; Nowlan 2005; Rivera 2005). British Columbia is just now beginning a general licensing or permitting system for groundwater withdrawals for non-domestic users. Until 2016, anyone could drill a well without regulatory permission, often in places with interconnected surface water systems. Indeed, even with the new law, the provincial government still permits this practice for domestic uses. Naturally, this creates significant challenges for a comprehensive or holistic approach to water management.

Notably, drought or limited supply have not yet heavily tested allocation systems in Canada. In addition, water managers have only rarely needed to curtail use among existing licensees (Nowlan 2012; Rivera 2005). Where shortages have occurred in the past, the responses of provincial regulators and local governments have emphasized voluntary temporary restrictions usually confined to irrigators or urban use. Other approaches, in more severe situations, include the use of temporary orders or enforcing existing licence conditions.

It most cases, management of drought across Canada relies heavily on ad hoc bureaucratic discretion. This approach generally lacks a structured approach for dealing with water scarcity—one that guides decisions and ensures fairness, economic efficiency, and environmental protection.

4.3 The Origins of Western Water Law—Foundational Principles in the Current Context

The North American tradition of water management and governance is premised on State or Crown ownership of the resource. This approach originated during a time when water was seen as "limitless" and the focus was on promoting settlement and creating certainty for economic development. The water management and law regime evolved out of a need to foster this frontier resource economy and serve the needs of the growing settler population with an emphasis on providing some level of certainty in the water supply for gold mining and agricultural development (Wilkinson 1992; Harris 2013).

"Western water law" is a term referring to the various freshwater allocation systems created to address the unique challenges of topography, climate, and precipitation patterns faced by settlers of the North American West in the late 1800s (Johnson and Dumars 1989).[5] In contrast, the eastern seaboard was similar to England's envi-

[5]Many water law scholars maintain that no true western water law exists in Canada, that Saskatchewan has not followed this kind of water law tradition for decades, and that all the other western provinces have modified it extensively; however, it is our viewpoint that these general

ronment, making it easier for early colonists to extend and apply the English water laws and customs that led to the riparian rights system still largely in effect in the East. The "arid" West challenged the colonial riparian rights laws governing water entitlements, forcing an initial evolution to address the need for certainty in investment, and for more significant water diversions further from source streams and rivers to support mining and agriculture, and to deal with the climatic variability in the western region (Hutchins 1971; Harris 2013).

The foundations of the current legal system emerged during this early pioneer period, and much of the original approach to water management remains intact even today. In the broader realm of environmental management and governance, this continued reliance on historical institutional and legal architecture is rare; but for water management in Canada many of the same legal principles and decision-making structures have remained intact since the enactment of the original provincial water legislation around the beginning of the twentieth century (Brandes et al. 2015; Curran 2014a; Brandes 2014). British Columbia's *Water Act* of 1909 codified the principals of the prior allocation doctrine for the province, which remains virtually unchanged in the various updates, including the most recent *Water Sustainability Act* (2014). A similar situation exists in Alberta and Manitoba under their updated water laws: the *Water Act* (RSA 2000), and the *Water Rights Act* (CCSM 2015; Percy 1988).

In this examination of the evolution of water law, it is useful to consider the foundational principles of the western water law system and to examine how they continue to have an impact on water management today. The four principles are (Percy 2004):

Principle 1: Crown asserts ownership over water resources

The primary foundation of the colonial legal system is the assertion of Crown (or State) ownership of water. In Canada, under western water law the Crown declares its ownership over water resources and then grants the beneficial use of a specific volume of water to landowners and to those in possession of infrastructure such as mines or hydroelectric facilities.

Principle 2: Water is allocated and diverted on a first-come, first-serve basis

British Columbia allocates water according to a prior allocation system,[6] also known as "first in time, first in right" (FITFIR). This is a priority ranking system based on date of licence issue and intended to provide certainty in water use. During times of scarcity, water licences with the earlier priority dates entitle their licensees (senior licence holders) to take their full water allocation over more junior licences, regardless of the purpose of water use. In principle, this system uses this dated prior-

principles and underpinnings do still exist and reveal important patterns in the evolution of law and governance, thus western water law still provides a useful framework.

[6] Prior *allocation* is the Canadian system and is distinguished from the western US system of prior *appropriation*. In Canada the priority date is based on the date at which the senior government granted permission, while the US system is predicated on the date of first use, or when the water was first appropriated.

ity approach all the way down the line until all the water is gone in fully allocated systems or during years of scarcity.

Principle 3: Requirement to "use it or lose it"

A common feature of western water law is the concept of "beneficial use." Western water law requires licensees to make beneficial use of the water they divert, to ensure they are not hording or creating speculative water licences. If water is not used for a specified period of time (for example, 3 years in British Columbia), the provincial government may cancel the rights. This provision is intended to offset concerns associated with FITFIR as it offers one of the few opportunities for government to cancel or alter water entitlements. It is an important counter measure to the relatively static and rigid nature of FITFIR.

In British Columbia, as in the rest of western Canada, legislators have not defined beneficial use, and both the courts and water-related tribunals have not interpreted the phrase in the water law context, which creates significant ambiguity. In practice, beneficial use has come to mean the payment of annual administrative fees associated with water licences and adherence to the terms of water use licences.

Principle 4: Water as a resource for economic development

Allocated water is first and foremost for human use and economic development, and the legal regime treats water as a resource with little attention to the cultural or ecological aspects of water as a socio-ecological condition. Ecological considerations, including baseline hydrological needs, are not explicitly considered in most decisions related to water use and licensing in Canada (Nowlan 2012; Brandes 2005; Brandes and Nowlan 2009). There are only minimal provisions to require formal consideration of ecosystem needs—sometimes called environmental flows—and the rigour with which ecological considerations factor into licensing decisions is highly variable. Also, scientific information about ecosystem needs, including data on even basic environmental flow thresholds, varies widely across Canada (NRTEE 2010).

Water law and allocations in Canada are becoming increasingly complex. Various legal and administrative amendments have been introduced to address emerging concerns such as water scarcity, bulk water exports, the need for shorter-term more flexible temporary permits, provisions to enable local water planning frameworks, and dated time limits on entitlements. In light of these changes, it is time to revisit the foundational principles of water law in Canada.

4.4 Critiques of Western Water Law

The most contentious of these foundational principles are the assertion of Crown ownership and the FITFIR allocation system, both of which continue to present challenges across western Canada. This is true even with the most recent legislation in British Columbia, the new *Water Sustainability Act*, which fails to address the historical colonial systems that privilege government ownership and management over traditional indigenous law. Many view that the authority and legitimacy of government action, including licensing and creating entitlements to water, rests on a legal fiction that the Crown "owns" the water (Phare 2009).

A recent Supreme Court of Canada decision clearly established that Aboriginal rights and title can no longer be ignored without significant legal implications (*Tsilhqot'in Nation v. British Columbia* 2014). It has been described as a legal "earthquake" and reinforces the notion that British Columbia cannot have a functional water law regime until First Nations are involved in a meaningful way that respects their constitutionally affirmed rights.[7]

Even within the confines of the existing system, Indigenous peoples have clearly been using water the longest and, based on legal recognition through the provincial FITFIR system, Indigenous peoples should be entitled to the oldest water rights. Yet, when British Columbia developed its water law regime in the early 1900s, the provincial government did not consistently grant Indigenous peoples the most senior water rights in quantities that would secure their continued livelihoods (Matsui 2009).

British Columbia missed the opportunity to address this inequality in the new *Water Sustainability Act* by not explicitly acknowledging Aboriginal water rights.[8] The Province continues to assert Crown ownership over all water in British Columbia, which is problematic for many Indigenous people and Aboriginal governments who have outstanding claims over the land and water of their traditional territories in the province (Wilson-Raybould 2013).

In addition, the FITFIR system offers the illusion of certainty for investment and development. The reality is that, as watersheds experience low flows due to shifting hydrology or periodic drought, the system has limited utility in allocating water that both keeps ecosystems functional and ensures robust local economic activity (Brandes et al. 2015). The most senior licence holders may not be beneficially diverting water for the most economically valuable uses. While administratively convenient, FITFIR is a rigid system that provides for senior water users to have priority over all other economic and environmental uses, unless the Crown intervenes, which historically is a rarity.

The more senior licences and their uses are often "locked in" because the government granted many of the oldest rights in perpetuity or with no end date. Only recently have governments begun granting newer licences with specified terms. These perpetual and fixed uses can be problematic as social and economic contexts and priorities change. Conflicts over water can also be amplified as changing hydrology creates more uncertainty and variability year to year, resulting in uncertainty for

---

[7] The need for the Province to work with First Nations in a meaningful way that respects their constitutionally protected rights is clearly illustrated in the British Columbia Environmental Appeal Board's September 2015 decision to revoke Nexen Inc's water licence in part because the Province failed to consult in good faith with the Fort Nelson First Nation. See *Chief Gale and the Fort Nelson First Nation v. Assistant Regional Water Manager* (2012).

[8] No colonial court has recognized a specific Aboriginal water right, but courts have acknowledged rights to conditions that support Aboriginal water rights, such as the right to fish.

even some senior licensees about how much water they can divert in the dry months of August and September.

Another concern associated with FITFIR is the lack of incentive for the efficient use of water. Hydrology, technology, best practices, and industry standards all change over time; however, it is not possible to change or modify how water is used or how much is used under licences. This inability to adapt coupled with the "use it or lose it" principle create a powerful incentive to protect licences since non-use can result in the loss of the rights to divert water. In effect the "use it or lose it" principle encourages, or even mandates, waste in the FITIR regime.

The overall lack of attention to nature's limits associated with the current approach to water law has significant impacts on watersheds, fish populations, and ecosystems. Failure to consider and protect environmental flows results in periodic (and often persistent) water shortages for fish and other species, as well as for overall watershed health and function. Overallocation is also a growing concern as water users continue to expect ongoing access, even in the face of changing hydrology and increasing scarcity. The result is mounting conflict between users, the environment and the economy when water is not reliably available.

An example of the impact of this lack of attention to the basic ecology of rivers, streams, and lakes is seen in the Cowichan River watershed on eastern Vancouver Island, British Columbia where salmon are regularly trucked upriver in the fall and released to spawn due to insufficient water. Unrestricted licensed water withdrawals continue in the area, and the capacity to store and holdback water for release at critical times is lost due to the opposition by private landowners claiming that requirements interfere with their riparian access (*Weir v. British Columbia* 2013).

This failure to attend to ecological needs is reinforced by the paucity of reliable information about the amount of water available in most watersheds, and when and how much of the water is being used (Woods and Pentland 2013). The lack of credible and reliable information extends to limited groundwater mapping, insufficient hydrometric information, and minimal climate projections. These major gaps in data are a Canada-wide problem that becomes more acute when attempting to make evidence-based decisions. The situation is reinforced by licences that do not require users to monitor and report actual water use, and by systematic cuts to provincial or federal comprehensive information management systems (NRTEE 2010).

4.5 The Evolution of Water Law—A Case Study in British Columbia

Although transforming the existing system—based on the legal foundations discussed above—will take time and a significant commitment to rearrange the static, top down approach to water management, interesting potential exists in British Columbia's new *Water Sustainability Act* with immediate opportunities for improvements.

British Columbia is an economically, ecologically, and socially diverse place that reflects a full suite of water-related issues. The province includes some of the driest places in the country, such as the Okanagan in the central interior; coastal rainforests; Canadian heritage rivers, such as the Cowichan and Kicking Horse; and globally significant rivers such as the Skeena, Fraser, and the headwaters of the Columbia and Mackenzie rivers.

British Columbia's hydrological and geographic diversity includes watersheds encompassing a range of ecosystems from arid grasslands and coastal temperate rainforest to high alpine source waters. Water uses in the province in many ways mirror those in other parts of Canada. This includes highly concentrated geographic water use in the south and along the coast and in urban watersheds that produce significant agricultural commodities. There is intensive water use for energy production, including large-scale hydroelectric generation, and run-of-the-river and hydraulic fracturing in the north- and south-east interiors. Roughly one-third of the population relies on groundwater, similar to the national average.

Although not facing the level of scarcity and water crisis unfolding in the western United States, recent drought events further challenge the current approaches to water management and law in the province (Christensen and Brandes 2015). For example, in the last two decades, British Columbia has faced an increasing number of droughts, floods, and other water issues, including conflicts over water use, streams running dry, declining aquifer levels, and degraded watersheds (Brandes et al. 2016).

In 2015 alone, several regions in British Columbia experienced pronounced water shortages (Pynn 2015; CBC 2015; Leighton 2015). Canadian heritage rivers such as the Cowichan River on Vancouver Island have had some of the lowest flows in recent history (Fumano 2015). Some watersheds faced fishing closures due to warm water temperatures and low-flow conditions, which put fish stocks in jeopardy (BC FLNRO 2015; Sienuic 2015). Throughout the summer of 2015, reservoir levels also rapidly declined in several areas: Metro Vancouver's reservoir storage dropped dangerously below its normal range in July (Metro Van 2015), and the reservoir levels in Campbell River reached historically low and critical levels, and the community was within weeks of running out of water (Campbell River Mirror 2015).

With this changing "waterscape" as backdrop, British Columbia's recent law reforms provide a useful opportunity to explore how the core principles of western water law are beginning to fracture and adapt to the changing world of water, offering insights into a path forward.

4.5.1 British Columbia's New Water Sustainability Act

In May 2014 the Province of British Columbia enacted the new *Water Sustainability Act* (WSA or "Act"), which replaced the 106-year-old *Water Act*. The new Act is the result of an extensive modernization process driven by numerous specific commitments in British Columbia's 2008 *Living Water Smart Plan* (BC MoE 2008). It came into force on February 29th, 2016, and is now the cornerstone of British Columbia's legal framework for water.

The *Water Sustainability Act* is a framework, or enabling legislation, which means that the critical details of the legislation will be determined in supporting regulations developed over the next several years. From a sustainability perspective, while many of the most promising elements of the Act are not yet in place, this new law portends a different kind of regime that emphasizes partnerships and begins to address some of the more serious challenges associated with the lingering foundations of western water law.

The Act has the potential to be a strong piece of environmental legislation, including features that promise to protect and restore British Columbia's fresh water and offer significant potential for improved outcomes on the ground (and in the water) (Curran 2014c; Gage 2014; WWF-Canada 2014; ORC 2014). Highlights include: regulating groundwater for the first time; environmental and critical flow protections; the establishment of "water objectives" that improve links between water quality and quantity and land and water management; incentives to increase water use efficiency; more detailed measurement and reporting; and improved planning and governance through localized and binding plans and provisions for shared or delegated decision-making.

Instead of reviewing the legal minutia of the new Act, the remainder of this chapter focuses on the broad strokes of how this new law begins the necessary evolution away from of the foundational principles associated with western water law, although significant historical influences remain, which may, ultimately, impede its effective implementation.

4.5.2 Contested Crown Ownership

The new Act reaffirms the assertion of ownership over water flowing in streams, lakes, and ground water. This is highly contested by Aboriginal peoples across Canada and especially in British Columbia where the majority of the province is unceded traditional lands and waters (i.e., where no treaties exist). This chapter does not discuss Aboriginal water rights in any detail; however, the importance of this issue and the fundamental flaw it highlights for the legitimacy of the western water law system cannot be overstated.[9]

The Provincial government has failed to acknowledge and quantify Indigenous water use in the new regime or in each watershed's water balance. This type of "exclusion solution" will not be legally feasible in the future as the law continues to

[9] The US Indian reserved rights doctrine (Winter's Doctrine) addresses this concern to some extent. It arises from a 1908 Supreme Court case that recognized an implied federal reserve right to water in a sufficient amount to fulfill the purpose of Indian reservations (see Brooks 2005) There is no legal principle that mirrors this protection for basic water entitlements for Indigenous people in Canada.

evolve in its recognition of Aboriginal rights and title. Canada's legal landscape is constantly changing, and the recent Supreme Court of Canada decision, *Tsilhqot'in Nation*, illustrates this dynamism. The Tsilhqot'in decision confirmed that any provincial laws that operate to extinguish Aboriginal title are illegal, including, by extension, those that relate to fresh water (*Tsilhqot'in Nation v. British Columbia* 2014).

The Province's failure to recognize Aboriginal water rights under the *Water Sustainability Act*, if unaddressed, will have legal and operational repercussions for the provincial water law regime as a whole. Courts could invalidate parts of the provincial licensing and permitting regime in regions where insufficient consultation and accommodation has occurred and Aboriginal title found. This lack of consideration of the extent of existing and future water entitlements amplifies uncertainty for the current and future licensing system.

4.5.3 First in Time, First in Right (FITFIR) "Off-ramps"

FIFFIR remains firmly entrenched in the British Columbian legal regime. This is problematic, especially as it now extends to groundwater and there is increasingly a need to adapt to changing hydrological cycles, economic demands, and social priorities.

As discussed, FITFIR creates a rigid system that locks in historical—often suboptimal—uses. Yet, the new Act does begin to erode this aspect of western water law by building on existing administrative remedies. Various short-term government interventions that prioritize water for fish and ecosystems over licence priority are possible and increasingly streamlined under the new regime. This includes temporary orders to deal with acute shortage or threats to fish populations. Also, a number of FITFIR "off-ramps" are now in place, but not yet tested, under the new regime. These off-ramps include the ability to prioritize domestic and critical environmental flows during periods of water shortage ahead of all other licensed uses, and the ability to initiate legally enforceable water sustainability plans that can suspend, add conditions to, and even claw back existing licences.

A crucial dimension of the Act in addressing the rigidity of FITFIR is the ability to review existing licences—even those with no termination dates. Such a review can include amending licence conditions and terms, for example, for the express purpose of water conservation, environmental flow protection or the more efficient use of water. The comptroller of water rights or a water manager may notify licensees of a review any time after February 29, 2046 (30 years after the date the Act came into effect) or 30 years after the granting of a new licence or review of the terms and conditions of their licence (*Water Sustainability Act* 2014, s.23).

Arguably these various alternatives and off-ramps have the potential to render FITFIR increasingly irrelevant in some of the more serious or persistent low-flow scenarios.

4.5.4 Incentives for Efficiency and Water Sustainability Planning

Incentives for efficiency, improved planning and, importantly, the potential for innovative forms of governance lie at the heart of the new legal regime. Beneficial use is better defined and clarified in the Act itself with explicit requirements to use water as efficiently as practicable. Further specificity can be developed through new regulations, and decision-makers can require licensees to undertake a water conservation audit to show efficiency of use and to take measures to meet a specified level of water use efficiency and conservation (*Water Sustainability Act* 2014, s.30).

A powerful change in the new legislative regime is the sweeping ability to enact legally enforceable water sustainability plans. These provisions are the best example of a new partnership approach to water management that veers away from a strictly top-down regulatory system. Plans have the potential to prevent or address conflicts between water users or users and the environment, and have significant scope to go well beyond water allocations to include water quality, drought planning, water sharing, and changes to exiting licences. Indeed, water sustainability plans may ultimately lay the watershed-specific foundation for moving past the problematic foundational principles associated with the western water law regime.

The provincial cabinet has far reaching powers to initiate such plans and make them binding across sectors and on a variety of decision-makers exercising jurisdiction over Crown and private lands, such as other land managers or local governments. Ultimately, the plan's ability to take an integrated watershed approach is significant, especially when coupled with explicit powers in the Act to delegate decision-making to more local watershed authorities. At this point the utility of water sustainability plans is more potential than real as the plans must be developed and tested with the support of industry and local and Aboriginal communities.

4.5.5 Water for Nature

The *Water Sustainability Act* provides a variety of methods to address ecological considerations, and specifically to protect environmental flows, or water for nature. This addition represents one of the most fascinating, and potentially significant, shifts in the approach to water management.

Groundwater is now integrated into the regulatory apparatus (starting with non-domestic groundwater users), which means that surface and groundwater are managed as one interconnected resource. Environmental flows are better protected. Key features now in force include protections for critical flows so ecosystems and fish can *survive* during periods of drought (*Water Sustainability Act* 2014, s.87), as well as formal requirements for decision-makers to consider environmental flow needs in future licensing decisions to ensure aquatic ecosystems can *thrive* over the long term (*Water Sustainability Act* 2014, s.15).

Other protections include sensitive stream designations that limit development and require explicit attention to protecting fish habitat (*Water Sustainability Act* 2014, s.17), the ability to set aside water in ecological reserves (*Water Sustainability Act* 2014, s.39) and water objectives, which are legal tools that can influence land use decisions affecting water quality and quantity (*Water Sustainability Act* 2014, s.43).

Taken together, this suite of tools represents a significant departure from the historical approach. It signals a powerful rebalancing of ecological and economic priorities associated with water and provides government decision-makers and even local governments and communities with considerable flexibility to better ensure that water protection and allocation match current ecological needs and conditions.

4.6 Changing Currents—Toward a Twenty-first Century Approach

The top-down approach to water management emphasizes the diversion of water for human uses and generally ignores ecosystem needs and ecological priorities. It does not align with the complexity of today's water issues. Water simply does not adhere to political boundaries as it flows across the landscape, with ecosystems that are dynamic, governed by uncertainty and continual change.

In the context of modern water challenges, some governments in Canada are realizing that they do not have the capacity to be the exclusive decision-makers or water managers. Neither can these senior governments alone possibly keep on top of the real-time information needed to make water management decisions as seasonal and annual hydrological variability increases, and local social demands and priorities change.

Consensus has emerged in the academic literature that "good" governance for water includes the principles of participation, legitimacy, transparency, and accountability (van der Valk and Keenan 2011; Matthews and Schmidt 2014; Rogers and Hall 2003; Cook and Bakker 2011). Walker and Salt (2006) indicate that an outcome of "good" or improved water governance would be that it provides the capacity to build social-ecological resilience—capacity of the watershed and the communities and businesses within them to withstand disturbances while maintaining their structure, function, identity, and ability to learn and/or transform as needed.

Twenty-first century water governance requires a more collaborative approach where all governments, rights holders, and stakeholders have roles and responsibilities, with creative integration of top-down and bottom-up planning and decision-making (Brandes and O'Riordan 2014; Shurbsole 2004; Nowlan and Bakker 2010).

A fundamentally important development in water law is the recasting of water allocation as a mechanism for planning and collaboration and not simply as an administrative task as it often is during "normal" times. The example of British Columbia demonstrates how this water law regime is "changing the current"—evolving gradually toward such a system with a promise (as yet unfulfilled) to successfully implement a strong *Water Sustainability Act*.

The potential of a modern water law regime is certainly significant. At the heart of the reforms in British Columbia is a move away from a rigidly controlled, state-run, and top-down approach toward a model that emphasizes new priorities around nature and a creative mix of local and government actors with partnership and shared decision-making at its core. This modified legal regime makes possible the transformation of how communities interact with water and how nature is fundamentally included.

Ultimately, the success of this new water law regime will depend on a vibrant partnership between the Province, Aboriginal peoples and First Nations, federal and local governments, water licence holders, and community and watershed organizations all coming together to take leadership in and responsibility for water stewardship.[10]

All partners will be required to contribute to the day-to-day management of freshwater resources, engage in long-term watershed planning, and provide appropriate local solutions to water-related problems as they emerge. More fundamentally, partners will need to work together in each watershed under a shared risk and responsibility approach to water management and governance.

The partnership model (see Fig. 4.1) illustrates the kind of relationships needed to fulfill the potential of this approach. In this conceptualization, the series of smaller tributaries represent key actors and their designated roles and responsibilities. These various tributaries flow into a larger metaphorical river, representing a modern management and governance approach. This approach ultimately changes the patterns, or currents, of water use to better balance the needs of nature and the multitude of uses that exist in the watershed. A complementary modern water law regime is needed to reinforce this vision and ensure that the appropriate incentives and institutional architecture exist to facilitate the paradigm shift.

Getting from the existing system to a truly modern and more sustainable system will require a herculean effort. In many ways the process is already underway as demonstrated in the British Columbia case study, and as western water law continues to evolve in the face of economic, social, and ecological challenges.

---

[10] British Columbia's *Northeast Water Strategy* articulates one approach to this partnership concept. Unified water stewardship is one of the Strategy's core principles. This includes co-stewardship of water resources with First Nations and other partners, and also sharing of knowledge, research, and data between partners and between other overlapping water management activities in the region. (British Columbia 2015).

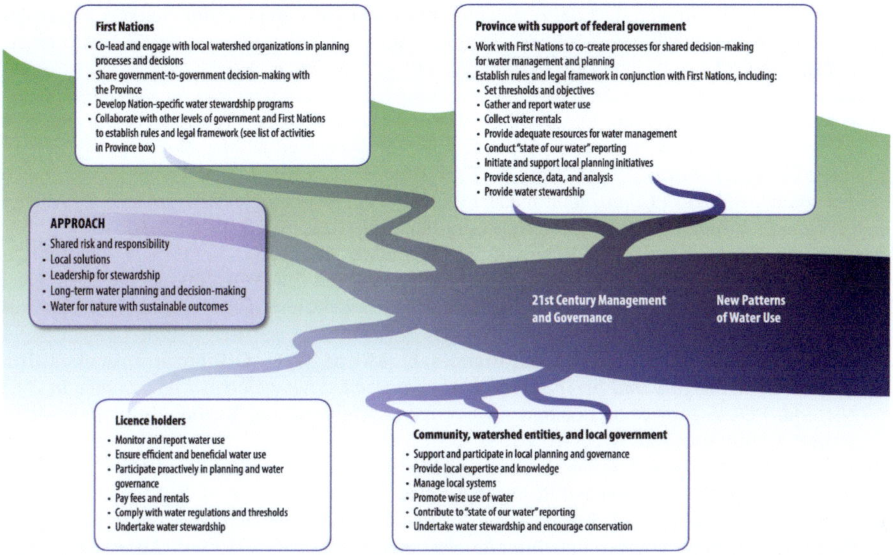

Fig. 4.1 Partnership model for British Columbia—roles and responsibilities (Brande s et al. 2015)

References

Acton v. Blundell. (1843). 12 M. W. 324.

Alberta Environment. (2003). *Water for life strategy*. http://www.waterforlife.gov.ab.ca. Accessed 20 Mar 2016.

Alberta Environment and Parks (AEP). (2016). *Water for life*. http://aep.alberta.ca/water/programs-and-services/water-for-life/default.aspx. Accessed 25 Mar 2016.

Bailey, A. (2008). *Federalism: Water law: The interjurisdictional context*. Centre for Constitutional Studies, University of Alberta. http://ualawccsprod.srv.ualberta.ca/ccs/index.php/constitutional-issues/federalism/50-water-law-the-interjurisdictional-context#_ftnref3. Accessed 24 Mar 2016.

BC Government. (2015). *Northeast water strategy*. http://www2.gov.bc.ca/gov/content/environment/air-land-water/water/water-planning-strategies/northeast-water-strategy. Accessed 20 Mar 2016.

BC Ministry of Environment (BC MoE). (2008). *Living water smart: British Columbia's water plan*. http://www.livingwatersmart.ca. Accessed 20 Mar 2016.

BC Ministry of Environment (BC MoE). (2016). *Modernization process. Development of living water smart and the water act modernization process*. https://engage.gov.bc.ca/watersustainabilityact/modernizationprocess/. Accessed 24 Mar 2016.

BC Ministry of Forests, Lands and Natural Resource Operations (BC FLNRO). (2015). Fishing closures announced for south coast. *BC Government News*, 20 July. https://news.gov.bc.ca/stories/fishing-closures-announced-for-south-coast. Accessed 20 Mar 2016.

Borrows, J. (2002). *Recovering Canada: The resurgence of indigenous law*. Toronto: University of Toronto Press.

Brandes, O. M. (2005). At a Watershed: Ecological governance and sustainable water management in Canada. *Journal of Environmental Law and Practice, 16*(1), 79–97.

Brandes, O. M. (2014, May 5). New BC water act generally gets things right. *The Province*.

Brandes, O. M., & Nowlan, L. (2009). Wading into uncertain waters: Using markets to transfer water rights in Canada—Possibilities and pitfalls. *Journal of Environmental Law and Practice, 19*(3), 267–287.

Brandes, O. M., & O'Riordan, J. (2014). *Decision-makers brief: A blueprint for watershed governance in British Columbia*. Victoria: POLIS Project on Ecological Governance, University of Victoria. http://poliswaterproject.org/publication/760. Accessed 18 Mar 2016.

Brandes, O. M., O'Riordan, J., O'Riordan, T., & Brandes, L. (2014). *A blueprint for watershed governance in British Columbia*. Canada: POLIS Project on Ecological Governance, University of Victoria. http://www.poliswaterproject.org/blueprint. Accessed 23 Mar 2016.

Brandes, O. M., Carr-Wilson, S., Curran, D., & Simms, R. (2015). *Awash with opportunity: Ensuring the sustainability of British Columbia's new water law*. POLIS Project on Ecological Governance, University of Victoria, Canada. http://poliswaterproject.org/awashwithopportunity. Accessed 20 Mar 2016.

Brandes, O. M., Morris, T., Archer, J. L., Brandes, L., Moore, M. L., O'Riordan, J., et al. (2016). *Illumination: Insights and perspectives for building effective watershed governance in B.C.* Victoria: POLIS Project on Ecological Governance, University of Victoria. http://poliswaterproject.org/illumination. Accessed 20 Mar 2016.

Brooks, N. (2005). *Indian reserved water rights: An overview*. CRS Report for Congress. Washington, DC: Congressional Research Service

Campbell River Mirror. (2015, June 23). *Record low flow forces down dam discharge*. Campbell River Mirror. http://www.campbellrivermirror.com/news/309362371.html. Accessed 22 Mar 2016.

CBC News. (2015, July 3). Vancouver Island fishing banned as drought hits level 4. *CBC News British Columbia*. http://www.cbc.ca/news/canada/british-columbia/vancouver-island-fishing-banned-as-drought-hits-level-4-1.3137599. Accessed 25 Mar 2016.

Chasemore v. Richards (1859) 7 HLC 349.

Chief Gale and the Fort Nelson First Nation v. Assistant Regional Water Manager. (2015, September 3). BC EAB Decision, 2012-WAT-013(c).

Christensen, R., & Brandes, O. M. (2015) *California's Oranges and B.C.'s Apples? Lessons for B.C. from California groundwater reform.* POLIS Project on Ecological Governance, University of Victoria, Canada. http://poliswaterproject.org/orangesapples. Accessed 20 Mar 2016.

Constitution Act (1867) (U.K.) 30 & 31 Vict., c. 3, s. 91 and 92, reprinted in RSC 1985, App. II, No. 5

Cook, C., & Bakker, K. (2011). Water governance in Canada: Innovation and fragmentation. *International Journal of Water Resources Development, 27*(2), 275–289.

Council of Canadian Academies. (2009). *The sustainable management of groundwater in Canada—Expert Panel on Groundwater*. Ottawa: Council of Canadian Academies.

Curran, D. (2014a). Can BC's New *Water Sustainability Act* Plug the Leaks? Seminar for University of Calgary Faculty of Law, Calgary, Canada.

Curran, D. (2014b). *British Columbia's Water Sustainability Act—A new approach to adaptive management and no compensation regulation*. The University of Calgary Faculty of Law Blog on Developments in Alberta Law. http://ablawg.ca. Accessed 28 May 2014.

Curran, D. (2014c). *British Columbia's New Water Sustainability Act—Waiting for the details*. Environmental Law Centre, University of Victoria. http://www.elc.uvic.ca/publications/british-columbias-new-water-sustainability-act-waiting-for-the-details/. Accessed 21 Mar 2016.

Curran, D. (2015). Water law as watershed endeavour: Federal inactivity as an opportunity for local initiative. *Journal of Environmental Law and Practice, 28*(1), 53–88.

Curran, D., & Brandes, O. M. (2012). When the water dries up: Lessons from the failure of water entitlements in Canada, the U.S. and Australia—Workshop Discussion Paper. University of Victoria, Canada. http://poliswaterproject.org/publication/478. Accessed 25 Mar 2016.

Curran, D., & Mascher, S. (2016). Adaptation in water law: Evaluating Australian (New South Wales) and Canadian (British Columbia) law reform initiatives. *McGill International Journal of Sustainable Development Law and Policy, 12*(1), Forthcoming.

de Loë, R., Varghese, J., Ferreyra, C., & Kreutzwiser, R. (2007). *Water allocation and water security in Canada: Initiating a policy dialogue for the 21st Century*. Guelph: Guelph Water Management Group, University of Guelph.

Fumano, D. (2015, June 10). Water experts urge B.C. to take a lesson from California drought. *The Province*. http://poliswaterproject.org/sites/default/files/2015-06-09_TheProvince_GroundwaterReport.pdf. Accessed 22 Mar 2016.

Furubotn, E., & Pejovich, S. (1972). Property rights and economic theory: A survey of recent literature. *Journal of Economic Literature, 10*(4), 1137–1162.

Gage, A. (2014). *The strengths and weaknesses of the new Water Sustainability Act*. http://wcel.org/resources/environmental-law-alert/strengths-and-weaknesses-new-water-sustainability-act. Accessed 23 Mar 2016.

Government of Northwest Territories (GNWT). (2016). *Strategy development timeline for northern voices northern peoples initiative*. NWT Water Stewardship. http://www.nwtwaterstewardship.ca/strategy_development_timeline. Accessed 21 Mar 2016.

Government of the Northwest Territories and Aboriginal Affairs and Northern Development Canada. (2010). *Northern voices, northern waters: NWT Water Stewardship Strategy*. http://www.nwtwaterstewardship.ca/introduction. Accessed 23 Mar 2016.

Harris, E. (2013). The evolution of water rights in the nineteenth century: The role of climate and asset type. *Natural Resources Journal, 53*, 217–263.

Hogg, P. (2007). *Constitutional Law of Canada*, student edition. Thomson Carswell, Scarborough Interprovincial Co-operatives Ltd. v. Manitoba (1975) [1976] 1 SCR 477.

Hutchins, W. A., Ellis, H. H., & DeBraal, J. P. (1971). *Water rights laws in the nineteen western states*. Washington, DC: United States Department of Agriculture.

Johnson, N. K., & DuMars, C. T. (1989). A survey of the evolution of western water law in response to changing economic and public interest demands. *Natural Resources Journal, 29*, 348–387.

Leighton, B. (2015, July 4). Heat wave forces burning ban, fishing closures as BC copes with drought. *The Province*.http://www.vancouverobserver.com/news/heat-wave-forces-burning-ban-fishing-closures-bc-copes-drought. Accessed 20 Mar 2016.

Lucas, A. R. (1990). *Security of title in Canadian water rights*. Calgary: Canadian Institute of Resources Law.

Matsui, K. (2009). *Native peoples and water rights: Irrigation, dams and the law in western Canada*. Montreal: McGill-Queens University Press.

Matthews, N., & Schmidt, J. J. (2014). False promises: The contours, contexts and contestation of good water governance in Lao PDR and Alberta, Canada. *International Journal of Water Governance, 2*(2–3), 21–40.

Mattison, J. (2016, February 1–2). Understanding the context: Environmental flow needs in British Columbia. Paper presented at the forum on environmental flow needs in British Columbia, WWF-Canada and POLIS Project on Ecological Governance, Vancouver, Canada.

Metro Vancouver. (2015). *Reservoir levels and daily consumption*. http://www.metrovancouver.org/services/water/conservation-reservoir-levels/reservoir-levels/Pages/default.aspx. Accessed 22 Mar 2016.

Muldoon, P. R., Lucas, A. R., Gibson, R. B., Pickfield, P., & Williams, J. (2015). *An introduction to environmental law and policy in Canada* (2nd ed.). Toronto: Emond Montgomery Publications.

Napoleon, V. (2007). *Thinking about indigenous legal orders*. National Centre for First Nations Governance. http://fngovernance.org/ncfng_research/val_napoleon.pdf. Accessed 20 Mar 2016.

National Round Table on the Environment and the Economy. (2010). *Changing currents: Water sustainability and the future of Canada's natural resource sectors*. Ottawa, Canada. http://nrt-trn.ca/water/water-sustainability-and-the-future-of-canadas-natural-resource-sectors-2/changing-currents. Accessed 21 Mar 2016.

Nowlan, L. (2005). *Buried treasure: Groundwater permitting and pricing in Canada*. Walter and Duncan Gordon Foundation. http://powi.ca/wp-content/uploads/2012/12/Buried_Treasure-Groundwater-Permitting-and-Pricing-in-Canada-2005.pdf. Accessed 22 Mar 2016.

Nowlan, L. (2012). CPR for Canadian rivers—Law to conserve, protect, and restore environmental flows in Canada. *Journal of Environmental Law and Practice, 23*, 203–252.
Nowlan, L, & Bakker, K. (2010). *Practising shared water governance in Canada: A primer.* UBC Program on Water Governance, Vancouver, Canada. http://watergovernance.ca/wpcontent/uploads/2010/08/PractisingSharedWaterGovernancePrimer_final1.pdf. Accessed 20 Mar 2016.
Ontario Ministry of the Environment and Climate Change (OMECC). (2016). *Source protection. Development of Ontario's comprehensive source protection program.* https://www.ontario.ca/page/source-protection. Accessed 21 Mar 2016.
Outdoor Recreation Council of British Columbia and Ducks Unlimited Canada. (2014, March 12). New water act will make BC leader in environmental stewardship: Environment Minister. *CTV News Vancouver.*http://bc.ctvnews.ca/new-water-act-will-make-b-c-leader-in-environmental-stewardship-environment-minister-1.1725666. Accessed 25 Mar 2016.
Pearse, P. H., & Quinn, F. (1996). Recent developments in federal water policy: One step forward, two steps back. *Canadian Water Resources Journal, 21*(4), 329–340.
Percy, D. R. (1988). *The framework of water rights legislation in Canada.* Calgary: Canadian Institute of Resources Law.
Percy, D. R. (2004). The limits of western Canadian water allocation law. *Journal of Environmental Law and Practice, 14*, 313–327.
Phare, M. S. (2009). *Denying the source: The crisis of first nations water rights.* Victoria: Rocky Mountain Books.
Pynn, L. (2015, September 17). Competition for precious water in drought-stricken Nicola Valley. *The Vancouver Sun.* http://www.vancouversun.com/Competition+precious+water+drought+stricken+Nicola+Valley/11372427/story.html. Accessed 20 Mar 2016.
Quebec Government. (2009). *An act to affirm the collective nature of water resources and provide for increased water resource protection, CQLR c C-6.2.* http://canlii.ca/t/52b2f
R. v. Crown Zellerbach Canada Ltd. (1988). 1 SCR 401
R. v. Hydro-Québec. (1997). 3 SCR 213
Rivera, A. (2005). How well do we understand groundwater in Canada? A science case study. In: Nowlan, L. (Ed.), *Buried treasure: Groundwater permitting and pricing in Canada.* Walter and Duncan Gordon Foundation (pp. 4–12). http://powi.ca/wp-content/uploads/2012/12/Buried_Treasure-Groundwater-Permitting-and-Pricing-in-Canada-2005.pdf. Accessed 20 Mar 2016.
Rogers, P., & Hall, A. W. (2003). *Effective water governance.* TEC Background Papers, No. 7. Global Water Partnership Secretariat, Stockholm, Sweden. http://water.worldbank.org/publications/effective-water-governance. Accessed 21 Mar 2016.
Saskatchewan Water Security Agency (SWSA). (2012). Area 1.3: Water allocation system. In: *25 year Saskatchewan water security plan action* (pp. 8–9). https://www.wsask.ca/About-WSA/Publications/25-Year-Water-Security-Plan/. Accessed 22 Mar 2016.
Schlager, E., & Ostrom, E. (1992). Property-rights regimes and natural resources: A conceptual analysis. *Land Economics, 68*(3), 249–262.
Shrubsole, D. (Ed.). (2004). *Canadian perspectives on integrated water resources management.* Cambridge: Canadian Water Resource Association.
Sieniuc, K. (2015, July 29). First Nations Tribal Council suspends Okanagan sockeye salmon fishery. *The Globe and Mail.* 29 July. http://www.theglobeandmail.com/news/british-columbia/first-nations-tribal-council-suspends-okanagan-sockeye-salmon-fishery/article25769695/. Accessed 22 Mar 2016.
Thayer v. California Dev. Co. (1912). 164 Cal. 117, 128 Pac. 21
Tsilhqot'in Nation v. British Columbia. (2014). SCC 44
van der Valk, M. R., & Keenan, P. (2011). Principles of good governance at different water governance levels. Papers presented at a workshop held on 22 March 2011 in Delft, The Netherlands. UNESCO, Netherlands.
Water Act, RSBC. (1996). c 483, s. 2. http://canlii.ca/t/52cdd
Water Act, RSA. (2000). c W-3, s. 30. http://canlii.ca/t/8220

Water Rights Act, CCSM. (2015). c W-80, ss. 8-9. http://canlii.ca/t/8gvc

Water Sustainability Act, SBC. (2014). c 15. http://canlii.ca/t/528ss

Weir v. British Columbia. (Deputy Comptroller of Water Rights). (2014). BC EAB Decision, 2013-WAT-013(b), 015(c), 016(b), 017(c), 018(c), & 019(c). 14–17 April 2014 and 3–5 June 2014.

Walker, B. H., & Salt, D. A. (2006). *Resilience thinking: Sustaining ecosystems and people in a changing world*. Washington, DC: Island Press.

Wilkinson, C. F. (1992). *Crossing the next meridian: Land, water, and the future of the west*. Washington, DC: Island Press.

Wilson-Raybould, J. (2013). *Re: Water sustainability act legislative proposal*. Submission to Minister Mary Polak on the Water Sustainability Act, BC Assembly of First Nations, 2 December 2013. http://engage.gov.bc.ca

Wood, C., & Pentland, R. (2013). *Down the drain: How we are failing to protect our water resources*. Vancouver: Greystone Publishing.

WWF Canada. (2014). *This year you rewrote history: A bold new BC water act protects water for nature and people*. WWF-Canada Annual Report 2014. http://awsassets.wwf.ca/downloads/2014_wwf_freshwatergcob_c__water_act.pdf. Accessed 22 Mar 2016.

Chapter 5
Reconciliation and Relationality in Water Research and Management in Canada: Implementing Indigenous Ontologies, Epistemologies, and Methodologies

Heather Castleden, Catherine Hart, Ashlee Cunsolo, Sherilee Harper, and Debbie Martin

Abstract Water-related issues disproportionately affect Indigenous communities in Canada. Despite millions in investment, Western-trained scientists, engineers, and other researchers as well as the government agencies that have constitutionally-mandated fiduciary responsibilities to address such issues have been rather unsuccessful in solving them. This has been due, in large part, to an overreliance on methods of Western science and management, ignoring the vast place-based wisdom of Indigenous knowledge systems and relational practices regarding water found across the country. The underlying reasons for this partiality are not innocuous; entrenched colonial and racist policies, programs, and practices have persisted across time and space. In recent years, there is increasing recognition of the

H. Castleden (✉)
Health, Environment, and Communities Research Lab, Department of Geography and Planning, Queen's University, 99 University Ave, Kingston, ON K7L 3N6, Canada
e-mail: heather.castleden@queensu.ca

C. Hart
Health, Environment, and Communities Research Lab, Queen's University, 99 University Ave, Kingston, ON K7L 3N6, Canada
e-mail: catherine.hart@queensu.ca

A. Cunsolo
Departments of Nursing and Indigenous Studies, Cape Breton University, 1250 Grand Lake, Sydney, NS B1P 6L2, Canada
e-mail: ashlee_cunsolowillox@cbu.ca

S. Harper
Department of Population Medicine, University of Guelph, 50 Stone Rd E, Guelph, ON N1G 2W1, Canada
e-mail: harpers@uoguelph.ca

D. Martin
School of Health and Human Performance, Dalhousie University, 6230 South St, Halifax, NS B3H 4R2, Canada
e-mail: debbie.martin@dal.ca

S. Renzetti, D.P. Dupont (eds.), *Water Policy and Governance in Canada*, Global Issues in Water Policy 17, DOI 10.1007/978-3-319-42806-2_5

importance of applying Indigenous approaches to water challenges in Canada. But strategies for successful implementation are only beginning to emerge. In an attempt to respond to this knowledge gap, our research has sought to systematically identify and assess how both Indigenous and Western ontologies, epistemologies, and methodologies have been implemented in water research and management. In doing so, this chapter identifies some of the most promising practices in Canada. We share these with the goal of contributing to processes of reconciliation and responsibility towards each other as well as our roles as water stewards across the country.

5.1 Introduction

Water is essential for all life on earth, but access to water is neither universal nor equitable; and – as we have learned from Algonquin Elders Barbara and Albert Hill and from many others in our own co-learning journeys – if the water stops flowing, so do we. Yet, Canada is generally seen as having vast reserves of water; access and availability are often taken for granted (Bakker 2007; Norman et al. 2010). But those who are actively engaged in water research and water management have begun to recognize that Canadians can no longer live with this false sense of security (Environment Canada 2004). Growing numbers are recognizing that current research, management practices, and policies, derived from Western science, technology, and ideology are not enough to address the problems we find ourselves facing at an increasing and alarming rate: contaminated drinking water sources, flooding, drought, and wastewater management (for example, at the time of our writing, there was an international debate underway as to whether the City of Montreal should be allowed to dump 8 billion litres of raw sewage into the St. Lawrence River which spans Canadian and American borders). If Western science and technology cannot 'fix' our problems, then surely we need to turn to other ontologies, epistemologies, and methodologies to explore how we can be better stewards of water.

The goal of this chapter is to identify, document, and understand some of the current methods and models that are being used to bridge/integrate/implement Indigenous ontologies, epistemologies, and methodologies with Western ones to inform water resource management in Indigenous (i.e., First Nations, Inuit, and Métis/Metis) communities, and to outline how such approaches would actually benefit all Canadians. With this goal in mind, we begin by reflecting on the stormy relationship that has permeated Indigenous-settler relations for 500+ years to contextualize our current reality in Canada. After laying this foundation, we look at the available literature for evidence of attempts to effectively and ethically implement Indigenous and Western knowledges in integrative ways with respect to water research, management, and policy in Canada. From there, we turn to a deeper exploration of the ways in which such implementation is taking place and some of the log jams that prevent us from realizing this praxis. We draw from expert interviews and exemplary case studies to illustrate these points. We conclude our chapter with implications for policy-makers as well as policy recommendations for three key audiences: academic and other institution-based researchers, policy-makers, and research funding agencies. We have intentionally not made recommendations that are specifically for Indigenous communities, peoples, and/or nations for a couple of reasons. First, to do so would actually contravene our other recommendations that call on researchers, policy-makers, and funding agencies to recognize Indigenous sovereignty (see Sect. 5.7). But also, recommendations arising from our research were, for the most part, focused on those who have usually held the decision-making power: researchers, funders, and policy-makers.

5.2 Storm Clouds on the Horizon: A Brief Introduction to 500+ Years of Indigenous-Settler Relations

Indigenous peoples[1] have occupied what is now known as Canada for millennia. Origin stories, whether of the Mi'kmaq, Maliseet, Anishinaabe, Dene, Haida, Mohawk, Nuu-chah-nulth, Inuit, or Métis/Metis[2] (and so on), often begin with a physical and social connection to the lands and waters around them (see, for example, Atleo 2004). Indigenous peoples have lived off the bounty of their traditional territories for much of their needs and comforts (i.e., diet, medicines, shelter, clothing, transportation, culture, spirituality, and the arts) for generations (Castleden and Skinner 2014). Knowledge and use of the surrounding landscape and waterways was and continues to be critical for physical and cultural survival for many (National Collaborating Centre for Aboriginal Health 2011). Complex social and cultural institutions and customary laws have traditionally helped govern the ways in which Indigenous peoples have lived as *part of* not *apart from* or having dominion over nature (Waldram et al. 2006). Water, for example, figures prominently in the spirituality of many Indigenous cultures, including water teachings and water ceremonies (Blackstock 2001); and for many, water is understood as more than just a necessity for life but as being alive and having agency (McGregor 2012). These ontologies, epistemologies, and methodologies have continued to nurture and

[1] We use the term 'Indigenous' to refer to and recognize the inherent rights of the original inhabitants of the land now known as Canada; the Canadian Constitution identifies three groups of Indigenous peoples collectively as 'Aboriginal': Indians (now commonly identified as First Nations), Inuit, and Métis/Metis.

[2] We use both 'Métis' and 'Metis' because (as per the National Aboriginal Health Organization) while the original 'Métis' is translated from French as 'mixed blood' and applied to the children of French fur traders and Cree women in the Prairies, of English and Scottish traders and Dene women in the North, and Inuit and British in Newfoundland and Labrador, many peoples and groups particularly from the North and western Canada have dropped the accent.

sustain Indigenous children, families, communities, and nations despite over five centuries of colonial disruption from the Settler population.[3]

The arrival of Settlers, and their accompanying colonial mentalities, severely compromised Indigenous peoples' social, political, economic, linguistic, educational, and cultural institutions across the continent. Once established as a colonial state, Canada's first Prime Minister, Sir John A. MacDonald, set out to:

> do away with the tribal system and assimilate the Indian people in all respects with the inhabitants of the Dominion, as speedily as they are fit for change (Royal Commission on Aboriginal Peoples 1996, p 165).

Settler-colonialism continues today in the form of multiple inequities seen in the lived experiences of Canada's Indigenous peoples (Adelson 2005); this includes inequitable relationships with water. For example, approximately 20 % of all drinking water advisories in Canada are found in Indigenous communities, yet the Indigenous population represents less than 5 % of the total population (Simeone 2009). Worse, many of these Indigenous communities' advisories have been issued for years, the longest being over 20 years in Neskantaga First Nation. Drinking water conditions in reserve communities[4] made national headlines in 2005 when E. coli in the water system caused the evacuation of over 800 members of Kashechewan. Phil Fontaine, former National Chief of the Assembly of First Nations noted:

> these conditions would not be tolerated in any other municipal setting and if they are to occur, swift and decisive action is the norm and is expected (Polaris Institute 2008).

In reality, nowhere in Canada are drinking water problems more apparent than in Indigenous communities, where water sources and standards are deplorable (Swain et al. 2006). As of January 2016, Health Canada (2016) reports 135 drinking water advisories for First Nations alone (this number does not include First Nations in British Columbia, any of the Territories, or Inuit and Métis/Metis communities). Five years ago, Health Canada reported 117 such advisories for these communities. Despite massive financial investments, things are clearly not getting better (Morrison et al.2015).

But the drinking water problem in many Indigenous communities is not the only inequitable progeny of the Indigenous-Settler-water trifecta. Indigenous peoples live with a lack of or improperly functioning wastewater treatment and/or storm water systems and they are more at risk for water-related threats including, for example, drought, flooding, and contamination (Castleden and Skinner 2014). All of these issues exist disproportionately among First Nations, Inuit, and Métis/Metis

[3] In Canada, the 'Settler population' refers to those non-Indigenous peoples whose ancestors, or themselves, have immigrated to Canada at any point over the past 500+ years and, through intentional or unwitting actions, have dispossessed Indigenous peoples from their land and now inhabit their territories.

[4] Reserve communities are a product of the colonial system whereby the Settler government forcibly relocated many First Nations to small tracts of land (often not arable or populated with game or medicines), and colonial agents instituted a 'pass' system to control and regulate movement on and off reserve; the pass system is no longer at play but the legacy of regulation continues.

communities in Canada. Indeed, many of these communities face significant water management challenges, and they are now also experiencing these challenges in the context of increasingly rapid climate change and its impacts, such as altered stream flows, rising sea levels and coastal flooding, as well as changes in precipitation patterns (Centre for Indigenous Environmental Resources 2011). These water issues cannot just be glibly attributed to the physical geographies of where Indigenous peoples live in Canada, but are rather the socio-political outcomes of colonial and racist policies, programs, and practices that have existed since the creation of the Indian Act in 1876 (Basdeo and Bharadwaj 2013). Moreover, Indigenous peoples continue to be excluded from important water-related policy-making that directly affects them and their inherent and Treaty rights to water (Phare 2009; Morrison et al. 2015).

5.3 Sandbagging Knowledge: Barriers to Integrative Approaches to Water Research and Management

Understanding the above-mentioned historic and current colonial context helps to situate why current water-related challenges and threats exist, particularly in Indigenous communities in Canada. The need to reduce these disparities has led to an increase in research support and activities aimed at finding ways to more safely and effectively manage water-related resources that affect Indigenous peoples. For example, the federal government has spent approximately $3 billion on water and wastewater services in Indigenous communities since 2006 (Hyslop 2014), with another $300 million promised by the end of 2016 (Health Canada 2015). Yet, in many cases, little progress has been made (Morrison et al. 2015). In fact, for decades, government agencies and researchers have relied on Western science both to deliver water-related resources, and also to improve water (and other) management challenges in Indigenous communities. These efforts have largely proven to be ineffective (White et al. 2012). The overall failure to correct water-related and other colonially-derived issues has spurred two trajectories. First, we are seeing a resurgence amongst Indigenous peoples to reclaim self-determination (see, for example, Alfred and Corntassel 2005; Coulthard 2014). Second, there is growing interest within the Western science community towards bridging/integrating Indigenous ontologies, epistemologies, and methodologies with Western approaches to create processes and outcomes that are respectful and responsive to Indigenous priorities and needs (see, for example, Lavalley 2006; McGregor 2012; Phare 2011; Walkem 2007). This means, briefly, using a variety of qualitative and quantitative methods guided by Indigenous Knowledge paradigms to privilege Indigenous voices, resist dominant discourses, respond to local contexts, decolonize investigator-driven leadership, honour cultural protocols (e.g., traditions, customs, ceremonies), and strengthen Indigenous communities (Wilson 2008; Kovach 2009).

These integrative approaches are indicative of a growing focus on the interface between Indigenous and Western knowledges, reflecting the increasing recognition of the need to conduct research 'by' and 'with' rather than 'on' or 'in' Indigenous communities, and a broader acceptance amongst the Western science community of the importance of Indigenous knowledge as a legitimate and valued way of knowing (Bartlett et al. 2012).[5] It has become clear that it is vital to create new pathways for stewardship of our shared water resources that honour (and, where needed, revitalize) the perspectives, experiences, understandings, knowledge, and wisdom of Indigenous peoples. Embedding Indigenous ontologies, epistemologies, and methodologies into the design and delivery of any water resource management system and practice alongside innovations in Western science, engineering, and policy holds great promise. There are instances where such approaches *are* working, where integrative Indigenous and Western knowledges have come together in respectful and responsible ways to challenge colonial policies and practices in the context of water research and management. And these instances must gain prominence in a milieu where attempts to do so have often fallen flat. Thus, we aim to draw readers' attention to examples of them here (see our Case Studies below; see also Lawless et al. 2016 as well as Candler et al. 2010).

Before we do, however, there are three overlapping and persistent facets of past and ongoing water research and management that serve as log jams to successfully practicing integrative approaches. We have learned about them directly from the Indigenous Knowledge-holders we have encountered throughout our research practice and particularly in our direct engagement on this topic over the past 2 years. We have become mindful of them in our own work and we encourage readers to refer back to these facets to help critique research and management practices they are exposed to now or involved with in the future. The first is that Indigenous peoples' emotional, relational, gendered, and spiritual ties to water are typically neglected in research and management processes. For example, the *Safe Drinking Water for First Nations Act* refers only to drinking water and wastewater safety and, although rights to water by Indigenous peoples in Canada flow from treaties, unceded lands, the *Canadian Constitution*, and inherent rights, these are often in conflict with economic development goals driven by the Settler-Canadian population and colonial government (von der Porten and de Loë 2013). An exception would be the Northwest Territories Water Stewardship Strategy, which emphasizes peoples' relationships with water in the NWT and acknowledges the need to recognize these relationships and values in decision-making about water (Fresque-Baxter 2015). Positivist approaches and the notion of objectivity, stemming from Western science, have

[5] Integrative science is now often termed "Two-Eyed Seeing" (translated from *Etuaptmumk*), an Indigenous concept first articulated by Mi'kmaq Elder Albert Marshall in 2004. It is a call to the research community to see the strengths of Indigenous knowledge (one eye) on equal terms as Settler knowledge (one eye), and to learn how to use both (two eyes) to answer pressing research questions in more wholistic ways (Bartlett et al. 2012). The term has seen relatively rapid uptake in Indigenous health research since 2012 due in large part to the decision taken by the Candian Institutes of Health Research-Institute of Aboriginal Peoples' Health to foster this concept by prioritizing projects for funding that show evidence of using this approach.

permeated most research designs and processes. With the cultural turn in the social sciences since the 1970s, we see this less so; but many in the natural sciences, engineering, and health sciences continue to operate within post-positivist and objectivist paradigms to the detriment of the research and the meaningful involvement of Indigenous peoples' perspectives in said research.

The second persistent trend is that Indigenous knowledge is often seen as some 'thing' to be collected; but those who share such knowledge know – and those who practice integrative science are beginning to understand – that such knowledge cannot be 'extracted' from the knowledge-holder or the place they are from (McGregor 2004). Given this 'complication' from a Western perspective, many researchers and managers are baffled as to how to truly perform bridging or integrative work (i.e., not 'add on' or 'tick the box' approaches) – as we shall see in the remainder of the chapter and there is little in the literature in the way of guidelines for how to do so (Rowan et al. 2015; Johnson et al. 2016). As a result, Indigenous knowledge is often overlooked, under-valued/discounted/not implemented/tokenized, and this dilemma manifests in research designs and the approaches Western-trained researchers take with their projects and the Indigenous communities with whom they work. Good bridging/integrative approaches begin from a position that values all knowledges *equally* and implements them *equitably* in research and management designs. When one does not know how to do this, having the humility to find out how is key. But such learning typically still takes place out of the academy, in Indigenous spaces with Indigenous peoples. At the same time, such activity has traditionally fallen outside of academic merit points and thus, scientists and others have dismissed Indigenous knowledge-holders. Even today, too often, the importance of privileging Indigenous knowledge and the holders of such knowledge are merely paid lip service, while any tangible implementation and valuation are starkly absent in designs and products.

The third continuing challenge is that water-related research and management is not typically situated within larger, wholistic (as opposed to holistic) understandings of water and its multiple uses, its intrinsic value, or its dynamic, trans-boundary, living nature. For example, in research water is uncritically divided into its uses (e.g. drinking, sanitation, recreation, ability to support aquatic life, treatment ponds, etc.), forms (e.g. groundwater, oceans, rivers, lakes, streams, ice, etc.), and by actions (rain, floods, snow, tsunamis, drought, etc.). By wholism, we draw from Pribram (2006) to mean two things. First, the whole is greater than, and different from, the sum of its parts – we need to consider cumulative effects and generational impacts, and we also need to seek governance mechanisms that incorporate these considerations (Moore et al. 2016). Research and management approaches tend to divide and compartmentalize water along socially constructed lines (e.g., political jurisdictions) and its value is predominantly measured as a resource for human use. Second, science and spirituality can (should) find congruence when it comes to water; an example can be found in the correspondence between astronomical explanations associated with the pull of the moon causing the ocean's tides and the full moon ceremonies that many Indigenous peoples have conducted for generations,

Fig. 5.1 Locations of integrative water research and water management activities identified in the literature through a modified systematic realist review (Taken from Castleden et al. 2015)

honouring what they refer to as the original instructions from the Creator.[6] These wholistic understandings about water are not commonplace across most research and management contexts, so we must challenge the status quo by bringing this narrative to the fore: that water flows everywhere, it is constantly moving and changing, it has agency, and if it stops, quite literally, so do we.

5.4 Skimming the Surface: What the Literature Tells Us

There is a growing array of literature emerging in Canada and worldwide that concerns the need to implement integrative approaches to engaging Indigenous and Western knowledge systems in water research and management. Our own combination of systematic and realist reviews of the published academic and grey (i.e., government and non-government reports, websites, etc.) literature draws primarily from the past 15 years; it has allowed us to observe key trends that are occurring in water research and management practices in Canada. Through this combination, we have a sense of what has been done, where it has been done and with whom in terms

[6] By original instructions, we refer to readers to, for example, the Haudenosaunee Thanksgiving Address for codes of behavior (e.g., http://www.naho.ca/documents/naho/publications/codeofBehaviour.pdf)

of integrative water research and management as well as what has worked. We also, importantly have an idea about *why* certain approaches, processes, and models have (or have not) worked out (see Fig. 5.1) (see Castleden et al. 2015 for detailed analysis).

From what we have found, the abundance of published research on integrative approaches is taking place in Ontario, British Columbia, Nunavut, and the Northwest Territories. Notably, much of that research has only been published in the last 15 years, which shows the downstream effects of Indigenous resurgence as well as the cultural turn in scholarly research mentioned earlier – a shift from a predominantly post-positivist research paradigm to a constructivist paradigm that recognizes how knowledge is socially constructed, privileged, and legitimized through processes of power. Around the same time we were also seeing an increased awareness amongst Western-trained social scientists and health researchers for the need to conduct ourselves 'in the field' with ethical considerations in mind. In Indigenous contexts this was being recognized through reference to the four 'Rs' of research involving Indigenous peoples in Canada. These 'Rs' were described by Kirkness and Barnhardt (1991) to include (1) **Respect**: respecting Indigenous peoples and worldviews; (2) **Relevance**: ensuring research is relevant and responsive to Indigenous peoples' needs and direction; (3) **Reciprocity**: engaging in research processes where the benefits of research are felt equally (but possibly differently) by all those involved; and, (4) **Responsibility**: acting on one's responsibility to conduct research 'in a good way'. These early signals were like openings for the flow of Indigenous scholarship on Indigenous epistemologies, ontologies, and methodologies, most notably recognized in the seminal work of Linda Tuhiwai Smith, a Māori scholar from Aotearoa/New Zealand, who wrote about the urgent need to decolonize research methodologies (Smith 1999). As well, there was growing recognition of the value and need for collaborative, integrative, community-based and participatory approaches to working with Indigenous peoples on issues that were important to them (Hart 2010; Castleden et al. 2012).

We found that most of the literature on integrative water research and management that is available is coming from the social sciences, with relatively little hailing from health, natural sciences or engineering. Until we see integrative approaches to science training in the academy, where Indigenous and Western ontologies, epistemologies, and methodologies are given equal space in lecture halls and laboratories (and getting out of the academy and onto the land), it will be difficult for researchers from these fields to learn how to embrace integrative, collaborative, or community-based participatory approaches to their work.[7] This trend has also been noted elsewhere (see, for example, Feldman 2004 as cited in Kershaw et al. 2014),

[7] In fact, some institutions are making headway in this upstream battle. For example, Lakehead University is introducing a mandatory degree requirement that all undergraduate students must complete the equivalent of a 'Half Course' containing at least 50% (approximately 18 h) of Indigenous knowledge and/or Indigenous content. The University of Winnipeg has also recently approved a requirement that all undergraduate students will be required to complete a course in Indigenous rights, traditions, histories, governance, and cultures. While this may not seem significant, it is a step in the right direction.

and while we are seeing change, we are also hearing from those who have participated in our national Water Gatherings that collaborative, and more specifically, community-based participatory research that involves Indigenous peoples and communities, is not (yet) regarded by many Western-trained natural scientists, health researchers, and engineers as a viable approach to the type of work that they do. At the same time, we recognize that the absence of this type of literature in the sciences could also be indicative of a publication bias; by this, we mean that collaborative and integrative approaches are not often reflected in journals (i.e., manuscripts written in this style do not make it through the peer-review process) (see Bradford et al. 2016). But even the presence of a publication bias represents – at a structural level – the reluctance of those within the academy to appreciate the importance and potential of strengthening their work through integrative approaches, the benefits of which include reconciliation and relationality across the Indigenous-Settler-water trifecta (discussed below).

Across the literature we have encountered, water is being studied primarily through the lens of improving safe drinking water provision. This siloed approach disengages water from its multi-faceted and omnipresent nature and boils it down to one purpose, which supports narratives that argue for technical solutions to allegedly technical problems. We are also beginning to see a proliferation of literature discussing water governance in recent years (Jackson 2006; McGregor 2010; von der Porten and de Loë 2013; Cave 2012; Moore 2013; Moore et al. 2016), indicating that an exciting and much-needed conversation is spreading about Indigenous governance and water co-governance in Canada. However, most of this literature is emerging from a particular (small) network of researchers. The conversation needs to expand and penetrate broader audiences.

Also worth noting is that within water research and management discourse, much of the literature is not yet acknowledging inherent Indigenous or Treaty rights. By this we mean the collective rights that flow from the use and occupation of land since time immemorial and Treaty rights, which refers to specific rights outlined in treaties, and includes historic treaties signed between Indigenous Nations and the British Crown as well as contemporary ones – the latter are also known as Modern Land Claims Agreements. Thus, while the conversation is certainly beginning to take hold, we need to amplify this shift in thought towards co-governance, in order to see new waves of policy and practice (more on this later).

The case study approach is most commonly used in published integrative research, and the methods of data collection emphasize conversation (i.e., interviews and focus groups) as well as document reviews. The integrative knowledge/bridging approaches that would be considered the most 'successful' have been those where Indigenous communities or Indigenous organizations were playing active roles in the research as partners, as co-learners, and as co-creators of knowledge (see Castleden et al. 2015). Indigenous ontologies, epistemologies, and methodologies that are being implemented during the research process in culturally appropriate ways include having a collaborative research team or through the participation of Indigenous partners at specific points in the research process. For example, in addition to interviews (e.g., Rizvi et al. 2013; Longboat 2012; Lemoine 2012; Joe-Strack 2012; Anderson 2010; Anderson et al. 2013; Armitage et al. 2011) and focus groups (e.g., Chilima et al. 2013; Szach 2013), Indigenous peoples' knowledge was being implemented through reported observations (e.g., Martin et al. 2007), analyses of survey data (e.g., Baird et al. 2013), Elders' teachings (e.g., Restoule et al. 2013), and oral histories (e.g., Sam 2013; Cave 2012). Other literature describes implementing Indigenous peoples' knowledge through a process of knowledge exchange (e.g., Woo et al. 2007; Water Policy and Governance Group 2010) where face-to-face communication leads to hands-on collaboration and transformative learning opportunities. New methods are also beginning to emerge (e.g., Bradford and Bharadwaj 2015; Strickert et al. 2015; and Lemelin et al. 2014).

Despite the above methods being useful as mechanisms to engage with Indigenous knowledge-holders (some more effectively than others), most of the literature is not distinguishing between Indigenous and Western approaches (for assistance in this area, see Smith 1999; Kovach 2009; and Wilson 2008). Similarly, much of the literature is not attempting to define Indigenous knowledge or Indigenous research methods, or even explicitly stating that Indigenous and Western ways of knowing are being used in integrative ways – even though they are. Moreover, few works describe mistakes made in science; perhaps not surprisingly, journal editors do not tend to publish articles where authors admit their errors. This perpetuates the notion that Settler science is foolproof. This presents a challenge for novice readers generally, who may be trying to learn how to do integrative research or manage their relationship with water 'in a good way', but who are unable to 'read between the lines' from processes described as such (one of a few notable exceptions is Findlay et al. 2011). When integrative processes are described (though not always using the term 'integrative' in the description), it is often within the context of co-management, with references also to co-learning and knowledge co-production (Castleden et al. 2015). In short, integrative processes are often regarded as never-ending, collaborative, respectful, requiring trust, and working towards a common goal (Castleden et al. 2015). The Northwest Territories Water Stewardship Strategy is an example of this and has become widely known for its transformative possibilities regarding our relationship to water; it is one of two exemplary case studies that illustrate what many see as successful implementation (see Case Study 5.1). Emphasis is often placed on equitable involvement, benefits, and responsibilities throughout the research process (e.g. Cave 2012; Gearheard et al. 2006; Armitage 2011), and we are hearing through our own conversations with Indigenous knowledge-holders and other leading figures in this area for the need to approach water management through nation-to-nation relationships as being the ideal way for those integrative processes to be actualized.

Case Study 5.1: Northern Voices, Northern Water: The NWT Water Stewardship Strategy
The NWT Water Stewardship Strategy began development in 2008 between the Government of Northwest Territories, Aboriginal Affairs and Northern Development Canada, and Aboriginal governments, as well as with regulatory boards, agencies, environmental organizations, industry, academic institutions, and members of the public and was released in 2010. The *Strategy's* vision is that "the waters of the Northwest Territories will remain clean, abundant and productive for all time" (AANDC/GNWT 2010, p. 10). The *Strategy* outlines a path for how those in the NWT can serve as stewards for water, including principles for the NWT to follow for how to use their water in the long run. An Action Plan for the initiative provides a timeline and tasks for partners of the Strategy until 2015 and an informative website accompanies the Strategy which provides information about what water means to the people living in the NWT (48 % of whom are Indigenous), the watersheds in the NWT, and what the most significant threats to water are in the NWT context. The Strategy takes an ecosystem-based approach and ensures "the values of Aboriginal governments and people, and that settled land claim and self-government agreements were respected as well as Aboriginal and treaty rights" (GNWT n.d.). The Action Plan and Strategy both acknowledge that Indigenous knowledge informs their development and they are accompanied by Indigenous knowledge protocols; there is a best practices summary as well as individualized protocols by the Gwich'in Tribal Council, Dehcho First Nations, Samba K'e Dene Band, and Inuvialuit Settlement Region. The implementation of the Strategy is guided by an Indigenous Steering Committee, which includes representatives of Dehcho First Nations, Gwich'in Tribal Council, Inuvialuit Regional Corporation, Katlodeeche First Nation, Northwest Territory Métis Nation, Sahtu Secretariat Incorporated, Salt River First Nation, and the Tłįcho Government. For more information about the Strategy see: http://www.nwtwaterstewardship.ca/.

5.5 Diving Deeper: Indigenous and Western Water Knowledge-Holders Tell Us More

Not surprisingly, the literature can only tell us so much about how to implement integrative processes in water research and management. As we alluded to earlier, often descriptive materials about integrative processes do not make their way into published products as a result of the peer-review process where value judgments are made about what are (and are not) legitimate or relevant aspects of research or management narratives. Thus, to enrich our analysis of the literature, we conducted semi-structured interviews with academic and other institution-based researchers as well as with community-based Indigenous Knowledge-holders who were identified through the literature and by our National Advisory Committee as knowledge leaders in integrative approaches to water research and management (for a full description of the methods, see Castleden et al. 2015). The interviews provided the opportunity to dive a little deeper into why integrative approaches are needed and how they are being operationalized. The research team collaborated with the National Advisory Committee to develop interview guides and once interviews were conducted, they were coded and analyzed using a constant comparative method (Guba and Lincoln 1994). The goals of the research were used to develop the first round of coding, but we also explored the data using data-driven coding through an iterative process. Four interconnected themes emerged from those conversations and, to help readers along in their own journeys towards integrative Indigenous and Western knowledge implementation, they are described herein.

5.5.1 We Need to Challenge the Dichotomy Discourse

We tend to generalize differences between Indigenous and Western ways of knowing as dichotomies: Indigenous knowledge is wholistic, Western knowledge is compartmentalized; Indigenous knowledge is relational, Western knowledge is objective. One might start to pick up on the subtle and not-so-subtle nuances of which is 'better'. Worse, these dichotomies can often have racist underpinnings: Indigenous knowledge is intuitive and unempirical, Western knowledge is systematic and rigorous; Indigenous knowledge is local and non-generalizable, Western knowledge is universal. In short, by creating such dichotomies, a privileging of one (Western knowledge) over the other (Indigenous knowledge) occurs. This is a reflection of who has predominantly controlled and reproduced knowledge and its accompanying narrative, and what that narrative has been used for (e.g. the process of 'Othering' and its use to justify the colonial oppression of Indigenous peoples). But maintaining this troublesome dichotomy does not acknowledge the fluid and permeable nature of any knowledge. Western science has expanded thought, brought about important insights, solved many problems, encouraged ambition, and compelled paradigm shifts; but it has also played a role in supporting colonial expansion and imperialism (Louis 2007; Pickerill 2009; Howitt and Suchet-Pearson 2006; Humphery 2001). By recognizing this history and naming it, we can begin to take steps towards overcoming the dichotomy discourse to one of mutual respect and value.

> There is a recognition of the value of [Western] science in protecting health in deciding quality of food and water that is outside the scope of traditional ability and on the flip side, there are a huge number of examples of where Inuit knowledge has informed research, identified trends or factors that scientists had overlooked that led to improved research in many different ways. (Indigenous Scientist/Knowledge-Holder)

> What we have done with Indigenous people of Canada, is that we have taken… a colonial approach to management, and we have imposed it on other people. And we have sold it or we have imposed it as the only way and the best way to manage their affairs… And then when those people repeatedly seem to fail to replicate our management structure that we have in our own communities, we say that, 'Well these people are not capable,' that they need us to help them set this up and train them. In reality, it may be that our management structure is not suitable, not just for their needs but also for our needs, as we are learning now. (Settler Scientist/Knowledge-Holder)
>
> [This dichotomizing approach] has the potential to cause problems, and I think it… is laden with power relationships that serve some people really well and serves some not so well. And I think the reality of any knowledge system is that there's adaptation and there's flux and fluidity… [W]e were talking about a similar issue with respect to the complementary aspects between Inuit knowledge and science, Western science, and [the Inuit leader] stood up and was like, 'Look, we've been doing science for our entire life. We just call it something else'. (Settler Scientist/Knowledge-Holder)

Progress in terms of challenging the dichotomy discourse has been made; for example Howitt and Suchet-Pearson have conceptualized the the notion of "ontological pluralism" as a:

> model of cross-cultural relations that acknowledges and respects Indigenous ontologies, or ways of being, and at the same time is attentive to the historical and current dominance of Eurocentric thinking within natural resource management (as cited in Zanotti and Palomino-Schalscha 2016).

Importantly, Zanotti and Palomino-Schalscha acknowledge that non-Indigenous researchers have a responsibility to decolonize their own and others' ontologies, epistemologies, and methodologies and explore ways that create spaces of 'plural co-existence' (2016).

5.5.2 We Need to Acknowledge That Power Dynamics Exist

Water research and management involving Indigenous peoples occurs in a colonial context and is therefore prone to reproducing the same inequities that manifest in the broader Canadian society. The issue of unequal decision-making power in collaborative research thus remains as seen, for example, between institutionally-bound researchers and their community partners under the guise of participatory research (see, for example, Coombes et al. 2014) or water management between Settler governments and Indigenous governments under the guise of co-management (see, for example Natcher and Davis 2007). As above, Johnson et al. (2016) recommend scientists "learn to see our own privilege, our own context, our own deep colonizing" (p. 3).

Recognizing this power imbalance is one thing but acting on it – resisting it – is another. This dynamic is closely connected to the first theme regarding the dichotomized valuation of Western science often being considered as a legitimate form of knowledge over Indigenous knowledge. The literature review and interview data helped to shape the three actions we have identified as opportunities to address power imbalances: (i) free, prior, and informed consent of Indigenous peoples in any water research and management in Canada; (ii) recognition of nation-to-nation relationships (Indigenous autonomy is often undermined in academic research projects and government-led management projects when the view taken is that Indigenous peoples are 'stakeholders', not 'rights holders'; see Sect. 5.5.4 below); and (iii) full and equal participation in the development and implementation of projects, programs, and policies.

> I mean, you know, who makes the rules, right? …I mean if we're supposed to have a partnership but I'm the one who decides what we're going to do and I'm the one who decides how we're going to do it and I'm the one who has the money, and I'm the one who picks the date, and I'm the one who picks the place, you know? And that's the nature of so much of the "collaborative experience", and why it doesn't work. And that's why so often First Nations don't participate in these processes – because even participating in it is reinforcing the existing power dynamics. (Settler Scientist/Knowledge-Holder)
>
> It's not only a difference of knowledge systems, but I also think it's a real issue of power differences, where [Indigenous] people and their knowledge simply don't have the respect from non-Indigenous decision-makers. And they don't have the same power and capacity and money and resources to implement their knowledge systems. There's a very, very dominant culture and a dominant way of thinking and a dominant knowledge, and that knowledge is [Western] science. (Indigenous Scientist/Knowledge-Holder)

5.5.3 We Need to Develop Awareness About Indigenous Rights and Indigenous Responsibilities

We are seeing a growing awareness of the distinctions between Indigenous rights (constitutionally-bound and legally derived) and Indigenous responsibilities for or about water (passed down from generation to generation). The latter is in contrast with the language of 'rights,' which implies entitlement whereas the notion of responsibility carries with it the understanding that humans are not above water, but are living because of it. We therefore have the responsibility to protect it and maintain its health in order to ensure the health of future generations.[8] Still, while the notion of 'rights' is conceptualized as a legal term of the colonial government, Indigenous peoples are increasingly adopting the concept behind this language as a mechanism in legal and policy challenges with the Crown as a platform to assert their responsibilities to water. 'Rights' language fits better within Canadian legal discourse and has the potential to achieve Indigenous goals of responsibility without disrupting the Western-derived notion of human rights. At the same time, the legal

[8] We also need to develop nuances around these distinctions in different Indigenous contexts; that is water rights and responsibilities are not generalizable across all First Nations, nor are they illustrative of the different relationships that Métis/Metis and Inuit have with water. Moreover, for example, gendered relationships with water play out differently in Inuit, First Nations Métis/Metis contexts, where some may have specific roles and responsibilities or spiritual relationships with water, others do not and instead see water for its pragmatic purposes: sanitation, travel, and sustenance.

recognition of Aboriginal and Treaty rights does not guarantee their implementation, and evidence as to how Indigenous water rights are being considered (i.e. existence or scope) have yet to be tested in Courts. But the right to use the land and adjacent waters as Indigenous peoples have from time immemorial can be seen as a fundamental right supported by common law and through the Constitution Act of 1982, for example, in the case of R. v. Sparrow in 1990.

> Really, if you go to an [Indigenous] community, people really don't talk about rights. You know, now they do because they have had to; their rights have been taken away in so many ways by removing their access to land and so on. But their relationship to water is far greater or more spiritual. So yeah, I don't really think about rights in that way. (Settler Scientist/Knowledge-Holder)
>
> I think that 'rights' is kind of a term that I often hear talked about kind of from a Western perspective, that people have a sense of entitlement to a resource, in this case water. From my understanding, what I'm learning… is that we really need to think in terms of responsibilities instead of entitlement so that all people have a responsibility to look after water, and that responsibility lies particularly with certain members of our [Indigenous] communities as well. (Indigenous Scientist/Knowledge-Holder)
>
> Inherent rights [tend to] come to mind [about water]… I'd rather say inherent responsibilities. That's what the Elders talk to me more about… we can sometimes be a little narrow once we start talking rights. (Indigenous Scientist/Knowledge-Holder)

5.5.4 We Need Innovative Strategies to Effectively Implement Integrative Water Research and Management

The looming question of how to achieve what we know must be done in Canada was expressed through three linked processes and actions. First, we (researchers, managers, and policy-makers, Indigenous and Settler alike) need to build on existing relationships; second, we need to come to new relationships with openness, a collaborative spirit, and transparency; and third, we need to be responsive and flexible to direction from Indigenous communities and governments (see Smith 1999; Wilson 2008; and Kovach 2009). A key principle of community-based participatory research is that relationships should begin before any research proposal or design has been conceived of, so that when it does (or if) it comes time to developing a research plan, trust has already been established and space has been created for Indigenous worldviews to be equally recognized in all aspects of the design. From there, we need to give equal value to the different knowledges involved. Participatory and decolonizing methodologies are often identified as central to integrative approaches. In writing about this, Johnson and colleagues (2016) point to ethnobotanists who have developed methodologies related to species identification and classification that use Indigenous and Western knowledge in integrative ways, and they note community-based monitoring as used by the Arctic Borderlands Ecological Knowledge Co-op. Effective and open communication (transparency, face-to-face communication as often as posssible, consistent project updates) as well as paying attention to power dynamics are also key; this includes being transparent about our positionalities and our decision-making processes. Put simply, there needs to be awareness about the fact that research and management decisions as well as policies and governance structures have the ability to either reproduce or eliminate inequitable processes.

> Perhaps this is really cliché but for me there is not a set of procedures, it just all comes down to respect and humility, and the ability to listen, and just recognition... recognition that you're entering someone's homeland and they have every right to define what they want and why and when and how. And so you know, showing up and listening is just the best thing you can do, and ensuring that you allow the time necessary and have the patience necessary to do that with care. (Settler Scientist/Knowledge-Holder)

Given this, the second case study we present (Case Study 5.2) is an ongoing process where an integrative approach to water management is being taken and where the above strategies are being applied.

Case Study 5.2: Bras d'Or Lakes Collaborative Environmental Planning Initiative (CEPI)

CEPI began in 2003 at the request of the five Unama'ki/Cape Breton First Nations Chiefs to develop a management plan for the Bras d'Or Lakes and watershed, and it is now coordinated by the Unama'ki Institute of Natural Resources (UINR). The Mi'kmaq Chiefs of Unama'ki recognized that the health of the Bras d'Or Lakes was deteriorating, despite various activities and calls to action by government agencies, Cape Breton University, and others, and continued deterioration would directly impact Mi'kmaw ways of life. Therefore, they "called on the provincial and federal governments to account for what was happening in the lakes" (CEPI 2015). Two workshops, the first bringing more than 100 senior government officials together, resulted in the Bras d'Or Lakes Charter, a charter based on the medicine wheel that has been signed by federal, provincial, municipal, and Mi'kmaq governments. Its uniquely collaborative approach involves a Steering Committee that meets quarterly (with federal, provincial, municipal, Mi'kmaq governments, industry, academia and NGO representation), a Management Committee, a Mi'kmaq Elder Advisor, an Elders Council, and a Senior Council. One of many outcomes that has emerged from CEPI is an innovative management process plan called *The Spirit of the Lakes Speaks. The Spirit of the Lakes Speaks* is a document that describes the guiding principles of the management process of the Lakes and "introduces the Lakes as a living entity that generates feelings in people, and supports them in many ways" (CEPI 2011, p. 11). It is a living document, and guides the integrative work done in the Bras d'Or Lakes. The CEPI is discussing the potential of developing twelve sub-watershed plans to address the local issues within the Bras d'Or Lakes watershed. With specific work plans, the CEPI, UINR, and Eskasoni Fish and Wildlife Commission employ a Two-Eyed Seeing approach to use "traditional ecological knowledge to inform modern science and management" (CEPI 2015).

5.6 Tacking Left, Tacking Right, and Coming About: Policy Implications

Just as all knowledge is fluid and ever changing, readers should not interpret 'knowledge integration' as having an endpoint – there isn't one. The principle behind integrative approaches to Indigenous and Western knowledge mobilization is that these approaches, in whatever form they take on the ground, embody a co-learning journey that is never-ending. This journey is why the active word, integrative, is the *process* of research and management done in a good way (i.e. embodying relationality and striving for reconciliation), not integration as an *outcome*. At the same time, there may be some knowledge that Indigenous knowledge-holders cannot or will not share – even within integrative approaches – because, for example, its holders may not want to (see, Battiste and Youngblood Henderson 2000), because it may not be translated across languages without losing some of its meaning, or because it may be embedded in place and taking it out of context would render it spiritless. Furthermore, because context-specificity is often a critical consideration, there is general agreement in the literature and in our circles that one-size-fits-all straightforward research or management models and policies do not work. Thus, integrative approaches are necessary precisely *because* every location and context is different, and because people whom policy and program decisions are going to impact are the most knowledgeable about their own experiences and what they need to achieve wellness (Gearheard et al. 2006; Daley et al. 2014; Harper et al. 2011).

Recall earlier in the chapter we refer to the 4Rs of research involving Indigenous peoples in Canada: respect, relevance, reciprocity, and responsibility. Emerging from the research described in this chapter, we recommend the addition of two more Rs: (1) relationality, and (2) reconciliation. By relationality, we mean taking the time to focus on relationships in research (before data are collected, while data are being collected and analyzed, and long after projects are done). Quite literally, we mean sit down and drink some tea (Castleden et al. 2012)! By reconciliation, we turn to the powerful opening statement made in the Final Report of The Truth and Reconciliation Commission of Canada (June 2015). While the focus of the Commission was on the experiences of thousands of Residential School survivors, not water – or our collective, albeit varied, relationship with water – the interconnectivity of both should become apparent vis-à-vis this chapter's introduction to Indigenous-Settler relations. In its opening statement, the report succinctly notes that:

> For over a century, the central goals of Canada's [Indigenous] policy were to eliminate [Indigenous] governments; ignore [Indigenous] rights; terminate the Treaties; and, through a process of assimilation, cause [Indigenous] peoples to cease to exist as distinct legal, social, cultural, religious, and racial entities in Canada.

Now that the 'truth' has been made known to all Canadians, to go forward in a good way, we must enter into a relationship of 'reconciliation'; that is, Settler-Canadians can no longer plead ignorance. To borrow from Indigenous scholar and story-teller, Thomas King, we hope that integrative approaches become the new normal, that researchers, policy-makers, funders, and the wider public let these approaches transform the way in which we live with water. Or, as King puts it, we could "forget it. But don't say in the years to come that you would have lived your life differently if only you had heard this story. You've heard it now" (2003, p. 60).

We call policy-makers and governance strategists to action. This action must include the 6 Rs; they are key to integrative research and management that will move us towards and beyond equitable relationships with water, but also with one another. To this end, not only will Indigenous peoples benefit, but all Canadians will benefit from integrative approaches to water-related problems (and beyond). We know with the advent of climate change impacts we need 'all hands on deck' so-to-speak in terms of putting all the knowledges available to us to use. We also know that reconciling relationships between Settler-Canadians and Indigenous peoples must be part of our individual and collective healing process and the healing of an increasingly fractured political, socio-economic, and environmental landscape.

5.7 When Two Waters Meet, Tides Can Turn: Policy Recommendations

There are three audiences for whom we have policy recommendations. These include academic and other institutionally-based researchers in Canada because as more researchers engage in collaborative, integrative, community-based, and participatory research, they will need resources to learn from and foundational knowledge to embed within their research practices. We include recommendations for policy-makers – the creators of the structures that guide the work on the ground – so that they can create the spaces needed to learn from Indigenous knowledge-holders and Indigenous ontologies, epistemologies, and methodologies, and to translate their learning into policy uptake, which will directly impact water. It is important for policy-makers to create possibilities that will benefit the Indigenous peoples of this land and all other Canadians, and it is important to be aware of how policies can reproduce existing inequities; doing so will help to avoid such harm. We have also identified recommendations for funders of water research and management to entreat them to create funding opportunities that will facilitate, not block, integrative approaches to research and integrative uses of knowledge. We begin by presenting recommendations that are applicable to all three audiences, followed by policy recommendations specific to each group.

Overarching Recommendations

1. Recognize and affirm that Western knowledge/science is one knowledge system of many;
2. Recognize and affirm the ways in which Western knowledge/science has played and continues to play a role in colonization;
3. Recognize that leadership (i.e. Chief and Council) is a First Nations-only model arising from the Indian Act and become familiar with other forms of leadership

(e.g. Hereditary) and that Inuit and the Métis/Metis Nation have their own leadership and organizational structures;
4. Recognize and affirm that local problems require local solutions, and build this into research, policy, and funding opportunities;
5. Become familiar with and a practitioner of integrative approaches to creating new knowledge;
6. Emphasize the importance of 'success' as measured and evaluated by non-conventional metrics from both Indigenous and western perspectives;
7. Familiarize yourself with the Royal Commission on Aboriginal Peoples (1996) and the Truth and Reconciliation Commission of Canada's Final Report (2015);
8. Conduct and support work that addresses the calls to action put forth by the Truth and Reconciliation Commission of Canada's Final Report (2015); and,
9. Recognize and affirm Indigenous peoples rights to Free, Prior, and Informed Consent as articulated in the United Nations' Declaration on the Rights of Indigenous Peoples (2008) with respect to water research and water management in Canada.

Researcher-Specific Recommendations

1. Recognize and express your positionality as well as your role and responsibilities to Indigenous lands;
2. Familiarize yourself with the Tri-Council Policy Statement: Ethical Conduct for Research involving Humans (Canadian Institutes of Health Research, Natural Sciences and Engineering Research Council, and Social Sciences and Humanities Research Council 2014), and the OCAP™ or OCAS principles (First Nations Centre 2007);
3. Recognize that research involving Indigenous peoples means a relational commitment to each other long before a project begins and long after a project ends and act on that awareness; and,
4. Recognize the institutional structures that you are working within are rooted in colonialism and work towards decolonizing them in ally-ship with others.

Policy-Maker Recommendations

1. Recognize and enact the need to take time to carefully and thoughtfully implement Indigenous ways of knowing in water policy, programs, and practice;
2. Reject water-related policies that directly affect Indigenous peoples without their Free, Prior, and Informed Consent;
3. Relinquish policy-making and decision-making power concerning water within the boundaries of Indigenous jurisdiction in Canada, and create governance structures that share these powers equally elsewhere; and,
4. Put integrative approaches to Indigenous and Western knowledge implementation into practice and programs that work to redress colonial legacies in Canada.

Funder Recommendations

1. Develop calls for research that require integrative Indigenous and Western ontologies, epistemologies, and methodologies;

2. Ensure that individuals with lived Indigenous experience are selected to conduct peer-review for those funding opportunities and research proposals;
3. Target training and research opportunities to increase the number of researchers with Indigenous ancestry doing work in this area; and,
4. Require that any funded water research or management being conducted in Indigenous communities is being led by Indigenous governments, organizations and communities or their designates.

Overall, it cannot be emphasized enough that what water researchers, water managers and integrative knowledge practitioners need to do most throughout their work to listen to Indigenous knowledge-holders and interact with respect. The importance of trust and respect is central to operationalizing integrative approaches, and as a group of Indigenous and Settler scholar-allies, we support the Truth and Reconciliation Commission's call to action to engage with Indigenous peoples on a nation-to-nation level. As evidenced by the literature and the stories we have been privileged to hear, it is going to take not only time but also political will to operationalize and normalize co-governance.

5.8 As Long as the Rivers Flow[9]: Concluding Comments

In concluding our chapter, we wish to throw a few more pebbles in the pond about how to approach and conduct integrative research and management that have emerged through the course of our learning in this area. The pebbles exemplify the intentions with which those who have chosen to move forward with integrative water research and management are embedding in their/our approaches and purposes: hope, humility, healing, and humour. Hope is what we feel is needed in order for us to want to move forward with integrative approaches embedded within research and management practices and policy; and it is springing up in various places across the country as demonstrated through the stories and cases presented herein. We are hopeful that we can redress colonial legacies in Canada, use Indigenous and Western knowledges together and, where relevant and necessary, privilege the application of Indigenous knowledge, to transform our responsibilities to each other and to water. In order to do so, we must approach our enterprise, whether it is research or management or policy-making and governance, and our relationships with humility. That is, while we all develop our own unique sets of expertise, we must also recognize that we have our own limits and that by working through problems in integrative ways we are often in a better position to reach wholistic solutions. Healing describes what integrative approaches to problems are attempting to achieve; such approaches are healing in and of themselves as we situate ourselves within what Mi'kmaw Elder Murdena Marshall has described as the 'healing tense'; one where we can put aspects of ourselves out in front of us, whether

[9] As quoted in Fumoleau 2004.

it be hurtful behaviour we have engaged in or our knowledge systems, and walk around it, critique it, and do the same for other knowledge systems so we can learn to see how multiple systems can be used together. We must engage in the healing tense if we are to heal relationships between Indigenous and Settler peoples in Canada and decolonize the Indigenous-Settler-water trifecta. The final pebble that we toss into the pond is humour; in our experience, again both individually and collectively, having a sense of humour, sharing in laughter, and with good spirits, goes a very long way in making integrative work happen (see hope, humility, and healing). So, on that note, we end with: What did the fish say when it hit the wall? Dam.[10] If you raised an eyebrow instead of groaning, giggling or guffawing, perhaps you might like to read the chapter again, sit down with an Elder and drink some tea, or come to our next Water Gathering…

References

Aboriginal Affairs and Northern Development Canada and Government of the Northwest Territories. (2010). *Northern voices, Northern waters: NWT water stewardship strategy.* Yellowknife: Environment and Natural Resources, GNWT.

Adelson, N. (2005). The embodiment of inequity: Health disparities in Aboriginal Canada. *Canadian Journal of Public Health, 96*(Supplement 2), S45–S60.

Alfred, T., & Corntassel, J. (2005). Being indigenous: Resurgences against contemporary colonialism. *Government and Opposition, 40*(4), 597–614. doi:10.1111/j.1477-7053.2005.00166.x.

Anderson, K. (2010). *Aboriginal women, water and health: Reflections from eleven first nations, Inuit, and Métis grandmothers.* Winnipeg: Prairie Women's Health Centre for Excellence.

Anderson, K., Clow, B., & Haworth-Brockman, M. (2013). Carriers of Water: aboriginal women's experiences, relationships, and reflections. *Journal of Cleaner Production, 60*(1), 11–17. doi:10.1016/j.jclepro.2011.10.023.

Armitage, D., Berkes, F., Dale, A., Kocho-Schellenberg, E., & Patton, E. (2011). Co-management and the co-production of knowledge: Learning to adapt in Canada's Arctic. *Global Environmental Change, 21*(3), 995–1004. doi:10.1016/j.gloenvcha.2011.04.006.

Atleo, R. U. (2004). *Tsawalk: A Nuu-chah-nulth worldview.* Vancouver: UBC Press.

Baird, J., Carter, B., Cave, K., Dupont, D., General, P., King, C., Plummer, R., & Varewyck, A. (2013). Gaining insights about water: The value of surveys in First Nations Communities to Inform Water Governance. *Indigenous Policy Journal, 23*(4). doi:10.1.1.659.8924.

Bakker, K. (Ed.). (2007). *Eau Canada: The future of Canada's water.* Vancouver: UBC Press.

Bartlett, C., Marshall, M., & Marshall, A. (2012). Two-eyed seeing and other lessons learned within a co-learning journey of bringing together indigenous and mainstream knowledges and

[10]The lead author doesn't know many good jokes but she likes a good laugh, even at her own expense; the one joke she does know just so happens to have a water reference and made it in here (if the Editors don't cut it). We hope you'll tease her about it when you bump into her along the way. The authors would like to acknowledge the Canadian Water Network for its financial support in 2014-2015 for funding our research in this area. The authors would also like to gratefully acknowledge and thank all those who participated in interviews with our team, those who participated in either/both Water Gatherings, as well as the National Advisory Committee who steered us in good directions over the course of our study and our fantastic research assistants: Rob Stefanelli, Lindsay Day, and Kaitlin Lauridson. Finally, this chapter is an abridged version of a full-length report, which is available at www.heclab.com.

ways of knowing. *Journal of Environmental Studies and Sciences, 2*(4), 331–340. doi:10.1007/s13412-012-0086-8.

Basdeo, M., & Bharadwaj, L. (2013). Beyond physical: Social dimensions of the water crisis on Canada's First Nations and considerations for governance. *Indigenous Policy Journal, 23*(4). http://www.indigenouspolicy.org/index.php/ipj/article/view/142

Battiste, M. A., & Youngblood Henderson, J. (2000). *Protecting indigenous knowledge and heritage: A global challenge*. Saskatoon: Purich.

Blackstock, M. (2001). Water: A first nations spiritual and ecological perspective. *BC Journal of Ecosystems and Management, 1*(1), 54–66.

Bradford, L. E., & Bharadwaj, L. A. (2015). Whiteboard animation for knowledge mobilization: A test case from the Slave River Delta, Canada. *International Journal of Circumpolar Health, 74*. http://dx.doi.org/10.3402/ijch.v74.28780

Bradford, L. E., Ovsenek, N., & Bharadwaj, L. A. (2016). Indigenizing water governance in Canada. In S. Renzetti & D. Dupont (Eds.), *Water policy and governance in Canada*. Cham: Springer.

Canadian Institutes of Health Research, Natural Sciences and Engineering Research Council of Canada, and Social Sciences and Humanities Research Council of Canada. (2014). *Tri-council policy statement: Ethical conduct for research involving humans*. Ottawa.

Candler, C., Olson, R., DeRoy, S., The Firelight Group Research Cooperative with the Athabasca Chipewyan First Nation, & The Mikisew Cree First Nation. (2010). *As long as the rivers flow: Athabasca river knowledge, use and change (Executive Summary)*. Edmonton: Parkland Institute, University of Alberta.

Castleden, H., & Skinner, E. (2014). Whitewashing indigenous water rights in Canada: How can we indigenize climate change adaptation in Canada if we ignore the fundamentals? In D. Stucker & E. Lopez-Gunn (Eds.), *Adaptation to climate change through water resource management: Capacity, equity, and sustainability*. London: EarthScan.

Castleden, H., Sloan Morgan, V., & Lamb, C. (2012). I spent the first year drinking tea: Exploring Canadian university researchers' perspectives on the ethical and institutional tensions of community-based participatory research involving indigenous peoples. *The Canadian Geographer, 56*(2), 160–179. doi:10.1111/j.1541-0064.2012.00432.x.

Castleden, H., Cunsolo Willox, A., Harper, S., Martin, D., Hart, C., Stefanellii, R., Day, L., & Lauridsen, K. (2015). *Living with water: Examining methods and models for integrative Indigenous and Western Knowledge to inform – And transform- water research and management in Canada*. Report prepared for the Canadian Water Network.

Cave, K. (2012). Exploring the influences of institutions on water governance and management: A first nation case study. Master's thesis, University of Waterloo.

Centre for Indigenous Environmental Resources. (2011). Climate change and adaptive capacity in Aboriginal Communities South of 60 Assessment Report. http://www.yourcier.org/uploads/2/5/6/1/25611440/160ina652_absos_final_reportfullcmprssd.pdf

CEPI (Collaborative Environmental Planning Initiative). (2011). *The spirit of the lakes speaks*. Bras D'Or Lakes Collaborative Environmental Planning Initiative Report. http://brasdorcepi.ca/wp/wp-content/uploads/2011/07/Spirit-of-the-Lake-speaks-June-23.pdf

CEPI (Collaborative Environmental Planning Initiative). (2015). *Bras D'Or Lakes*. Collaborative Environmental Planning Initiative. www.brasdorcepi.ca

Chilima, J., Gunn, J., Noble, B., & Patrick, R. (2013). Institutional considerations in watershed cumulative effects assessment and management. *Impact Assessment and Project Appraisal, 31*(1), 74–84. doi:10.1080/14615517.2012.760227.

Coombes, B., Johnson, J. T., & Howitt, R. (2014). Indigenous geographies III methodological innovation and the unsettling of participatory research. *Progress in Human Geography, 38*(6), 845–854. doi:10.1177/0309132513514723.

Coulthard, G. (2014). *Red skin, white masks. Rejecting the colonial politics of recognition*. Minneapolis: University of Minnesota Press.

Daley, K., Castleden, H., Jamieson, R., Furgal, C., & Ell, L. (2014) Municipal water quantities and health in Nunavut households: An exploratory case study in Coral Harbour, Nunavut, Canada. *International Journal of Circumpolar Health, 73*(1). doi:10.3402/ijch.v73.23843

Environment Canada. (2004). *Threats to water availability in Canada*. National Water Research Institute, Burlington, Ontario. NWRI Scientific Assessment Report Series No. 3 and ACSD Science Assessment Series No. 1, pp 128.

Findlay, I., Ray, C., & Basualdo, M. (2011). Research as engagement: Rebuilding the knowledge economy of the Northern Saskatchewan trappers association co-operative. In P. V. Hall & I. MacPherson (Eds.), *Community-university research partnerships: Reflections on the Canadian social economy experience* (pp. 141–158). Victoria: University of Victoria.

First Nations Centre. (2007). OCAP: Ownership, control, access and possession. Sanctioned by the First Nations Information Governance Committee, Assembly of First Nations. National Aboriginal Health Organization, Ottawa.

Fresque-Baxter, J. A. (2015). Water is life: Exploring the relationship between place identity, water and adaptive capacity in fort resolution, northwest territories, Canada. Dissertation, Wilfred Laurier University.

Fumoleau, R. (2004). *As long as this land shall last. A history of treaty 8 and 11, 1870–1939* (2nd ed., p. 247). Calgary: University of Calgary Press.

Gearheard, S., Matumeak, W., Angutikjuaq, I., Maslanik, J., Huntington, H. P., Leavitt, J., & Barry, R. (2006). It's not that simple: A collaborative comparison of sea ice environments, their uses, observed changes, and adaptations in Barrow, Alaska, USA, and Clyde river, Nunavut, Canada. *Ambio, 35*(4), 203–211. doi:10.1579/0044-7447(2006)35[203:INTSAC]2.0.CO;2.

Gupa, E. G., & Lincoln, Y. S. (1994). Competing paradigms in qualitative research. *Handbook of Qualitative Research, 2*, 163–194.

GWNT (Government of the Northwest Territories). (n.d.). Aboriginal Steering Committee. NWT Water Stewardship. http://www.nwtwaterstewardship.ca/ascommittee

Harper, S. L., Edge, V. L., Schuster-Wallace, C. J., Berke, O., & McEwen, S. A. (2011). Weather, water quality and infectious gastrointestinal illness in two Inuit communities in Nunatsiavut, Canada: Potential implications for climate change. *EcoHealth, 8*(1), 93–108. doi:10.1007/s10393-011-0690-1.

Hart, M. A. (2010). Indigenous worldviews, knowledge, and research: The development of an indigenous research paradigm. *Journal of Indigenous Voices in Social Work, 1*(1), 1–16. E-ISSN 2151-349X.

Health Canada. (2015). *Drinking water advisories in first nations communities*. http://www.hc-sc.gc.ca/fniah-spnia/promotion/public-publique/water-dwa-eau-aqep-eng.php

Howitt, R., & Suchet-Pearson, S. (2006). Rethinking the building blocks: Ontological pluralism and the idea of 'management'. *Geografiska Annaler Series B: Human Geography, 88*, 323–335. doi:10.1111/j.1468-0459.2006.00225.x.

Humphery, K. (2001). Dirty questions: Indigenous health and "Western research.". *Australian and New Zealand Journal of Public Health, 25*(3), 197–202.

Hyslop, K. (2014). A fresh fix for unsafe water on First Nations reserves. *The Tyee*. http://thetyee.ca/News/2014/01/13/Water-on-Reserves/

Jackson, S. (2006). Compartmentalising culture: The articulation and consideration of indigenous values in water resource management. *Australian Geographer, 37*(1), 19–31.

Joe-Strack, J. A. (2012). *Respect, resilience and prosperity: Recommendations for the Yukon's water strategy*. Walter and Duncan Gordon Foundation. http://gordonfoundation.ca/sites/default/files/publications/RespectResilienceProsperity_JJS.pdf

Johnson, J., Howitt, R., Cajete, G., Berkes, F., Louis, R. P., & Kliskey, A. (2016). Weaving indigenous and sustainability sciences to diversify our methods. *Sustainability Science, 11*, 1–11. doi:10.1007/s11625-015-0349-x.

Kershaw, G. G. L., Castleden, H., & Laroque, C. P. (2014). An argument for ethical physical geography research on indigenous landscapes in Canada. *The Canadian Geographer, 58*(4), 393–399. doi:10.1111/cag.12092.

King, T. (2003). *The truth about stories*. Minneapolis: University of Minnesota Press.
Kirkness, V. J., & Barnhardt, R. (1991). First nations and higher education: The four R's-respect, relevance, reciprocity, responsibility. *Journal of American Indian Education, 30*(3), 1–15.
Kovach, M. (2009). *Indigenous methodologies: Characteristics, conversations, and context*. Toronto: University of Toronto Press.
Lavalley, G. (2006). *Aboriginal traditional knowledge and source water protection: First Nations' views on taking care of water*. Gatineau: Environment Canada.
Lawless, J., Taylor, D., Marshall, R., Nickerson, E., & Anderson, K. (2016). Meaningful engagement: Women, diverse identities, and Indigenous water and wastewater responsibilities. *Canadian Woman Studies; Les cahiers de la femme, 30*(2,3), 81–88.
Lemelin, H., Dampier, J., Makin, D., & Cross, J. (2014). Aboriginal erasure or aboriginal historical exclusion? Using video interviews to recognize the role of Aboriginal peoples on Kitchi-Gami (Lake superior). *Journal of Rural & Community Development, 9*(3), 176–185.
Lemoine, N. (2012). Exploring water governance in northern Saskatchewan: Opportunities for a watershed council. Master's thesis, University of Saskatchewan.
Longboat, S. A. (2012). First Nation water security and collaborative governance: Chippewas of Kettle and Stony Point First Nations, Ontario, Canada. Dissertation, Wilfrid Laurier University.
Louis, R. P. (2007). Can you hear us now? Voices from the margin: Using indigenous methodologies in geographic research. *Geographical Research, 45*(2), 130–139. doi:10.1111/j.1745-5871.2007.00443.x.
Martin, D., Belanger, D., Gosselin, P., Brazeau, J., Furgal, C., & Dery, S. (2007). Drinking water and potential threats to human health in Nunavik: Adaptation strategies under climate change conditions. *Arctic 60*(2), 195–202. doi:10.14430/arctic244
McGregor, D. (2004). Coming full circle: Indigenous knowledge, environment, and our future. *The American Indian Quarterly, 28*(3), 385–410. doi:10.1353/aiq.2004.0101.
McGregor, D. (2010). Traditional knowledge: Considerations for protecting water in Ontario. *International Indigenous Policy Journal, 3*(3). http://ir.lib.uwo.ca/cgi/viewcontent.cgi?article=1080&context=iipj
McGregor, D. (2012). Traditional knowledge: Considerations for protecting water in Ontario. *The International Indigenous Policy Journal, 3*(3): Article 11. doi:10.18584/iipj.2012.3.3.11
Moore, M.-L. (2013). Perspectives of complexity in water governance: Local experiences of global trends. *Water Alternatives, 6*(3), 154–172.
Moore, M.-L., Shaw, K., Castleden, H., & Reid, J. (2016). Patchy resources for the governance of Canada's resource patches: How hydraulic fracturing is illuminating the need to improve water governance in Canada. In S. Renzetti & D. Dupont (Eds.), *Water policy and governance in Canada*. Cham: Springer.
Morrison, A., Bradford, L., & Bharadwaj, L. (2015). Quantifiable progress of the First Nations Water Management Strategy, 2001–2013: Ready for regulation? *Canadian Water Resources Journal/Revue Canadienne des Ressources Hydriques, 40*(4), 352–372. doi:10.1080/07011784.2015.1080124.
Natcher, D. C., & Davis, S. (2007). Rethinking devolution: Challenges for Aboriginal resource management in the Yukon territory. *Society & Natural Resources, 20*(3), 271–279. doi:10.1080/08941920601117405.
National Collaborating Centre for Aboriginal Health. (2011). *Health land, health people: Forging international connections*. www.nccah-ccnsa.ca/264/Healthy_Land__Healthy_People_-_International_Gathering.nccah
Norman, E., Bakker, K., Cook, C., Dunn, G., & Allen, D. (2010). *Water security: A primer*. Ottawa: Library and Archives Canada Cataloguing in Publication.
Phare, M. A. S. (2009). *Denying the source: The crisis of first nations water rights*. Surrey: Rocky Mountain Books Ltd.
Phare, M. A. S. (2011). *Restoring the lifeblood: Water, first nations and opportunities for change*. Toronto: Walter and Duncan Gordon Foundation.

Pickerill, J. (2009). Finding common ground? Spaces of dialogue and the negotiation of indigenous interests in the environmental campaigns in Australia. *Geoforum, 40*, 66–79. doi:10.1016/j.geoforum.2008.06.009.

Polaris Institute, Assembly of First Nations. (2008). *Boiling point! Six community profiles of the water crisis facing first nations within Canada*. Ottawa: Polaris Institute.

Pribram, K. (2006). Holism vs. Wholism. *World Futures, 62*(1), 42–46. doi:10.1080/02604020500406255.

Restoule, J., Gruner, S., & Metatawabin, E. (2013). Learning from place: A return to traditional Mushkegowuk ways of knowing. *Canadian Journal of Education, 36*(2), 68–86.

Rizvi, Z., Adamowski, J., & Patrick, R. J. (2013). First nation capacity in Québec to practice integrated water resources management. *International Journal of Water, 7*(3), 161–190. doi:10.1504/ijw.2013.054859.

Rowan, M., Poole, N., Shea, B., Mykota, D., Farag, M., Hopkins, C., Hall, L., Mushquash, C., Fornssler, B., & Dell, C. A. (2015). A scoping study of cultural interventions to treat addictions in Indigenous populations: methods, strategies and insights from a Two-Eyed Seeing approach. *Substance Abuse Treatment, Prevention, and Policy, 10*(26). doi:10.1186/s13011-015-0021-6.

Royal Commission on Aboriginal Peoples (RCAP). (1996). *Report of the Royal Commission on Aboriginal peoples: Looking forward, looking back*. Ottawa: Minister of Supply and Services Canada.

Sam, M. G. (2013). Oral narratives, customary laws and indigenous water rights in Canada. Master's thesis, University of British Columbia.

Simeone, T. (2009). Safe drinking water in First Nations Communities (No. PRB 08-43E). Ottawa: Library of Parliament.

Smith, L. T. (1999). *Decolonizing methodologies: Research and indigenous peoples*. New York: Zed Books.

Strickert, G., Chun, K. P., Bradford, L., Clark, D., Gober, P., Reed, M. G., & Payton, D. (2015). Unpacking viewpoints on water security: Lessons from the South Saskatchewan river basin. *Water Policy, 18*(1), 50–72. doi:10.2166/wp.2015.195.

Swain, H., Loutitt, S., & Hrudey, S. (2006). *Report of the expert panel on safe drinking water for first nations* (Vol. 1). Ottawa: Government of Canada.

Szach, N. J. (2013). Keepers of the water: Exploring Anishinaabe and Metis women's knowledge of water and participation in water governance in Kenora, Ontario. Master's thesis, University of Manitoba.

Truth and Reconciliation Commission of Canada. (2015). *Honouring the truth, reconciling for the future: Summary of the final report of the truth and reconciliation commission of Canada*. http://www.trc.ca/websites/trcinstitution/File/2015/Findings/Exec_Summary_2015_05_31_web_o.pdf

United Nations. (2008). United Nations declaration on the rights of indigenous peoples. http://www.un.org/esa/socdev/unpfii/documents/DRIPS_en.pdf

von der Porten, S., & de Loë, R. (2013). Water governance and indigenous governance: Towards a synthesis. *Indigenous Policy Journal, 23*(4), 1–12.

Waldram, J., Herring, D., & Young, T. (2006). *Aboriginal health in Canada: Historical, cultural, and epidemiological perspectives* (2nd ed.). Toronto: University of Toronto Press.

Walkem, A. (2007). The land is dry: Indigenous people, water, and environmental justice. In K. Bakker (Ed.), *Eau Canada: The future of Canada's water* (pp. 303–319). Vancouver: University of British Columbia Press.

Water Policy and Governance Group. (2010). Water challenges and solutions in first nations communities: Summary findings from the workshop sharing water challenges and solutions: Experiences of first nations communities, 15–16 April 2010., Ontario: Kitchener-Waterloo.

White, J., Murphy, L., & Spence, N. (2012). Water and indigenous peoples: Canada's Paradox. *The Indigenous Policy Journal, 3*(3): Article 3. doi:10.18584/iipj.2012.3.3.3

Wilson, S. (2008). *Research is ceremony: Indigenous research methods*. Winnipeg: Fernwood Publishing.

Woo, M. K., Modeste, P., Martz, L., Blondin, J., Kochtubajda, B., Tutcho, D., Gyakum, J., Takazo, A., Spence, C., Tutcho, J., Di Cenzo, P., Kenny, G., Stone, J., Neyelle, I., Baptiste, G., Modeste, M., Kenny, B., & Modeste, W. (2007). Science meets traditional knowledge: Water and climate in the Sahtu (Great Bear Lake) region, Northwest Territories, Canada. *Arctic, 60*(1), 37–46. doi:10.14430/arctic263

Zanotti, L., & Palomino-Schalscha, M. (2016). Taking different ways of knowing seriously: Cross-cultural work as translations and multiplicity. *Sustainability Science, 11*, 139–152.

Part II
International and Transboundary

Chapter 6
Placing Canada's Water Policies in an International Context

Zafar Adeel

Abstract Canada is a country rich in water resources that has mastered the management of its abundant water resources through effective policies. Similarly, effective and relatively conflict-free management of shared waters with its southern neighbor is also cited as a success story in popular discourse, with over a century of joint management since the Boundary Waters Treaty of 1909. This chapter undertakes a critical review of these commonly held notions in the context of evidence, particularly incorporating the challenges stemming from a limited federal role in water management and a non-existing national water policy. A key part of this evidence comes in the form of a comparative analysis of the policies adopted by other countries. There are three comparative approaches utilized in this chapter: similarity of geographic scale and economic development; similarity of economic development but not scale; and, contrast of economic development. Selection of countries for comparative analysis is based on: fit with the comparison criteria, availability of comparable data, and relative success in managing their respective circumstances. First, the chapter compares Canadian approaches with those applied in developed countries of similar geographic scale; the most comparable examples in this case are that of United States of America and Brazil. Second, a comparison is made with the policies utilized in developed countries that have demonstrated success in managing their own water resources as well as shared ones; the examples used here are Germany and the Netherlands. Third, the Canadian approach is contrasted with approaches deployed in water-scarce developing countries; the examples used in this case are Jordan and South Africa. This multi-faceted analysis is intended to contextualize Canadian successes and failures, while also charting the potential for further enhancements and modifications, learning from experiences of other countries presented in this chapter.

Z. Adeel (✉)
Pacific Water Research Centre, Simon Fraser University,
TASC2 Room 8800, 8888 University Drive, Burnaby V5A 1S6, BC, Canada
e-mail: zadeel@sfu.ca

S. Renzetti, D.P. Dupont (eds.), *Water Policy and Governance in Canada*,
Global Issues in Water Policy 17, DOI 10.1007/978-3-319-42806-2_6

6.1 Context of Water Management

International dialogues around development issues have increasingly recognized the importance of water in achieving sustainable human, economic and social development. Successive analyses over the last 3 years by the World Economic Forum has depicted water insecurity as one of the top five major concerns and threats, indicating that the world leadership is duly recognizing its relevance (Vörösmarty et al. 2015; WEF 2016). The world leaders gathered in New York in September 2015 and approved a transformative development agenda that will guide nations over the next 15 years; water received major attention within the so-called Sustainable Development Goals (SDGs) – both in the form of a dedicated goal (SDG 6) for universal water security, and through a number of specific targets embedded under other SDGs (UN 2015).

The reasons for this level of attention are obvious. The availability of adequate drinking water supplies to a large number of individuals is not assured: Over 660 million people do not have access to an "improved water sources" (UNICEF and WHO 2015); approximately 1.8 billion people do not have access to "safe water" on a regular basis; and, an additional 1.2 billion of the world's population does not receive safe water as a regular supply in sufficient amounts (Onda et al. 2012). While the number of people in Canada without access to safe drinking water is approximately 1 % of the rural population (UNICEF and WHO 2015), these numbers dominantly represent First Nations communities for which over 100 communities remain under boil water advisories (Dupont et al. 2015). The consequences of poor service record worldwide are devastating on societies. It is estimated by the World Health Organization that about 3.5 million people, including 1.5 million children under the age of five, die each year from gastro-intestinal diseases such as diarrhea because they do not have access to safe water free from microbial and chemical pollution.

This narrative around access to safe water, however, has a much bigger underlying challenge that relates to management of water as a resource. Numerous countries across the world regularly withdraw water for various uses that far exceeds the capacity of natural water systems, prompting the need for policies and management approaches to cope with this exceedance. It is estimated that if our consumption patterns do not change, water demand will exceed supply by about 55 % by the year 2050, mainly because of much higher water demand from key sectors in this period: manufacturing (400 % increase), thermal electricity generation (140 % increase) and domestic use (130 % increase) (WWAP 2014). The largest user of water in this context is agriculture – the global average withdrawals by agriculture exceed 70 % of total water withdrawals (WWAP 2014). This figure is higher in water-scarce countries, and could go as high as 90 % in the least developed countries, in an effort to assure food security (FAO 2011). But agriculture is not the only consumer, almost all forms of energy generation also require considerable volumes of water – particularly for cooling systems in thermal and nuclear plants; it is estimated that approximately 15 % of the world's water withdrawals are meant for energy production

(WWAP 2014). The water-energy nexus for Canada links to hydropower on one end of the spectrum (approx. 59 % of the nationwide energy generation; UNEP 2012), and to exploitation of oil sands and resulting water contamination on the other (WWAP 2014).

6.1.1 Growth of IWRM

As a result of the multiple dimensions of water, the so-called "global water crisis" is considered multi-faceted and cutting across sectors of economy. This realization has taken considerable time to be noticed at the policy formulation level. The notion of Integrated Water Resources Management has been around for about 25 years, it was presented at the 1992 International Conference on Water and the Environment (Dublin, Ireland) where it was offered as a means of introducing system-wide water management approaches.[1,2] The resulting Dublin Principles were incorporated into the Agenda 21 later in the same year, and incorporated into the goals included in Chap. 18 of the Agenda 21 (UN 1992). It has since made considerable inroads into the policy formulation arena at the national level. A study by the United Nations has shown that about 80 % of countries in the world have undertaken policy reforms to create a more enabling environment for application of IWRM approaches (UNEP 2012). It offers a mechanism through which allocation of water resources at various scales and to various uses can be undertaken, while considering conservation and protection of key aquatic and land-based ecosystems. Other authors offer a more detailed analysis of the history and evolution of the IWRM concept, which is beyond the scope of this chapter (Norman et al. 2012; Solanes and Gonzalez-Villarreal 1999).

Much has been written about the principles of IWRM; many authoritative treatises exist on the subject (Rahaman and Varis 2005; Jeffrey and Gearey 2006; Lautze et al. 2011). Its success is driven in part by governmental policies and legislation that enable its implementation, and in equal part by engagement of the full spectrum of stakeholders (Solanes and Gonzalez-Villarreal 1999). The concept, however, still holds little appeal outside the circle of water managers, experts, and researchers; critics like Molle (2008) argue that it is a "nirvana concept" that embodies a reconciliation of opposing view and re-appropriated by various constituencies to their benefit. There is limited evidence that IWRM concepts are integrated into national development planning; a few examples of relatively good integration include water-

[1]"Although the conceptual approach remains fuzzy, central elements related to IWRM typically include (1) a combined consideration of all water uses, including social, economic, and ecological dimensions; (2) cross-sectoral water management based on integrated planning; and (3) participation and good governance." (Kramer and Pahl-Wostl 2014).

[2]"IWRM is a process which promotes the co-ordinated development and management of water, land and related resources, in order to maximize the resultant economic and social welfare in an equitable manner without compromising the sustainability of vital ecosystems" (GWP 2000).

scarce countries like Egypt, Jordan, and Pakistan. Similarly, there is an argument that better water management and provisioning of safe water can yield multiple health benefits through better nutrition (i.e., enhanced food production) and reduced morbidity and mortality through safe water (i.e., improved provisioning of safe water). However, there is also very limited buy-in from the health sector, despite the obvious benefits – about 80 % of hospital beds in Sub-Saharan Africa are occupied by those afflicted with water-borne diseases. Nonetheless, up to 80 % of countries report that they have made improvements to the institutional and governance frameworks – particularly in terms of better policies, laws and management systems (UNEP 2012). Some have also attempted to measure the social and economic benefits of these policy transitions.

Another dimension of IWRM deals with reducing risks to society from extreme events like floods and droughts. Increased awareness and heightened level of preparedness for these extreme event requires cutting across several sectors – building sufficient infrastructure and transport mechanisms, connecting with mass-communication channels, streamlining of government response mechanisms, and linking up with national and regional meteorological monitoring systems. It may be argued that only limited success has been achieved in using IWRM principles to achieve disaster-risk reduction. This could be partly explained by the fact that practitioners and policy makers who do not deal with water management on a day-to-day basis are less likely to be familiar with the IWRM constructs and principles. As a result, much decision making – even when its results may bear direct impacts on water resources – is divorced from IWRM principles.

Finally, tackling financial flows in a manner that aligns with IWRM principles also remains a challenge. In development planning and policy formulation, it is ultimately the allocation of budgetary resources and other fiscal measures (such as taxes, subsidies, and other incentives) that determine the strength of governmental uptake of these principles. The same UN report indicates that there is a 'gradual but positive trend' in receiving better budgetary allocation for water management and an increasing diversity of sources for this finance (UNEP 2012). At the same time, almost no appreciable progress is apparent around the notion of payment for water resources and related ecosystem services.

Implementation of IWRM principles in Canada also encountered a number challenges and obstacles: difficulties in integration with other resources management approaches for land, forest and fish; governance challenges when catchments cross political and administrative boundaries; knowledge gaps in fully understanding inter-relationship between ecosystems, sectors, and stakeholders – leading to "silo-ed" decision making; and, competing economic interests (Mitchell 2014). It can be argued that the severity of these challenges was not extreme enough to bring about complete dysfunction in water management, while also noticing that the nature of these challenges is no different than those encountered in developing countries.

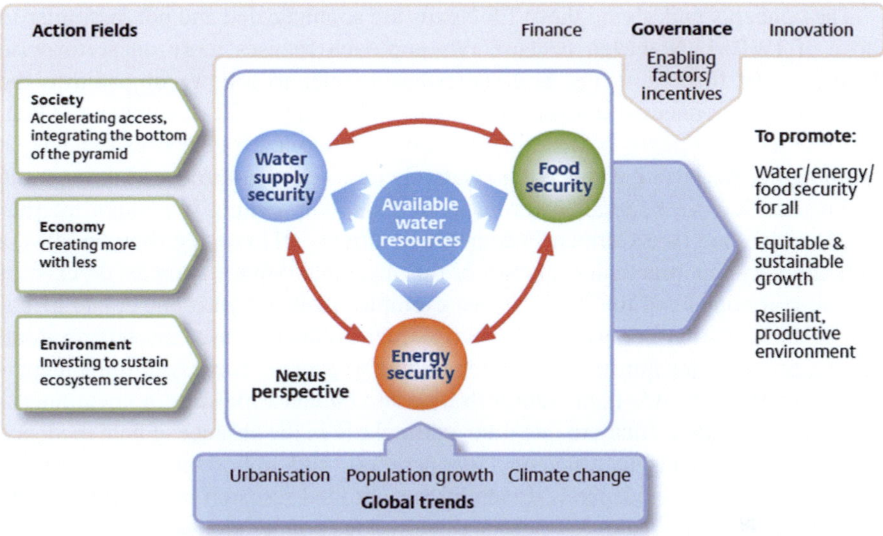

Fig. 6.1 The dimensions and interlinkages of the water-energy-food security nexus (Hoff 2011)

6.1.2 Emergence of Water-Energy-Food Nexus

The complexity of IWRM concepts, coupled with limited uptake in the last two decades, has led to discussions and debates about other approaches that might yield a more pliable policy uptake. As a result of these debates, the notion of Water-Energy-Food Nexus has emerged in the past decade (Hoff 2011). The concept essentially cross-links three services critical to economic and social development. It is argued that linking water management to two issues – food and energy – that are at the forefront of public awareness and political debates, would lead to better uptake in policy formulation.

The concept of the WEF Nexus is more complicated than a simple triangular relationship between three services or sectors. As is shown in Fig. 6.1, the WEF Nexus is influenced actors in the societal, economic and environmental domains, and is impacted by "external" drivers such as urbanization patterns, population growth and dynamics, and climate change-related impacts. Successful implementation of the WEF Nexus requires appropriate governance structure that enable collaboration across sectors, adequate financial resources to facilitate joint actions, and innovation and out-of-the-box thinking to devise mechanisms that align with this approach. At the end of the day, a successful implementation of the WEF nexus should lead to achievement of water, energy and food security for all people. In turn, this would foster equitable and sustainable economic growth as well as a robust and dynamic economy that is able to withstand external shocks.

The concepts underlying the WEF Nexus are sophisticated and not dissimilar to those of IWRM. However, because this approach focuses more on sector-wise integration of three domains, and somewhat simpler to follow compared to the 'everything-is-connected-to-everything' approach used in IWRM. Having said that, translation of this nexus-based thinking into active policy frameworks remains an uphill battle. Because the concept is relatively new – it was introduced through the lead of the German Federal Government in 2009, culminating in a major international conference (see German Federal Government 2011) on the topic in 2011 – there are very few practical examples of this available. Some examples do exist of the sub-components of the WEF Nexus: examples include hydro-power and irrigation schemes (water-energy nexus) and biofuel production systems (energy-food nexus). Some major hurdles must be overcome to achieve successful implementation of the WEF Nexus. First, administrative and bureaucratic silos in governments and institutions are difficult to integrate, particularly in the absence of hard evidence that integration leads to efficiency in operation, and importantly, cost-savings. Second, the lack of evidence is driven by very limited research and empirical data available outside the realm of traditional hydropower generation schemes.

6.2 A Historical Perspective on Canada's Water Management

The Canadian Constitution Act (1867) does not include an explicit mention of water management, although provincial public works in the form of "canals, with lands and Water Power connected therewith" and "Rivers and Lake Improvements" in provinces became property of Canada. As a natural resource, water management is delegated to provinces and other territorial governments as their responsibility (Brandes 2005). It has been argued that there is a complex web of normative, institutional, legal, legislative and jurisdictional processes that in itself becomes a roadblock to effective and efficient water management policy at the national level (Brandes 2005). That recent governments have made no attempts to untangle this situation or move towards a more cohesive and collaborative approaches in tandem with the provinces has not led to any significant improvements. A nationwide and national narrative on water management or provisioning of water services to citizens remains largely missing. In particular, for drinking water quality guidelines, Canada's 13 provinces and territories retain a decentralized and discretionary approach – in contradiction with the recommendations from the World Health Organization to deploy national, legally binding standards (Dunn et al. 2014). A variety of regulatory approaches and standards remain in practice despite a non-binding federally-mandated process undertaken through the Federal-Provincial-Technical Committee on Drinking Water (FPTCDW).

6.2.1 National Water Management in a Federal Environment

A more in-depth analysis of the national water management and governance regime is provided elsewhere in this volume. The Canadian Water Act (1986, and most recently revised in 2014) provides the framework for management of water resources in Canada, including research and the planning and implementation of programs relating to the conservation, development and utilization of water resources. The Canadian Federal government retains primacy in jurisdictions related to fisheries, navigation, federal lands, and international relations, including responsibilities related to the management of boundary waters shared with the US. At the level of the Federal government, there is a multiplicity of ministries and agencies that deal with various aspects of the national water management. Environment Canada is responsible for monitoring and measuring water quality across the board in freshwater aquatic ecosystems; it incorporates some regulatory and reporting functions. Health Canada is responsible for defining national standards water quality from a health perspective. Natural Resources Canada and Fisheries and Oceans view aquatic ecosystems as essentially resource base. For international engagement, most notably with the US, the recently minted Global Affairs Canada (GAC) also plays an increasingly important role.

As is discussed in Sect. 6.2.3, US-Canada relations are an important and central element of the Canadian national water discourse. The policy interface is largely governed at the Federal level even though many of the interfaces are "localized." The Laurentian Great Lakes are a case in point in which the primary interface on the Canadian side is with the province of Ontario. The Federal Government appoints commissioners and staff to the International Joint Commission (IJC), and also provides the interface to the political processes in Ottawa.

6.2.2 Provincial Lead in Water Management

A more in-depth analysis of water management at the provincial level in Canada is provided elsewhere in this volume. Water management in generally delegated to provincial government, particularly for water resources that are included entirely within a province's boundaries. As an exception, the Indian and Northern Affairs Canada (INAC) has the mandate to manage water resources in the Northwest Territories and Nunavut. Provincial legislative powers *inter alia* include the following areas: water flow regulation; authorization of water use development; water supply; pollution control; thermal and hydroelectric power development. Boundary Waters, shared between Canada and United States, are fall under Federal water management.

Norman and Bakker (2009) undertook a quantitative analysis of transboundary water governance regimes at different jurisdictional scales and argued that enhanced governance of transboundary resources is not taking place at the local scale (read:

provincial scale for Canada), despite ostensible attempts to devolve governance mechanisms. In particular, they found a disconnect between the governance trend of devolution – in which managerial responsibilities are delegated to more local or regional scales of governance – and the corresponding adequate institutional support to make their participation effective or agents of change.

6.2.3 Canada's Interface with US

In the history of transboundary water sharing arrangements, perhaps the United States of America and Canada share a unique relationship, in two major ways. First, it is unique in that a legal agreement has existed since 1908, one of the oldest such treaty that is sufficiently robust to be still actively enacted upon and implemented today. The need for this treaty was recognized in response to disputes arising around hydropower development as well as health impacts as a result of poor water quality. Second, it is also unusual in that the underlying relationship between the two countries has been cordial to a large extent – based on evidence, scientific knowledge, and sharing of a common purpose. This is notwithstanding the legacy in which the First Nations were excluded from the transboundary water sharing dialogue (Norman 2015). By the same token, there have been numerous disputes over shared water and natural resources, none of which have spun into unmanageable problems.

The establishment of IJC as part of the Boundary Waters Treaty has further facilitated the shared water management in the form of an institutionalized mechanism for addressing common concerns and issues. A parallel track also emerged around the fisheries industries, given their obvious commercial significance. The introduction of sea lamprey – a top predatory fish – into the Great Lakes and resulting ecological and economic impacts prompted the signing of the Convention of the Great Lakes Fisheries in 1954. This was followed by the establishment of the Great Lakes Fishery Commission in 1955. These two treaties, combined with a number of interstate agreements and an agreement between the Province of Ontario and the Canadian federal government create a legally-binding governance realm around the Great Lakes.

Cooperation between the US and Canada around the Columbia River basin is another example of long-standing joint management. Dialogue between the two countries took off in earnest in the aftermath of a major flooding event in 1948, which caused extensive damage in British Columbia and Oregon. After a long gestation period involving detailed discussions around hydropower generation and flood control, the Columbia River Treaty was signed in 1961 and came into force in 1964. A corresponding agreement is also in place between the Federal government and British Columbia, which defines rights and obligations of the parties. In essence, the treaty provided Canada the opportunity and resources (partly coming from the US) to build three reservoirs for flood control, although some were modified later for hydropower generation purposes as well. In addition to direct monetary benefits, flood protection and hydropower generation, the treaty also created opportunities for creating employment in significant numbers. Nonetheless, there are ongoing arguments and some disputes about inaccuracies in the calculation of costs/benefits and long-term usefulness of the treaty.

Over the past two decades, increasing trade relationships between Canada and the US – particularly in the form of North American Free Trade Agreement (NAFTA) – has offered increasing incentives for mutual collaboration.

6.3 Comparative Analysis of Water Management and Governance

The purpose of this section is to juxtapose Canadian governance regime with those in similar and contrasting situations. The thesis being tested is that whether economic and social development can help arrive at effective governance mechanisms despite the geographic scale or the relative abundance of the resource, water in this case. In other words, are states with a higher level of economic development – and thus with greater economic, human and technological resources at their disposal – more likely to effectively manage their water resources and overcome any scarcity?

For the sake of expediency, the testing of this thesis is through comparisons that are constrained to three categories of pairings, with the realization that there can be many more matching and contrasting scenarios: similarity of geographic scale and economic development; similarity of economic development but not scale; and, contrast of economic development. Because the ultimate objective is to be able to develop some critique of Canadian approaches and suggest ways to further improve the situation in a Canadian context, it does not make sense to draw comparisons that do not yield usable results. Selection of countries for comparative analysis is based on: fit with the respective comparison criteria (size, economic development, water endowment), availability of comparable data, and relative success in managing their respective circumstances. As result, there are six countries selected for this comparative analysis; an overview of their respective national water governance approaches is provided in Fig. 6.2. This data synthesis, and the descriptive evaluation of national approaches and policies provided in the subsequent sub-sections, are based on the responses provided by the respective national governments to a 2012 global survey conducted by UN-Water.[3] Canada did not respond to this survey, and hence the corresponding data for Canada are not available for comparison purposes.

---

[3] UN-Water is the United Nations inter-agency coordination mechanism for all freshwater related issues, including sanitation. Established in 2003, it provides the platform to 32 UN organizations and agencies to address the cross-cutting nature of water and maximize system-wide coordinated action and coherence.

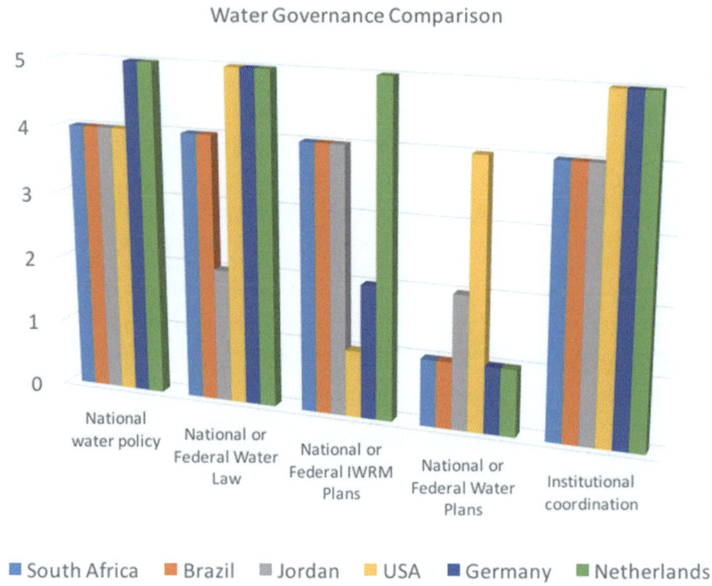

Fig. 6.2 A comparison of water governance approaches in selected countries. Description of scores: *1* = Not relevant; *2* = Under consideration; *3* = In place but not yet implemented; *4* = In place but partially implemented; *5* = Fully implemented. (UN-Water Survey of Water Resources Management 2012, unpublished data; UNEP 2012)

The formulation of legislative and legal frameworks that enable adequate water governance is correlated to development of cross-sectoral institutes that are crucial to understanding water management in a complex and diversified environment. As shown in Fig. 6.3, data from countries with varying degree of social and economic development (measured through the level of Human Development Index, or HDI) show a clear link between the two processes/systems.

6.3.1 Water Management in Large Developed Countries

This section compares Canadian water governance approaches with those applied in developed countries of similar geographic scale and similar water endowment. The most comparable examples in this case are that of United States of America and Brazil.

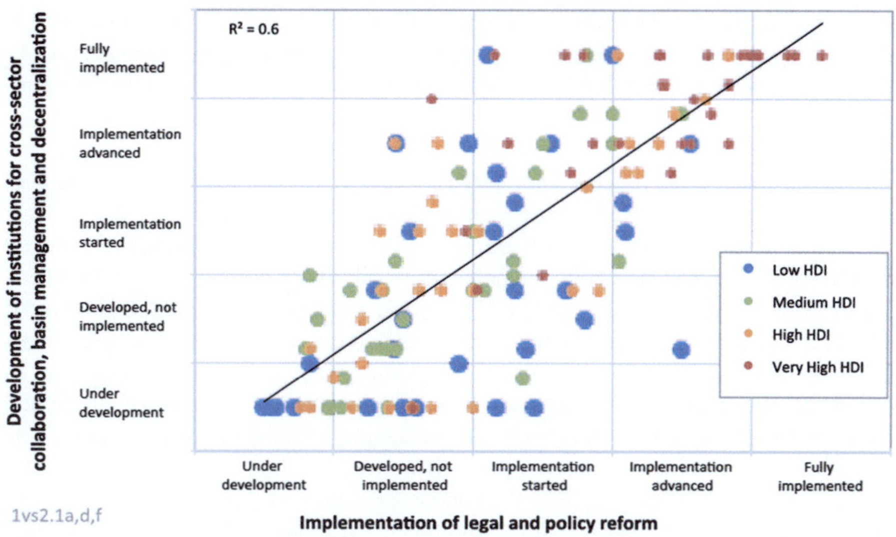

Fig. 6.3 The relationship between progress in implementation of legal and policy reforms, and the development of cross-sectoral institutions (UNEP 2012)

6.3.1.1 United States of America

The US has made significant advances in incorporating water policies and approaches for water resources management at the national level. A number of control measures at the national level, such as the Clean Water Act (1948, with the most recent revision in 1987) or the Water Resources Planning Act (1965, with the most recent amendment in 2007), provide national scale legal frameworks (Veiga and Magrini 2013). As presented in Fig. 6.2, these approaches (e.g., national water policies and national water plan) are in place but their implementation is not yet complete. At the same time, the US has not made attempts to develop a national IWRM plan explicitly. Much of this progress is driven by the national and sub-national demands, as well as a growing level of concern about environmental impacts and ecosystem conservation. The US government has also argued that other extraneous factors also drive movement in this direction, particularly the emergence of a governance system that encourages transparency, accountability, and public participation in government decision making. While these principles may not be specific to water, they nevertheless support ideals that create the conditions for a sound water policy that reflects the needs, wants and desires of the public.

At the national level, the National Science and Technology Council (which coordinates science and technology policy within the executive branch) has a Subcommittee on Water Availability and Quality. This subcommittee consists of 25 federal agencies that collectively are responsible for all aspects of federal water research and water resource management. The subcommittee has 7 elements of coordinated strategy designed to bridge the water demand/availability gap. They include elements that aim: (i) to implement a National water census; (ii) to develop a new generation of water monitoring techniques; (iii) to develop and expand technologies for enhancing reliable water supply; (iv) to develop innovative water-use technologies and tools to enhance public acceptance of them; (v) to develop collaborative tools and processes for U.S. water solutions; (vi) to improve understanding of the water-related ecosystem services and ecosystems needs for water; and, (vii) to improve predictive hydrologic models and their applications.

Waters in the United States are monitored by a broad range of stakeholders that include state, federal, and local agencies, universities, dischargers, and volunteers. Taken collectively, these data are used for water management and allocation under the many laws, regulations, compacts, treaties and many other types of agreements in existence in the U.S. These data are also central to planning and management of water resources for diverse purposes including irrigation, hydropower, urban and industrial water supplies, flood control management and drought mitigation. The United States Geological Survey (USGS) and the United States Environmental Protection Agency (USEPA) have the two largest water monitoring databases. This, however, masks the wide diversity of monitoring mechanisms at state, county, and municipal level, some of which are directly or indirectly supported by the US EPA.

Overall, the fundamental and constitutional principles of federalism, decentralization, and separation of powers make the implementation of an explicitly designed national IWRM plan very unlikely. It also remains challenging to address a range of needs – which include recreation, ecosystem conservation, hydropower generation, irrigation, municipal water supply and flood control – in a transparent manner, while accounting for changing hydrologic cycle as a result of global climate change. Canada finds itself in a similar situation. Nevertheless, it may be argued that the combination national and sub-national mechanisms lead to satisfactory coordination of development and management of water resources in the US.

6.3.1.2 Brazil

Brazil, a country similar in size to Canada albeit with six times larger population, is also a federated country in which states carry considerable delegated authority for water management. Since the promulgation of the Brazilian Constitution in 1988, two levels of water governance are at play: "federal waters" are those rivers and water bodies that cross through one or more state boundaries or into the territory of other countries and "state waters" are groundwater aquifers and rivers that are located entirely within the territory of a single state. A major transition in national water governance took place in 1997 when Law 9433 established the National Water Resources Policy (NWRP) and the National Water Resource Management System (NWRMS). These mechanisms introduced "important changes in the administrative, legal and institutional aspects of water resources" (Veiga and Magrini 2013). Brazilian water management takes place in an abundance of water resources in the aggregate; the average per capita water availability is in excess of 43,000 m^3 per year. It should also be noted that this average masks very diverse endowment of water from state to state, with some of the northeastern states operating in semi-arid conditions.

The Law 9433 created a shared and decentralized water resources framework that comprises the following nested components: National Water Resources Council (NWRC), State Water Resources Councils (SWRCs), River Basin Committees (RBCs), State Water Resources Management Institutions (SWRIs), and Water Agencies (WAs) (Magrini and Santos 2001). The water resource management hierarchy should be noted, within which the NWRC resolves disputes in federal waters and guides the implementation of the instruments provided in NWRP. The function of SWRCs is analogous to the NWRC, but operates for basins at the state level (i.e. "state waters"). In many cases, the jurisdiction of the RBCs – which are considered as water parliaments – transcend those of the states. Over the years, some attempts have been made to bring further consolidation and coherence through legislation. For example, Law 9984 in the year 2000 created the National Water Agency (ANA), which brought some consolidation to the Brazilian water sector (Braga et al. 2006). Similarly, Law 10881 in the year 2004 further clarified the relationship between ANA and nongovernmental organizations and allowed sharing of revenues between respective basin-level WAs.

Despite considerable progress in legislation and the creation of institutional frameworks such as a centralized ANA, challenges remain in coordinating diverse stakeholders and integrating state and national level water resource management (Veiga and Magrini 2013). This is in part the reason why scores shown in Fig. 6.2 are level 4 for national water policy, national water law, and national IWRM plans. Regional differences in the water endowment, particularly in the water-rich regions in the North and Midwest, and varying levels of water-related legislation and governmental prioritization has led to uneven development of water governance mechanisms.

6.3.1.3 Analysis

Both the US and Brazil are more populous than Canada, with comparable size of land mass. Both are also federated countries, with considerable autonomy exercised at the state level that could be considered similar to the Canadian provincial situation. However, in contrast to Canada, both countries have proactively promulgated legislations to bring about cohesion in approaches and policies. In both cases, apex water governance institutions – Subcommittee on Water in the US and ANA in Brazil – offer a platform at which continuous dialogue takes place. Presence of such a platform *per se* affords the possibility of analyzing emerging water challenges and reframing legislative and institutional frameworks in response to such challenges. Such an apex mechanism does not exist in Canada.

6.3.2 Advanced Water Management in European Context

In this section, a comparison is made with the policies utilized in developed countries that have demonstrated success in managing their own water resources as well as shared ones. These success stories have to do with management of water resources, particularly those of transboundary water resources in Germany and the Netherlands. Both countries also fall under European Commission's Water Framework Directive (WFD), which was established in the year 2000 to create a new paradigm to manage water resources throughout the European Union (EU 2003). It comprises the following steps that take place in a cyclical manner: (a) estimating demands and resources accounting; (b) defining objectives, which is reviewed and revised every 6 years; (c) environmental assessment of aquatic ecosystems; (d) control and monitoring; and, (e) measures to improve ecological status. The underlying approach is that aquatic ecosystems are providers of ecosystem services and must be managed in a sustainable manner. The two countries are also contiguous and share water resources, most notable is the Rhine river basin which has had both a history of major pollution problems and more recent successes in collaborative management.

6.3.2.1 Germany

It is important to note at the outset that the overall annual water resources availability in Germany is 188 billion m^3, whereas the aggregate water demand is 36 billion m^3; this essentially implies that there is a very significant surplus in water resources to the tune of about 152 billion m^3. Water governance regimes are indeed colored through this abundance of resource.

The German water governance has evolved through a combination of national and state legislation developed under the framework of WFD. Most recently, the new Federal Water Act (WHG) entered into force at national level in 2010; it represents a key milestone through the creation of the first set of comprehensive national regulations for the management of water bodies in Germany, which is in a result of the Federalism Reform (Irmer and Kirschbaum 2010). This is being further supplemented through a range of national ordinances; for example, ordinances that deal with protection of groundwater and surface waters and on facilities for the handling of substances potentially hazardous to water. In parallel, implementation of the Integrated River Basin Management (IRBM) of the WFD prompted the German federal government to appoint competent authorities – which include the German federal states (Laender) – which cooperate within the river basins and serve a coordination function. Active Stakeholder participation is guaranteed both under the WFD and the WHG, especially within the process of the establishment of an Integrated River Basin Management Plans.

Germany has a long history and tradition of monitoring water quality and quantity in surface waters as well as groundwater. There are established monitoring systems and networks which enable the competent authorities to recognize relevant detrimental changes, particularly major and persistent trends, in water quality and quantity just in time, and if needed, respond with appropriate measures. The competent authorities for the establishment and management of the monitoring systems are the Federal states (Laender). Germany adopted implementation processes that meet the WFD obligations for monitoring and reporting of chemical, biological and hydro-morphological parameters for surface waters and for chemical and quantitative parameters for groundwater.

Despite having a history of successes and implementation of various elements of the WFD, some challenges remain in achieving complete success. Some of these challenges stem from upstream/downstream relationships, which make it difficult to determine who pays for and who benefits from water management measures. Equally, it is difficult to resolve issues related to conflicts arising from competing interests and uses, particularly when there are strong financial interests of the stakeholders involved. The polluter pays principle is only fully implemented for discharges from point sources, and in all the other areas it is still the state which takes most of the financial burdens to finance the necessary measures.

At the international level, Germany interfaces with neighboring countries through a number of international organizations and conventions focused on water management for transboundary resources. Germany is party to numerous international and regional environmental protection conventions, including the various Conventions for the protection of the marine environment (London Convention, MARPOL, OSPAR and Helsinki Conventions) and the river basin commissions for transboundary river basins, namely Danube, Elbe, Oder, Rhine.

One interesting and instructive example of transboundary collaboration is around the Rhine river basin, which is managed through the International Commission for the Protection of the Rhine against Pollution (IKSR) and a number of subsidiary mechanisms. The cooperation between IKSR member countries – Germany, France, Luxembourg, the Netherlands and Switzerland – spans over 60 years; the European Community also joined in 1976. A new Rhine Convention (1999) superseded the international treaty dating from 1963 and a new program for the sustainable development of the Rhine was adopted in 2001, entitled "Rhine 2020." The success of the IKSR has been recorded; in particular, the remediation measures have resulted in a significant improvement in the water quality of the Rhine and its tributaries (Irmer and Kirschbaum 2010). Increasing demands and some deficits in water resource necessitates additional long-term measures that bring together key stakeholders such as agriculture, industry, local authorities, and shipping industry, which need to be undertaken in concert with the use of water for hydropower generation and flood protection.

6.3.2.2 The Netherlands

The Dutch geography is somewhat unique in the world in that slightly less than a third of the country lies below the sea level, and about a quarter is prone to flooding. In a major way, the ultimate objective of managing water in the Netherlands is to keep 'Dutch feet dry.' Consequently, the engineering and infrastructure solutions as well as the institutional arrangements are rather unique as well, most notably the systems of dykes and pumps that aim to prevent flooding while keeping seawater at bay. Management of these systems is integrally linked to the economy of the Netherlands, which relies heavily on export of agricultural and food products.

Water governance in the Netherlands has evolved over the years in accordance with emerging environmental, political, and economic conditions. The overall trend in this evolution has been towards consolidation in which, for example, Regional Water Authorities (RWAs) have shrunk in number from 2650 (traditionally known as water boards) to 24 and public drinking water companies reduced from 200 to 10 (OECD 2014). In 2009, eight water laws were combined into the National Water Act to create an integrated water management legal framework. This trend was further reinforced by the creation of the Ministry of Infrastructure and the Environment in 2010, which provides a consolidated entry point for water management at the national level. It also leads development of the national water policy, which ties in closely to spatial planning and flood protection at the national level. It is no surprise the Netherland ranks the highest in nearly all the water management categories presented in Fig. 6.2.

The international dimension of the Netherlands' water management is complicated by the fact that it is the downstream riparian for four international rivers (Scheldt, Meuse, Rhine, and Ems). Many river basin commissions provide the framework for transboundary water management. Like Germany, water governance and management approaches in the Netherlands are also subservient to the European Union's WFD. There is vertical nesting of these mechanisms as they go from European level to national, provincial, and watershed levels. These systems operate within unique Dutch water governance and financing characteristics, which include a well-defined taxation regime, cost recovery mechanisms and decentralized democracies (OECD 2014). Coordination amongst the water management bodies is handled by the National Administrative Council on Water (LBOW). There is also annual reporting to the Dutch Parliament (called Water in Beeld, or "water in focus"), which *inter alia* includes social and economic aspects.

6.3.2.3 Analysis

Both Germany and the Netherlands are similar to Canada in terms of their economic and social development, even if their economic throughput is less diverse than Canada and driven primarily by industrial output and agriculture, respectively. However, water governance approaches stand in stark contrast. In both cases, water management is taken as central to their economic and social development. As a result, water governance paradigms constantly undergo intense scrutiny and are well integrated into national political discourse. In the Netherlands, the Ministry of Infrastructure and the Environment provides an institutional umbrella at the national level. The German Federal Water Act (WHG) plays a somewhat similar role in national consolidation of policy perspectives.

6.3.3 Management of Water in Water-Scarce Countries

As noted earlier, the rich water endowment of Canada is usually taken in combination with general perception that Canada has no water problems. The reality of water availability in Canada, however, is somewhat different in which population pressures on water resources do not always co-locate with the rich water endowment and many municipalities struggle with stress on water resources. Over 60 % of Canada's renewable water supply flows north to the arctic or subarctic regions, which leaves a much smaller amount (about 2.6 % of the world supply of renewable) available for southern part of the country where most people reside (Sprague 2007). Therefore, it is instructive to compare water governance utilized in countries which are water scarce at the national scale. Accordingly, in this section the Canadian approach is compared with those deployed in water-scarce developing countries; the examples used in this case are Jordan and South Africa.

6.3.3.1 Jordan

The Hashemite Kingdom of Jordan is perhaps one of the top-ten water scarce countries, with annual water availability of 140 m^3 per capita that fall well below the water scarcity threshold of 1000 m^3 per capita (Molle and Berkhoff 2006; Nortcliff et al. 2008). To make matters worse, there is significant increase in the water consumption for municipal uses, which are forecasted to further increase due to population growth. This water scarcity has led to some remarkable choices for water management. One of the most recent examples is a trilateral agreement signed between Israel, Jordan, and the Palestinian Authority to undertake large-scale transfer of water from the Red Sea to the Dead Sea. This venture, if completed successfully, would serve multiple functions: reducing freshwater scarcity through desalination of water being pumped from through Red-to-Dead canal; hydropower generation by taking advantage of roughly 500 m net drop in hydraulic head between Red and Dead Seas; and, reversing the rapidly dropping water levels in the Dead Sea.

The water governance at the national level is managed by the Ministry of Water and Irrigation (MWI), which carries the responsibility of monitoring the water sector and managing freshwater supply and wastewater flows (Nortcliff et al. 2008). The MWI overseas two critical water management institutions: the Water Authority of Jordan, which deals with freshwater supply and wastewater management; and, the Jordan Valley Authority which deals with social and economic development in the valley including irrigation allocations. The MWI also instigated the development of the National Water Master Plan (NWMP) of Jordan in 2004, even if the plan did not contain a well-defined legal framework. The NWMP is adaptive in nature and can be iteratively revised, and accommodates the fact that the national water resources are *de facto* managed by a combination of public authorities and private organizations.

The overall governance approach in Jordan embraces: resource management strategies (e.g., sustainable use, international negotiation and treaties, and recycling wastewater); water allocation strategies (e.g., agricultural usage, incorporating treated wastewater, groundwater allocation); and, water quality control (e.g., enforcement of national health standards for water quality, ensuring groundwater quality in shallow aquifers). As in many arid countries, agriculture remains the largest consumer of freshwater while being linked to food security and rural livelihoods. As per the NWMP, any increases in irrigation demands should ideally be met through reclaimed wastewater, desalination, and water harvesting.

6.3.3.2 South Africa

Water distribution and availability in South Africa is uneven, and yet the country is classified as an arid (read: water-scarce) country. Social impacts of this water scarcity are serious, including for example the burden of water management in rural areas resting primarily on women. There is a vast number of South Africans were left unserved for safe water and adequate sanitation under the apartheid regime, a situation that has not been fully corrected (Van Koppen et al. 2011). Historic neglect and mismanagement of water resources in South Africa has resulted in serious food security challenges coupled with adverse environmental impacts (Molobela and Sinha 2011). Increasing water scarcity as a result of global climate change is forecasted to further exacerbate the water security situation, unless there is a significant paradigm shift in water resources management.

Water governance in South Africa has been undertaken in the context of poverty reduction and minimizing inequalities, problems that mostly dated back to the apartheid era. Since 1994, this process has entailed development of new water legislation (i.e., National Water Act 1998), decentralization of water management, and creation of local and regional level institutions (Molobela and Sinha 2011). South Africa is one of the few countries in the world where access to freshwater is recognized as a human right constitutionally. This places high political premiums on the one hand, and incorporates unique approaches to pricing of water supply on the other. The latter is designed to ensure basic water supply to everyone in the country through a tiered pricing system based on economic conditions in households. While these supply systems are broadly hailed as successful, anecdotal evidence suggests that their ability to meet the quality and quantity standards in rural settings are quite limited (Majuru et al. 2012). In the same vein, a strong sense of entitlement supported by constitutional declaration of water as a human right has led many residents to initiate litigation and street protests as a way achieving better water security (Manzungu et al. 2013); the success of these actions in achieving their stated objectives remains open to question.

6.3.3.3 Analysis

Severe water scarcity can often trigger proactive approaches to management of water resources as well as water demand. This is amply obvious in the cases of Jordan and South Africa; one might argue that similar institutional innovation emerges in the Canadian context in cases where water scarcity becomes a serious concern. The response in South Africa to social inequities stemming from the apartheid era in terms of tiered water pricing is most interesting. This approach incorporates real cost pricing for water supply systems while providing social security to those who have been disadvantaged historically.

6.4 Conclusions

The comparison of Canada's water management and governance approaches shows limited successes and a greater number of challenges. It may be argued that, on the average, the relatively high water endowment of the Canadian provinces and territories tends to mask and minimize challenges, or in some cases outright failures, of water management. Some of the salient gaps in Canadian water governance regime can be described as follows:

First, and foremost, there has been limited dialogue and discourse at the national level to define a common vision and pathways to achieving that mission. Attempts to undertake such dialogues in the absence of explicit support from the Canadian federal government has meant that the uptake from such dialogues into policy formulation is limited at best. Examples of such dialogues include the Blue Water Initiative (Crane 2013) and the future of Canada's water supply undertaken by the National Round Table on the Environment and the Economy (NRTEE 2010). Successful water management, such as the case of Germany and the Netherlands, is demonstratively linked to proactive national dialogues on the role of water in social and economic development. Such dialogues typically culminate in legislative frameworks and institutional mechanisms for their implementation.

Second, pricing of water services remains a controversial topic. IWRM principles promote the notion of true-cost pricing to enable revenue generation for successful service delivery and operation. South Africa offers an interesting model of tiered pricing to enable access to disadvantaged household at a subsidized rate, whereas those who can afford it are required to pay the full cost of water usage (which proportionately goes up with water consumption). There is a need to earnest discussion of such models for application in Canadian urban settings. The main argument for moving in this direction is the dire need for fresh capital to undertake renewal and retrofitting of Canada's aging water infrastructure.

Third, it is repeatedly demonstrated in all the six examples of water governance presented in this chapter that creation of an apex institution is crucial to successful water governance. Additionally, the apex institutions are typically the ones that lead the public discourse on water management issues and are central to creation of legislative and policy frameworks that are coherent and integrated. It would stand to reason if Environment and Climate Change Canada (ECCC) is accordingly reinforced to take on this central role in the Canadian context.

References

Braga, B. P. F., Flecha, R., Penna, D. S., & Kelman, J. (2006). A Reforma Institucional do Setor de Recursos Hídricos. In *Águas Doces no Brasil: capital ecológico, uso e conservação* (3rd ed., Vol. 1, pp. 639–675). São Paulo: Escrituras Editora.

Brandes, O. M. (2005). At a watershed: Ecological governance and sustainable water management in Canada. *Journal of Environmental Law & Practice, 16*(1), 79–97.

Crane, D. (2013). *Canada as the water solutions country: Defining the opportunities* (The blue economy initiative). Toronto: Canadian Electronic Library.

Dunn, G., Bakker, K., & Harris, L. (2014). Drinking water quality guidelines across Canadian provinces and territories: Jurisdictional variation in the context of decentralized water governance. *International Journal of Environmental Research and Public Health, 11*, 4634–4651.

Dupont, D., Waldner, C., Bharadwaj, L., Plummer, R., Carter, B., Cave, K., & Zagozewski, R. (2015). Drinking water management: Health risk perceptions and choices in first nations and non-first nations communities in Canada. *International Journal of Environmental Research and Public Health, 11*, 5889–5903.

EU. (2003). *Common implementation strategy for the water framework directive – Carrying forward the common implementation strategy for the water framework directive*. European Union, 17 June 2003. http://ec.europa.eu/environment/water/water-framework/objectives/pdf/strategy2.pdf. Accessed 23 Nov 2015.

FAO (Food and Agriculture Organization of the United Nations). (2011). *The state of the world's land and water resources: Managing systems at risk*. London: Earthscan/FAO.

German Federal Government. (2011). The water energy and food security Nexus – Solutions for the green economy. In *Bonn2011 Nexus Conference, 16–18 November 2011*. Bonn, Germany.

GWP (2000). *Integrated water resources management*. TAC Background Papers No. 4. Stockholm: Global Water Partnership.

Hoff, H. (2011). Understanding the Nexus. In *Background paper for the Bonn 2011 conference: The water, energy and food security Nexus*. Stockholm Environment Institute, Stockholm, Sweden.

Irmer, U., & Kirschbaum, B. (2010). *Water resource management in Germany: Part 1 – Fundamentals*. Dessau-Roßlau: Federal Environment Agency (UBA)

Jeffrey, P., & Gearey, M. (2006). Integrated water resources management: lost on the road from ambition to realisation? *Water Science & Technology, 53*(1), 1–8.

Kramer, A., & Pahl-Wostl, C. (2014). The global policy network behind integrated water resources management: Is it an effective norm diffusor? *Ecology and Society, 19*(4), 11

Lautze, J., de Silva, S., Giordano, M., & Sanford, L. (2011). Putting the cart before the horse: Water governance and IWRM. *Natural Resources Forum, 35*, 1–8.

Magrini, A., & Santos, M. (2001). River basin management and the national water resources policy in Brazil. In: C. A. Brebbia, P. Anagnostopoulos, K. Katsifarakis, A. H-D. Cheng (Eds.), *Water resources management* (Vol. 4, pp. 79–88). Southampton: WIT Press.

Majuru, B., Jagals, P., & Hunter, P. R. (2012). Assessing rural small community water supply in Limpopo, South Africa: Water service benchmarks and reliability. *Science of the Total Environment, 435–436*, 479–486.

Manzungu, E., Jonker, L., Madaka, E., Naka, Z., Sithole, E., & Dzingirai, V. (2013). Emerging forms of social action in urban domestic water supply in South Africa and Zimbabwe. *Journal of Sustainable Development, 6*(3), 70–82.

Mitchell, B. (2014). Addressing implementation deficits related to IWRM in Canada. In V. R. Squires, H. M. Milner, & K. A. Daniell (Eds.), *Basin management in the twenty-first century*. Boca Raton: CRC Press.

Molle, F. (2008). Nirvana concepts, narratives and policy models: Insight from the water sector. *Water Alternatives, 1*(1), 131–156.

Molle, F., & Berkhoff, J. (2006). *Cities versus agriculture: Revising intersectoral water transfer, potential gains and conflicts*. Comprehensive Assessment Report 10, International Water Management Institute, Colombo, Sri Lanka.

Molobela, I. P., & Sinha, P. (2011). Management of water resources in South Africa: A review. *African Journal of Environmental Science and Technology, 5*(12), 993–1002.

Norman, E. S. (2015). *Governing transboundary waters: Canada, the United States and indigenous communities*. London: Routledge.

Norman, E. S., & Bakker, K. (2009). Transgressing scales: Water governance across the Canada--U.S. Borderland. *Annals of the Association of American Geographers, 99*(1), 99–117.

Norman, E. S., Bakker, K., & Cook, C. (2012). Introduction to the themed section: Water governance and the politics of scale. *Water Alternatives, 5*(1), 52–61.

Nortcliff, S., Carr, G., Potter, R. B., & Darmame, K. (2008). *Jordan's water resources: Challenges for the future*. Geographical Paper No. 185, University of Reading, UK.

NRTEE. (2010). *Changing currents: Water sustainability and the future of Canada's natural resource sectors*. National Round Table on the Environment and the Economy (NRTEE), Ottawa, Canada.

OECD. (2014). *Water governance in the Netherlands: Fit for the future?* Paris: OECD Studies on Water, Organisation for Economic Co-operation and Development (OECD) Publishing. doi:http://dx.doi.org/10.1787/9789264102637-en.

Onda, K., LoBuglio, J., & Bartram, J. (2012). Global access to safe water: Accounting for water quality and the resulting impact on MDG progress. *International Journal of Environmental Research and Public Health, 9*(3), 880–894.

Rahaman, M. M., & Varis, O. (2005). Integrated water resources management: Evolution, prospects and future challenges. *Sustainability: Science, Practice, & Policy, 1*(1), 16.

Solanes, M., & Gonzalez-Villarreal, F. (1999). *The Dublin Principles for water as reflected in a comparative assessment of institutional and legal arrangements for integrated water resources management*. TAC Background Papers No. 3. Stockholm: Global Water Partnership.

Sprague, J. B. (2007). Great wet North? Canada's myth of water abundance. In K. Bakker (Ed.), *Eau Canada: The future of Canada's water*. Vancouver: UBC Press.

UN. (1992). Agenda 21, United Nations conference on environment & development. Rio de Janerio, Brazil, 3–14 June 1992. United Nations, New York, USA.

UN. (2015). *Transforming our world: The 2030 agenda for sustainable development*. A/RES/70/1. United Nations, New York, USA. https://sustainabledevelopment.un.org/content/documents/21252030%20Agenda%20for%20Sustainable%20Development%20web.pdf. Accessed 28 Mar 2016.

UNEP. (2012). *The UN-water status report on the application of integrated approaches to water resources management*. Nairobi: United Nations Environment Programme.

UNICEF and World Health Organization. (2015). *Progress on sanitation and drinking water – 2015 update and MDG assessment*. Geneva: World Health Organization.

Van Koppen, B., Schriener, B., & Fakir, S. (2011). The political, social and economic context of changing water policy in South Africa post-1994. In B. Schreiner & R. Hassan (Eds.), *Transforming water management in South Africa – Designing and implementing a new policy framework*. Dordecht: Springer.

Veiga, L. B. E., & Magrini, A. (2013). The Brazilian water resources management policy: Fifteen years of success and challenges. *Water Resources Management, 27*, 2287–2302.

Vörösmarty, C. J., Hoekstra, A. Y., Bunn, S. E., Conway, D., & Gupta, J. (2015). Fresh water goes global. *Science, 349*(6247), 478–479.

WEF. (2016). The Global Risks Report 2016 (11th ed.). World Economic Forum, Geneva, Switzerland. http://www3.weforum.org/docs/GRR/WEF_GRR16.pdf. Accessed 28 Mar 2016.

WWAP (United Nations World Water Assessment Programme). (2014). The United Nations World Water Development Report 2014: Water and energy. United Nations Educational, Scientific and Cultural Organization (UNESCO), Paris, France.

Chapter 7
Water Security and Adaptation to Climate Extremes in Transboundary Rivers of North America

Dustin Evan Garrick

Abstract Three basins in North America are used to examine how transboundary water governance arrangements have developed and performed in the face of recent severe droughts: the Colorado and Rio Grande Rivers on the US-Mexico Border and the Columbia River on the Canada-US border. The chapter delivers insights about water governance responses to the key problems in each basin, as well as the opportunities and limits to transfer policy lessons across basins. The findings illustrate the: (a) importance of proportional resource sharing mechanisms that spread risk and benefits in ways that are more likely to be perceived as fair; (b) potential for economic instruments and fiscal decentralization to reduce risks of natural hazards by enabling more localized responses; and (c) the need to establish, and strengthen, coordination mechanisms (e.g. river basin authorities, joint monitoring, conflict resolution venues) that are well matched to local conditions, including informal institutions (e.g. working groups, networks, joint studies). The chapter concludes with lessons about adaptation to extreme climate events in transboundary rivers of North America, including governance insights and practices that have enhanced (or reduced) freshwater security.

D.E. Garrick (✉)
Smith School of Enterprise and the Environment, University of Oxford,
School of Geography and the Environment, South Parks Road, Oxford OX1 3QY, UK
e-mail: dustin.garrick@smithschool.ox.ac.uk

S. Renzetti, D.P. Dupont (eds.), *Water Policy and Governance in Canada*,
Global Issues in Water Policy 17, DOI 10.1007/978-3-319-42806-2_7

7.1 Introduction

Climate variability and extreme events, such as droughts and floods, are superimposed on the chronic pressures associated with freshwater scarcity. Together, water stress and climate hazards pose a growing threat to freshwater security for people and ecosystems. The world's federal countries confront such water security challenges in a situation of institutional fragmentation: authority for managing transboundary rivers and aquifers is splintered across independent territorial and national jurisdictions, which can impede integrated water management and contribute to geopolitical tensions and instability.

North America is an ideal laboratory to investigate water security and adaptation to climate extremes in three contrasting federal political systems: Canada, Mexico and the United States. Transboundary water management varies across and within the three countries in terms of the level of centralization and institutional mechanisms for power sharing, conflict resolution, information gathering and planning, and fiscal arrangements. This raises fundamental questions about the evolution, design and performance of transboundary water governance arrangements at multiple scales – both within and across national borders.[1]

Three basins in North America are used to examine how transboundary water governance arrangements have developed and performed in the face of recent severe droughts. The case studies include the Colorado and Rio Grande Rivers on the US-Mexico Border and the Columbia River on the Canada-US border. The chapter delivers insights about water governance responses to the key problems in each basin, as well as the opportunities and limits to transfer policy lessons across basins. The findings illustrate the: (a) importance of proportional resource sharing mechanisms that spread risk and benefits across governance arrangements in ways that are more likely to be perceived as fair (b) potential for economic instruments and fiscal decentralization to reduce risks of natural hazards by enabling more localized responses; and (c) the need to establish coordination mechanisms (e.g. river basin authorities, joint monitoring, conflict resolution venues) that are well matched to local conditions, including informal institutions (e.g. working groups, networks, joint studies). The chapter concludes with lessons about adaptation to extreme climate events in transboundary rivers of North America, including governance insights and practices that have enhanced (or reduced) freshwater security.

---

[1] Governance has been defined by Young (1992) as "the structures and processes through which people in societies make decisions and share power" as cited by Davidson and de Loe (2014). A focus on governance is distinct from government by accounting for formal and informal institutions and actors. Transbounary governance accounts for the special coordination challenges when resources cross political borders.

7.2 Climate Change and Difficult Hydrology in North America

The observed historic and projected impacts of climate change are highly dependent on context. The 5th Assessment of the Intergovernmental Panel on Climate Change (IPCC) characterized the impacts and adaptation needs in North America (Romero-Lankao et al. 2014). It reviewed the observed climate changes relevant for water resource governance in North America, including the increasing prevalence of severe hot weather events (affecting outdoor water use) in the USA and heavy precipitation events throughout North America, which influences flooding extremes and strains stormwater infrastructure. The attribution of individual extreme weather and climate events to anthropogenic remains complex (Trenberth et al. 2015). However, climate change has been linked with observed changes to earlier snowmelt and declining spring snowpack in cold mountain rivers of the Western USA and Canada (Romero-Lankao et al. 2014). Future climate change is projected to cause a series of water-related climate hazards in North America associated with risks of shortage and flooding (low snow years, earlier runoff, intense droughts and increased precipitation variability, storm surges and higher sea levels). Climate change impacts also combine with other factors including urbanization, poor governance and mounting resource demands for food, energy and water security (Grafton et al. 2013).

7.2.1 Difficulty Hydrology and Extreme Climate Events

Extreme climate and weather events are a prominent feature of this observed climate record in North America. The IPCC's Special Report on Climate Extremes, known as the SREX report (Field 2012), defined climate extremes as the 'occurrence of a value of a weather or climate variable above or below a threshold value near the upper or lower ends of the range of observed values of the variable.' Droughts and floods are two prominent types of water-related extreme climate events; drought is the prime focus of this chapter.

It is important to note that droughts and floods are a natural feature of hydroclimatology and there is a long history of adaptation and maladaptation to their impacts. David Grey and Claudia Sadoff (2007) introduce the concept of 'difficulty hydrology' to describe the water security challenges posed by climate variability – both seasonal and inter-annual (year over year) fluctuations in runoff. Such variability has been linked with persistent poverty through the recurrent economic losses and impacts associated with droughts, floods and unpredictable timing of monsoonal events. Maladaptation to existing climate variability and extremes is viewed as an impediment to sustainable development; hence, enhanced adaptive capacity to deal with climate extremes is considered a prerequisite for addressing future climate change. These arguments have since been bolstered by empirical analysis and econometric modeling linking variability and economic growth (Brown and Lall 2006; Hall et al. 2014).

The North American context is noteworthy for its examples of regions confronting the challenges associated with difficulty hydrology, including the Colorado and Rio Grande Rivers. Grey and Sadoff cite the Colorado as an example of a 'harnessed hydrology' for its potential to decouple, or buffer, its economic development from the water-related risks posed by climate variability and extremes. They acknowledge that water security in regions with harnessed hydrology is precarious and requires ongoing vigilance and innovations in institutional reform, infrastructure development and information systems. They also note that past water security has come with a high social and environmental cost. Therefore, the lessons from adaptation in North America are also of increasing interest internationally. Past drought and flood adaptation provide an analog for both future climate change adaptation and may offer lessons from both successes and failures for many of semi-arid and subtropical rivers and regions in the developing world.

7.3 Adaptation

Climate change adaptation in human systems refers to 'adjustment to actual or expected climate and its effects' and 'seeks to moderate or avoid harm or exploit beneficial opportunities' (Agard et al. 2014).

This chapter highlights the *institutional options* for adaptation in the multi-layered context of transboundary rivers shared by multiple political jurisdictions. Dovers and Hezri (2010) identify three types of 'institutional resilience strategies'[2] for adapting to climate change: resistance and maintenance, change at the margins and openness and flexibility. Each can be deployed with a range of positive and negative outcomes, or strengths and weaknesses. **Resistance and maintenance** characterizes the 'no action' or status quo based on a reluctance to adapt prematurely or pursue maladaptive strategies. This can lead to efforts to prioritize stability and optimization of resource use (positive outcomes) or resistance to change and delay until crisis triggers change (negative outcomes). Strategies based on incrementalism favor **change at the margins**. Recognition of the need for learning and iterative adaptation can favor gradual approaches (positive outcome) but miss opportunities to address major changes and adopt long-term, strategic approaches (negative outcome). Finally, strategies based on **openness and adaptability** may embrace uncertainty and the need for rapid changes by preserving flexibility (positive outcome) while also permitting rash changes and maladaptation (negative outcome). These three dominant strategies also align with the recent distinctions between incremental and transformational adaptation with the latter restricted to

[2] For a thorough treatment of the concept of resilience applied to social-ecological systems, please see Martin-Breen and Anderies (2011).

conditions when vested interests and path dependencies can be overcome due to the stakes (i.e. risks and benefits) motivating reform (Kates et al. 2012).

7.3.1 *Water-Related Adaptation in a Transboundary Context*

Pittock (2013) adapts the approach developed by Dovers and Hezri (above) to examine adaptation options in a transboundary river context. He identifies at least five prevalent strategies:

1. **Iterative or cyclical water planning**. Coordinated and multi-layered planning processes incorporate opportunities for adaptation and learning, as well as integration of knowledge and stakeholders within (horizontal) and across (vertical) tiers of governance.
2. **Cap-and-trade water allocation reforms**. Water markets enable allocation in response to changing supply and demand conditions by using price to signal the scarcity value of water and to cue reallocation. In large transboundary rivers, such reforms involve the development of a nested set of caps on water diversions and trading rules to account for social and physical connections between water users.
3. **Reallocation of water for the environment**. In overallocated rivers, efforts to address environmental water requirements involve reallocation to preserve or restore ecosystem functions and enhance resilience to climate change impacts.
4. **Expansion or modification of water infrastructure**. Physical and natural water infrastructure, particularly different forms of storage (including reservoirs) has been proposed as a means of buffering the impacts of climate variability by addressing shortage and flood risk.
5. **Ecosystem conservation**. Following # 3 and #4, ecosystem conservation enhances the resilience of rivers as ecosystems and their capacity to provide ecosystem services related to floods and water storage.

In practice, these elements are almost always pursued in combinations to adapt to the impacts of 'difficult hydrology' and extreme climate events associated with droughts and floods. Almost all major rivers traverse political borders. The roles and responsibilities in water governance and climate adaptation are divided across multiple levels of governance and between the public and private sector at a range of scales. This has led to calls for decentralization or retention of adaptation tasks at the lowest level of governance possible (Field 2012), following the principle of subsidiarity (Marshall 2008). This raises the challenge of balancing decentralization and subsidiarity with complementary mechanisms for coordination and conflict resolution (Marshall 2008, Pahl-Wostl and Knieper 2014). These challenges are particularly pronounced in federal political systems where authority and capacity are divided and shared across national and sub-national levels, and include nation-to-nation relationships with indigenous communities.

7.3.2 The Special Challenge of Adaptation in Federal Political Systems

Water management poses unique coordination challenges in federal countries because authority over water is divided and shared between national and sub-national governments (Garrick et al. 2016). Extreme climate events exacerbate these governance dilemmas by creating situations ripe for opportunism: shirking of responsibility and burden-shifting by states or provinces and encroachment and crowding out by the federal government.

Federalism affects adaptation to climate extremes in a wide range of geographic and political economic contexts. Garrick et al. (2013) document the global influence of federalism in over 300 of the world's 554 major river basins, including approximately half of the world's international rivers. These include iconic freshwater bodies as varied as the Colorado, Nile, Indus and the Great Lakes. Prior research has examined the character and evolution of state-federal relations in water management, particularly in the older federations such as the US, Canada and Australia (Heinmiller 2009). However, comparative research remains surprisingly rare. Existing research also emphasizes policy responses to chronic water stress (e.g. competition or pollution), although there has been growing attention over the past 15 years to the management of climate variability and extreme weather events (Schlager et al. 2011; de Loë et al. 2001; de Loë and Plummer 2010).

7.4 North America as a Laboratory for Transboundary Adaptation

North America is an ideal laboratory to investigate the relationship between federalism, climate extremes and adaptive capacity. The federal approaches to adaptation to climate extremes and the relative roles and responsibilities across the central, sub-national and local levels vary across the three countries in terms of the level of centralization and of institutional mechanisms for power sharing, fiscal arrangements and conflict resolution in the management of water resources and climate extremes. In the broadest terms, federal political systems in Canada and Mexico form a spectrum from more decentralized to more centralized with the US occupying an intermediate point with high levels of internal diversity, although these attributes vary over time and across policy issues. Below, the review looks for commonalities that apply and differences across these diverse contexts.

This section reviews lessons from adaptation experiences in three North American rivers – the Colorado (US-Mexico), Columbia (US-Canada) and Rio Grande (US-Mexico) – which together traverse all three countries – Canada, Mexico and the United States (Table 7.1). The basins share conditions of water stress and overallocation, and are also prone to extended drought (from a year to decade) and periodic flooding. After briefly introducing the geographic context of each basin below, lessons from recent severe droughts in each basin are elaborated based on the adaptation options and insights outlined above.

7.4.1 Colorado

The Colorado River straddles seven states in the US and two in Mexico (637,100 km^2), as well as several Indian tribes, cities and irrigation districts. It has supported extensive irrigation development (4.5–5.5 million acres of irrigated agriculture), hydropower production and rapid urban growth for up to 40 million people in the major population centers of the Western US (U.S. Bureau of Reclamation 2012). There is a chronic imbalance in lower basin deliveries from Lake Mead, which has been described as a 'structural deficit' with total outflows and losses of approximately 12.6 billion m^3 and inflows of only approximately 11.1 billion m^3. The annual imbalance is buffered by reservoir storage, but this cushion has been depleted during sustained drought.

Drought is a prominent feature of the observed and paleoclimate records. The 1922 Colorado River 'Compact' – an interstate apportionment scheme – was famously negotiated after an unusually wet period, causing an overestimate of supply and a system of fixed volumetric allocations dividing water between the Upper and Lower Division states and between the US and Mexico. The four Upper Division states in the US (Wyoming, Colorado, Utah and New Mexico) devised a proportional allocation scheme after recognizing the variability of inflows and the limited likelihood of developing the full allocations under the 1922 agreement. Subsequent severe drought (in the late 1940s and 1950s) was followed by tree-ring studies in the 1970s identifying the prevalence of mega-droughts in the paleoclimate record and also revising the long-term annual average flows available (Stockton and Jacoby 1976; Woodhouse et al. 2006).

Drought conditions interact with demand pressures to shape contemporary transboundary adaptation efforts. Long-term supply and demand intersected in the late 1990s after nearly a century of infrastructure development and population and economic growth. In the context of this river basin closure, the Basin has experienced an unprecedented sequence of dry years since 2000, which has left the Basin's vast storage (4:1 ratio of storage: annual runoff) at 50 % capacity as of September 2015. This has triggered a spate of transboundary adaptation efforts, including the development of shortage rules (2007) for reducing interstate downstream deliveries from Lake Mead – the Lower Basin's primary storage at specified reservoir levels.

A comprehensive Basin Study in 2012 assessed supply-demand imbalances under projected climate change impacts through 2060 and examined vulnerability of different states sharing the river. The 1944 international treaty allocating Colorado River water to Mexico specified the sharing of shortages under 'extraordinary drought' conditions but did not define this term. In 2012, the US and Mexico also agreed to an international shortage sharing rule to include Mexico in shortages as part of "Minute 319" – a form of amendment – to the 1944 international water treaty governing the Colorado and Rio Grande Rivers (Gerlak 2015).

7.4.2 Columbia

The Columbia Basin is one of the most developed rivers in Western North America with more than 200 dams supporting approximately 5 million acres of irrigation; 16 000 MW of hydropower; a population of more than 7 million with increasing development in the rural, semi-arid interior; and a salmon fishery with high ecological, cultural and economic significance. Like the Colorado, the Columbia straddles an international border, but between the US and Canada. The basin drains almost 700,000 km^2 across seven US states, one Canadian province and a number of First Nations and tribal nations. The Columbia River Basin is comparable in size with the

Table 7.1 Selected transboundary rivers in North America

	Colorado	Columbia	Rio Grande
Political geography			
Size	637,137 km^2	668,000 km^2	471,900–870,000 km^2
Jurisdictions (Sub-national)	7 US, 2 Mexico	7 US, 1 Canada	3 US, 5 Mexico
Transboundary framework			
International	1944 US-Mexico Treaty	1964 Columbia River Treaty	1944 US-Mexico Treaty; 1906 International Convention
Drought provisions	Proportional reduction under "extraordinary drought"	N/A	Proportional reduction under "extraordinary drought"
Interstate	1922 Colorado River Compact	N/A	1938/9 Rio Grande Compact
Drought provisions	Tiered Shortage (Lower Basin and Mexico); Proportional Rule (Upper Basin)	N/A	Proportional Reduction

All three rivers have several indigenous First Nations (Canada) and Tribal Nations (US/Mexico); A range is provided for the size Rio Grande due to the endorheic (closed) sub-basins within the Basin.

Colorado in terms of drainage area but not in volume. It has an average volume at the Dalles Dam of 165 billion m^3, an order of magnitude higher than the Colorado. However, like the Colorado, stream flow is characterized by spatial and seasonal variability due to a snowmelt dominated hydrograph. Tributaries still experience seasonal variability and exhibit scarcity conditions characteristic of semi-arid irrigation regions. Chronic seasonal water deficits in the tributaries occur in late summer (August, September), when peak agricultural use coincides with natural low flows after snowmelt.

Semi-arid conditions, irrigation development and recurrent drought affect the tributaries of the interior basin with drought years causing economic losses in the 1930s, 1977 and early 2000s (Xiao et al. 2014). Most recently, droughts have tested water allocation institutions in 2001, 2005 and 2015. The Columbia lacks interstate or international allocation agreements for water quantity, which concentrates adaptation actions at the local, state and federal level within each country. Transboundary basin planning and conflict resolution during extreme events is restricted primarily to flood control, hydropower production, and ecosystem restoration. States and provinces issue drought orders and associated water use restrictions. Adaptation to climate extremes is therefore concentrated *within* individual states on the US and Canadian sides of the basin. In the US, shortage conditions have typically been managed at the level of tributaries (where water scarcity is prevalent) because the main 'stem' of the river is relatively abundant and its vulnerability to climate extremes is limited primarily to fluctuations in hydropower production. In the Yakima River of Southern Washington, for example, federal and state agencies in cooperation with irrigation districts and tribal governments have developed drought plans, undertaking climate change studies under the 2009 Water Smart Act program, and developed reverse auctions for dry-year leasing to acquire water from farmers during dry years to protect salmon fisheries.

The Okanagan River of British Columbia and Washington and the Walla Walla of Southwestern Washington and Northeastern Oregon are among the only major international and interstate tributaries, respectively. The Walla Walla has been subject to a US Supreme Court dispute in the 1930s. More recent programs in Washington promote "Flow from Flexibility" to work outside the priority based system of water allocation ('prior appropriation' or 'first in time, first in right' in the US) by entering into collaborative agreements, the development of water markets and mitigation bank agreements. These have been concentrated within Washington due to the lack of interstate agreements (Siemann and Martin 2007). Changing snowpack and winter precipitation under climate change promises to increase the transboundary nexus within the US (across states) and internationally (between the US and Canada) as changing streamflow patterns affect hydropower production, salmon recovery and other water uses and infrastructure operations on the main stem (Cosens and Bankes 2014).

7.4.3 *Rio Grande*

The Rio Grande is shared by the US and Mexico and forms the international border between them. The river originates in the San Juan Mountains in Colorado and flows through New Mexico (43 $m^3 s^{-1}$) until it reaches the border cities of El Paso, Texas and Ciudad Juarez, Chihuahua where the river is reduced to irrigation return flows and wastewater. The Rio Conchos is a major tributary originating in Mexico that joins the border reach upstream of the border cities of Presidio, Texas and Ojinaga, Chihuahua increasing average flows by an order of magnitude from 3 $m^3 s^{-1}$ to 30 $m^3 s^{-1}$ (Woodhouse and Stahle 2012).

The geography of the Basin creates an ideal setting for comparative institutional analysis. Both countries occupy an upstream position on a major tributary – the US on the Rio Grande, Mexico on the Rio Conchos – which creates mutual obligations for downstream deliveries and involves internal, interstate dynamics associated with meeting international commitments. The Rio Grande is a snowmelt-dominated hydrograph with runoff peaking in the spring. The Rio Conchos is heavily influenced by the North American monsoon with over half of the annual precipitation occurring between the middle of June and middle of September. This creates a situation where drought can affect one or both parts of the basin.

Drought in the Rio Conchos and Rio Grande is not strongly coupled in the instrumental record with the exception of the 1930s and 1950s (Woodhouse and Stahle 2012). The continental scale drought in North America (2012) is another recent example (Cook et al. 2013). However, the drought of record in the Upper Rio Grande is the most recent drought, which did not extend to the Rio Conchos. The drought patterns differ partly due to the contrasting hydrology of the snowmelt-dominated Upper Rio Grande and the monsoon-driven Rio Conchos. Since the 1990s there have been at least three major events, one in the US, two in Mexico, which overlap only partially. In the Upper Rio Grande, flows at the Del Norte Colorado gauge, which are used to establish Colorado's downstream water delivery requirements, have been below the 1895–2010 average for at least two of every 3 years since 2000. The diminished flows have affected the major water users in New Mexico and Texas (cities of Albuquerque, Santa Fe, and Las Cruces, New Mexico; irrigation districts of the San Luis Valley of Colorado Middle Rio Grande and Elephant Butte, and the El Paso Irrigation District in Texas). In the Rio Conchos, there are two events of drought and/or shortage: 1994–2003 and 2010–2014. Droughts that affect part of the basin may create asymmetries in the compliance issues as both countries are upstream on at least one major tributary, while basin-wide drought is likely to create compliance problems in both directions. However, because drought interacts with water demand characteristics, even minor anomalies have led to compliance problems.

Two international agreements govern the Rio Grande/Bravo: the 1906 Convention[3] and 1944 Treaty[4] between the US and Mexico. Both international agreements provide fixed volumetric commitments from the US to Mexico (60,000 acre feet per year) and Mexico to the US (350,000 per year assessed over a cycle), respectively. Both agreements also referred to "extraordinary droughts" as the basis for reducing deliveries proportionally. However, neither agreement defined such droughts in operational terms (e.g. triggers or thresholds). Under the 1906 Convention, deliveries to Mexico can be reduced in proportion to reductions in the US. Deliveries to Mexico have been reduced approximately one in 3 years on average since 1939, and the US is not required to "repay" the deficit. The reductions have been severe in recent years with the US delivering 39 % and 6 % of Mexico's allotment in 2012 and 2013, respectively (Carter et al. 2013). Under the terms of the 1944 Treaty, Mexico has accrued water debts during two consecutive drought cycles during the drought from 1994–2003. As of the cycle concluding on October 2, 2002, Mexico's water debt had reached 1.5 MAF over two consecutive cycles (Marin 2003). The accumulated deficits prompted political negotiations between Presidents G.W. Bush and Vicente Fox in spring 2002. Despite these intense negotiations and mutual suspicions of opportunism in the interpretations of 'extraordinary drought', the drought prompted a period of coordination between the US and Mexico sections of the International Boundary and Waters Commission.

Within both countries, different transboundary approaches are taken for interstate adaptation, governed by the Rio Grande Compact and its proportional allocation rules (US) and through central government and river basin council deliberations (Mexico). Internal arrangements within both countries have been marked by high degrees of conflict and Supreme Court cases (US) or threats of legal action (Mexico). Limited pockets of water trading in the Upper Rio Grande and recent sub-basin studies of climate change vulnerability under the 2009 Water Smart program have aimed to reduce impacts of drought and water shortage in the US portion of the Basin. Groundwater development and infrastructure efficiency investments have been the focus of adaptation efforts on both sides of the border causing challenges for meeting downstream compliance obligations.

7.5 Lesson from Adaptation to Climate Extremes in Transboundary Rivers

The three rivers of North America exhibit contrasting roles for water users, other stakeholders and different levels of governance, including local, regional, and federal bodies. This section reviews the institutional mechanisms and strategies proposed by Pittock. In Canada, the decentralized system of water governance vests allocation and many related planning and adaptation decisions at or below the

[3] Equitable Distribution of the Waters of the Rio Grande.

[4] Utilization of Waters of the Colorado and Tijuana Rivers and of the Rio Grande.

provincial level. In Mexico, water governance and climate adaptation remain comparably centralized despite over 20 years of decentralization since the 1992 National Water Law. In the US, context matters greatly, but transboundary adaptation is shaped by strong state governments and the existence of horizontal coordination through interstate river 'compacts.' International agreements between the US and Canada (administered by the International Joint Commission) and between the US and Mexico (administered by the International Boundary and Water Commission) are therefore shaped by these underlying and internal agreements. While the agreements between the US and Mexico reference 'extraordinary drought', and the agreements between the US and Canada on the Columbia address flood control benefit sharing, the scope of international agreements for climate adaptation remains limited to data exchange and allocation agreements with mixed downstream compliance (relatively high in the Colorado from the US to Mexico, relatively low, or at least disputed, in the Rio Grande). Despite these contrasting systems for transboundary adaptation, the Canadian, Mexican and US experiences generates some broad lessons.

7.5.1 Proportional Allocation Rules

The allocation of water based on proportional shares, rather than fixed volumes, is expected to be viewed as fair and hence more likely to be well-matched to drought prone areas (Schlager and Heikkila 2011). All three rivers include instances of proportional rules for sharing water across political borders. In the Colorado River, interstate agreements include both fixed (between Lower Division US States) and proportional (between Upper Division US States) rules, while shortage sharing measures have been expanded to include tiered reductions in downstream deliveries to Mexico when official shortages are triggered in the US. The Rio Grande is governed by a nested set of interstate and international agreements with both governed by proportional rules related to water allocation (US Rio Grande Compact) and triggers related to 'extraordinary drought' that reduce international deliveries from the US to Mexico by a proportional amount when the Upper Rio Grande experiences shortages within the US. Contrasted with fixed allocation rules (e.g. Lower Division US Colorado River states), proportional allocation rules enhance flexibility.

7.5.2 Water Markets and Efficiency Improvements

Water markets and associated cap-and-trade allocation reforms have been applied unevenly within all three rivers, although principally in the US portions. Pockets of water leasing have developed in portions of the Colorado (e.g. Arizona, California, Colorado), Columbia (all major US states) and the Upper Rio Grande (New Mexico and Colorado), including the development and use of dry-year leasing for both urban and ecosystem water security. Water allocation flexibility via water markets is projected to enhance adaptive capacity by containing or managing conflicts at lower levels (Olmstead 2010). However, the development of such cap-and-trade-mechanisms has depended heavily on multi-layered planning and associated monitoring/metering, information sharing and modeling at higher levels of governance through coordinated state, federal and tribal government actions (Garrick 2015).

7.5.3 Multi-Layered Planning and Conflict Resolution

New physical infrastructure (storage) programs constitute a popular supply side strategy for adaptation to both drought and flood extremes. However, all three rivers feature commitments to soft infrastructure through transboundary and multi-layered planning efforts to, inter alia, conduct water supply and demand studies, undertake drought and climate change vulnerability assessments and coordinate infrastructure operations. The Colorado River Basin is perhaps the most successful example with new rules passed to coordinate reservoir operations under a range of supply conditions (2007) followed by a 50-year basin study under the Water Smart program. The potential renegotiation of the Columbia Basin Treaty has opened potential for similar initiatives across the US-Canada border with a range of international networks and the International Joint Commission fostering data sharing and a platform for dialogue.

7.5.4 Subsidiarity

Subsidiarity, or the notion that governance tasks should be assigned at the lowest level possible, offers a potential guiding principle for transboundary adaptation in the North American context. Subsidiarity involves the assignment of tasks (planning, monitoring, conflict resolution) as close to the water user level as possible, but also implies its corollary, namely complementary higher-level coordination institutions for tasks that span jurisdictions (Marshall 2008). Following the work by Schneider (2008), the transboundary frameworks in the three rivers share a focus on local and horizontal coordination until stakeholders ‘pull’ in additional funding, conflict resolution and related institutional mechanisms for coping with climate variability and the impacts of extreme climate events. When local (water user, state/provincial) measures have proven insufficient, international and interstate frameworks offer the binding conflict resolution mechanisms, although these decisions have often been met with high transaction costs, low compliance and lingering disputes.

7.6 Pathways to Water Security and Transboundary Adaptation in Federal Rivers: Implications for Canada and Future Research

Natural hazards are projected to cost the Canadian economy up to $43 billion in losses annually by 2050 (TD Economics 2014). Drought conditions across Canada in 2001–2002 reduced GDP by almost $6 billion (Canada 2015). Droughts pose special governance challenges in federal political systems like Canada due to the division of powers and functions between national and sub-national governments and the blurring of roles and responsibilities during shortages. These governance challenges elevate the importance of conflict resolution and other institutional mechanisms to share risks and enhance resilience to extreme climate events.

The lessons from this chapter are relevant because Canada is not alone in facing these challenges: during the summer of 2015, the impacts of droughts and water shortages were felt from Brazil and British Columbia to California, South Africa and Saskatchewan. In this context, sharing knowledge, experiences and best practices developed across a spectrum of federal countries is a powerful way to build capacity to address present and future challenges posed by droughts and other extreme climate events.

What are the pathways to climate resilience in Canada's international and interprovincial rivers? The IPCC's Working Group II report on adaptation defines the concept of 'climate-resilient' adaptation pathways as a:

> Continuing process for managing changes in the climate and other driving forces affecting development, combining flexibility, innovativeness, and participative problem solving with effectiveness in mitigating and adapting to climate change (Denton et al. 2014).

In this context, the institutional resilience strategies and options adopted in the Colorado, Columbia and Rio Grande form part of multi-dimensional, dynamic and path dependent decision-making to build adaptive capacity in the face of uncertain changes to climate (Haasnoot et al. 2013). The elements, design and sequencing of these pathways depends on context but also exhibits potential for building adaptive capacity over the long term through a portfolio of the elements noted above.

Ongoing and future work aims to build on a growing tradition of comparative research by responding to the challenge set out by James Wescoat who noted the need to harness the global circulation of ideas and water policy expertise to support mutual learning and policy transfer across diverse contexts facing similar challenges (Wescoat 2009). Directions for future research include systems-based, interdisciplinary analysis of triggers, sequencing and portfolios of investment in institutions, information and infrastructure to achieve water security and build capacity to adapt to climate variability, change and extremes. It also requires networks of interdisciplinary researchers and practitioners working with a common framework and set of coding procedures to diagnose the risks and governance challenges of climate extremes in transboundary rivers (Armitage et al. 2015; Srinivasan et al. 2012). Doing so will involve establishing a global, multi-scale data architecture and set of institutional and governance indicators that can be used in risk assessment, planning and evaluation, and can be analyzed in combination with established datasets for international and shared waters.

References

Agard, J, Schipper, E. L. F., et al. (Eds.). (2014). Annex II: Glossary. Intergovernmental Panel on Climate Change (IPCC). In: V. R. Barros, C. B. Field, D. J. Dokken, M. D. Mastrandrea, K. J. Mach, T. E. Bilir, M. Chatterjee, K. L. Ebi, Y. O. Estrada, R. C. Genova, B. Girma, E. S. Kissel, A. N. Levy, S. MacCracken, P. R. Mastrandrea, L. L. White (Eds.), *Climate change 2014: Impacts, adaptation, and vulnerability. Part B: Regional aspects*. Contribution of working group II to the fifth assessment report of the intergovernmental panel on climate change (pp. 1757–1776). Cambridge/New York: Cambridge University Press.

Armitage, D., de Loë, R. C., Morris, M., Edwards, T. W. D., Gerlak, A. K., Hall, R. I., Huitema, D., Ison, R., Livingstone, D., MacDonald, G., Mirumachi, N., Plummer, R., & Wolfe, B. B. (2015). Science–policy processes for transboundary water governance. *Ambio, 44*(5), 353–366. doi:10.1007/s13280-015-0644-x.

Brown, C., & Lall, U. (2006). Water and economic development: The role of variability and a framework for resilience. *Natural Resources Forum, 30*(4), 306–317.

Canada, Agriculture and Agri-Food. (2015). *Lessons learned from the Canadian drought years 2001 and 2002*. http://www.agr.gc.ca/eng/?id=1326987176314. Accessed Dec 2015.

Carter, N. T., Seelke, C. R., & Shedd, D. T. (2013). *U.S.-Mexico water sharing: Background and recent developments*. Washington, DC: US Congressional Research Service.

Cook, B. I., Smerdon, J. E., Seager, R., & Cook, E. R. (2013). Pan-Continental droughts in North America over the last millennium. *Journal of Climate, 27*(1), 383–397. doi:10.1175/JCLI-D-13-00100.1.

Cosens, B. A., & Bankes, N. (2014). *The Columbia River treaty and the dynamics of transboundary water negotiations in a changing environment: How might climate change alter the game?* Available at SSRN.

Davidson, S. L., & de Loë, C. (2014). Watershed governance: Transcending boundaries. *Water Alternatives, 7*(2), 367–387.

de Loë, R., & Plummer, R. (2010). Climate change, adaptive capacity, and governance for drinking water in Canada. In *Adaptive capacity and environmental governance* (pp. 157–178). Berlin/London: Springer.

de Loë, R., Kreutzwiser, R., & Moraru, L. (2001). Adaptation options for the near term: Climate change and the Canadian water sector. *Global Environmental Change, 11*(3), 231–245.

Denton, F., Wilbanks, T. J., Abeysinghe, A. C., Burton, I., Gao, Q., Lemos, M. C., Masui, T., O'Brien, K. L., & Warner, K. (2014). Climate-resilient pathways: Adaptation, mitigation, and sustainable development. Intergovernmental Panel on Climate Change (IPCC). In C. B. Field, V. R. Barros, D. J. Dokken, K. J. Mach, M. D. Mastrandrea, T. E. Bilir, M. Chatterjee, K. L. Ebi, Y. O. Estrada, R. C. Genova, B. Girma, E. S. Kissel, A. N. Levy, S. MacCracken, P. R. Mastrandrea, & L. L. White (Eds.), *Climate change 2014: Impacts, adaptation, and vulnerability. Part A: Global and sectoral aspects* (Contribution of working group II to the fifth assessment report of the intergovernmental panel of climate change, pp. 1101–1131). Cambridge/New York: Cambridge University Press.

Dovers, S. R., & Adnan, A. H. (2010). Institutions and policy processes: The means to the ends of adaptation. *Wiley Interdisciplinary Reviews: Climate Change, 1*(2), 212–231.

Economics, T. D. (2014). *Special Report: Natural catastrophes: A Canadian economic perspective*. TD Bank. http://www.td.com/document/PDF/economics/special/NaturalCatastrophes.pdf

Field, C. B. (2012). *Managing the risks of extreme events and disasters to advance climate change adaptation: Special report of the intergovernmental panel on climate change*. New York: Cambridge University Press.

Garrick, D. E. (2015). *Water allocation in rivers under pressure: Water trading, transaction costs and transboundary governance in the western US and Australia*. Cheltenham: Edward Elgar Publishing.

Garrick, D., De Stefano, L., Fung, F., Pittock, J., Schlager, E., New, M., & Connell, D. (2013). Managing hydroclimatic risks in federal rivers: A diagnostic assessment. *Philosophical Transactions of the Royal Society A: Mathematical, Physical and Engineering Sciences, 371*(2002), 20120415.

Garrick, D., Schlager, E., & Villamayor-Tomas, S. (2016). Governing an international transboundary river: Opportunism, safeguards and drought adaptation in the Rio Grande. *Publius: The Journal of Federalism (Oxford Journals) 46*, in press.

Gerlak, A. K. (2015). Resistance and reform: Transboundary water governance in the Colorado River Delta. *Review of Policy Research, 32*(1), 100–123.

Grafton, R. Q., Pittock, J., Davis, R., Williams, J., Fu, G., Warburton, M., Udall, B., McKenzie, R., Yu, X., & Che, N. (2013). Global insights into water resources, climate change and governance. *Nature Climate Change, 3*(4), 315–321.

Grey, D., & Sadoff, C. W. (2007). Sink or swim? Water security for growth and development. *Water Policy, 9*(6), 545.

Haasnoot, M., Kwakkel, J. H., Walker, W. E., & ter Maat, J. (2013). Dynamic adaptive policy pathways: A method for crafting robust decisions for a deeply uncertain world. *Global Environmental Change, 23*(2), 485–498.

Hall, J. W., Grey, D., Garrick, D., Fung, F., Brown, C., Dadson, S. J., & Sadoff, C. W. (2014). Coping with the curse of freshwater variability. *Science, 346*(6208), 429–430.

Heinmiller, T. B. (2009). Path dependency and collective action in common pool governance. *International Journal of the Commons, 3*(1), 131–147.

Kates, R. W., Travis, W. R., & Wilbanks, T. J. (2012). Transformational adaptation when incremental adaptations to climate change are insufficient. *Proceedings of the National Academy of Sciences, 109*(19), 7156–7161.

Marshall, G. (2008). Nesting, subsidiarity, and community-based environmental governance beyond the local scale. *International Journal of the Commons, 2*(1), 75–97.

Marin, C. (2003). Bi-national border water supply issues from the perspective of the IBWC. *United States-Mexico Law Journal, 11*, 35.

Martin-Breen, P., & Anderies, J. M. (2011). *Resilience: A literature review*. Rockerfeller Foundation. Institute of Development Studies and The Resource Alliance. Available at https://opendocs.ids.ac.uk/opendocs/bitstream/handle/123456789/3692/Bellagio-Rockefeller%20bp.pdf?sequence=1

Mumme, S. P. (2005). The international boundary and water commission under fire: Policy prospective for the 21st century. *The Journal of Environment & Development, 14*(4), 507–524. doi:10.1177/1070496505282123.

Olmstead, S. M. (2010). The economics of managing scarce water resources. *Review of Environmental Economics and Policy, 4*(2), 179–198. doi:10.1093/reep/req004.

Pahl-Wostl, C., & Knieper, C. (2014). The capacity of water governance to deal with the climate change adaptation challenge: Using fuzzy set Qualitative Comparative Analysis to distinguish between polycentric, fragmented and centralized regimes. *Global Environmental Change, 29*, 139–154. doi:http://dx.doi.org/10.1016/j.gloenvcha.2014.09.003

Pittock, J. (2013). Lessons from adaptation to sustain freshwater environments in the Murray-Darling Basin, Australia. *Wiley Interdisciplinary Reviews: Climate Change, 4*(5), 429–438. doi:10.1002/wcc.230.

Romero-Lankao, P., Smith, J. B., Davidson, D. J., Diffenbaugh, N. S., Kinney, P. L., Kirshen, P., Kovacs, P., & Villers-Ruiz, L. (2014). North America. Intergovernmental Panel on Climate Change (IPCC). In V. R. Barros, C. B. Field, D. J. Dokken, M. D. Mastrandrea, K. J. Mach,

T. E. Bilir, M. Chatterjee, K. L. Ebi, Y. O. Estrada, R. C. Genova, B. Girma, E. S. Kissel, A. N. Levy, S. MacCracken, P. R. Mastrandrea, & L. L. White (Eds.), *Climate change 2014: Impacts, adaptation, and vulnerability. Part B: Regional aspects* (Contribution of working group II to the fifth assessment report of the intergovernmental panel of climate change, pp. 1439–1498). Cambridge/New York: Cambridge University Press.

Schlager, E., & Heikkila, T. (2011). Left high and dry? climate change, common-pool resource theory, and the adaptability of western water compacts. *Public Administration Review, 71*(3), 461–470.

Schlager, E., Engel, K., & Rider, S. (2011). *Navigating climate change policy: The opportunities of federalism*. Tucson: University of Arizona Press.

Schneider, S. (2008). Who's to blame? (Mis)perceptions of the intergovernmental response to disasters. *Publius: The Journal of Federalism, 38*(4), 715–738.

Siemann, D., & Martin, S. (2007). *Managing many waters: An assessment of capacities for implementing water and fish improvements in the Walla Walla Basin*. William D. Ruckelshaus Center.

Srinivasan, V., Lambin, E. F., Gorelick, S. M., Thompson, B. H., & Rozelle, S. (2012). The nature and causes of the global water crisis: Syndromes from a meta-analysis of coupled human-water studies. *Water Resources Research, 48*(10). doi:10.1029/2011WR011087

Stockton, C. W., & Jacoby, G. C. (1976). Long-term surface-water supply and streamflow trends in the Upper Colorado river basin based on tree-ring analyses. *Lake Powell Research Project Bulletin, 18*, 1–70.

Trenberth, K. E., Fasullo, J. T., & Shepherd, T. G. (2015). Attribution of climate extreme events. *Nature Climate Change, 5*, 725–730.

US Bureau of Reclamation. (2012). Colorado River Basin Water Supply and Demand Study. Boulder City: Bureau of Reclamation, Lower Colorado Region.

Wescoat, J. L. (2009). Comparative international water research. *Journal of Contemporary Water Research & Education, 142*(1), 61–66.

Woodhouse, C. A., & Stahle, D. W. (2012). Rio Grande and Rio Conchos water supply variability over the past 500 years. *Climate Research, 51*(2), 147.

Woodhouse, C. A., Gray, S. T., & Meko, D. M. (2006). Updated streamflow reconstructions for the Upper Colorado River basin. *Water Resources Research, 42*(5)

Xiao, M., Nijssen, B., Mao, Y., & Lettenmaier, D. P. (2014). *Drought in the Pacific Northwest, 1920–2013*. AGU Fall Meeting Abstracts.

Young, O. R. (1992). *The effectiveness of international institutions: Hard cases and critical variables governance without government*. Cambridge/New York: Cambridge University Press.

Chapter 8
Transcending Borders Through Postcolonial Water Governance? Indigenous Water Governance Across the Canada-US Border

Emma S. Norman and Karen Bakker

Abstract This chapter explores the evolution of transboundary water governance along the Canada-US border. We examine two key examples in two eras of water management across the Canada-US border, separated by more than a century. First, we examine the Boundary Waters Treaty (a bi-national agreement between the federal governments of Canada and the United States), as an emblematic example of the dominant concerns that underpinned (colonial settler) water governance at the turn of the twentieth century, creating the framework in which nation-state governance mechanisms were dominant. Second, we examine the development of Indigenous-led transboundary governing bodies, focusing on the Yukon River Intertribal Watershed Council. We argue that the YRITWC is emblematic of a new era of transboundary water governance: participatory, and (in an increasing number of cases) Indigenous led – which implies new principles for water governance, involving an expanded network of actors beyond the nation-state.

8.1 Introduction

This chapter examines two eras of water management across the Canada-US border, separated by more than a century. First, we examine the Boundary Waters Treaty (BWT, a bi-national agreement between the federal governments of Canada and the United States), as an emblematic example of the dominant concerns that underpinned water governance at the turn of the twentieth century. Second, we examine

E.S. Norman (✉)
Department of Native Environmental Science, Northwest Indian College,
2522 Kwina Rd, Bellingham 98226, WA, USA
e-mail: enorman@nwic.edu

K. Bakker
Department of Geography, Program on Water Governance, University of British Columbia,
1984 West Mall, Vancouver V6T 1Z2, BC, Canada
e-mail: karen.bakker@ubc.ca

S. Renzetti, D.P. Dupont (eds.), *Water Policy and Governance in Canada*,
Global Issues in Water Policy 17, DOI 10.1007/978-3-319-42806-2_8

the development of Indigenous-led transboundary governing bodies, focusing on the Yukon River Intertribal Watershed Council (YRITWC). We argue that the YRITWC is emblematic of a new era of transboundary water governance: participatory, and (in an increasing number of cases) Indigenous-led – which implies new principles (such as postcolonial governance) and an expanded network of actors beyond the nation-state.

The chapter thus builds on earlier discussions in this volume related to international and transboundary water. As a complement to these other contributions, we offer a constructivist perspective. Simply put, a constructivist perspective holds that water is governed through the social construction and application of political jurisdictions and associated regulatory frameworks – which are influenced by underlying worldviews. This, in turn, highlights a well-recognized and foundational challenge: the mismatch between multiple scales of jurisdiction (which arise from the social construction of (geo)political boundaries) and water's biophysical characteristics – a flow resource that transgresses these boundaries at multiple scales. This mismatch is underwritten by power dynamics, in which some worldviews have more perceived legitimacy than others. In our view, the study of changing patterns of water governance must thus be attentive not only to the content of regulations, but also to the underlying worldviews and power dynamics in which they are embedded. Moreover, these power dynamics evolve over time, as illustrated (in this chapter) by the resurgence of Indigenous participation in transboundary water governance across the Canada-US border.

Our chapter begins with a summary of the historical context of the foundational transboundary treaties and frameworks, which provides important context for our examination of their strengths and limitations. In particular, we emphasize what was *excluded* from the original treaties and frameworks – particularly Indigenous rights and representation (as well as other issues, such as ecological integrity). A historically situated analysis of the legal framework for transboundary governance (which was established more than a century ago) is important for understanding the subsequent evolution of transboundary water governance (Table 8.1). This also underpins our analysis, later in the chapter, of the constraints and context in which new patterns of postcolonial governance are emerging.

8.2 The Boundary Waters Treaty, the International Joint Commission, and Binational Cooperation

8.2.1 The Boundary Waters Treaty: The Tug of War between Sovereign Reach and Duty to Cooperate

On 11 January 1910, in Washington D.C., the United States Secretary of State, Elihu Root, and the United Kingdom's (His Majesty's) ambassador to the United States, James Bryce, signed the BWT – more formally known as "Treaty between

Table 8.1 Eras of Canada–US transboundary water governance, 1900–2015

Era	Period	Role	Example
Binational	1900–1910	Establish legal framework for management of transboundary Canada-US waters	Boundary Waters Treaty
Cooperative development	1945–1965	Projects of mutual benefit; Federal government encourages hydroelectric development	Columbia River Treaty; St. Lawrence Seaway and hydropower development; Niagara Treaty comprehensive management
Comprehensive management	1965–1985	Issue-based comprehensive river basin planning and more 'environmentally conscious' framework; Water expertise built up at federal-level	Great Lakes Water Quality Agreement
Sustainable development	1985–2005	Linking of economy and environment; Issues more integrative, anticipatory and preventive	Great Lakes Annex
Participatory/ post-colonial	2000–current	Indigenous – led transboundary governance mechanisms organized by culturally connected communities; Opening of space for Indigenous leadership roles in existing or new mechanism	IJC International Watersheds Initiative; Coast Salish Gathering/Yukon River Inter-tribal Watershed Council

Adapted from Norman and Bakker (2014)

the United States and Great Britain relating to Boundary Waters, and Questions arising between the United States and Canada". The BWT represented the culmination of years of diplomatic negotiations aimed to secure mutually-agreed upon terms related to the shared waters that span the 7000 km Canada-US border. Initially, this push led to the creation of the International Waterways Commission (IWC), which was established by the Rivers and Harbors Act of 1902. Although the IWC's focus was regional, mainly concerned with the Great-Lakes-St Lawrence system, it provided a platform for the development of a commission that would address water issues across the entire Canada-US Border.

The negotiation of the Boundary Waters Treaty was heavily influenced by the political climate of the time. A central component of the BWT was the creation of the International Joint Commission (IJC), the central governing body established to mediate conflict between the two federal governments. Given the extraordinary length of the international border, this mediation was considered to be desirable (indeed, inevitable) and long-term. As US Secretary of State, Elihu Root, stated prior to signing the treaty (quoted in Munton 1981):

> I do not anticipate that the time will ever come when this Commission will not be needed. I think that as the two countries along this tremendous boundary become more and more thickly settled the need for it will increase.

Despite the relatively widespread agreement on the need for a formal cooperation arrangement, the settlement of the treaty was delayed as government representatives debated a critical point in the negotiation: the relative strength of the IJC. A key issue for the negotiators was the question of whether to assign the commission:

> real arbitral power or merely to assign a more limited mandate to recommend courses of action, which the governments could accept or reject as they saw fit (Carroll 1981, p 44).

Two competing philosophies played out at the turn of the century during the Treaty's negotiation. On the Canadian side, growing political power in the US made the Canadian negotiators advocate for the strongest, most binding Treaty possible; this was represented in the Gibbons-Clinton draft. In this version, the IJC had binding authority over the final decision and the creation of a supranational court to adjudicate boundary water disputes. Root articulated this preference in a letter to British ambassador James Bryce in 1908:

> The difficulty of the United States in assenting to an agreement that all questions within the broad field described by the Gibbons-Clinton draft shall be referred for final determination to such a commission as is proposed, is in the main that such questions necessarily involve, not merely questions of fact and of law suitable for the determination of a commission or arbitral tribunal, but many questions of policy, of mutual concession and of the give and take which is in so great a number of cases the efficient means of reaching possible settlement of difficult controversies. Such questions of policy, of concession, of discretion make it impossible for the Government of the United States to commit to any commission under our system of government. (Carroll 1981)

On the US side, the negotiators leaned towards an agreement with less authoritative power. The preference was for mechanisms that would make "recommendations" rather than have any real arbitral and binding power. Growing concern regarding Mexican unrest and protests was at the forefront of American thinking during the time of negotiation. Concern over this unrest led to the Harmon Doctrine – named after US Attorney General Judson Harmon, who noted that:

> the fundamental principle of international law is the absolute sovereignty of every nation, as against all others, within its own Territory.

This underlying philosophy ultimately provided the guiding context for BWT.

The BWT achieved two crucial goals. It established a core of legal principles to govern the management of internationally shared waters between Canada and the United States, and it established an institutional framework to supervise the implementation of the principles. The historical context of this negotiation is important context in the contemporary discussion of 'sovereign reach', and the balance between nation-state and non-nation-state actors in water governance. At the time of the signing of the BWT, the tension focused on where the power of adjudication lay – in the commission, or in the country itself. The fact that the power balance was seen as an 'either-or' rather than a 'both-and' is an important point to consider for contemporary transboundary water governance mechanisms. Thus, for both Canada and the US, the politics of the southern neighbor (in Canada – the US – and in the US – Mexico) influenced the binational treaty between Canada and the US.

One hundred years later, tensions have shifted, and much has changed in terms of the make-up of the actors involved in the governance and the expectations for public involvement. Although the balance between sovereign rights has continued to evolve, Secretary Root's prediction held true: along the Canada–US border, the need remains for federal governments to address issues of shared concern, communicate intentions, and mediate potential as well as actual conflicts. However, some key factors have changed. First, a greater range of types and numbers of actors have become involved in these discussions. At the time of the Treaty signing, the governance structure was envisaged as nation-to-nation. Although the six commissioners (three from each country) are asked to act in their professional capacity, rather than representing 'national interests', the structure of the IJC remains distinctly binational. However, a number of other sub-national actors have become increasingly important actors in transboundary water governance processes (such as the Great Lakes Fisheries Council); in some cases, the IJC is thus playing a reduced role (Norman and Bakker 2009). Second, the trend towards rescaling has caused the IJC – which remained largely federally oriented until the past decade – to rethink its approach and role beyond the nation-state.

8.2.2 The IJC's Transition from Binational to Participatory

The IJC's International Watershed Initiative (IWI) is an example of how the IJC has evolved to include more inclusive governance models – but still within a binational framework. Generally, the IJC proposed the following roles for the Boards:

- Coordinate with existing institutions and agencies within the watershed;
- Report on the state of the watershed to the IJC;
- Serve as liaison between the community and the IJC; develop monitoring indicators; run water-related studies (as directed by the IJC);
- Facilitate the prevention of disputes; foster the development of an "informed transboundary community" regarding watershed management;
- Field comments and complaints regarding the watershed; interface between different jurisdictional and ecological systems (i.e., freshwater, terrestrial and marine) (IJC 1997, 30).

Following IJC protocol, in order for the Watershed Boards to become initially established, political support from both countries and subnational stakeholders is necessary. In 2005, the Commission identified three existing boards to apply the IWI concept: The St. Croix River (New Brunswick, Maine), The Red River (Minnesota, North Dakota and Manitoba), and Rainy River (Minnesota and Ontario). In 2007, the Commission added a fourth pilot international board: The Souris River (Saskatchewan, Manitoba, and North Dakota). In 2007, the International St. Croix

River Watershed Board became the first ‘official’ International Watersheds Board and, in 2013, the International Rainy Lake of the Woods Watershed Board became the second official board. The International Red River and the International Souris River Board remain pilot IWI boards.

Although Boards have long been part of the IJC’s governing process, the IWI marks a departure from the previous IJC governance model. The watershed role moves away from the legal and engineering framework – which dominated the Commission during its first 50 years – to embrace a broader governance approach. During the earlier years of the IJC, the governance structure had a narrower scope and, subsequently, operated with much less infrastructure. Watershed Board is an attempt to broaden the participation within the constructs of the binational approach – and to take a *proactive* rather than *reactive* approach to water governance. In addition, the Board approach provides the opportunity to diversify its membership, open up space for multiple worldviews, and respond to and plan for contemporary water issues. Whether the Boards will fulfill this is yet to be seen; however, the structure is in place for ongoing flexibility.

The evolving approach adopted by the IJC discussed above is one example of a more general trend, in which new issues and concerns have come to occupy a central place in discussions of water governance, including: groundwater, water quality, aquatic species (e.g., salmon), and Indigenous (American Indian or First Nations) communities. As explored below, this exclusion has continuing impacts on water governance frameworks; shifts in governance models increasingly include First Nations communities and actors – and in fact – as discussed in the following sections, legal precedents both in the US and in Canada have the extraordinary opportunity to reshape governance frameworks where Indigenous rights are central to, rather than excluded from, water politics.

8.3 Indigenous-Led Transboundary Water Governance: Legal and Historical Context

For most of the twentieth century, Indigenous communities were systematically excluded from water governance on both sides of the Canada-US border. The BWT did not include or reference relationships to Indigenous or First Nations communities. This was particularly problematic for communities whose traditional territories were and are bifurcated by the international border (Norman 2015).

In recent years, as explored below, there has been a resurgence of Indigenous involvement in water governance processes on both sides of the border (Battiste 2000; Norman 2015; Phare 2009; Thom 2010; Von der Porten and de Loë 2013a, b). Although a full discussion of the many reasons for this resurgence is beyond the scope of this chapter, three issues merit discussion: definition and extent of rights, legitimacy of Indigenous law, and Indigenous rights in International Law. The first issue is the definition and extent of Indigenous water rights, particularly given that although state recognition of Indigenous water rights varies widely, international jurisprudence links water rights to land rights, which may include the right to determine development for lands, territories, and resources, including water (Boelens 2003, 2014; Boelens and Doornbos 2001; Getches 2005; Goodman 2000; Phare 2009; Perreault 2005, 2008; Roth et al. 2005; Thorson et al. 2006; United Nations Declaration on the Rights of Indigenous Peoples, Article 32). The response to this issue hinges, in part, on a second key issue: the degree of recognition of the legitimacy of Indigenous law, which emphasizes the importance of the distinction between inherent rights (based on historic, reciprocal and often spiritual as well as material relationships between Indigenous Peoples and water) as opposed to established ("Aboriginal law") rights (negotiated or defined in and through courts and/or Treaties) (Borrows 2002, 2010). As inherent Indigenous rights continue to grow in importance (Beccar et al. 2002; Boelens 2008, 2009; Boelens et al. 2010; Borrows 1997a, b), they will continue to be subject to critique regarding their comprehensiveness and co-optation (Corntassel 2008; Coulthard 2007, 2014; Mascarenhas 2007). Nonetheless, we believe that this will be an important trend in the next few decades, and that the distinction between Indigenous law and Aboriginal law will become increasingly central to water governance in traditional territories of Indigenous peoples along the Canada-US border. This is also consistent with the third issue often invoked in this debate: recognition of Indigenous Rights in international law, as exemplified by Article 6 and Article 36b of the UN Declaration on the Rights of Indigenous Peoples (United Nations 2007).

Underlying debates about Indigenous water rights is a key issue that must be addressed: another important issue: the articulation between Indigenous and Western worldviews of water. Asserting a dichotomy between Indigenous and Western approaches runs the risk of essentialization, and threatens to obscure diversity within these traditions. With these caveats in mind, scholars have observed that whereas Western views often frame water as a resource available for human exploitation (Bakker and Cook 2011; although see Strang 2004, 2009 re: the diversity of Western sociocultural relations to water), Indigenous views may frame water as an animate (living) entity which is imbued with socio-cultural and spiritual meaning, is inter-related with all aspects of the environment, is the subject of values, use practices and rituals transmitted across generations, and is thus constitutive of Indigenous law, knowledge, and identity (Barbera-Hernandez 2010; Basdeo and Bharadwaj 2013; Battiste and Henderson 2000; Berkes 2008; Blackstock 2001; Boelens et al. 2010; Craft 2014; McGregor 2012; Perreault 2005, 2008; Phare 2011; Richard 1999; Rizvi et al. 2013; Singh 2006; Toussaint et al. 2005; UBCIC 1991; Wilson 2014). In other words, our worldviews of water influence our approaches to governance (e.g., Craft 2014; Sandford and Phare 2011; Wong 2011; Wong and Christian 2013). The incorporation of Indigenous law (and hence worldviews) into water governance processes – and the concomitant evolution of these processes – thus raises a set of important conceptual issues that are relevant to a broad range of disciplines (Fresque-Baxter 2015). As explored below, these issues are reflected – albeit in varying degrees and with distinct emphases – in the evolving approach to Indigenous water governance in Canada and the United States.

8.3.1 Indigenous Water Governance in Canada

Water is absent in the Supreme Court of Canada's definition of Aboriginal title, and Aboriginal rights to water have never been explicitly established or disproven through a Supreme Court ruling in Canada (Phare 2009). This situation contrasts to the United States, where the Winters doctrine and subsequent Cappaert decision (Cappaert v. United States 1976; Winters v. United States 1908) state that surface and ground water rights are implied by the federal establishment of American Indian reservations, and (further) set standards to which the US government must adhere in ensuring sufficient water for reservations, thereby recognizing the centrality of water-land interactions to American Indian communities (Shurts 2000). In Canada, comparatively, Indigenous water rights and participation in water governance have, with few exceptions, been treated implicitly within land-focused legal claims; as a result, historical inequalities have often constrained Indigenous communities' access to water and exercise of Aboriginal rights (Phare 2009; Simms 2015; von der Porton 2012; von der Porten and de Loë 2013a, b). As an example, Bill S-8 – the Safe Drinking Water for First Nations Act – has been criticized for its treatment of Aboriginal rights and title, particularly a clause that permits Aboriginal and treaty rights to be overridden to ensure drinking water safety (AFN 2012; Simeone and Troniak 2012).

Another example is the case of First in Time, First in Right (FITFIR) water rights regime in British Columbia and Alberta. Although First Nations are undeniably "First in Time," they are usually not the first rights-holders (BCAFN 2010; Simms 2015). This situation has led to litigation (Tsuu T'ina Nation v. Alberta 2010; Piikani Nation v. Alberta 2002) in which the legal argument has been made that Aboriginal rights and title extend to water. There are several bases to this argument. First, Aboriginal water rights are inherently necessary to fulfill the purpose and intent for which reserves were created (Bartlett 1998; Matsui 2009; Phare 2009; Walkem 2007). Second, insofar as water and land are integrally interconnected, Aboriginal title gives Indigenous peoples the right to lands submerged by waters, and entitles them to make use of waters (Passelac-Ross and Buss 2011). Third, Indigenous peoples have inherent water rights that stem from their historical, ongoing claim to and relationships with their traditional territories, including the customs, practices, and traditions of a given community (BCAFN 2010; Norman 2015; Walkem 2007). As noted by the British Columbia Assembly of First Nations:

> For our Nations, ownership of water, or title to water, is considered an aspect of Aboriginal title. We maintain that our Nations have Aboriginal title to water, and therefore the right to use it, and to govern its use (BCAFN 2010, p 445).

Recent jurisprudence is noteworthy in this regard; there have been several recent key court cases that specifically address First Nations' water access and rights (Halalt First Nation v. British Columbia 2011). For example, the BC Supreme Court's recent *Halalt First Nation vs. BC Environment* found that the Halalt had a proprietary interest in the groundwater beneath their reserve, although the Court's decision also emphasized the fact that the proceeding was not conclusive.

8.3.2 Indigenous Water Rights in the US

In the United States, federally recognized Indigenous communities have a sovereign status, which provides them with the right to negotiate with the federal government as equals. In essence, there are hundreds of 'nations' within the United States. The nation-to-nation relationship is a complex legal relationship (Deloria and Lytle 1998). Although in the US, tribes have many treaty rights (including rights for self-governance), in practice, the relationship is fraught with power-dynamics that consistently privilege federal and state governments over the tribes themselves. These asymmetrical power dynamics have led tribes to need to reassert treaty rights over-and-over again.

One example of the jurisdictional complexity—and the wrought power dynamics included—is the example of the challenges Indigenous communities are facing in energy extraction. The Ojibwa communities who reside along the coastline of North America's Lake Superior (and whose traditional territory spans into Canada and the United States) face a number of water-related issues that compromise their rights to clean water. One of the biggest threats currently facing Ojibwa communities (similar to many Indigenous communities throughout the world) is mining activity, compromising both surface and groundwater sources. In one case, in Keweenaw Bay Indian Community of northern Michigan, mining permits being issued through the state (avoiding the US Environmental Protection Agency as a loop-hole for conducting federally-run Environmental Impact Assessments) are posing significant and long-term threats to the community's water sources.

In particular, the Eagle Mine in the Upper Peninsula of Michigan has become a point of controversy, as the mine location has disturbed a spiritually and culturally significant place. Although the American Indian Religious Freedom Act was passed in1978, Indigenous communities still struggle to protect sacred places in the face of resource extraction. As Jessica Koski of the Keweenaw Bay Indian Community states:

> it is a shame that the United States of America, proudly founded on values of religious freedom, has trouble guaranteeing this right to all of its nation's first people.

Furthermore, a press release put out by the Keweenaw Bay Indian Community in 2014, stated:

> The Anishinaabeg band has opposed the Eagle Mine development, located on Treaty of 1842 ceded homeland, since it was first permitted by the Michigan Department of Environmental Quality in 2006. Unsettled concerns involve the mining regulatory process, improper permitting and inadequate assessment of impacts to the area environment, cultural resources and water quality, including groundwater contamination and the potential for perpetual acid mine drainage upstream from Lake Superior.

However, the 'loopholes' of gaining permits through a state (rather than US entity) has allowed a fast-track for the company to move forward with the extraction. The mining companies have surveyed parcels throughout northern Michigan, Wisconsin, and Ontario (throughout Ojibwa and Anishinaabe territory) and have plans to continue to mine throughout this region. The mining techniques that are proposed pose the risk of compromising water quality for generations to come. Could attempting to join The United Nations' Convention on the Protection and Use of Transboundary Watercourses and International Lakes (The Convention) as the "Ojibwa Nation" help to keep the mining companies in check? The Keweenaw Bay Indian Community has already sent letters to the United Nations General Assembly citing the United Nations Declaration of the Rights of Indigenous Peoples (Koski 2014).

8.4 Emerging Trends in Indigenous-Led Transboundary Water Governance Across the Canada-US Border

In conjunction with the shifting legal landscape described above, Indigenous peoples are increasingly reclaiming transboundary water governance processes along the Canada-US border. A growing number of Indigenous-led mechanisms have emerged along the Canada–US borderland, which continues to contribute to the growing rescaling and the new era of governance discussed in the previous section. Specifically, Indigenous transboundary water governance mechanisms have become established in the Great Lakes basin (Ontario and Minnesota, Wisconsin, Michigan, New York), the Yukon River basin (Yukon Territory and Alaska) and the Salish Sea basin (British Columbia and Washington). In this new era, not only are Indigenous communities working collaboratively to address environmental issues facing their communities, but they are also working on strengthening traditional lifeways and promoting self-governance.

In essence, Indigenous-led initiatives are serving as mechanisms to both decolonize governance and provide means for self-determination. Arguably, the sociocultural and political impacts of colonial occupation are exacerbated at the borderland, where Indigenous homelands were bifurcated by the demarcation of the international border and the subsequent creation of Native reserves – reservations, in the United States (Norman 2014). The emergence of an Indigenous-led era in North American transboundary water governance has international relevance, because it gives question to the context of sovereignty and cooperation. For example, the recent move to designate membership of First Nations, Metis, and Tribal members in the Rainy Lake board is precedent-setting for the IJC. Although there has been an increasing move to include representation of Indigenous actors on other IJC boards, this practice has not been consistently applied. The explicit inclusion of Indigenous actors in transboundary water governance activities – which had previously been reserved for federal actors – points to the shift that is occurring.

The rescaling of water governance to account for greater inclusion of Indigenous Peoples in its process occurs in at least two fundamentally different ways. First, through the restructuring of diplomatic relations from nation-to-nation to multinational – and, the subsequent renegotiations of Treaty terms. Second, when the restructuring occurs through a greater emphasis on Indigenous-led governance mechanisms.

The first approach, which requires a shift from bi-national to multi-national understandings of Canadian-US water politics is critical, but will be structurally difficult and require long-term changes either when Treaties come to terms (through Treaty renegotiations) or during the establishment of entirely new treaties. Given, the "Treaty era" of water governance has largely been usurped by more "informal mechanism" (see Table 8.1 and Norman and Bakker 2009) such as MOUs, MOAs and other informal agreements, the Treaty approach would likely occur as Treaty's come through a renegotiation process – such as the Columbia River Treaty renegotiation process.

However, the ability to renegotiate the terms of all Treaties signed as it relates to Canada-US water is likely an unrealistic goal, given the expense and lack of political movement from the federal governments to implement such a sweeping change, let alone engage in Treaty negotiations at a rate more common in the "Treaty era" of early last century. More likely, treaties that are up for renegotiation – such as the Columbia River Treaty – can be "modernized" – to reflect current political sensibilities and priorities. The Columbia River Treaty is an example, where, through negotiations, greater inclusion of First Nations is leading to a fundamental rethinking of the treaty terms (Cohen and Norman forthcoming; Bakker et al. 2014).

The second approach (which is the focus of the following section of the chapter), in which Indigenous governance mechanisms are established by and for indigenous governments, is less structurally and temporally fraught. In this case, the restructuring occurs through inter-tribal processes and other governments (federal, state, provincial and city, etc.) are invited as partners or *guests*, to the process. This is the case with the Coast Salish Gatherings (Norman 2013) as well as with the Yukon River Inter-tribal Watershed Council (Wilson 2014).

8.4.1 The Yukon River Inter-Tribal Watershed Council

The Yukon River Inter-Tribal Watershed Council (YRITWC) is a leading example in North America of an Indigenous grassroots organization whose purview spans multiple borders (Canada, US provincial, state, and Indigenous nations) and whose governance processes are explicitly multi-national (see Fig. 8.1).

The YRITWC represents a collective initiative of 70 First Nations and tribes that reside within the massive watershed. The aim of the Council is to improve the health and wellbeing of the watershed and the people who live within it. Simply put, the Council's vision is "to be able to drink water directly from the Yukon River" (YRITWC 2013). The mission lays out its mandate:

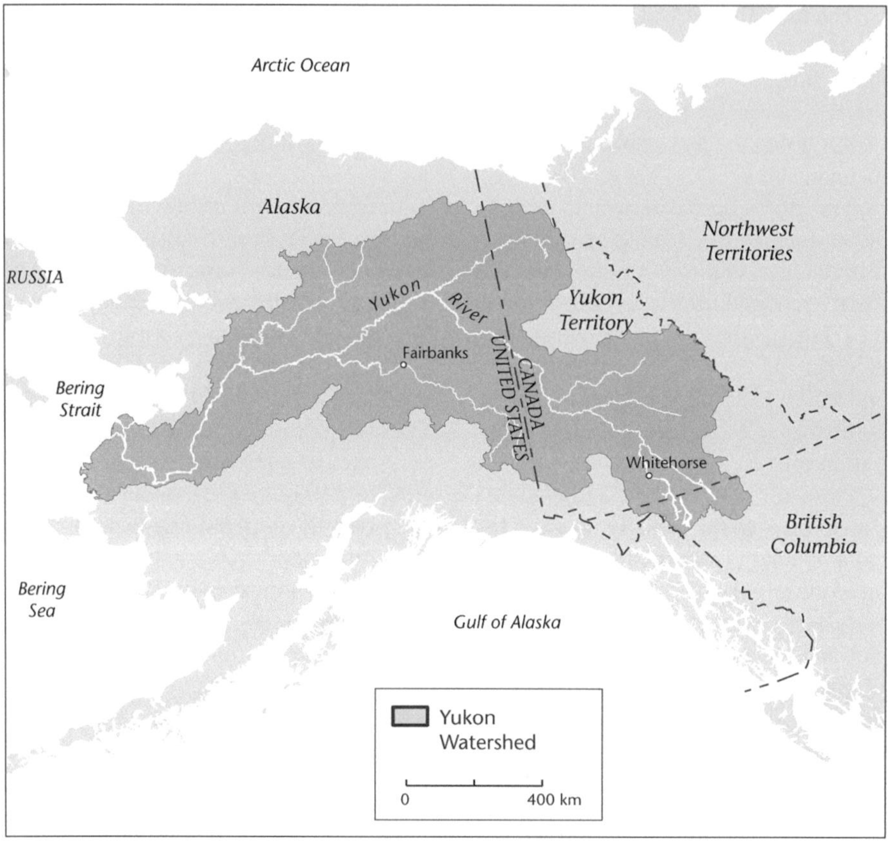

Fig. 8.1 The Yukon watershed

> We, the Indigenous Tribes/First Nations from the headwaters to the mouth of the Yukon River, having been placed here by our Creator, do hereby agree to initiate and continue the clean up and preservation of the Yukon River for the protection of our own and future generations of our Tribes/First Nations and for the continuation of our traditional Native way of life.

Foundational to the Council's governance model is a place-based, basin wide approach that accounts for traditional governance mechanisms. The YRITWC's governance structure represents a commitment to inter-basin cooperation; consensus based decision-making, and integrated resource governance, which is consistent with Indigenous governance mechanisms throughout the region (YRITWC 2013). In addition, as former Director Jon Waterhouse, notes, the YRITWC:

> understands the need to move forward with the modern technological advances available to them, but are also committed to cultural preservation and grounding in tradition (Waterhouse 2007).

This is particularly important for preparing the youth to serve as the next generation of leaders, of stewards of the Yukon – multigenerational education is a priority for the YRITWC.

Governing the Yukon River is no small feat. The size of the watershed requires a dedicated effort by staff, volunteers, and partner agencies to continue to work towards their goals. With offices in Anchorage, Fairbanks, and Whitehorse, the organization plays a leading role in advocating for the River (both regionally and internationally) and communicating between its upstream and downstream member tribes.

To carry out its mission, the YRITWC relies on a staff of 14 individuals, and an Executive Committee that is selected through a process of consensus at the biennial Summits. Consistent with collective governance models, members of the Executive Committee represent the geographic area of those governments, rather than the First Nation or Tribes themselves, including: Yukon Flats, Tanana River, Innoko Confluence, Middle Yukon, Innoko Confluence, Dahka Tlingit, Vuntut Gwich'in/Tr'ondek Hwech'in, Northern Tutchone, Kaska, and Southern Tutchone. The representatives provide steering for the staff and volunteers to meet its mission – they are also instrumental in communicating the work of the Council back to its constituents.

A central part of the YRITWC's governance structure is to host biennial summits that represent the signatory Tribes and First Nation governments. The Summit is a place to reinvigorate its members and allies and to (re)align them to goals and missions of the YRITWC. The Summit also provides an opportunity to exchange information about activities, voice concerns, celebrate successes, realign priorities, and strengthen allies. The Summit provides the intellectual and administrative space to reflect on the YRITWC program activities (as executed through the five departments), it also provides the staff and director an opportunity to prioritize project foci and assures the continual realignment of activities to the mission of the Council.

Through these reflective, consensus-based governance mechanisms, five priorities of YRITC have been identified through inter-departmental cooperation: sustainable lands, solid waste, science, energy, and drinking water. In the area of sustainable lands, the YRITWC developed a brownfield information system database, which can be targeted for cleanup and redevelopment. The projects include identifying contaminant sites, organizing a corrective action plan, including cleanup planning and reuse of lands. In this process, they document and map a range of contaminates including, asbestos, lead, mercury, PCBs, sewage petroleum, (among others).

The Backhaul Program was established to address the solid waste issues in the Yukon River Watershed – and arose out of the prioritization in the governance process. Between 2004 and 2008 alone, the Backhaul program removed 10 million pounds of solid waste material out of the Watershed.

Water Quality assessment is also a fundamental component of the goals of the YRITWC. Water Quality monitoring occurs consistently throughout the entire River Basin (and is collected through an educational program).

Not only are the accomplishments that directly related to water quality important to consider – the governance process itself is an important component to consider.

Reclaiming traditional governance structures is a key component to Indigenous-led governance. Like the Coast Salish Gathering, equally important as the goals of increasing water quality are revitalizing traditional governance mechanisms and fostering Indigenous self-determination.

In addition, the development of the Yukon River basin-wide Water Management Plan is largely viewed as one of the most significant tangible accomplishments of the YRITWC. The plan incorporates years of discussions and dialogues between member tribes and First Nations and highlights the need for ongoing comprehensive and basin-wide planning.

8.4.2 Connection to Place

Lastly, (re)connecting to place is also seen as an important contribution of the YRITWC and a central mission of The Summit. For example, the YRITWC Summit location changes sites every year – the participants experience the River from different vantage points and explore the diverse ecological habitats along the River. In addition, the changing locations provide an opportunity for the host communities to highlight their culture (for example through food preparation that is unique to the region, storytelling, or dance). Another way the YRITWC connects people to the River is through the annual Healing Journey, where people physically connect on the waterway on multi-day canoe trips. This embodiment of the River – getting to know the place in which you work – is starkly different from other approaches, where management is largely disconnected from the place. For example, offices in Washington D.C. and Ottawa, where the locus of power for international water management are far removed from the physical places that are on their "file". Officials in charge of managing transnational fresh water governance rarely have the chance to physically visit the locales for which they have oversight responsibility. Interviews and field work with managers across Canada and the US found that this lack of physical connection to waterways proved to be a significant barrier to its management (Norman and Bakker 2009, 2015). This is in stark contrast to Indigenous governance – where knowing the place – in body, mind, and spirit – is integral to the practice of water governance. In fact, it would seem unfathomable that one could make decisions about a place without knowing the place intimately – and have the historic knowledge passed down through generations that would guide the relationship with the river.

8.5 Conclusion

Water governance is a reflection of dominant values and worldviews. This is an important observation. This chapter has sought to provide some empirical depth to this observation, through using two examples to illustrate the evolution of transboundary water governance frameworks along the Canada-US border. We believe that the contemporary, participatory-era governance described above provides an opportunity to rethink transboundary water governance. Indeed, we believe that it signals a coming era of post-colonial, Indigenous-led transboundary water governance.

We offer this historical analysis in order to provide context for contemporary opportunities to renegotiate and reimagine these transboundary governance Treaties, policies and governing frameworks; specifically, we claim that it is important to remember the original historical context in which these agreements emerged in order to more effectively negotiate their future. In short, we argue it is essential to look backward (in time) in order to move forward with the goal of enhancing social and environmental equity in transboundary water governance across the Canada-US border and beyond. These arguments are, we feel, relevant to more general debates about water governance, and in particular the need for "social learning" to support adaptive management. Within this broader context, scholars have called for the evaluation of frameworks under which water laws, policies, and regulations were developed and are operationalized (Hirsch 2001; Jonas and Gibbs 2003; Kramsch and Mamadouh 2003; Kramsch 2002; Maddock 2004; Paehlke 2001; Parson 2001; Verchick 2003; Wismer and Mitchell 2005). This reevaluation provides opportunities to strengthen mechanisms and approaches and also remedy previous omissions.

Moreover we claim that hallmarks of this era will include active attempts to decolonize the borderland by creating indigenous-led mechanisms that rely on traditional (and non-nation-state) networks and protocols. In cases along the Canada–US borderlands, this is already occurring. Indigenous-led transboundary water governance is being enacted, with very significant impacts on nation-state-centric governance mechanisms. For example, in the Pacific Coast region, Coast Salish Gatherings and intertribal Canoe Journeys are raising public awareness of water-related issues – particularly ocean acidification, declining marine habitat and resources such as salmon, and impacts of water pollution. In the Great Lakes, the Grandmother Water Walkers and the Great Lakes Indian Fish and Wildlife Commission are working on issues related to declining water quality and fish contamination due to extraterritorial pollutants, such as PCBs and mercury. And in the Yukon Basin, the Yukon River Inter-Tribal Watershed Council is working towards protecting one of North America's last 'wild rivers' and to continue to maintain the priorities of the Indigenous peoples who rely on the river for subsistence. In all three cases, the preservation of these waterways is also explicitly linked to cultural preservation, economic vitality, self-determination and self-governance (Norman 2015).

Thus, in rethinking environmental governance – and transboundary water governance more specifically – we suggest that there is a need to support communities and activities that have historically been absent from formal binational treaties. Indigenous peoples had little presence in the original treaty-making process in North America related to transboundary waters. Court cases and treaties between tribes and governments have had increased presence in water governance. The North American example provides an opportunity to learn from retrofitting treaties or, as in the Columbia River renegotiation, 'modernizing' the treaty. Thus, for countries and regions that are currently creating or updating their transboundary water governance mechanisms, a key consideration is to look at the cultural politics associated with the act of creation of the border. The design of the mechanism should reflect historical 'upstream' and 'downstream' relationships and provide avenues for regional mechanisms to take leading roles in long-term governance (Cohen et al. 2014).

References

Assembly of First Nations. (2012). Bill S-8: Safe drinking water for First Nations Act. Submission to the Standing Senate Committee on Aboriginal Peoples. http://www.afn.ca/uploads/files/water/senatestandingcommitteebill_s-8.pdf. Accessed 5 Nov 2015.

Bakker, K., & Cook, C. (2011). Water governance in Canada: Innovation and fragmentation. *International Journal of Water Resources Development, 27*, 275–289.

Bakker, K., Cohen, A., Norman, E., & Mattson, M. (2014, January 28). Opinion: Transboundary water governance needs new approach. *The Vancouver Sun*. http://www.vancouversun.com/technology/Opinion+Transboundary+water+governance+needs+approach/9436407/story.html?__lsa=33d9-cf86

Barbera-Hernandez, L. (2010). Indigenous peoples, human rights and natural resource development: Chile's Mapuche peoples and the right to water. *Annual Survey of International Comparative Law, 11*, 1–28.

Bartlett, R. (1998). *Aboriginal water rights in Canada: A study of aboriginal title to water and Indian water rights*. Calgary: Canadian Institute of Resources Law.

Basdeo, M., & Bharadwaj, L. (2013). Beyond physical: Social dimensions of the water crisis on Canada's first nations and considerations for governance. *Indigenous Policy Journal, 13*(4), 1–14.

Battiste, M. (2000). *Maintaining aboriginal identity, language, and culture in modern society. Reclaiming indigenous vice and vision* (pp. 192–208). Vancouver: UBC Press.

Battiste, M. (Ed.). (2000). *Decolonizing education: nourishing the learning spirit*. Toronto: Purich Publishing, Ltd.

Battiste, M., & Henderson, J. Y. (2000). *Protecting indigenous knowledge and heritage: A global challenge*. Saskatoon: Purich Publishing Ltd.

Beccar, L., Boelens, R., & Hoogendam, P. (2002). Water rights and collective action in community irrigation. *Water Rights and Empowerment*, 1–21, Netherlands: Van Gorcum, Assen.

Berkes, F. (2008). *Sacred ecology* (2nd ed.). New York: Routledge.

Blackstock, M. (2001). Water: A first nations' spiritual and ecological perspective. *BC Journal of Ecosystem Management, 1*, 54–66.

Boelens, R. (2003). Local rights and legal recognition. The struggle for indigenous water rights and the cultural politics of participation. Paper presented at the Third World Water Forum, Kyoto, March 2003. http://www.cepal.org/drni/proyectos/walir/doc/walir15.pdf

Boelens, R. (2008). *The rules of the game and the game of the rules: Normalization and resistance in Andean water control*. Wageningen: Wageningen University.

Boelens, R. (2009). The politics of disciplining water rights. *Development and Change, 40*(2), 307–331.

Boelens, R. (2014). Cultural politics and the hydrosocial cycle: Water, power and identity in the Andean highlands. *Geoforum, 57*, 234–247.

Boelens, R., & Doornbos, B. (2001). The battlefield of water rights: Rule making amidst conflicting normative frameworks in the Ecuadorian highlands. *Human Organization, 60*(4), 343–355.

Boelens, R., Getches, D., & Guevara-Gil, A. (2010). *Out of the mainstream: Water rights, politics and identity*. London: Earthscan.

Borrows, J. (1997a). Frozen rights in Canada: Constitutional interpretation and the trickster. *American Indian Law Review, 22*(1), 37–64.

Borrows, J. (1997b). Living between water and rocks: First nations, environmental planning and democracy. *University of Toronto Law Journal, 47*(4), 417–468.

Borrows, J. (2002). *Recovering Canada: The resurgence of indigenous law*. Toronto: University of Toronto Press.

Borrows, J. (2010). *Canada's indigenous constitution*. Toronto: University of Toronto Press.

British Columbia Assembly of First Nations (BCAFN). (2010). Section 3.31: Water. In: *BCAFN governance toolkit: A guide to nation building* (pp. 443–462), Vancouver, British Columbia.

Cappaert v. United States. (1976). 129 U.S. 426 http://nnsa.energy.gov/sites/default/files/nnsa/inlinefiles/united%20states%20v%20cappaert.pdf

Carroll, J. E. (1981). Patterns old and new. In R. Spencer, J. J. Kirton, & K. R. Nossal (Eds.), *The international joint commission seventy years on* (p. 44). Toronto: University of Toronto.

Cohen, A., & Norman, E. (2016). Beyond critique in political ecology: Re-imagining the border in the Columbia Basin. Forthcoming.

Cohen, A., & Norman, E. S., & Bakker, K. (2014, February). Renegotiating boundaries: Political ecologies of shared waters. Paper presented at the dimensions of political ecology conference, Lexington, Kentucky.

Corntassel, J. (2008). Toward sustainable self-determination: Rethinking the contemporary indigenous-rights discourse. *Alternatives: Global, Local, Political, 33*(1), 105–132.

Coulthard, G. S. (2007). Subjects of empire: Indigenous peoples and the 'politics of recognition' in Canada. *Contemporary Political Theory, 6*(4), 437–460.

Coulthard, G. S. (2014). *Red skin, white masks: Rejecting the colonial politics of recognition*. Minneapolis: University of Minnesota Press.

Craft, A. (2014). Anishinaabe Nibi Inaakonigewin Report. Reflecting the Water Laws: Research Gathering conducted with Anishinaabe Elders, 20–23 June 2013 at Roseau River, Manitoba. University of Manitoba, Centre for Human Rights Research (CHRR) and Public Interest Law Centre (PILC). http://static1.squarespace.com/static/54ade7ebe4b07588aa079c94/t/54ec082ee4b01dbc251c6069/1424754734413/Anissinaabe-Water-Law.pdf

Deloria, V., Jr., & Lytle, C. M. (1998). *The nations within: The past and future of American Indian sovereignty*. Austin: University of Texas Press.

Fresque-Baxter, J. A. (2015). Water is life: Exploring the relationship between place identity, water and adaptive capacity in fort resolution, Northwest Territories, Canada. Dissertation, Wilfrid Laurier University. http://scholars.wlu.ca/cgi/viewcontent.cgi?article=2792&context=etd

Getches, D. (2005). Defending indigenous water rights with the laws of dominant culture. In D. Roth, R. Boelens, & M. Zwarteveen (Eds.), *Liquid relations* (pp. 44–65). New Jersey: Rutgers University Press.

Goodman, E. J. (2000). Indian tribal sovereignty and water resources: Watersheds, ecosystems, tribal co-management. *Journal Land Resources and Environmental Law, 20*(2), 185–221.

Halalt First Nation v. British Columbia (Ministry of Environment). (2011). 2011 BCSC 945.

Hirsch, P. (2001). Globalisation, regionalisation and local voices: The Asian development bank and rescaled politics of environment in the Mekong region. *Singapore Journal of Tropical Geography, 22*(3), 237–251. doi:10.1111/1467-9493.00108.

International Joint Commission. (1997). *The IJC and the 21st Century* (pp. 30)., Washington, DC/Ottawa: International Joint Commission. http://www.ijc.org/php/publications/pdf/ID1011.pdf

Jonas, A. E. G., & Gibbs, D. C. (2003). Changing local modes of economic and environmental governance in England: A tale of two areas. *Social Science Quarterly, 84*(4), 1018–1037. doi:10.1046/j.0038-4941.2003.08404011.x.

Koski, J. (2014, April 10). *Personal Interview with author*. Baraga, Michigan.

Kramsch, O. (2002). Reimagining the scalar topologies of cross-border governance: Eu(ro)regions in the post-colonial present. *Space & Polity, 6*(2), 169–196. doi:10.1080/1356257022000003617.

Kramsch, O., & Mamadouh, V. (2003). Crossing borders of political governance and democracy. *Journal of Borderlands Studies, 18*(1), 39–50. doi:10.1080/08865655.2003.9695600.

Maddock, T. A. (2004). Fragmenting regimes: How water quality regulation is changing politicaleconomic landscapes. *Geoforum, 35*, 217–230. doi:10.1016/j.geoforum.2003.09.004

Mascarenhas, M. (2007). Where the waters divide: First nations, tainted water and environmental justice in Canada. *Local Environment, 12*(6), 565–577.

Matsui, K. (2009). *Native peoples and water rights: Irrigation, dams, and the law in Western Canada* (Vol. 55). Montreal: McGill-Queen's University Press.

McGregor, D. (2012). Traditional knowledge: Considerations for protecting water in Ontario. *International Indigenous Policy Journal, 3*, 1–21.

Munton, D. (1981). Dependence and interdependence in transboundary environmental relations. *International Journal, 36*(1), 139–184.

Norman, E. S. (2014). *Governing transboundary waters: Canada, the United States and Indigenous communities*. London: Routledge.

Norman, E. S. (2015). *Governing transboundary waters: Canada, the United States and Indigenous Communities*. London: Routledge.

Norman, E. S., & Bakker, K. (2009). Transgressing scales: Water governance across the Canada–US borderland. *Annals of the Association of American Geographers, 99*(1), 99–117.

Paehlke, R. (2001). Spatial proportionality: Right-sizing environmental decision-making. In E. A. Parson (Ed.), *The environment: Persistent challenges, uncertain innovations* (pp. 73–125). Toronto: University of Toronto Press.

Parson, E. A. (2001). Environmental trends: A challenge to Canadian governance. In E. A. Parson (Ed.), *The environment: Persistent challenges* (pp. 3–30). Toronto: Uncertain Innovations. University of Toronto Press.

Passelac-Ross, M., & Buss, K. (2011). Water stewardship in the lower Athabasca river: Is the Alberta government paying attention to aboriginal rights to water? *Journal of Environmental Law and Practice, 23*(1), 69–83.

Perreault, T. (2005). State restructuring and the scale politics of rural water governance in Bolivia. *Environmental Planning A, 37*(2), 263–284.

Perreault, T. (2008). Custom and contradiction: Rural water governance and the politics of Usos y Costumbres in Bolivia's irrigators' movement. *Annals of the Association of American Geographers, 98*(4), 834–854.

Phare, M.-A. S. (2009). *Denying the source: A crisis of First Nations water rights*. Surrey: Rocky Mountain Books.

Phare, M. (2011). *Restoring the lifeblood: Water, first nations and opportunities for change*. Toronto: Walter and Duncan Gordon Foundation.

Piikani Nation v. Alberta. (2002). settled out of court.

Richard, G. (1999). When the ditch runs dry: Okanagan natives, water rights, and the tragedy of no commons. *B.C. Historical News, 32*(2), 10–17.

Rizvi, Z., Adamowski, J., & Patrick, R. J. (2013). First nations capacity in Quebec to practice integrated water resources management. *International Journal of Water, 7*(3), 161–190.

Roth, D., Boelens, R., & Zwarteveen, M. (2005). *Liquid relations*. New Brunswick: Rutgers University Press.

Sandford, R. W., & Phare, M. (2011). *Ethical water*. Vancouver: Rocky Mountain Books.

Shurts, J. (2000). *Indian reserved water rights: The winters doctrine in its social and legal context, 1880s–1930s*. Norman: University of Oklahoma Press.

Simeone, T., & Troniak, S. (2012). *Legislative summary: Bill S-8: The safe drinking water for First Nations Act*. Publication No. 41-1-S8E. Ottawa: Library of Parliament.

Simms, B. R. (2015). *All of the water that is in our reserves and that is in our territory is ours: Colonial and Indigenous water governance in unceded Indigenous territories in British Columbia*. Master's Dissertation, University of British Columbia. http://circle.ubc.ca/handle/2429/51475

Singh, N. (2006). Indigenous water management systems. *Society & Natural Resources, 19*(4), 357–366.

Strang, V. (2004). *The meaning of water*. New York: Berghahn Books.

Strang, V. (2009). *Gardening the world*. New York: Berghahn Books.

Thom, B. (2010). The anathema of aggregation: Towards 21st-century self-government in the Coast Salish world. *Anthropoligica, 52*, 33–48.

Thorson, J. E., Britton, S., & Colby, B. G. (2006). *Tribal water rights: Essays in contemporary law, policy, and economics*. Tuscon: University of Arizona Press.

Toussaint, S., Sullivan, P., & Yu, S. (2005). Water ways in aboriginal Australia: An interconnected analysis. *Anthropologial Forum, 15*(1), 61–74.

Tsuu T'ina Nation v. Alberta (Minister of Environment). (2010). 2010 ABCA 137.

Union of British Columbia Indian Chiefs. (1991). *Indian Water Rights in British Columbia: A handbook*. http://d3n8a8pro7vhmx.cloudfront.net/ubcic/legacy_url/549/ubcic_indianwaterrights_1991.pdf?1426350418. Accessed 5 Nov 2015.

United Nations. (2007). *UN declaration on the rights of indigenous peoples*. Geneva: United Nations. http://www.un.org/esa/socdev/unpfii/documents/DRIPS_en.pdf

Verchick, R. R. M. (2003). Why the global environment needs local government: Lessons from the Johannesburg summit. *Urban Lawyer, 35*(3), 471–494.

von der Porten, S., & de Loë, R. C. (2013). Collaborative approaches to governance for water and indigenous peoples: A case study from British Columbia. *Geoforum, 50*, 149–160.

von der Porten, S., & de Loë, R. C. (2013a). Water governance and Indigenous governance: Towards a synthesis. *Indigenous Policy Journal, 23*(4), 1–12.

von der Porton, S. (2012). Canadian indigenous governance literature: A review. *AlterNative: An International Journal of Indigenous Peoples, 8*(1), 1177–1801.

Walkem, A. (2007). The land is Dry: Indigenous peoples, water, and environmental justice. In K. Bakker (Ed.), *Eau Canada* (pp. 311–329). Vancouver: UBC Press.

Waterhouse, J. (2007, March 4). Phone interview with E. Norman.

Wilson, N. J. (2014). Indigenous water governance: Insights from the hydrosocial relations of the Koyukon Athabascan village of Ruby, Alaska. *Geoforum, 57*, 1–11.

Winters v. United States. (1908). 564 U.S. 207. https://supreme.justia.com/cases/federal/us/207/564/

Wismer, S., & Mitchell, B. (2005). Community-based approaches to resource and environmental management. *Environments, 33*(1), 1–4.

Wong, R. (2011). What would restitution and regeneration look like from the point of view of water. In A. Mathur, J. Dewar, & M. DeGagné (Eds.), *Cultivating Canada: Reconciliation through the lens of cultural diversity* (pp. 81–90). Ottawa: Aboriginal Healing Foundation.

Wong, R., & Christian, D. (2013). Untapping watershed mind. In C. Chen, J. MacLeod, & A. Neimanis (Eds.), *Thinking with water* (pp. 232–253). Montreal: McGill-Queen's University Press.

Yukon River Inter-Tribal Watershed Council. (2013). http://www.yritwc.org. Accessed 5 Nov 2015.

Chapter 9
The Great Lakes, Water Quality and Water Policy in Canada

Carolyn M. Johns

Abstract This chapter focuses on water policy in the Great Lakes region to illustrate the complexities of transboundary, multi-level water governance and policy regimes in Canada. The chapter begins with an introduction of the Great Lakes as a natural resource that provides multiple ecological and human uses. The evolution of policy challenges and responses are introduced along with the various policy stakeholders. The second section reviews the transboundary water policy regime that has evolved to govern this complex system with a particular emphasis on water quality and the Great Lakes Water Quality Agreement. The third section focuses on analyzing what the Great Lakes case illustrates about water policy in Canada from a historical and comparative perspective. The focus in this section is on answering two key questions: How does the Great Lakes case help us understand water policy in Canada and how does this regional transboundary water policy regime impact water policy in Canada? The final section of the chapter focuses on the lessons and insights that can be drawn from the case of the Great Lakes for water policy in Canada and beyond.

9.1 Introduction

The Great Lakes – St. Lawrence is the largest freshwater basin on earth, covering 767,000 miles^2, and straddling a combined shoreline of 10,210 miles or 17,017 km on the border between Canada and the United States. The basin contains approximately one-fifth of the world's surface freshwater. It sustains a complex set of ecosystems, flora, fauna and human uses. Collectively, the five Great Lakes and their draining river systems span two countries, two provinces, eight US states, hundreds of Indigenous communities and numerous cities and towns. The region is home to 105 million people, including more than one-tenth of the population of the United States and one-quarter of the population of Canada (Austin et al. 2010;

C.M. Johns (✉)
Department of Politics and Public Administration, Ryerson University,
350 Victoria Street, Toronto, ON M5B 2K3, Canada
e-mail: cjohns@politics.ryerson.ca

S. Renzetti, D.P. Dupont (eds.), *Water Policy and Governance in Canada*,
Global Issues in Water Policy 17, DOI 10.1007/978-3-319-42806-2_9

Kavcic 2015). It is the largest and most populated megalopolis in North America with 60 million people living in the largest cities around the lakes.

The basin has played a major role in the economic development of the US and Canada. Water resources from the basin have supported fishing and shipping, agriculture, power generation and a wide range of industries. Some of the world's largest concentrations of industrial activity are located in the Great Lakes region totaling some $4 trillion in Gross Domestic Product annually and providing 47 million jobs (Austin et al. 2010; Kavcic 2015). The region accounts for 30 % of combined Canadian and US economic activity and would rank as the third largest economy in the world if it were a country (Kavcic 2015).

In addition to the significant economic uses, the basin provides water for numerous human uses such as drinking water, waste disposal and a variety of domestic, recreational and spiritual uses. In a survey of Canadian and US Great Lakes residents in 2013, 71 % stated that the Lakes have a significant impact on their daily life and 94 % believe the Lakes are a valuable economic resource for their state or province. Some 98 % reported that they had visited at least one of the Lakes; 54 % reported visiting a Great Lakes beach and another 54 % reported having eaten fish from one of the Lakes in the past year (Maack et al. GLPRN 2014). In a 2015 public opinionc survey, some 85 % of residents felt it was important to protect the Great Lakes (IJC 2016).

The number and variety of water uses has changed dramatically and increased over time. Industrialization, urbanization and globalization have all resulted in a range of uses that cumulatively have had negative impacts on surface water, wetlands, and groundwater. Despite their large size, the Great Lakes are sensitive to the effects of all of these various uses and sources of pollution. To manage all these multiple uses of water a complex set of policies has evolved to manage this valuable, shared resource.

Given water in the Great Lakes is shared by Canada and the US, a complex set of water policies and governance arrangements involving numerous governments and a variety of users and stakeholders has evolved over the past 100 years to collectively manage water quantity and quality issues. There have clearly been some important water policy success stories related to acid rain, toxic substances and endangered species. However, despite the existence of long-standing policies, institutions and agreements between Canada and the US, enduring challenges remain and new challenges have emerged. Harmful algal blooms that were once thought to have been address by policy regimes have reappeared, particularly in Lake Erie. New challenges such as flood and water level concerns related to climate change; new energy developments related to nuclear waste; shale gas fracking and offshore wind; and new concerns about aquatic invasive species such as Asian carp that have found their way into Lake Ontario; all have significant implications for the Great Lakes water policy regime.

This chapter focuses on the Great Lakes as an excellent case to examine water policy in Canada in the broader context of a complex, transboundary, multi-level policy regime. The next section outlines the water policy regime and governance arrangements that have evolved to address the enduring and new water policy challenges in the region, with a particular focus on transboundary policies and institutions to collectively address water pollution and water quality issues. The third section analyzes the evolution and state of Canadian water policy in this broader transboundary context. The final section outlines the insights and lessons that can be learned from this case about water policy in Canada.

9.2 Great Lakes Water Policy and Governance: The Transboundary Policy Context

9.2.1 Historical Context

Water policy in the Great Lakes region has a long history and foundations that predate Confederation in Canada. Indigenous peoples across several different tribes that inhabited the Great Lakes region had governance principles based on a fundamental and spiritual respect for water for thousands of years (McGregor 2014; Rashidi 2014). With immigration and colonization, property and water rights became embodied in constitutional and statutory law related to both water quantity and quality. Laws and policies related to water rights, use and quantity preceded laws and policies for water quality resulting in two, overlapping, yet distinct water policy regimes (Brooks 2003; Boyd 2003; Sproule-Jones et al. 2008; Johns and Thorn 2015). Since the 1800s, numerous laws and court decisions in Canada and the US have shaped the evolution of a multi-level law and policy regime. As outlined in other chapters in this book, the constitutional division of powers and asymmetrical responsibility for water as a natural resource resulted in the development of water policies related to the Great Lakes at all levels of government – federal, provincial and municipal. In addition, a regime of international soft law in the form of international agreements with limited, legally binding features developed at the transboundary level.

Transboundary governance and management of the vast watersheds shared by Canada and the United States began over 100 years ago with the signing of the Boundary Waters Treaty in 1909. The treaty established the International Joint Commission (IJC) as a unique, binational, transboundary institution for the resolution of binational water disputes and cooperation on water policy priorities. The IJC, headed by six commissioners (three appointed by the US president and three appointed by the Canadian prime minister), acts on references by both governments to cooperatively address disputes and other issues related to transboundary water resources. Initially, water policies focused on water quantity impacts related to shipping and navigation, industrial uses, hydro-electricity development and water diversions (Heinmiller 2003, 2007; MacFarlane 2015). Some of the earliest water policies evolved at the transboundary level related to the development of the Great Lakes-St. Lawrence seaway for navigation and industrial purposes involving both national and subnational diplomacy (MacFarlane 2015). Since that time, other water quantity policies such as the 2008 Great Lakes Compact and Agreement between provinces and states which bans water diversions outside the basin and the IJC's Plan 2014 to update regulations related to water level fluctuations and management in the Lake Ontario-St. Lawrence River area have increased the focus on water quantity and the impacts of climate change in the region. The water quantity regime is it itself being tested by the recent approval of the City of Waukesha Wisconsin's application to divert water from Lake Michigan. Both these developments clearly indicate that over time there has also been increasing recognition that water quantity and water quality regimes have evolved and need to be more fully integrated in the region (Johns and Thorn 2015).

In terms of water quality, as early as 1912, water pollution issues were referred to the IJC for study. Scientific studies began to emerge in the 1940s and 1950s related to local concerns about water pollution and fish kills in several of the lakes. However, it wasn't until serious water pollution events and national news coverage of the Cayuga River fires, toxic contamination in the Love Canal area of New York State, and other pollution events mobilized political action and got the attention of policy makers in the late 1960s and early 1970s. Growing scientific research related to the impacts of water pollution culminated in a policy push for basin-wide efforts focused on point source controls, toxic chemical regulations and effluent limits for industries and municipal sewage treatment systems.

In 1971 the federal government and the Government of Ontario signed the first intergovernmental Canada-Ontario Agreement (COA) related to Great Lakes issues. A year later in 1972, joint fact-finding and diplomacy at the federal led by the IJC resulted in the signing of the first Great Lakes Water Quality Agreement (GLWQA) between Canada and the United States. This non-binding, good-faith agreement was a form of international soft law committing the federal governments in Canada and the US to take action to clean up the lakes.

In the 1970s and 1980s governments in Canada and the US focused on developing policies, programs and their bureaucratic capacity to implement the agreement ushering in a period of significant water policy development. Many of the foundational water laws and policies were developed during this period (see Chap. 4) and there were clearly some successes related to point source water pollution. However, by the late-1980s, an estimated 57 million tonnes of liquid waste were still being poured into the Great Lakes annually by its inhabitants, their industries, and their municipalities and the degraded state of ecosystems was well known (Colborn et al. 1990, p 64). In 1987, the two national governments renewed the GLWQA with a more focused approach on 43 Areas of Concern (AOCs) that had been identified by the IJC as particularly problematic watersheds with serious pollution and governance challenges (see Fig. 9.1).

AOCs highlighted various types of water pollution—heavy metals, pathogens, contaminated sediments, and toxic chemicals in the basin—that were impairing various ecosystem uses, such as the health and reproduction of various species, including fish populations; wildlife habitat; human health; and other human uses. Twenty-six AOCs were located entirely within the United States, twelve entirely within Canada, and five shared by both countries. In each AOC multi-level and

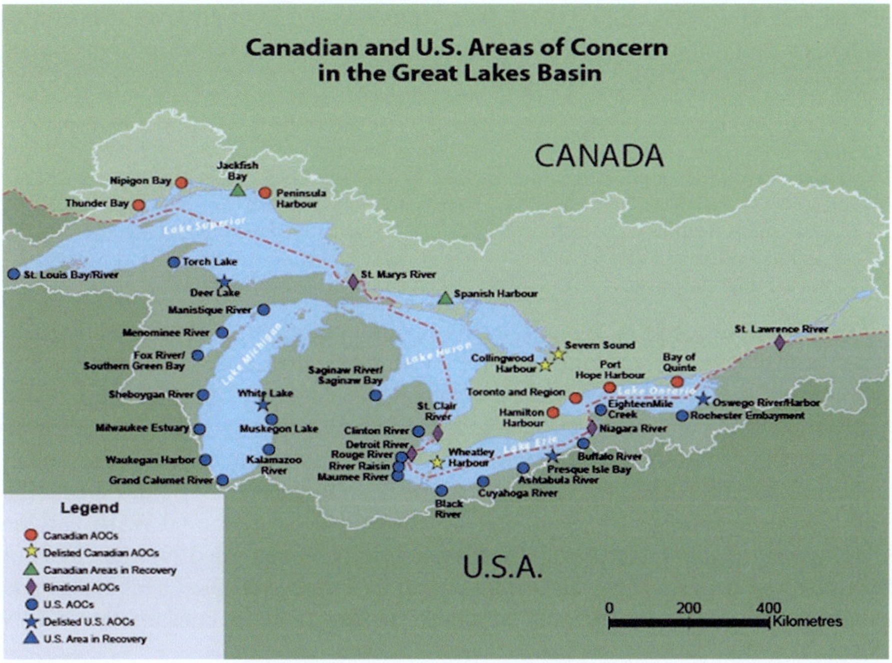

Fig. 9.1 Map of Great Lakes Basin and Areas of Concern (Environment Canada, Areas of Concern 2015, http://www.ec.gc.ca/raps-pas/)

multi-stakeholder governance institutions were established to develop and implement Remedial Action Plans (RAPs) to clean up and improve the human and ecological uses in these watersheds. Although signatories to the GLWQA were the federal governments in Canada and the US, the eight US states, province of Ontario along with local governments and a wide range of local stakeholder had central roles in preparing and implementing RAPs. There was a gradual shift in focus from IJC-level efforts to a more decentralized approach related to RAPs (Hartig and Zarrell 1992; Sproule-Jones 2002a) with the IJC playing a "permanent watchdog role" over the implementing parties (Schwartz 2005).

Watersheds became the natural boundaries around which networks of state and societal stakeholders came together to address environmental problems in the 43 AOCs. Non-government actors included local citizens, scientists, industrial and agricultural users, local environmental groups, water system professionals, academics and scientists. The involvement of this wider range of stakeholders from all levels of government, a wide range of government agencies, and key societal actors and organizations resulted in various levels of success (Sproule-Jones 2002a, b; Botts and Muldoon 2005; Johns 2009; Krantzberg and Manno 2010; Greitens et al. 2012).

There were clearly some success stories such as habitat restoration, return of species on the brink of extinction, improvements in ecological function, and the declining presence of some of the most toxic chemicals, including PCBs and dioxins

(Botts and Muldoon 2005, p 137). The focus on AOCs, multi-level cooperative action, as well as the supporting role of the IJC was credited for some of the policy successes. According to the IJC itself:

> No other institution has the IJC's broad mandate or its successful track record in preventing and resolving transboundary disputes around environmental and water-resource issues, and no other institution provides the opportunities for officials from all levels of government, scientists, stakeholders and interested citizens to work together on these issues. The Commission's flexibility and historic emphasis on consultation, joint fact-finding, objectivity and independence, and its ability to engage local governments and serve as a public forum are important assets in meeting the challenges of the 21st century. (IJC 1997)

Despite these accomplishments, the basic water policy goals of swimmable, fishable, drinkable water remained elusive in many Great Lakes communities. The complexity of water pollution problems were also being realized. By the mid-1990s, the US Environmental Protection Agency (EPA) and state governments began to scientifically document that non-point sources and runoff from agriculture and other land uses constituted the major causes of water pollution in many watersheds (US EPA 1995). Despite policy goals of 'virtual elimination' in the GLWQA and the development of the Great Lakes Binational Toxics Strategy in 1997, hundreds of thousands of tons of chemicals continued to be released in the Great Lakes and concerns remained about contaminants, beach closings and fish consumption (GLIN 2006; GLU 2007).

9.2.2 Decline and Re-Engagement?

By the late 1990s, public attention and the general sense of community between researchers and stakeholder groups in the Great Lakes began to decline (Botts and Muldoon 2005, p 161). Although the early 1990s saw some increased activity by industry associations, particularly related to the toxic chemicals and the debate about "virtual elimination," in many AOCs, industries were not active participants (Sproule-Jones 2002a, b) and there was a "lack of industry attention reflecting the sector's limited participation in the Great Lakes community in the first two decades of the agreement's history" (Botts and Muldoon 2005, p 101). At the same time, there was a shift away from point source regulatory instruments to voluntary instruments as governments on both sides of the border tried to address non-point sources of water pollution (Johns 2000).

The Great Lakes fell off the agendas of governments in Canada and the US and a period of public indifference was evident (Botts and Muldoon 2005). Even significant water pollution events, such as those in Milwaukee, Wisconsin, in 1993, and Walkerton, Ontario, in 2000, where people died from drinking water pollution did not generate a public or policy response in connection with broader water quality issues in the Great Lakes (Johns 2009). Water policy and environmental policy were not federal priorities under the Chretien Liberal government. The federal Commissioner of Environment and Sustainable Development noted that the "limited use of federal powers, weakness in basic management and accountability, and the politics of federal–provincial relations have all played a part" in diminishing the federal water policy role generally and in the Great Lakes specifically since the mid-1990s (Canada OAG, CESD 2001a).

During this same period, the COA lapsed between 2000 and 2002 and although agreements were signed and progress reporting continued from 2002 to 2010 these intergovernmental agreements were a low priority for both the federal government and Ontario. In Ontario, water policy did become a priority after the Walkerton tragedy and Inquiry but consecutive governments were internally focused on drinking water policy. During this period the IJC continued to identify issues that were becoming more pressing such as uncertainty related to the impacts of climate change, aging water infrastructure, chemical use and release, habitat loss and biological diversity and invasive species (IJC 2006) but the IJC itself lost capacity (Canada OAG, CESD 2001b).

The declining transboundary and domestic policy effort from the late 1990s to 2000s was particularly evident in the slow progress cleaning up the most polluted sites on the Great Lakes (Sproule-Jones 2002b). Some 20 years after various policy initiatives, only three of the 43 AOCs had been delisted, and two designated as areas in recovery. The progress in the remaining AOCs, and many other watersheds in the Great Lakes, remained slow on both sides of the basin. Although water policy and efforts have been comparatively higher and more consistent in the US than in Canada, most Great Lakes watersheds in the United States were still not meeting water quality objectives (US EPA 2007). Binational State of the Lakes Ecosystem Conference (SOLEC) reports have consistently shown progress has plateaued or declined for many physical, biological and chemical indicators (SOLEC 2011). An IJC report focusing on a 25-year assessment of scientific, ecosystem and performance indicators highlighted some important successes but clearly outlined many outstanding challenges (IJC 2011).

It wasn't until the mid-2000s that the Great Lakes were placed back on the policy agenda, not by governments in Canada but by the federal government in the US. The US *Great Lakes Legacy Act in 2002* allocated $270 million over 5 years to partnership projects with states, federal agencies, municipalities, non-government organizations (NGOs) and the private sector to focus on remediation and cleaning up contaminated sites (US EPA 2014). Similar funding was reauthorized under this statute in 2008. The emphasis in the US was not on developing new water laws and policies but implementing the wide range of existing policies that already existed. This policy initiative paralleled the mobilization of U.S. NGOs and other levels of government in a review of the GLWQA as part of the lead up to the IJC's 100th anniversary in 2009. The election of President Obama in 2008 was followed by the announcement of the Great Lakes Restoration Initiative (GLRI), a major investment of $500 million (2010–2014) creating considerable momentum for state and societal actors to re-engage in Great Lakes efforts for both economic and environmental protection reasons – the largest investment in the Great Lakes in two decades (US EPA 2015). In 2009 U.S. Secretary of State Hillary Clinton and Canadian Minister of Foreign Affairs Lawrence Cannon announced that the US and Canada would renegotiate the Great Lakes Water Quality Agreement last signed in 1987.

This water policy effort in the US was not matched in Canada. Indeed the decade covering 2000–2010 was a period of significant policy inaction and ambivalence related to the Great Lakes (Johns and Sproule-Jones 2015). Neither the Chretien nor Harper governments had water policy or environmental policy on their agendas, despite the policy initiatives south of the border and the context of economic prosperity and federal budget surpluses for most of this period. In Ontario it was not until the late 2000s that broader connections were made to the Great Lakes through the Clean Water Act (2006) and later with provincial involvement in the renegotiation of the GLWQA (Johns and Thorn 2015). Although water policy was a significant policy priority in Ontario, the lack of intergovernmental collaboration, federal leadership and transboundary activity did not result in re-engagement until the renegotiation period of the GLWQA was underway.

The past 5 years has been a very interesting period to examine the implications of transboundary water policy related to the Great Lakes for water policy in Canada. After 3 years of renegotiation, the new GLWQA was signed in 2012 renewing interest in policy efforts and collective action. Implementation by the 'parties' (led by US EPA and Environment Canada) began in 2013 with an emphasis on partnerships and broad engagement. No new legislation was required on either side of the basin. In the US, the GLRI has been the main program to achieve GLWQA objectives with the Great Lakes National Program Office leading federal and intergovernmental efforts. The US placed a high priority on getting some AOCs delisted through targeted GLRI funding and concerted intergovernmental effort resulting in the first four US AOCs being delisted (see Fig. 9.1). In Canada the Canadian Federal Great Lakes Program led by Environment Canada in collaboration with several other federal departments provides the framework for working on Canada's commitments in the GLWQA. The Canada-Ontario Agreement is also a cornerstone of Great Lakes water policy work and was renegotiated and signed in December 2014 for a 5-year period. At the transboundary scale two institutions play a key role. The IJC plays the watchdog and accountability role and is responsible for progress reporting every 3 years under the new agreement. The other binational institution established under the Agreement is the Great Lakes Executive Committee (GLEC) which leads and coordinates policy efforts at all levels of government, including Indigenous organizations, NGOs, and private sector organizations and actors engaged in implementing programs and measures related to GLWQA and its 10 issue-focused Annexes.

The new 2012 GLWQA reflects a binational consensus that existing policies and institutions are sufficient and that the focus needs to be on improving the performance of existing transboundary, national, subnational and watershed-based policies and institutions to improve outcomes. There is a renewed focus on implementation by the Parties. The IJC's role in reporting on progress has move from a biennial to triennial reporting cycle with some interest in incorporating program effectiveness, governance and public engagement indicators to supplement the ecosystem and human health indicators in SOLEC. The focus is on making the existing institutions at all levels work better and increasing stakeholder and public engagement as policy makers realize that government's alone cannot collectively manage this complex system and there is a need to focus more explicitly on the 'implementation deficit' that exists (McClaughlin and Krantzberg 2011). There is also additional emphasis on partnerships and recognition that Indigenous peoples and communities need to be more engaged in GLWQA implementation. This is also reflected in the most recent Canada-Ontario Agreement and the province of Ontario's 2015 Great Lakes Protection Act. However, how to incorporate traditional ecological and indigenous knowledge, and increase engagement of Indigenous peoples and communities remains an important policy and governance challenge.

In general, Canadian water policy aligns well with this approach as the Canadian federal government seems quite content with the status quo related to the GLWQA, happy to have the provinces be the leaders on managing water as a natural resource, and supportive of the partnership approach, as neither water quantity nor quality are political priorities. At the same time the GLWQA was being signed and launched, the Harper government introduced omnibus legislation aimed at consolidating and considerably reducing the scope of a suite of federal statutes that have historically been used as the basis of action in the Great Lakes including the Environmental Assessment Act, Fisheries Act, Environmental Protection Act and Species at Risk Act. Clearly these legislative changes were not made with the Great Lakes in mind but with energy and natural resources development in the West and North at the forefront of the government's second mandate. Although some investments have been made in intergovernmental water infrastructure projects since the 2008–2009 economic recession, the climate of austerity has not supported the implementation of existing water policies in the region. With the election of the new Liberal government under Justin Trudeau it remains to be seen if the Great Lakes and water quality will become policy priorities. In order to more fully assess the impact of domestic water policy in Canada on the Great Lakes, and the impact of transboundary water policy in the Great Lakes on Canadian water policy in the region and beyond it is important to historically and comparatively examine several dimensions of Canadian water policy more closely.

9.3 Assessing Canadian Water Policy Using the Great Lakes Case

Over the past several decades' Canada's water policy related to the Great Lakes has evolved into a complex, multilevel policy regime. For over 25 years, the state of domestic water policy has been used to partially explain transboundary outcomes and environmental performance in the Great Lakes (Hartig and Zarrell 1992; Rabe and Zimmerman 1995; Rabe 1999; Sproule-Jones 2002a, b; Botts and Muldoon 2005; Hill et al. 2007; Heinmiller 2003; Johns 2009, 2010). However, examining the implications of the transboundary water policy regime for water policy in Canada is a less common approach. This section highlights some of the key dimensions that can be used to analyze the implications of the transboundary context for water policy in the region and more broadly in Canada. It starts with the broad economic and political context, covers some of the important institutional dimensions, highlights the significance of political and bureaucratic leadership and capacity and ends with a focus on the significance of private sector organizations and actors, NGOs and public engagement.

As mentioned in the introduction, Canadian water policy in the Great Lakes transboundary context has to be broadly understood in the broader historical, geographic, demographic, economic, social and political context. The Great Lakes has played a very important role in Canadian history, economy and culture. The Great Lakes is a massive water system with a growing and changing demographic base, multiple uses and a changing economy within the broader context of North American integration (Clarkson 2008). Indeed the significance and transformation of the economy in the region has been one of the important policy frames for recent investment in ecosystem services and infrastructure (Austin et al. 2010; Kavcic 2015).

Politically, the region is governed in the context of Canada-US relations amidst a general political culture of trust and cooperation which are strong foundations and the underpinnings of what some argue is a model of transboundary cooperation (Linton and Hall 2013). Indeed, a highly integrated and developed governance regime exists that has had some impact on water policy in Canada, in the region and beyond. However, deeply vested water use interests, the pervasiveness of a politics of abundance, and a federal government which has not viewed the region as central to Canada's economic future, prevents the region from being on the top of the political agenda. The next section thus focuses on the impact of the transboundary water policy on Canadian water policy in the region while keeping in mind that the impact on Canadian water policy outside the region is much more limited.

The transboundary governance regime rests on international soft law, diplomacy and a complex set of transboundary institutional arrangements and web of transboundary networks working on a wide range of water policy issues. Although the Parties are the signatories to the GLWQA and responsible for implementation through the GLEC, Annex committees and working groups and domestic laws and policies, the IJC still has important responsibilities under Article 7. These responsibilities include collecting, analysing and disseminating data related to achievement of the General Objectives, reporting on progress and effectiveness, engaging the public and making recommendations to the Parties. However, some analysts have observed that the level of commitment and support from the Canadian and US governments for the IJC and its role has diminished over the last two decades (Canada OAG, CESD 2001b; Brooks 2010). The IJC has been active related to its roles under the GLWQA and the first triennial reporting process. The first triennial report is forthcoming in 2016. The role and significance of the IJC as a central institution in the Great Lakes water policy and governance system has changed but it is not clear if it will remain central as a transboundary institution or if a more decentralized policy and governance model has diminished its role in terms of leadership and accountability related to the Agreement.

The GLWQA illustrates the challenges the IJC faces as a transboundary institution and the limitations of a non-binding international agreement. No politicians or governments in the US or Canada face serious political fallout if the commitments are not achieved or ignored. Although the longer-term implications of environmental degradation and economic impacts are recognized, the vested interests in current water uses are quite powerful (see Chap. 8). There is little recourse if the goals and objectives of the GLWQA are not achieved and few political incentives to alter existing economic and citizen uses and behaviours. This is compounded by the reality that the policy goals are long-term, beyond the political time frame of any government.

The accountability regime at the transboundary level consists primarily of government and public reporting by the Parties and the IJC. Under Article 5 of the GLWQA the Parties and the IJC are mandated to have a Great Lakes Public Forum fill a review and accountability role. The IJC does have many forums for public engagement on various issues related to water quantity, quality and the GLWQA but continues to be criticized, along with Great Lakes Executive Committee (GLEC) meetings and Annex work, for engaging organized groups and stakeholders and not effectively engaging the public, citizens and Indigenous communities. However, efforts are underway in the various Annex groups to improve public engagement and measure this engagement. The IJC has also been using webinars and various technologies to reach out beyond policy experts, scientists and traditional stakeholders. A true test of public engagement efforts of the Parties and the IJC's role in accountability will be the first triennial report on the new GLWQA expected in 2016. However, in a soft law, transboundary regime, accountability is an enduring issue and the most coercive elements come in the form of naming, shaming and blaming through public reporting and pressures for action in implementation networks. Political leadership and engagement is also important. Not surprisingly, the accountability regime has been the target of criticism for some time (GLU 2007; McLaughlin and Krantzberg 2011), particularly in Canada where the domestic Great Lakes policy and accountability regime is weaker.

As outlined in several chapters in this volume, the constitutional and jurisdictional foundations of water policy in Canada set the stage for decentralized water policy regimes across the country. However, as outlined above, water quality policy in the Great Lakes is more highly coordinated and centralized because of the Canada-US, transboundary, and intergovernmental pull of the GLWQA. As noted by other scholarly observers, beyond the federal-provincial dynamic in Canadian water policy "perhaps the most influential aspect, in terms of shaping policy-making, is the need to coordinate policy with the United States" (Bakker and Cook 2011, p 278).

Although the GLWQA does provide several coordinating mechanisms through the IJC, the GLEC, and numerous issue-specific policy networks that exist across the basin, coordination has also been noted as an enduring issue and one that varies by water policy issue and across several other related policy areas.

Despite fragmentation and coordination challenges, the transboundary water policy regime does seem to exhibit a vertically integrating force that is evident in the text of the GLWQA and in the institutional arrangements designed for implementation. Although comparative policy research indicates water policy in the US federal system has been more centralized and concerted (Hoberg 1992; Johns 2000, 2009, 2010), the longstanding federal-provincial diplomacy and COAs create direct policy linkages between the transboundary, federal and provincial water policy regimes. In addition, over the past 30 years subnational diplomacy, agreements and institutional arrangements have also evolved related to transboundary water quantity and quality governance (Norman and Bakker 2010; Johns and Thorne 2015). Indeed there is some evidence of policy learning and lesson drawing through this complex transboundary and intergovernmental policy regime (Johns 2009).

Nonetheless, there is also evidence that as a result of different political cultures and institutions, Canada and the U.S. have distinct legal approaches and different 'policy styles' when it comes to environmental policy and water policy (Hoberg 1992; Howlett 1994). This is evident in the Great Lakes case. In contrast to the US which has specific legislation related to the Great Lakes and a range of detailed federal statutes with legislated appropriations to Great Lakes programs and initiatives, the Canadian federal government, while officially adopting similar goals for the Great Lakes, does not have any "hard law"' devoted to articulating goals and dedicating resources related to transboundary or domestic efforts and the GLWQA. Although long-standing statutes exist outlining authorities and mandates related to the Great Lakes, there has been no federal legislative action related to the Great Lakes or water for decades. In Ontario, Great Lakes water policy falls under a number of existing and recent statutes such as the Ontario Water Resources Act; the Safe Drinking Water Act; the Nutrient Management Act; and the Clean Water Act, all new or recently amended statutes in the past 10–15 years. This policy development has been positive for water quality in the Great Lakes. In addition, Ontario recently passed a new Great Lakes Protection Act in 2015 to increase the focus and connection of these various policies to the Great Lakes and mobilize resources for Great Lakes efforts.

Another important dimension of analysis related to the implications of the Great Lakes transboundary policy regime on water policy in Canada is on the bureaucratic machinery designed to implement transboundary policy goals. The two main parties to the Agreement are the US EPA and Environment Canada. Policy history indicates that the transboundary water policy regime does have an impact on this front in Canada but that the impact is limited. There is evidence that bureaucratic leadership and capacity has ebbed and flowed along with the priority of the Great Lakes and the GLWQA. There have been periods when the transboundary policy regime had significant implications for bureaucratic leadership and capacity with the establishment and resourcing of the National Water Resources Institute and Canada Centre for Inland Waters in the 1970s and 1980s. However, this leadership and capacity has declined significantly in the past few decades. A Senate committee characterized the federal role in the area of water management and research as in "retreat" and urged the federal government, "once a well-respected leader in advancing the scientific study of water", to reinvest and "take up that leadership role once again" (Canada 2005).

Bureaucratic leadership and capacity at the federal and provincial levels relates directly to the priority of the Great Lakes by various governments. Here a comparison with the US is arguably not a fair comparison given the sheer size of the US bureaucracy and the lack of direct evidence, but there is some related case study research indicating that US governments have exerted more effort and developed more substantial capacity in the area of scientific capacity, measurement and reporting on water quality through the US Environmental Protection Agency (US EPA) and performance indicators under Great Lakes Restoration Initiative (GLRI). There is some evidence that policy and technical expertise from the US EPA has also been comparatively more collaborative with subnational governments in the intergovernmental context and integrated with university extension programs (Johns 2000). Some have attributed this capacity to more adversarial policymaking focused on best-available technologies and reporting and a demand for scientific information by non-governmental actors (Bocking 2006). In the US, policy and scientific capacity has been enhanced significantly through the GLRI, GLWQA and the Great Lakes National Program Office. The impact of the GLWQA on Canadian water policy has been much more limited. Although science has also been central to Environment Canada's role in the Great Lakes, and EC does have a Great Lakes Office and role in SOLEC reporting, its policy and technical capacity is more limited. The federal government's capacity to report on the state of waters in Canada has declined in the past two decades and there is less intergovernmental collaboration on water science and technical reporting. However, scientific and technical capacity has developed through university-based research networks in the past several years and Canada does have some scientific and technical capacity at the watershed scale through Ontario's Conservation Authorities.

The implications of transboundary Great Lakes water policy at the provincial level are less clear. Policy research indicates leadership and policy capacity at the subnational level partially explains the plateauing and declining levels of progress on water quality policy in the Great Lakes (Rabe and Zimmerman 1995; Botts and Muldoon 2005; Johns 2009, 2010). Despite evidence that subnational actors are increasingly engaged in transboundary environmental policy networks and cross-border regional activities (VanNijnatten 2006; Friedman and Foster 2011; Norman et al. 2013), there is very little evidence of the impact of the transboundary water policy regime on water policies in Ontario from the late 1990s to the late 2000s, the same period when there was little federal action. However, provincial engagement with the federal government and other subnational governments during the GLWQA renegotiation period, and the recent proposal for a new Great Lakes Protection Act, are clear indications that the transboundary water policy regime has had an impact at the provincial level in Ontario in the past 5 years.

At the local scale, the transboundary water policy regime has also had some positive implications, particularly for AOCs. There is evidence that the 1987 GLWQA did mobilize local governments and NGOs in problem-definition and watershed planning in the Stage 1 phase of RAPs but that engaging key stakeholders and the private sector became challenges at the Stage 2 implementation stage when significant resources and changes to the status quo were required to remediate and restore these local areas (Sproule-Jones 2002b). Research shows there has also been a higher percentage of subnational and local government participation in those AOCs that have been delisted (Greitens et al. 2012). The other positive development that evolved from the transboundary policy regime was the creation of the Great Lakes-St. Lawrence Cities Initiative in 2003 that engages several local governments in water policy related to the GLWQA, water quantity issues and a range of other water policy initiatives.

The Great Lakes case also illustrates that water policy and governance does not just involve state actors and agencies. Like other water policy regimes in Canada, the Great Lakes contain multiple state and societal actors that are highly dependent on subnational actors and government-led networks for implementation. Networks of government and non-government actors and organizations are involved in a wide range of water quantity and quality policy networks. In the Great Lakes there is evidence that subnational and regional cross-border environmental linkages and networks have become more numerous, dense and regionally concentrated over time, but there is no evidence they are more productive in terms of implementation and policy outcomes (Vannijnatten 2006; Friedman and Foster 2011). Although there is empirical evidence that these networks exist related to the GLWQA and other issues in the region, (Johns 2014; VanNijnatten et al. 2016), there are some important water users and actors missing.

Non-governmental and private sector actors and organizations are also significant to water policy and governance in the region. The transboundary policy regime has clearly had mobilizing effects yet the sheer size and complexity of the water system and the domestic foundations of Great Lakes water policy has made it difficult for environmental organizations to form effective cross-border alliances (Botts and Muldoon 2005). The rise and fall of Great Lakes United is a recent example of this. The sustained engagement and capacity of Canadian environmental groups in Great Lakes efforts is also comparatively low (Botts and Muldoon 2005). Canada does not have Great Lakes specific environmental groups such as Great Lakes Alliance or large, well-resourced NGOs with long-standing Great Lakes agendas. Some of the environmental law and policy organizations do have Great Lakes agendas such as the Canadian Environmental Law Association and Lake Ontario Waterkeepers but their interests and advocacy span many other environmental issues, not just water, and there are challenges related to mobilizing citizen interest and engagement related to Great Lakes issues.

In the past few decades there have been growing opportunities for public engagement in the implementation of water policies at the international, national and local levels particularly related to increasing accountability of the two federal governments in the Great Lakes (Hall 2007). In theory this has helped create a more informed and engaged citizenry and an increased role for public engagement on both sides of the border, and in water policy in Canada, but empirical research indicates this engagement has been very limited. Research on AOCs and RAPs indicates citizen engagement in efforts to develop RAPs but far less in implementation, monitoring and accountability efforts (Hartig and Zarrell 1992; Sproule-Jones 2002b). Research shows that engagement in Canadian AOCs has primarily been government actors constituting 78.5 % of stakeholders while only 49 % of stakeholders in U.S. AOCs are government actors (Greitens et al. 2012). Although there is evidence that the IJC's review of the GLWQA re-engaged both state and societal stakeholders, the 3-year re-negotiation of the agreement has been criticized for the lack of authentic engagement of stakeholders and the public (Krantzberg 2011; Grover and Krantzberg 2015).

Despite the existence of the Council of Great Lakes Industries, which plays an active role in the transboundary water policy regime and is based in the US, the transboundary policy regime has also not led to any serious engagement of the private sector in Canadian water policy in the region. Some companies such as the Royal Bank have invested significantly in corporate social responsibility related to water in Canada but only indirectly related to the Great Lakes and are not involved in the policy community. As outlined in other chapters in this book, this is not surprising given the economic 'vested interests' in the continued uses of water in the basin that can be threatened by changes in use and water quality policies.

The other significant engagement challenge that has not been addressed, even with explicit goals in the new GLWQA, is the engagement of Indigenous peoples and communities. Water is at the heart of Indigenous culture and the 75 First Nations communities in the basin (Rashidi 2014). Articles 3 and 4 in the GLWQA explicitly state that tribal governments, First Nations and Metis will be represented and consulted in Great Lakes water governance. The 2014–2019 COA also states engagement of Indigenous peoples is a priority. However, serious issues remain related to engaging Indigenous peoples, their communities and their traditional ecological knowledge in the transboundary and domestic water policy regimes (McGregor 2012). This is particularly problematic given that values and views of jurisdiction and borders Indigenous peoples hold related to water policy (Phare 2013) are fundamentally in tension with many of the foundations of both the transboundary and domestic water policy regimes in Canada. Although there is growing awareness and recognition of the need for engagement and some important Indigenous-led initiatives such as the Great Lakes Water Walk, there is little convincing evidence that there has been rescaling and a new era of post-colonial water governance since 2000 (Norman 2015; Norman and Bakker 2015, 2016). Serious engagement challenges and power asymmetries remain. This in part is related to broader citizen awareness of both Indigenous peoples and Great Lakes issues.

Water policy in the Great Lakes has witnessed periods of increased salience, awareness and engagement but this has not been the case for decades. Public opinion research consistently shows Canadians value water as a resource (RBC 2016) and this is reflected in public opinion research on the Great Lakes. Some 60 % of residents surveyed feel that their personal actions impact the Lakes' health and only 39 % believe that the lakes are in better health today than they were 20 years ago. A majority of residents (73 %) rank pollution and water quality as one of their top three concerns in the region. There is near-unanimous agreement among residents in the Basin that responsibility for protecting the Great Lakes should be widely shared by all levels of government, business and industry, and individual citizens with 95 % of residents saying that each of these actors has at least some responsibility for lake protection and 70 % expressing that "a great deal of responsibility" lies with governments at all levels and the private sector and a smaller majority (56 %) saying that individual citizens have "a great deal of responsibility" for protecting the Lakes (Maack et al. GLPRN 2014). Although 78 % of residents indicate individuals have responsibilities for protecting the lakes, some 30 % of those do not know waht actions they could take to make a difference (IJC 2016).

Public opinion research clearly indicates that when asked, citizens value the Great Lakes and recognized the challenges. But in the context of many other policy issues facing them and as a political priority, water and the Great Lakes are not at the top of either citizen or political agendas. Thus in terms of what the Great Lakes means for water policy in Canada, it likely means a lot more for those that live in the region and beyond the region is bound-up with the broader politics and economics of abundance.

9.4 Conclusion

The Great Lakes- St. Lawrence basin is unique in Canada but like many other complex transboundary watersheds requires water policies and governance at multiple scales involving a wide range of users and actors to ensure its sustainability. Several decades ago, the policy and institutional complexity make the overall structure and functioning of Great Lakes water governance difficult to fully comprehend and analyze (Francis 1987; Caldwell 1994). Research on water policy and governance in the region over the past 20–30 years has clearly improved our understanding. Nonetheless, this case illustrates the challenge of teasing out Canada's water policy related to the Great Lakes independent of the transboundary water policy regime. This brief history, evolution and comparative analysis of the Great Lakes case does shed some light on what transboundary policy in the region has meant in terms of Canadian water policy.

Compared to water policy in Canada which has been characterized by decentralization, fragmentation, lack of coordination, no national standards, and limited information sharing, the Great Lakes is an interesting case. Although not a political priority in Canada there is a lot of intergovernmental, coordination, and information sharing. There is a lot of good water policy work going on related to water quantity and quality issues. However, this clearly ebbs and flows and this does not mean there are not conflicts and serious governance challenges. As the above chapter illustrates, there are clearly tensions between governments and Indigenous communities in the region; water quantity and quality regimes, government and a wide range of private sector, NGO and citizen interests; the IJC and the parties; and the interface of water policy with other policy domains in the region such as energy.

Water pollution and degradation remain significant environmental problems in the basin and it is clear from both scientific and policy research to date that bilateral and transboundary efforts to date have resulted in only limited success. Although the 2012 GLWQA has incorporated new policy priorities and reconfirmed the focus on lake-wide action, AOCs and watersheds, the focus on new approaches to governance are noticeably absent. Most agree on the water policy goals and transboundary priorities for the basin and the policy document is sprinkled with language of partnerships, community engagement and public participation, however, the policy regime remains state-centered and does not focus on the need to seriously examine the multiple uses and vested interests in the region and the need to change the values and behaviour of citizens, users and government decision makers. These are not easy policy fixes and require a serious rebalancing of uses and more fundamental governance and institutional reform to meet enduring and future challenges (Sproule-Jones et al. 2008).

This case illustrates that even with clear transboundary and domestic water policy goals and transboundary and domestic policies at every level, policy implementation is critical and requires significant political will and resources. It also requires significant changes in values, organizational and individual behavior. Current water policy regimes are not adapting to address the fundamental challenges of institutional reform, vested interests and engagement. This will likely become more evident as new challenges such as climate change test the existing policy regime.

What does exist is a politically legitimate and coordinated governance framework compared to other transboundary water systems and regions in Canada and the world. Although the approach is in some ways a model for water policy in other transboundary systems and for water policy in Canada, there are several important political and policy ingredients missing. Political will and the desire to be a regional leader in balancing economic and environmental systems; lack of broader and more authentic public engagement beyond the usual policy actors; lack of private sector leadership and engagement; lack of openness to new perspectives and new information tools; lack of a rigorous accountability regime, and lack of embracing of new values and a different water ethic are just some of the missing policy ingredients that are highlighted as challenges in this case.

What is clear from this case, and others in the volume, is that multi-level, multi-sector, multi-actor policy approaches are central to improving our understanding and outcomes in these complex systems. Only with new research approaches, new engagement and new tools related to making policies more adaptive and proactive will water policy in this region become a true model of Canadian water policy.

References

Austin, J., Bradely, J., & Vey, J. (2010). *The next economy: Economic recovery and transformation in the Great Lakes region*. Washington, DC: Brookings Institute.

Bakker, K., & Cook, C. (2011). Water governance in Canada: Innovation in the context of fragmentation. *International Journal of Water Resources Development, 27*(2), 275–289.

Bocking, S. (2006, June). *Scientific expertise and environmental politics: Cross-border contrasts*. Paper presented at the Canadian Political Science annual conference.

Botts, L., & Muldoon, P. (2005). *Evolution of the Great Lakes water quality agreement*. East Lansing: Michigan State University Press.

Boyd, D. R. (2003). *'Water' in unnatural law: Rethinking Canadian environmental law and policy* (pp. 13–53). Vancouver: University of British Columbia Press.
Brooks, S. (2003). Water Policy. In S. Brooks & L. Miljan (Eds.), *Public policy in Canada: An introduction* (pp. 262–279). Toronto: Oxford University Press.
Brooks, S. (2010). The International Joint Commission: Convergence, divergence or submergence? In S. Brooks, B. Rabe (Eds.), *Environmental governance on the 49th parallel: New century, new approaches* (pp. 16–32). Washington, DC: Woodrow Wilson Centre. http://wilsoncenter.org/topics/pubs/IJC%20FINAL.pdf
Canada, Office of the Auditor General, Commissioner of the Environment and Sustainable Development. (2001a). A legacy worth protecting: Charting a sustainable course in the Great Lakes and St. Lawrence River basin. In *Report of the commissioner of the environment and sustainable development*. Office of the Auditor General. Minister of Public Works and Government Services, Ottawa, pp. 301–314
Canada, Office of the Auditor General, Commissioner of the Environment and Sustainable Development. (2001b). *The International Joint Commission: A key binational organization*. http://www.oag-bvg.gc.ca/internet/docs/c101sec8e.pdf
Canada, Office of the Auditor General. (2000). *Report of the commissioner of the environment and sustainable development*. Ottawa: Office of the Auditor General of Canada.
Canada, Standing Senate Committee on Energy, the Environment and Natural Resources. (2005). *Report on Federal Water Policy*.
Caldwell, L. K. (1994). Disharmony in the Great Lakes Basin: Institutional jurisdictions frustrate the ecosystem approach. *Alternatives, 20*(3), 26–31.
Clarkson, S. (2008). Transboundary water governance. In *Does North America exist? Governing the continent after NAFTA and 9/11* (pp. 138–162). Toronto: University of Toronto Press.
Colborn, T., Davidson, A., Green, S., Hodge, R. A., Jackson, C. I., & Liroff, R. A. (1990). *Great lakes, great legacy?* Washington, DC/Ottawa: Conservation Foundation/Institute for Research on Public Policy.
Environment Canada. (2015). *Great Lakes areas of concern*. http://www.ec.gc.ca/raps-pas/
Francis, G. (1987). Toward understanding Great Lakes "organizational ecosystems". *Journal of Great Lakes Research, 13*, 233–245.
Friedman, K., & Foster, K. (2011). *Environmental collaboration: Lessons learned about cross boundary collaborations*. IBM Centre for the Business of Government.
Great Lakes Information Network. (2006). *Water pollution in the Great Lakes*. www.greatlakes.net/teach/pollution
Great Lakes United. (2007). *The Great Lakes blueprint: A Canadian vision for protecting and restoring the Great Lakes and the St. Lawrence River ecosystem*. Toronto: Environmental Defence Canada.
Greitens, T. J., Strachan, C. J., & Welton, C. S. (2012). The importance of multi-level governance participation in the 'Great Lakes areas of concern'. In I. Roberge, D. Jesuit, & D. Cepiku (Eds.), *Making multi-level public management work: Cases from the EU and North America*. New York: Taylor and Francis.
Grover, V., & Krantzberg, G. (2015). Transboundary water management: Lesson learnt from North America. *Water International, 40*(1), 183–198.
Hall, N. D. (2007). The evolving role of citizens in United States-Canadian international environmental law compliance. *PACE Environmental Law Review, 24*, 131–160.
Hartig, J. H., & Zarrell, Z. (1992). *Under RAPs*. Ann Arbour: University of Michigan Press.
Heinmiller, B. T. (2003). Harmonization through emulation: Federalism and water export policy in Canada. *Canadian Public Administration, 46*(4), 495–513.
Heinmiller, B. T. (2007). Do intergovernmental institutions matter? The case of water diversion regulation in the Great Lakes basin. *Governance, 20*(4), 655–674.
Hill, C., Furlong, K., Bakker, K., & Cohen, A. (2007). A survey of water governance legislation and policies in the provinces and territories. In K. Bakker (Ed.), *Eau Canada: The future of Canada's water* (pp. 370–376). Vancouver: University of British Columbia Press.
Hoberg, G. (1992). Comparing Canadian performance in environmental policy. In R. Boardman (Ed.), *Canadian environmental policy: Ecosystems, politics and process* (pp. 246–262). Toronto: Oxford University Press.

Howlett, M. (1994). The judicialization of Canadian environmental policy, 1980–1990: A test of the Canada-United States convergence thesis. *Canadian Journal of Political Science, 27*, 99–127.
International Joint Commission. (1997). *The IJC and the 21st century: Response of the IJC to a request by the governments of Canada and the United States for proposals on how to best assist them to meet the environmental challenges of the 21st century*. www.ijc.org
International Joint Commission. (2006). *Advice to the government on their review of the Great Lakes water quality agreement*. A special report to the governments of Canada and the United States.
International Joint Commission. (2011, October). *Assessment of progress made towards restoring and maintaining Great Lakes water quality since 1987*. (Draft Report).
International Joint Commission. (2013). *Sixteen Biennial report on the Great Lakes Water Quality Agreement.*
International Joint Commission. (2016). *2015 Binational great lakes basin poll*. http://ijc.org/files/tinymce/uploaded/WQB/WQB_GreatLakesPollReport_March2016.pdf
Johns, C. (2000). *Non-point source water pollution management in Canada and the US: A comparative analysis of institutional arrangements and policy instruments*. Dissertation, McMaster University, Hamilton.
Johns, C. (2009). Water pollution in the Great Lakes basin: The global-local dynamic. In C. Gore & P. Stoett (Eds.), *Environmental challenges and opportunities: Local-global perspectives on Canadian issues* (pp. 95–129). Toronto: Emond Montgomery.
Johns, C. (2010). Transboundary water pollution efforts in the Great Lakes: The significance of national and sub-national policy capacity. In B. Rabe & S. Brooks (Eds.), *Environmental governance on the 49th parallel: New century, new approaches* (pp. 63–82). Washington, DC: Woodrow Wilson International Center for Scholars, Canada Institute.
Johns, C. (2014). *The significance of networks in understanding governance & public administration: Using social network analysis to analyze governance and policy implementation challenges in the Great Lakes Region*. Paper presented at the Canadian Political Science Association Annual Meeting, May 28, 2014.
Johns, C., & Sproule-Jones, M. (2015). Water policy in Canada and the Great Lakes case. In D. VanNijnatten & R. Boardman (Eds.), *Environmental policy in Canada* (4th ed.). Toronto: Oxford University Press.
Johns, C., & Thorn, A. (2015). Subnational diplomacy in the Great Lakes region: Toward explaining variation between water quantity and quality regimes. *Canadian Foreign Policy Journal, 21*(3), 195–211.
Kavcic, R. (2015). *North America's economic engine*. Bank of Montreal Special Report
Krantzberg, G. (2011). *Renegotiating the 1987 Great Lakes water quality agreement: From confusion to promise*. Paper presented at the 1st World Sustainability Form. www.wsforum.org
Krantzberg, G., & Manno, J. (2010). Renovation and innovation: It's time for the Great Lakes regime to respond. *Water Resources Management, 24*, 4273–4285.
Linton, J., & Hall, N. (2013). The Great Lakes: A model of Transboundary Cooperation. In E. Norman, A. Cohen, & K. Bakker (Eds.), *Water without borders? Canada, the United States and shared waters* (pp. 221–246). Toronto: University of Toronto Press.
Maack, E., Mills, S., Borick, C., Gore, C., Rabe, B., & Great Lakes Policy Research Network (GLRPN). (2014). *Environmental policy in the Great Lakes region: Current issues and public opinion*. http://www.greatlakespolicyresearch.org/wp-content/uploads/2014/05/Public-Opinion-and-Enviromental-Policy-in-the-Great-Lakes-Region.pdf
MacFarlane, D. (2015). Subnational diplomacy and the Great Lakes St. Lawrence Seaway. *Canadian Foreign Policy Journal, 21*(3), 212–223.
McGregor, D. (2012). Traditional knowledge: Considerations for protecting water in Ontario. *The International Indigenous Policy Journal, 3*(3), 1–21.
McGregor, D. (2014). Lessons for collaboration involving traditional knowledge and environmental governance in Ontario, Canada. *AlterNATIVE. An International Journal of Indigenous Peoples, 10*(4), 340–353.
McLaughlin, C., & Krantzberg, G. (2011). An appraisal of policy implementation deficits in the Great Lakes. *Journal of Great Lakes Research, 37*, 390–396.

Mitchell, B. (2016). The hydrological and policy contexts for water in Canada. In S. Renzetti, D. Dupont (Eds.), *Water policy & Governance in Canada*. Cham: Springer.
Norman, E. S. (2015). *Governing transboundary waters: Canada, the United States and Indigenous Communities*. New York: Routledge.
Norman, E., & Bakker, K. (2015). Do good fences make good neighbors? Canada-United States Transboundary Water Governance, the Boundary Waters Treaty, and 21st century challenges. *Water International, 40*(1), 199–213.
Norman, E., & Bakker, K. (2016). Transcending borders through postcolonial water governance? Indigenous water governance across the Canada-US border. In S. Renzetti, D. Dupont (Eds.), *Water policy and governance*. Cham: Springer.
Norman, E., & Karen, B. (2010). Governing water across the Canada-US Borderland. In M. Gattinger & G. Hale (Eds.), *Borders and bridges: Canada's policy relations in North America* (pp. 194–212). Toronto: Oxford University Press.
Norman, E., Cohen, A., & Bakker, K. (Eds.). (2013). *Water without borders? The future of Canada-US Transboundary Water Governance*. Toronto: University of Toronto Press.
Phare, M.-A. (2013). Indigenous peoples and water: Governing across borders. In E. Norman, A. Cohen, & K. Bakker (Eds.), *Water without borders? The future of Canada-US Transboundary Water Governance* (pp. 27–46). Toronto: University of Toronto Press.
Rabe, B. G. (1999). Federalism and entrepreneurship: Explaining American and Canadian innovation in pollution prevention and regulatory integration. *Policy Studies Journal, 27*(2), 288–306.
Rabe, B. G., & Zimmerman, J. B. (1995). Beyond environmental regulatory fragmentation: Signs of integration in the case of the Great Lakes basin. *Governance, 8*(1), 58–77.
Rashidi, M. (2014). Indigenous peoples and the Great Lakes, Great Lakes policy research network policy brief. http://www.greatlakespolicyresearch.org/wp-content/uploads/2013/09/Indigenous-Peoples-and-the-Great-Lakes-Policy-Brief.pdf
Royal Bank of Canada. (2016). *RBC Canadian water attitudes study*. http://www.rbc.com/community-sustainability/environment/rbc-blue-water/water-attitude-study.html
Schwartz, A. M. (2005). The Canada-US environmental relationship: Calm waters but slow sailing. *International Journal, 60*(2), 437–448.
SOLEC. (2011). *State of the lakes ecosystem conference report*. http://binational.net/solec/sogl2011/sogl-2011-technical-report-en.pdf.
Sproule-Jones, M. H. (2002a). Institutional experiments in the restoration of the North American Great Lakes environment. *Canadian Journal of Political Science, 35*(4), 835–857.
Sproule-Jones, M. H. (2002b). *Restoration of the Great Lakes: Promises, practices, performance*. Vancouver: UBC Press.
Sproule-Jones, M., Johns, C., & Heinmiller, B. T. (Eds.). (2008). *Water politics in Canada: Conflicts and institutions*. Montreal/Kingston: McGill-Queen's University Press.
US, EPA. (2007). *National water quality report to congress*. https://www.epa.gov/sites/production/files/2015-09/documents/2007_10_15_305b_2002report_report2002305b.pdf
US, EPA. (2014). *Great Lakes legacy act*. https://www.epa.gov/great-lakes-legacy-act/about-great-lakes-legacy-act
US, EPA (2015). *Great Lakes restoration initiative report to congress and the president 2010–2014*. https://www.epa.gov/sites/production/files/2015-09/documents/21050720-report_to_congress-2.pdf
United States Environmental Protection Agency. (1995). *EPA progress report*. Washington, DC: Environmental Protection Agency.
VanNijnatten, D. L. (2006). Towards cross-border environmental policy spaces in North America: Province-state linkages on the Canada-U.S. Border. *AmeriQuests: The Journal of the Center for the Americas, 3*(1).
VanNijnatten, D., Johns, C., Friedman, K., & Krantzberg, G. (2016). Transboundary governance capacity in the Great Lakes basin. *International Journal of Water Governance, 4*(1), 7–32.

Part III
Economics

Chapter 10
Water Valuation

Diane P. Dupont and Wiktor L. Adamowicz

Abstract Canadians reap the benefits of having an abundance of good quality water. In some cases, however, decisions regarding water resource allocation, quantity allocation decisions or investments in quality improvement, need to be made. A better understanding of the economic value of water would be helpful in these decision contexts. However, most of the values associated with water are not easily calculated since the benefits arise from goods/services for which there are no explicit markets or market prices, and values vary over space and time and differ across people. The goals of this chapter are threefold: (1) briefly outline and provide a critical assessment of the approach and methods associated with the Total Economic Value (TEV) framework for obtaining non-market water values; (2) illustrate a number of case studies of whether or not (and how) market and non-market values have been used for the purposes of allocating water or making investments in improved water management; and (3) identify important gaps in both available data and information for applying valuation information to support policy analysis. In particular, the chapter focuses on policy around improvements in water quality and aquatic ecosystems. The chapter concludes with an evaluation of the availability of Canadian water value estimates. This reveals both data and information gaps. The chapter discusses the importance of filling these gaps, along with the need to develop a systematic approach to their use in policy discussions around water quality, in particular.

D.P. Dupont (✉)
Department of Economics and Environmental Sustainability Research Centre, Brock University, PL 441, 1812 Sir Isaac Brock Way, St. Catharines, ON L2S 3A1, Canada
e-mail: diane.dupont@brocku.ca

W.L. Adamowicz
Department of Resource Economics and Environmental Sociology, University of Alberta, 515 General Services Building, Edmonton, AB T6G 2H1, Canada
e-mail: vic.adamowicz@ualberta.ca

S. Renzetti, D.P. Dupont (eds.), *Water Policy and Governance in Canada*,
Global Issues in Water Policy 17, DOI 10.1007/978-3-319-42806-2_10

10.1 Introduction

The allocation of water resources, and the development of policies and regulations to address water quality concerns, requires information on tradeoffs. Determination of minimum instream flows in the rivers in irrigated areas, or in oil sands mining areas, requires assessment of the costs and benefits of different levels of flow. In the oil sands context, the benefits of maintaining instream flows include the maintenance of ecological function and the provision of ecosystem services such as fishing by Aboriginal and non-Aboriginal people. The costs of increasing instream flows include the increased costs to industry if water is unavailable in times of low flow (Mannix et al. 2014). Economic values provide measures of tradeoffs, and provide a way to compare costs and benefits in the same units. But determining the economic value of the various components of water is challenging in part because water has many dimensions and provides several services or functions for people. Its value is associated with drinking water, sanitation, recreation, habitat, and other elements. The objective of this chapter is to outline the concept of total economic value of water, and provide illustrations or examples of how this information on value can be used in policy and/or decision making.

Economic value is the monetary value that accrues to people from a good or service. Value may be expressed for a unit of irrigation water for crop growth, or for a change in water quality that improves recreational fishing. As described above – the relative values of competing uses of water can help inform resource allocation decisions between uses like irrigation and recreation or in-stream flows. The economic value of a change in water quantity, water quality or other related changes is the amount of money that needs to be given to (or taken away from) an individual (or a firm) to make them as well off as before that change (Freeman et al. 2014). It can also be defined by the person's (or firm's) willingness to pay or accept to be as well off as after the change. Furthermore, goods and services that are not traded in a market, and do not have prices can have measurable economic values.

One of the most popular misconceptions about economic value is that the price of a good or service is its value. That is not the case. Prices, when they exist for market goods and services, can help inform valuation. But they are not values. The price paid by a consumer for the last unit of the good reflects their *marginal* value for the last unit of the good. It does not, however, represent the value of changes involving more than the last unit, as there will generally be a difference in what the consumer would be willing to pay for units of the good and the price that they pay. This distinction between marginal value and non-marginal value gives rise to the famous "water – diamonds" paradox. Water typically has a price that is much smaller than the price of diamonds, yet water is essential to life and diamonds are not. But the price reflects the value of the last unit of each. If one valued a change that removed all water from an individual that value would undoubtedly be higher than the value of diamonds! (See Hanemann 2005).

10.1.1 *Types of Values: Market and Non-Market Values*

10.1.1.1 Market Value

In some cases, water is traded on markets (water rights, bottled water, etc.) and is referred to as a "market good". In southern Alberta, for example, owners of water rights have the opportunity to sell them. The prices received reflect the scarcity value of the water resources. In a landmark trade, the Western Irrigation District (just east of Calgary) sold the right to use a quantity of water to a development just north of Calgary for what was then one of the highest per unit values of water rights trade in the world (Martel 2011).

Market values for products related to water can also be used to help assess the value of water. The economic value of agricultural commodities, for example, is often used to develop estimates of the value of additional irrigation water. In this case the crops can be considered a market ecosystem service and the value of changes in water quality or quantity can be determined at least in part by the assessment of the market value for these commodities.

10.1.1.2 Non-Market Values

Many goods and services are not bought or sold on markets, yet they have economic value. Recreational fishing is not priced in a market, yet consumers are willing to pay for access to the activity and may be willing to pay for improvements to the activity, such as water quality improvements. Similarly, water may contribute to scenic amenities or views that result in higher property values, or health risk reductions that may lower drinking water treatment costs, or provide other goods/services that are not traded on markets. Measurement of economic value for non-market goods and services is also derived from the maximum willingness to pay or the minimum willingness to accept for a change in that good or service. But in the non-market case there are no market prices to inform value.

There are two main types of non-market values: use values and passive (or non-use) values. Use values are related to activities, or items that have a behavioral trail. Water related examples include outdoor recreation, fishing, swimming and other activity based goods that are not typically priced in a market. Values arising from other observable activities, like the value of scenery that may be embedded within property values, are also considered use values because there is a behavioral trail associated with the purchase of a home. Passive use values are the values that an individual has for goods/services that are not expressed through any market behavior. The value an individual may have for the protection of a unique habitat or ecosystem, for example, cannot easily be observed through any market purchases or behaviors. The sum of use and passive values is called **total economic value (TEV)** (Freeman et al. 2014). When evaluating the economic value of a change in water quantity or quality, the most comprehensive measure is the total economic value associated with the change.

10.1.2 Non-Market Valuation Methods

Non-market valuation methods also fall into two categories: revealed preference (RP) and stated preference (SP) methods. RP analysis examines actual behavior like the choice (or frequency of choice) of a fishing site. An individual is observed to spend money on water-based recreation trips to sites and thus a trade-off between money and the non-market activity can be identified. Note that water-based recreation is typically a non-market activity because there is usually no "per-trip" entrance fee or price for the activity like there would be for a market good. A measure of the value of water quality improvements can be determined by examining the additional costs of travel that individuals may incur to access higher quality sites. The expenditures on travel are not the values of the activities; they reflect the prices paid to access the site. The willingness to pay to improve water quality can be measured by the additional travel cost that an individual would be willing to pay to access the higher quality site. Using travel costs as a proxy for the price of a non-market good is called the travel cost method (Freeman et al. 2014). An example of the valuation of water-based recreation is the research by Adamowicz et al. (1994) that measured the value of changes in water quality. Recreation site choices made by individuals in southern Alberta were used to assess the trade-offs that individuals would make to experience higher quality recreational experiences (better fish catch, better water quality).

As mentioned above, prices of properties can be used to identify information on the value of non-market goods. This so-called Hedonic Price method decomposes the price into attributes such as the size of the home, the number of bedrooms, and other structural and neighborhood characteristics. In addition, environmental quality attributes like scenery or water quality in a nearby lake may also affect property value. For example, Kim et al. (2015) estimate the value of improved water management on property values located near a reservoir close to the city of Calgary. They identify the impact that water management (stabilization of water levels and water quality improvements) had in one location by statistically comparing it with similar reservoir based properties in other nearby regions. A related method (Hedonic Wage) decomposes wages into various characteristics, including mortality risks. The value of mortality risk reduction, or willingness to pay to reduce the chances of dying, can be used to assess the benefits of improving water quality and correspondingly reducing health risks (Chestnut and de Civita 2009).

The second category of valuation methods (SP) involves the careful development of questionnaires to elicit trade-offs from individuals (Mitchell and Carson 1989; Freeman et al. 2014). This method describes the baseline situation and then attempts to identify how much an individual would be willing to pay (or accept) for a change from the baseline. Goods or services are presented as scenarios (referred to as contingent valuation) or as a set of attributes (choice experiments) (Freeman et al. 2014).

While SP methods can estimate use values, they are also the only methods that can be used to estimate passive use values, such as the willingness to pay to reduce the risk of the loss of threatened aquatic species (Forbes et al. 2015). Forbes et al. asked a national sample of individuals to vote on whether they would accept an increase in income tax payments in exchange for an improvement in the status of threatened species of rockfish. These monetary values are used in regulatory impact assessments of the listing of a species at risk. While stated preference methods are very flexible, there is concern that the responses to survey questions may not be the same as the actual behavioural responses that may arise if a change occurs. There has been a great deal of investment in developing methods to attempt to address this potential difference between responses on surveys and in real choice situations (e.g. Vossler et al. 2012).

Conceptually when both market and non-market values are affected by a project or policy they both can be measured and used in an analysis such as a benefit cost analysis or regulatory impact analysis. But quite often the non-market values are not measured. Since the information required for non-market valuation is not systematically collected, these values are usually only measured as "one off" occurrences. Also, as there are few regulatory settings in which formal benefit cost analysis or regulatory impact analysis is required, there is little "demand" for the generation of estimates of non-market values.

This situation is somewhat different in the United States where regulatory impact assessments are often mandated by law, and there are requirements for the measurement of all relevant values (see e.g. Mansfield et al. 2012). Furthermore, systematic collection of various forms of recreation data, and property value data, in the United States allows for more continuous analysis of at least some types of non-market value. In Canada there are few such systematic data collection efforts, and those that have existed in the past have either disappeared or occur infrequently (e.g. The Value of Nature to Canadians survey). Given these challenges to the measurement of non-market values, these more environmentally oriented aspects of the value of water are often not incorporated into policy or regulatory decision making. In the sections below we outline a few examples where policy analyses did include the broader set of values, and also discuss some of the continuing challenges and limitations in this area.

10.1.3 How Is Total Economic Value Used?

Economic values can provide information in a number of ways. Benefit cost analysis – the systematic examination of the benefits and costs of a project or policy change – employs market and non-market value to identify the most economically efficient option available to the public. Historical examples of benefit cost analysis include cases such as the Old Man River Dam and the Meridian Dam in Alberta. These historical examples included market values, but tended not to include non-market values in a rigorous quantitative fashion as the field was not as well developed at that time. Examples of benefit cost analyses that include market and non-market values for water services will be presented below. Essentially, benefit cost analysis provides information on whether the "project" or which of a set of possible "projects" provides benefits that exceed the costs. This information is also used in understanding who benefits and who loses if a project is implemented. Benefit cost analysis should not be viewed as the only tool for decision making – there are many other considerations that should be included in such social decisions. Plus benefit cost analysis is based on principles of identifying economic efficiency and is not focused on distributional or fairness principles. The technique uses methods such as the discounting of future benefits and costs and the monetary valuation of environmental and health risks – practices that not all individuals agree with (Griffin 1998). Thus not all individuals view benefit cost analysis as a useful tool for social decision making. But benefit cost analysis provides a rigorous and transparent framework for the description of possible outcomes of project based decisions.

Benefit cost analysis can be extended to consider policy changes. This is often referred to as a regulatory impact assessment (Government of Canada 2012). This method examines the benefits and costs of a policy change, relative to without the change. Examples include an analysis of the benefits and costs of changes in pollution regulation, or the impacts of a change in regulations protecting an endangered species. Non-market values associated with health risk reductions, for example, could be used in a regulatory impact assessment that measures the benefits and costs of improved water treatment to protect human health.

Finally, economic values, and particularly non-market values, can be used to measure the amount of compensation that is required as payment in the case of environmental damages. Such an approach, which is common in the U.S. but only now emerging in Canada, requires that parties who are responsible for environmental damage must compensate the public for the monetary value of these damages. The values are often non-market in nature. High profile cases of such damage assessment include the Exxon Valdez case (Duffield 1997) and the American Trader case (Chapman and Hanemann 2001). The former is often highlighted as the first major case involving the use of passive use values in damage assessment and compensation.

10.1.4 Water and Values

Before turning to Canada's experience with economic valuation and water resource decision making, it is worth discussing the complexity of water values. As Michael Hanemann points out in his 2005 essay on the value of water – water is "different". It is at times a private good (e.g. bottled water) and a public good (scenery, habitat). Its value changes over time because of water flow regimes and because of changing preferences and information. It is an emotional "commodity" and thus there is concern about treating water as a market commodity when that may mean limited access for poor people. It is also mobile and thus has a spatial component to its value. It is used in both production (firms) and consumption (consumers). Thus water is perhaps one of the most complex goods/services to understand and value in economic terms.

10.2 Canada's Experience with Using Water Values

As The Politics of Water Policy Development in Canada illustrates, there is no single government agency at any level in Canada with sole responsibility for issues relating to water. Instead, at various times and in various circumstances each of the three levels of government in Canada (Federal, Provincial, and Local/Municipal) has had sometimes overlapping jurisdictions over water. Provinces are primarily responsible for management and protection of water bodies that lie within their geographic boundaries and, for the most part, have delegated authority to local (municipal) governments over drinking water treatment and distribution, as well as wastewater management. The federal government is charged with protection of oceans, fisheries, navigation, waters with boundaries shared by two countries, and water management on federal lands (including First Nations reserves). In matters relating to agriculture and health, the federal and provincial governments work together.

It is noteworthy that none of the most important pieces of federal legislation such as the *Fisheries Act* (1868), the *Canada Water Act* (1970), the *Canadian Environmental Protection Act* (1988) or the *Canadian Environmental Assessment Act* (1999) require the explicit calculation of the costs and benefits of proposed projects that might impact water quality or quantity. A potential exception to this general conclusion includes the requirement found in the *Canadian Environmental Protection Act* that the government must "take preventative and remedial measures to protect, enhance and restore the environment" (s.2.1(a.1); Government of Canada 2003) and, in deciding what actions to undertake, the government is required to consider, amongst other factors:

> the positive economic impacts arising from the measures including those cost-savings arising from health, environmental and technological advances and innovation, among others and… any other benefits accruing from the measure. (s.2.1.1(b); Government of Canada 2003).

Another potential exception is found in the *Canadian Environmental Assessment Act* which is concerned with the application of environmental assessments to projects under federal jurisdiction. The Act allows the government to specify the issues that an Environmental Assessment Review Panel must consider in its appraisal of a project. A new Canadian Act outlining compensation for damage from oil spills (Energy Safety and Security Act, S.C. 2015, c. 4) may require the use of nonmarket valuation in the case of oil spills and related environmental injuries. This act enables the inclusion of use and passive-use values into determination of fines and penalties (see Olszyinski and Boxall 2014). However, they note that the use of environmental valuation for the purposes of quantifying harm in a sentencing case has never been successfully carried out in Canada. This is in spite of the requirement that programs or projects with significant enough impacts need to submit a Regulatory Impact Analysis Statement (Treasury Board of Canada 2015).

At the provincial level, the situation is largely the same with most provinces' major water-related and environmental laws eschewing economic analysis of proposed projects or government actions in the past (Percy 1988). Virani and Graham (1998) note that there is a significant variance in the application of economic analysis of environmental legislation across provinces. As the authors conclude "some provinces do not use these tools at all, while others apply CBA and CEA extensively to environmental policy" (p A1–100).

Alberta's Environmental Protection and Enhancement Act includes in the description of an environmental impact assessment report:

> a description of potential positive and negative environmental, social, economic and cultural impacts of the proposed activity, including cumulative, regional, temporal and spatial considerations (49 (d));

which appears to enable benefit cost analysis.

The remainder of this chapter discusses a number of case studies to illustrate how water values have or have not been employed within Canada to support policy and/or decision-making that may impact water.

10.2.1 International and Federal Level Experiences

Under the Great Lakes Water Quality Agreement signed in 1978, Canada and the United States agreed to work with local state and provincial governments to develop and implement Remedial Action Plans (RAP) for a number of locations identified as Areas of Concern (AoCs). The purpose of the RAPs was to develop a concrete set of remedial action steps to address impairments associated with poor water quality for a number of identified beneficial uses (both ecological and cultural). While RAP committees were instructed to calculate the capital and operating costs of the chosen remedial actions and demonstrate in principle the expected ecological changes, they were not charged with determining the economic value of the anticipated benefits. In the absence of these values, a cost-benefit analysis could not be conducted to determine whether these policies provided positive net benefits to Canadian society, as a whole, and, therefore, be worthwhile pursuing.

Dupont and Renzetti (2005) used the limited amount of data available showing the willingness-to-pay (benefit) values associated with both current and potential future direct uses (swimming, fishing, boating, and birdwatching), as well as some non-use values in the Hamilton Harbour AoC. These values were obtained using SP methods for the specific Hamilton Harbour case (Dupont 2003). Assuming a 20 year time horizon (1995–2015) and a social discount rate of 10 %,[1] they calculated

[1] In the current context, this is a very high discount rate. However, it was the value suggested by the Treasurer Board of Canada in 1976 in the Benefit Cost Analysis Guide (Treasury Board Canada Secretariat 2002). This was changed in 2007 to 8 %.

the present value of RAP costs to be $C240 million with present value of benefits of only $68 million. Assuming Canada-wide estimates of population increases (that would lead to increases in benefits), this net benefit rose to $77 million, still only less than 1/3 of the actual costs spent at that time period. This analysis illustrates how actions/policies relating to water in Canada were undertaken without evaluating whether they provided net benefits to Canadians. By Dupont and Renzetti's calculations, Canadians would have been better served had the money directed towards remedial actions been invested in some other, higher yielding, way.[2]

The need to better incorporate values (particularly, those relating to uses) of the Great Lakes has recently found a champion in the Great Lakes Blue Accounting initiative. This initiative came from Great Lakes governors and the premier in Ontario (in 2013). The focus was upon treating the Great Lakes as a set of regional water assets, requiring both information and data and understanding of metrics for evaluating policies and actions, as well as agreement upon desired goals for optimal water resources management (Seelbach et al. 2014). Access to water values in a Canadian context is a crucial input into any such effort and requires a concerted effort to collect the relevant information.

In the late 1990s two federal ministries (Statistics Canada and Environment Canada) joined forces to examine the value of Canada's water resources in all uses. Their goal was to develop a framework that would value natural capital (associated with water) in Canada for the purpose of integration with traditional national income accounts. In order to support this initiative, Environment Canada worked in collaboration with US EPA and the Office of Water on the Environmental Valuation Reference Inventory (EVRI). This is an online database of environmental values launched in 1999 (EVRI 2015). Its purpose is to serve as a repository of valuation estimates from both the academic and grey literature and to support the use of benefit transfer techniques as a support for policy making (Brouwer 2000). The Final Report on natural capital valuation and integration was released in 2002 (Gardner Pinfold 2002). It was guided by principles of the SEEA (System of Integrated Environmental and Economics Accounting) proposed by the United Nations Statistical Division in 1993. Specifically, the measurement of natural capital would make it possible to calculate "an environmentally adjusted measure of net domestic product which accounts for resource depletion and environmental degradation" (Gardner Pinfold 2002, p 67).

The process of integration (for the purposes of informed policy decision-making) is very slow, in part due to continued reliance on regulations for policy purposes and to a lack of water management capacity and expertise within government (NRTEE 2010). This has been exacerbated by complexity of jurisdictional oversight, large

[2]The Council of Great Lakes Industries (1996), an organization representing common interests of industries operating in the Great Lakes Basin, identified their positions on public policy, specifically, on that relating to remedial action plans. A key recommendation was that effort be made on both sides (Canada and the United States) to develop "methodologies for analyzing the … costs [..] and benefits of addressing any given use impairment". In particular, they noted that the process used to decide on which projects should proceed "lacks quantitative technical and/or economic data".

gaps in both water quantity and valuation data, and a lack of resources devoted to finding solutions. Such challenges make it virtually impossible to undertake effective and timely policy-decisions. By way of example, a proposal made by Environment Canada in 2002 to develop natural capital measures finally resulted in a Statistics Canada project in 2011 called the MEGS Project (Measuring Ecosystem Goods and Services). This particular project reported in 2013 (Statistics Canada 2013). Most of the work was focused on land based values, however, it did develop some national wetlands indicators. While encouraging, this project reveals the challenges associated with first measuring water values and then incorporating them into a decision-making framework for the purposes of policy development.[3]

Recent plans note that economic valuation is important and that research efforts to develop and apply models for economic valuation of natural capital are to be supported (Sustainable Development Office, Environment Canada 2013):

> Provide scientific expertise, guidance and advice to decision makers, and develop and apply models for social, cultural and economic valuation of ecosystem services to support sustainable development decision making…

Three years out, this still has not been advanced in concrete terms. Compared to other countries, Canada has relatively abundant water that supports the provision of high quality ecosystem services. However, increasing pollution and demand for high quality water will severely challenge the status quo approach to water policy decision-making. Policy-makers need to be building capacity and expertise now in order to be ready to meet these future challenges. Otherwise, the current fortuitous position in which Canada finds itself may rapidly be eroded and irreversible damages may occur.

In 2011 the National Round Table on the Environment and the Economy (a round table group of sustainability leaders active in business, universities, environmentalism, labor, public policy and community life from across Canada and appointed by the federal government that has since been disbanded) published a report on how best to ensure "sustainable water use by the natural resource sectors" in Canada (NRTEE 2011, p 16). The goal was to have NRTEE:

> advise… inform governments as they develop water strategies, as well as industrial and agricultural sectors in their practices and management plans. Achieving sustainable water use depends upon getting strategies and policy right. …[to] ensure ecosystem protection, allow jurisdictions to be better prepared in time of water shortage, provide a safety margin that may help avert water shortages, and promote best management practices to access an uncertain water future (p 17).

They noted that economic instruments (such as water charges or tradable water permits) would allow Canada to improve water conservation and efficiency but that such efforts are hampered by a …"lack of reliable, publicly available data on water

[3] The lack of progress at the federal level to value water for the purposes of decision-making stands in contrast to the use of EVRI by both Environment Canada and Health Canada to value human health benefits from better air quality and make regulatory decisions about ambient air quality. These agencies developed and have used the Air Quality Benefits Assessment Tool (AQBAT) since 2006.

use". In particular, they noted the need for national water-use forecast data that would be updated on a regular basis and for governments to develop capacity to forecast such water use.[4] The most recent data available at the municipal level are from 2009 and Environment Canada has ended the survey that collected these data. The most recent data on industrial water use is from 2013, collected on an irregular basis. There is no systematic collection of irrigation data or water quality change data across Canada.

10.2.2 Provincial Experiences

In the mid-1990s Alberta Public Works, Supply and Services submitted a water management project that involved diversion and storage of water from the Highwood River. This was influenced by low flows in the Highwood River in the later 1970s that had led to stressed in-stream uses (fisheries, water quality, and aesthetics). This project consisted of four interrelated components. These were first a dam and reservoir on the Little Bow River to contain diverted water from the Highwood River, as well as natural runoff in the Little Bow River. A second component was a canal and diversion works to allow water to be diverted during high water flow periods. Third, a second diversion and canal in Clear Lake was proposed to hold diverted water from the Highwood River during peak flows. Last, an expansion of an existing reservoir (Squaw Coulee) was proposed. Along with the proposed infrastructure, the agency proposed an operating plan that would service existing and projected license demands in the Highwood and Little Bow river basins, as well as improve in-stream flows on the Highwood River in order to protect aquatic resources.

Under both federal (Canadian Environmental Assessment Act of CEAA) and provincial (Natural Resources Conservation Board Act or NRCBA) legislation this project was required to undergo an environmental impact assessment. The NRCB and Canadian Environmental Assessment Agency agreed to convene a Joint Review Panel (NRCB/CEAA 1998). The report considered not only a variety of environmental effects (including water quality, habitat and fish, prairie vegetation, and soil and land capability) but also social and economic effects. Of these, enhanced water-based recreational opportunities (boating, fishing, beach, and swimming) that would arise from the project were considered. The benefit-cost analysis undertaken included irrigation benefits, as well as benefits from increased recreational opportunities estimated using stated and revealed preference valuation techniques (Adamowicz et al. 1994). The ecosystem service values associated with improvements to recreational opportunities were estimated to be between \$0.38 and \$0.55 million per year.[5] The first three components of the project were ultimately approved and opened in 2004.

---

[4] In 2013, the federal government announced the elimination of the NRTEE.

[5] This project's explicit inclusion of non-market benefits stands in contrast to the more typical approach that discusses the possibility of using non-market values and then does not include them due to lack of data (e.g., Alberta's Meridian dam project (Golder Associates 2002)).

A second example of the important role that is played by the inclusion of non-market benefits from water quality improvements in a provincial setting is found in a study undertaken for the Canadian Council of Ministers of the Environment (CCME 2007).[6] The study undertook a cost-benefit analysis of a number of scenarios involving changes to wastewater treatment and discharge. As the report stated:

> …the policy objective becomes to not only improve surface water quality per se but rather to improve surface water quality in an economically efficient manner (CCME 2007, page ii).

The scenarios were identified as falling under CCME's development of a Canada-wide Strategy for the Management of Municipal Waste Water Effluent. The report examined the potential net benefits to wastewater treatment plants in two provinces (Newfoundland/Labrador and New Brunswick). While the report identified a large number of potential benefits using the Total Economic Value framework that would arise from infrastructure improvements (e.g., benefits such as better water-based recreational experiences and habitat support) the authors were challenged by a lack of available data. To illustrate the potential benefits they used values from the EVRI database to calculate values associated with households' willingness to pay for water quality improvements linked to different rungs on the Resources for the Future water quality ladder (Vaughan 1986).[7] As an example the estimated present value of benefits for households in Newfoundland and Labrador for the period 2011–2035 were calculated to be $308 million (in 2005$ using a 4 % discount rate). The same calculation for New Brunswick produced an estimate of $443 million. A second potential benefit in the form of increased property values from water quality improvements was also determined from available data. The present value of these benefits was estimated to be $335 million for Newfoundland and Labrador and $255 million for New Brunswick. The sum of these values is set against the present value of estimated costs (incremental capital requirements and operating and maintenance) for the same time period. For Newfoundland and Labrador, these present value costs are $507 million and for New Brunswick, they are $222 million. Thus, a comparison of the present value of net benefits for the two provincial case studies suggests that predicted benefits far exceed the costs of water quality improvements. It should be recalled that the benefits are only a subset of the entire suite of benefits flowing from improved water quality. The values of improved water quality that would be associated with reduced risks to human health for example or reductions in operating costs (from having better quality intake water) to industrial firms are not included in these calculations.[8]

---

[6] This is a council comprised of environment ministers from provincial and territorial governments, along with the federal minister of the Environment. This group aims to work collectively to discuss national environmental priorities and to commission research in support of positive environmental results.

[7] This was developed in the mid-1980s as an effort to connect scientific measurements in water quality with environmental outcomes that are easily understood by the general public (e.g., water quality that is excellent such as for drinking to water quality that is extremely poor and only suitable for boating).

[8] Adamowicz et al. (2011) employ a stated preference survey to obtain the willingness to pay for changes to municipally supplied water in order to reduce the illnesses and deaths from microbial

Recently, the Ontario government commissioned several studies of economic value of protecting the Great Lakes watershed as part of its responsibilities under the Great Lakes Compact.[9] It was guided by a similar study undertaken by researchers at the Brookings Institute for the United States (Austin et al. 2007). The first Ontario report identified a list of ecosystem goods and services and provided estimates for a number of use and passive use (including ecosystem services) values within the watershed (Marbek 2010a). As the report notes, many of the values are not site-specific (not even Canada-specific), so benefit transfer techniques are used (Brouwer 2000). The final report (Marbek 2010b) undertook a cost-benefit of a proposed watershed management plan for the Rouge River designed to provide a suite of both use and non-use benefits related both to improvements in water quality and quantity. The present value of net benefits estimated is approximately $687 million. Recognition of these potential benefits may have encouraged the province of Ontario to sign the 2014 entitled Canada-Ontario Agreement on Great Lakes Water Quality and Ecosystem Health (Environment Canada 2014). However, it is too early to tell whether this agreement will employ information on non-market values associated with water quality as the basis for future decision and policy making.[10]

10.2.3 Municipal Governments (or Local Agencies)

In 1946 the province of Ontario (the largest province in Canada) created a number of corporate bodies called Conservation Authorities charged with developing programs to enhance resource management of lands, water, and natural habitats on a watershed basis. Their goal was to "promote an integrated watershed approach balancing human, environmental and economic needs" (Conservation Ontario 2015). Currently, 36 such bodies are in operation.

One of the largest conservation authorities is the Grand River Conservation Authority. In 1982 it undertook a water management study designed to help it meet three water management objectives that included: reduction in flood damages, provision of an adequate water supply to local communities, and the maintenance of

---

contaminants (e.g., such as E Coli). They find that the average household in Canada is willing to pay between $175 and $237 per year (in 2004 $) to reduce microbial risks. According to Vander Ploeg (2011) of the Canada West Foundation, in 2004 the average monthly charge for water supply in Canada was $23 (for an annual payment of $276). Thus, the Adamowicz et al, finding reveals that Canadians are willing to pay a very large premium on top of existing water bills in order to obtain a reduction in health risks associated with tap water consumption.

[9] The compact is a legally binding agreement signed by the United States government and eight states that border the Great Lakes with Canada being a voluntary signatory. The goal of the compact is to create a framework for future water resource management in the Great Lakes St. Lawrence River basin.

[10] A recently released report by the office of the environmental commissioner of Ontario notes the need to adopt both full-cost pricing for water and a user-pay (polluter pay) approach to improve water management (Environmental Commissioner of Ontario 2015).

adequate water quality within the watershed (Grand River Implementation Committee 1982). At the time, the report noted that future wastewater management planning required a full evaluation of costs and benefits. However, the analysis in the report included an explicit monetary evaluation of only a limited number of benefits (specifically, those associated with a better water supply and flood control). The report, while noting that benefits such as aesthetics, improved aquatic life and public health were important, made no attempt to put a dollar value upon benefits of improved water quality because of the difficulty of estimating the monetary value (Grand River Implementation Committee 1982).

The recent water management plan from the Grand River Conservation Authority (Grand River Watershed Water Management Plan 2014) is intended to address management of both surface and ground water resources through to the year 2031. Amongst the stated water objectives, it identifies rivers such as the Grand as "an amenity in communities through which they pass", "rivers are aesthetically pleasing to support recreational, cultural and destination tourism uses", and "water quality and quantity needs of sports fish populations are met, such that angling opportunities and community benefits are realized" (Grand River Watershed Water Management Plan 2014, p 3–4).

The document contains a large number of water quality milestones for sampling purposes, however, these are not determined with reference to the net benefits to members of society from achieving these milestones. Moreover, the report notes that:

> Additional work to identify and describe appropriate indicators and targets that measure the full suite of cultural, recreational and tourism aspects associated with the two broad water objectives is required and has been initiated by the Grand Strategy's Heritage Working Group. However, this additional work is beyond the scope of the Water Management Plan (p 3–15).

Sadly, this is in contrast to the widespread use of techniques to establish non-monetary water values for the management purposes by other agencies (e.g., Environmental Protection Agency in the United States (Griffiths et al. 2015) and the United Kingdom National Ecosystem Assessment[11] (United Kingdom Department for Environment, Food and Rural Affairs 2014).

10.3 Gaps/Weaknesses

In the mid-1980s an Inquiry on Federal Water Policy was undertaken with the goal of informing a federal Water Policy (Pearse et al. 1985). In the preface to the policy, then-Minister of the Environment Tom McMillan noted "We must manage water like any other natural resource – with care." This policy was an ambitious effort to:

---

[11] The goal of this long-term assessment project is analyze the natural environment in the United Kingdom in order to determine the benefits to society. In particular, the 2014 assessment examined the cultural and economic value of nature and identified tools and methods for assisting decision-makers (United Kingdom, Department for Environment, Food, and Rural Affairs 2014).

encourage the use of freshwater in an efficient and equitable manner consistent with the social, economic, and environmental needs of present and future generations (Campbell 2013).

Three of the five goals that were cited as helping achieve this objective included legislation and water pricing, as well as integrated planning. One report written for the Inquiry assessed the then value of water to the Canadian economy using the TEV framework (Muller 1985). He estimated this to be between $7.5 and $23 billion per year (1985 dollars), although he noted that this was likely to be an underestimate since some values were not available, specifically many of the passive use values associated with the provision of ecosystem goods and services. He included non-market estimates of only two components: waste assimilation and sports fishing. Interestingly, the combined non-market values of these two items were larger in total estimated value than the so-called withdrawal use values (municipal, irrigation, thermal power and industrial uses) and comprised 23 % of the total WTP calculated using lower end estimates and 36 % of the total WTP calculated using higher end estimates.[12] He noted some important limitations of his study. These included the absence of additional recreational instream water values, potentially large passive use values relating to maintenance of wetlands, and other values associated with aesthetic enjoyment of water (Muller 1985). Renzetti et al. (2011) used the same methodology to update the values to the current period. They estimated the 2011 values to be between $7.8 and $22.9 billion per year (in 2011 dollars). Their estimates were also stymied by the continual problem of very large data gaps in available values for the Canadian context. The largest gaps are to be found in the myriad of values associated with water's role in providing non-market social (ecosystem) benefits from, for example, the use of water as a sink into which effluent from agricultural and manufacturing, as well as municipal wastewater, flows. Such activities may not reduce the quantity of water available, however, they worsen the quality of water available for alternative beneficial uses.

An underlying reason for the lack of information on water's non-market contribution to the economy is that there are few instances in Canada that dictate that such information must be used as part of a government decision-making process. Taylor and Kennedy (2008) point out that an important boost for CBA came from United States' Executive Orders 12866 and 12291 which required a formal economic analysis of any federal rules or regulations expected to have an annual economic impact exceeding $100 million. An NRC report points out that the Executive Order was subsequently revised as EO 13258 in 1996. The revisions weakened the requirement that proposed projects pass a CBA test and, instead "required that the benefits of the regulation *justify* the costs" (Committee on Assessing and Valuing the Services of Aquatic and Related Terrestrial Ecosystems 2004, p 41). While Canada does have a Regulatory Impact Assessment process, the quality of assessment is variable and little effort is made to estimate environmental benefits. As a review by the OECD of

---

[12] Similarly, non-market values were a substantial portion of the total benefits associated with the Highwood River dam project discussed earlier.

regulatory reform in 1997 notes "half of the RIAS examined did not quantify benefits at all." (OECD 1997, p 45).

It is important to recognize that decision-making regarding water resources should not be guided solely by economic efficiency considerations that exclusively incorporate the value of water. Most analysts believe that policies related to water resources can and should be assessed from a variety of perspectives using many possible objectives. These include: sustainability, economic efficiency, fairness, transparency, etc. Traditional CBA focuses upon efficiency but may be modified to incorporate fairness/equity concerns. In a multi-criteria approach, the rigorous need to obtain monetary values for every component is relaxed, however, some form of weighting or ranking of the various objectives must be undertaken in order to make a recommendation. Arrow et al. (1996) note the value of a cost-benefit framework for its ability to organize and compare different types of information for the purposes of undertaking policy analysis. They identify its important potential role in government development of environmental regulations, in particular. Importantly, such a framework requires that analysts include the widest range of values possible. For the case of water-related policy, many of these are necessarily non-market values.

As this chapter highlights, there are a large number of instances in which Canadian policies and regulations related to water at different levels have mostly failed to incorporate water value information aimed at promoting more efficient decision-making. In particular, the absence of information on the value of water in its alternative uses means that we are not maximizing the wellbeing of Canadian society from its valuable water resources. Ultimately, this can lead to misallocation of water resources and inefficient estimates in water quality.

References

Adamowicz, W. L., Louviere, J., & Williams, M. (1994). Combining stated and revealed preference methods for valuing environmental amenities. *Journal of Environmental and Economics and Management, 26*, 271–292.

Arrow, K. J., Cropper, M. L., Eads, G. C., et al. (1996). Is there a role for benefit-cost analysis in environmental, health, and safety regulation? *Science, 272*(5259), 221–222.

Austin, J., Anderson, S., Courant, P., & Litan, R. (2007). *America's North Coast: A benefit-cost analysis of a program to protect and restore the Great Lakes*. http://www.brookings.edu/research/reports/2007/09/04gleiecosystem-austin. Accessed 20 Sept 2015.

Brouwer, R. (2000). Environmental value transfer: State of the art and future prospects. *Ecological Economics, 32*(1), 137–152.

Campbell, I. (2013). *Feature article – Toward a national freshwater policy framework*. Policy Research Initiative, Government of Canada. http://www.horizons.gc.ca/eng/content/feature-article-%E2%80%93-toward-national-freshwater-policy%C2%A0framework. Accessed 10 Sept 2015.

Canadian Council of Ministers of the Environment. (2007). *A protocol for the derivation of water quality guidelines for the protection of aquatic life*. http://www.ccme.ca/files/Resources/supporting_scientific_documents/protocol_aql_2007e.pdf. Accessed 28 July 2016.

Chapman, D. J., & Hanemann, M. W. (2001). Environmental damages in Court: The American trader case. In A. Heyes (Ed.), *The law and economics of the environment* (pp. 319–367). Cheltenham: Edward Elgar.

Chestnut, L., & De Civita, P. (2009, March). "Economic Valuation of Mortality Risk Reduction: Review and Recommendations for Policy and Regulatory Analysis" Research paper. Government of Canada Policy Research Initiative Project, Research Strategy. http://www.horizons.gc.ca/sites/default/files/Publication-alt-format/2009-0012-eng.pdf. Accessed 27 July 2016.

Committee on Assessing and Valuing the Services of Aquatic and Related Terrestrial Ecosystems. (2004). *Valuing ecosystem services: Toward better environmental decision-making*. Washington, DC: Water Science and Technology Board. Division on Earth and Life Studies; National Research Council of the National Academies. The National Academies Press.

Conservation Ontario. (2015). *Natural Champions*. http://www.conservation-ontario.on.ca/. Accessed 27 Sept 2015.

Council of Great Lakes Industries. (1996). *Remedial Action Plan*. http://cgli.org/wp-content/uploads/2013/11/1996-Remedial-Action-Plan.pdf. Accessed 7 Sept 2015.

Duffield, J. (1997). Nonmarket valuation and the courts: The case of the Exxon Valdez. *Contemporary Economic Policy, 15*, 98–109.

Dupont, D. P. (2003). CVM embedding effects when there are active, potentially active and passive users of environmental goods. *Environmental & Resource Economics, 25*(3), 319–341.

Dupont, D. P., & Renzetti, S. (2005). Cost-benefit analysis of the Remedial Action Plan to improve water quality in the Great Lakes in Canada. In R. Brouwer & D. Pearce (Eds.), *Cost-benefit analysis and water resources management* (pp. 195–222). Cheltenham: Edward Elgar.

Environment Canada. (2014). *Canada-Ontario agreement on Great Lakes water quality and ecosystemhealth*.http://www.downloads.ene.gov.on.ca/envision/env_reg/er/documents/2014/011--9290_d.pdf. Accessed 15 Oct 2015.

Environmental Commissioner of Ontario. (2015). *Small things matter*. Annual Report 2014/2015. Toronto, Ontario. http://eco.on.ca/wp-content/uploads/2015/07/2014_2015-AR.pdf. Accessed 24 Nov 2015.

Environmental Valuation Reference Inventory. (2015). https://www.evri.ca/Global/HomeAnonymous.aspx. Accessed 10 Sept 2015.

Forbes, K., Boxall, P. C., Adamowicz, W., & de Maio Sukic, A. (2015) Recovering Pacific Rockfish at risk: The economic valuation of management actions. *Frontiers in Marine Science*, *2*, Article 71, 1–10. http://dx.doi.org/10.3389/fmars.2015.00071

Freeman, A. M., Herriges, J. A., & Kling, C. L. (2014). *The measurement of environmental and resource values theory and methods* (3rd ed.). Washington, DC: Resources for the Future Press.

Gardner Pinfold Consulting Limited et al. (2002). *Monitoring the value of natural capital: Water final report*. Prepared for Environment Canada and Statistics Canada.

Golder Associates. (2002). *Report on Meridian Dam preliminary Feasibility study*. Submitted to Alberta Environment and Saskatchewan Water Corporation. Pub T/642. http://www.assembly.ab.ca/lao/library/egovdocs/2002/alen/133264.pdf. Accessed 15 Oct 2015.

Government of Canada. (2003). *Consolidated statutes and regulations*. http://lawslois.justice.gc.ca/eng/annualstatutes/2003%5F9/. Accessed 28 July 2016.

Government of Canada. (2012). Analyzing the benefits and costs of regulation. In *Cabinet directive on regulatory management*. http://www.tbs-sct.gc.ca/hgw-cgf/priorities-priorites/rtrap-parfa/guides/cdrm-dcgr-eng.asp#cha67. Accessed 25 Nov 2015.

Grand River Implementation Committee. (1982). *Grand River basin water management study*. http://www.grandriver.ca/WaterPlan/1982_BasinStudy.pdf. Accessed 17 Sept 2015.

Grand River Watershed Water Management Plan. (2014). Grand River Conservation Authority, Cambridge, Ontario, p 137, appendices. http://www.grandriver.ca/waterplan/2014_WMP_Final.pdf. Accessed 18 Sept 2015.

Griffin, R. C. (1998). The fundamental principles of cost-benefit analysis. *Water Resources Research, 34*(8), 2063–2071.

Griffiths, C., Klemick, H., Massey, M., et al. (2015). U.S. Environmental Protection Agency valuation of surface water quality improvements. *Review of Environmental Economics and Policy, 6*(1), 130–146.
Hanemann, W. M. (2005). *The value of water*. Berkeley: University of California.
Justice Laws Website. (2015). *Energy safety and security act*. Government of Canada. http://laws-lois.justice.gc.ca/eng/AnnualStatutes/2015_4/page-1.html Accessed 14 Oct 2015.
Kim, H. N., Boxall, P. C., & Adamowicz, W. (2015). The demonstration and capture of the value of an ecosystem service: A quasi-experimental hedonic property analysis. *American Journal of Agricultural Economics, 98*, 819–837.
Mansfield, C., Van Houtven, G., Hendershott, A., Chen, P., Porter, J., Nourani, V., & Kilambi, V. (2012). Klamath River basin restoration: Nonuse value survey. Final Report. Prepared for the U.S. Bureau of Reclamation. Research Triangle Park, North Carolina: RTI International. https://klamathrestoration.gov/sites/klamathrestoration.gov/files/DDDDD.Printable.Klamath%20Nonuse%20Survey%20Final%20Report%202012%5B1%5D.pdf. Accessed 28 July 2016
Mannix, A., Adamowicz, W., & Dridi, C. (2014). Solutions to the high costs of future water restrictions for new oil-sands industry along the Athabasca river. *Canadian Water Resource Journal, 39*(4), 395–408.
Marbek Resource Consultants. (2010a). *Economic value of protecting the Great Lakes*. Literature Review Report. Submitted to the Ontario Ministry of Environment. http://www.blue-economy.ca/sites/default/files/reports/resource/stdprod_086944.pdf
Marbek Resource Consultants. (2010b). *Assessing the economic value of protecting the Great Lakes: Rouge river case study for nutrient reduction and nearshore protection*. Final Report. Submitted to the Ontario Ministry of Environment. http://www.greeninfrastructureontario.org/sites/greeninfrastructureontario.org/files/Final%20Rouge%20Report%20Nov%2030.pdf. Accessed 25 Sept 2015.
Martel, L. (2011). Cash flow: Buying and selling priceless water. *Albertaviews* 14(8):28–34. https://albertaviews.ab.ca/2014/07/16/cash-flow/. Accessed 25 Nov 2015.
Mitchell, R., & Carson, R. (1989). *Using surveys to value public goods: The contingent valuation method*. Washington, DC: Resources for the Future.
Muller, A. (1985). *The socioeconomic value of water in Canada* (Canada Inquiry on Federal Water Policy Research paper, Vol. 5, p. 104). Ottawa: Inquiry on Federal Water Policy.
National Round Table on the Environment and the Economy. (2010). *Changing currents. Water sustainability and the future of Canada's natural resource sectors*. http://www.blue-economy.ca/sites/default/files/reports/resource/changing-currents-water-report-eng-1.pdf. Accessed 23 Nov 2015.
National Round Table on the Environment and the Economy. (2011). *Charting a course. Sustainable water use by Canada's resource sectors*. http://www.blue-economy.ca/sites/default/files/reports/charting-a-course-eng.pdf. Accessed 10 Sept 2015.
Natural Resources Conservation Board, Alberta, Canadian Environmental Assessment Agency. (1998). *Little Bow Project/Highwood diversion plan application to construct a water management project to convey and store water diverted from the Highwood River*. Report of the NRCB/CEAA Joint Review Panel Application #9601 – Alberta PublicWorks, Supply and Services. NRCB/CEAA Joint Review Panel. https://nrp.nrcb.ca/Portals/1/Documents/Decisions/Little-Bow/Decision-report.pdf. Accessed 10 Sept 2015.
Olszynski, M., & Boxall, P. (2014). *The law and economics of environmental harm: A primer and updating for environmental sentencing*. Paper presented at a symposium on Environment in the Courtroom (III): Sentencing and Environmental Offences, University Canadian Institute of Resources Law Working Paper, Dalhousie, 21–22 February 2014. https://www.cirl.ca/files/cirl/peter_boxall-en.pdf. Accessed 15 Oct 2015.
Organization for Economic Co-operation and Development. (1997). *Regulatory impact analysis: Best practices in OECD Countries*. http://www.oecd.org/gov/regulatory-policy/35258828.pdf. Accessed 15 Oct 2015.

Pearse, P., Bertrand, F., & MacLaren, J. W. (1985). *Currents of change. Final report of the inquiry on federal water policy*. Ottawa: Environment Canada.

Percy, D. R. (1988). *The framework of water rights legislation in Canada*. Calgary: The Canadian Institute of Resources Law.

Renzetti, S., Dupont, D. P., & Wood, C. (2011). *Running through our fingers. How Canada fails to capture the full value of its top asset*. Report for the Canadian Water Network Blue Economy Initiative. http://www.blue-economy.ca/report/running-through-our-fingers-0. Accessed 7 Sept 2015.

Seelbach, P. W., Read, J. G., Buckner, K. A, et al. (2014). *Great Lakes Blue accounting: Empowering decisions to realize regional water values*. A Report to the Council of Great Lakes Governors. http://glc.org/files/docs/2014-blue-accounting-recommendations-glc.pdf. Accessed 8 Sept 2015.

Statistics Canada. (2013). *Human activity and the environment. Measuring ecosystem goods and services in Canada*. Ministry of Industry. Catalogue no. 16-201-X. http://www.statcan.gc.ca/pub/16-201-x/16-201-x2013000-eng.pdf. Accessed 10 Sept 2015.

Sustainable Development Office, Environment Canada. (2013). *Planning for a sustainable future: A federal sustainable development strategy for Canada 2013–2016*. http://ec.gc.ca/dd-sd/default.asp?lang=En&n=CD4179F6-1. Accessed 10 Sept 2015.

Taylor, A., & Kennedy, M. (2008, June). Scoping report on Canadian and International Water Valuation. Prepared by Pembina Institute for Appropriate Development. Prepared for the Canadian Council of Ministers of the Environment Project 423-008

Treasury Board of Canada. (2002). *Benefit-cost analysis guide*. http://www.tbs-sct.gc.ca/fm-gf/tools-outils/guides/bca2-gaa-eng.asp. Accessed 8 Sept 2015.

Treasury Board of Canada. (2015). *Guide to the federal regulatory development process* https://www.tbs-sct.gc.ca/rtrap-parfa/gfrpg-gperf/gfrpg-gperf02-eng.asp. Accessed 15 Oct 2015.

United Kingdom Department for Environment, Food and Rural Affairs. (2014). *UK National Ecosystem Assessment Follow-on – NR0150*. http://randd.defra.gov.uk/Default.aspx?Menu=Menu&Module=More&Location=None&Completed=2&ProjectID=18081#Description. Accessed 29 Sept 2015.

Vander Ploeg, C. G. (2011). Water pricing. Seizing a public policy Dilemma by the horns. In *Canadian water policy Backgrounder 7*. Canada West Foundation. http://cwf.ca/pdf-docs/publications/Water_Backgrounder_7_Sept_2011.pdf. Accessed 27 Sept 2015.

Vaughan, W. (1986). Appendix B: The RFF water quality ladder. In R. Mitchell & R. Carson (Eds.), *The use of contingent valuation data for benefit/cost analysis in water pollution control, final report*. Washington, DC: Resources for the Future.

Virani, S., & Graham, S. (1998). *Economic evaluation of environmental policies and legislation*. Prepared for the European Commission Directorate General III (Industrial Affairs). http://europa.eu.int/comm/environment/enveco/others/environmental_policies_and_legislation.pdf. Accessed 27 Sept 2015.

Vossler, C. A., Doyon, M., & Rondeau, D. (2012). Truth in consequentiality: Theory and field evidence on discrete choice experiments. *American Economic Journal: Microeconomics, 4*(4), 145–171.

Chapter 11
Water Pricing in Canada

Steven Renzetti

Abstract Water pricing has historically not played an important role in most facets of water policy in Canada. Specifically, policies relating to the allocation of water across major water-using sectors, water quality and even the provision of potable water and sewerage services have not relied heavily on prices or other economic instruments (Conference Board of Canada. How Canada performs: Water withdrawals. http://www.conferenceboard.ca/hcp/details/environment/water-consumption.aspx. Accessed 3 Apr 2014, 2014). For example, in many provinces, self-supplied water users face minor administrative fees to access raw water supplies. At the municipal level, inadequate pricing has meant consumers have not faced the full cost of water supply and sewage treatment and has left water agencies with aging infrastructure. The first half of this chapter discusses this feature of past and current Canadian water policies at the provincial and municipal levels and critically assesses its implications for the efficiency and sustainability of Canadian water use. The second half of the chapter presents evidence of recent changes to Canadian water pricing policies, assesses their implications and discusses opportunities and challenges for future pricing reforms. Interestingly, there are a number of initiatives underway that signal provincial and local governments' greater reliance on prices and, more generally, market forces. In a number of provinces, fees for water withdrawals permits are rising. At the municipal level, some water agencies are developing charges associated with the cost of stormwater management. These trends have not been without controversy, however, and raise questions about a shifting in attitudes towards the potential 'commodification' of water as well as the distributional consequences of these policy reorientations (Bakker K. Ann Rev Environ Resour 39:469–494, 2014).

S. Renzetti (✉)
Department of Economics and Environmental Sustainability Research Centre,
Brock University, 1812 Sir Isaac Brock Way, L2S 3A1 St. Catharines, ON, Canada
e-mail: steven.renzetti@brocku.ca

S. Renzetti, D.P. Dupont (eds.), *Water Policy and Governance in Canada*,
Global Issues in Water Policy 17, DOI 10.1007/978-3-319-42806-2_11

11.1 Introduction

A decade ago I reviewed the state of water pricing in Canada and came to the following set of conclusions:

> Simply put, water prices at the municipal and provincial levels are in pretty bad shape, and they fail in each of the four areas set out at the beginning of this chapter. In other words, they do not generate the revenues needed to support water agencies; they do not inform consumers of the full costs of their water use decisions; they do not contribute to protecting environmental ecosystems; and they do not satisfy basic principles of fairness. These shortcomings have not arisen out of a lack of technology or know-how; rather, they reflect not only the water supply industry's historic disinterest in using water prices for any purpose other than raising revenue but also a general provincial and federal neglect of this important part of water management. (Renzetti 2007, p 277)

This chapter provides an update on the state of water pricing at the provincial and municipal levels. As will be seen in the discussion that follows, there have been some advances in water pricing at the provincial and municipal levels in Canada but many of the concerns expressed in 2007 remain.

The next section provides a brief overview of the role of the different levels of government in Canada in water management and, more specifically, water pricing. The third section critically assesses the state of water pricing at the provincial level and examines two issues; the setting of fees for permits to withdraw water and the pricing of discharges to water bodies. The fourth section provides a critical assessment of the state of pricing for water supply and sewerage treatment services at the municipal level. In both the third and fourth sections, I identify several innovative policy directions being undertaken regarding water pricing. In a number of provinces, for example, fees for water withdrawals permits are rising and being applied to more water-using sectors. At the municipal level, some water agencies are developing charges associated with the cost of stormwater management. The chapter concludes by pointing to a number of emerging trends and possible directions for future water pricing in Canada.

11.2 Background

There are a variety of types of water pricing considered in this chapter. These include pricing for potable water supply and provision of sewerage services at the municipal level as well as charges levied by provincial governments for permits to withdraw water. The pricing of irrigation water is, for the most part, under the control of privately-owned irrigation districts in Canada's prairie provinces. This topic is not discussed here but is covered in Renzetti and Dupont (2015).

As indicated in The Hydrological and Policy Contexts for Water in Canada, the responsibility for the regulation of water use and water quality is divided between the Canadian federal and provincial governments. This division of responsibilities has implications for the evolution and current state of water pricing in Canada. The federal government has responsibility over water pricing on federal lands (e.g. First Nations communities) but this responsibility has rarely been exercised. Thus, this chapter focuses on water pricing at the provincial and municipal levels.

Provincial governments are responsible for setting fees for permits to withdraw water and for creating the regulations that govern municipal rate setting. There is little coordination across provinces in approaches taken to either fees for withdrawal permits or regulation of municipal rate setting. This is in part due to historical differences in provinces' approaches to the allocation of water (Water Policy in Canada). In the absence of regulations or policy direction from provincial governments, a major source of guidance and coordination has come from industry associations such as the American Water Works Association which have published guidelines for pricing, cost accounting and other water agency activities. In the absence of provincial regulations, these have often become *de facto* standards for the industry.

A final form of water pricing that is, in theory, available to provincial governments is the pricing of discharges to receiving water bodies. As has been well established in the research literature, pricing can also be used to regulate changes in water quality (following the 'polluter pay' principle) but, aside from the pricing of municipal sewerage services, this approach has largely been eschewed by Canadian provincial governments (Kenny et al. 2011; Canada West Foundation 2011). Instead, provincial governments have relied almost exclusively on setting quantitative limits on discharges to meet water quality objectives.

11.3 Provincial Water Pricing

All Canadian provincial governments have the authority to regulate the allocation of water (Horbulyk 2016). The principle way this is done in most provinces is through a permitting system whereby firms, irrigation districts, farms and municipalities wishing to withdraw water from surface or underground sources must first obtain a licence allowing them to do so from the province government. There are several features of this permitting system that are noteworthy. First, the regulations governing the issuance of permits vary by province. Most provinces' regulatory frameworks include the allocation of groundwater but not all do (Alberta does not and B.C. has only recently changed its regulations to do so). This is a potentially important omission as it means significant volumes of water use (and potential reductions in water quality) are not regulated. Second, in all provinces, the issuing of the permit does not convey legal ownership of the water – only the right to withdraw a given volume under stipulated conditions. This is an important point because, if the issuance of a water permit did convey some form of ownership right, then it is possible that the provisions of the North American Free Trade Act would allow holders of water permits to undertake large-scale exports of water from Canada to the United States. It is now widely accepted amongst legal researchers that neither the issuance of a provincial permit to take water nor the pricing of these permits allows permit holders to undertake large-scale exports of water (Coffin et al. 2011).

Third, for the most part, the permits are for a limited term (usually less than 10 years) and are not transferable. One province (Alberta) is the exception to the last point as it has introduced a limited market for water permits in which licence holders operating in the same water basin may sell or temporarily lease permits Bjornlund (2010). The fourth feature of the permit system concerns the criteria provincial governments employ to decide whether to issue a licence. While these vary by province, governments generally require applicants to demonstrate the absence of hydrologic impacts on existing permit holders. Interestingly, while all provinces require applicants to identify the purpose of the proposed withdrawal, no province requires the applicant to provide information on the economic value of the proposed use of water. Thus, provinces rely almost exclusively on hydrologic rather than economic information in making their decisions regarding water allocations.

The fifth feature of the provincial permitting system concerns the fees that are charged for permits. Most provinces charge some type of fee for a water permit and these fees usually vary by sector (agricultural, manufacturing, mining and natural resource extraction, etc.). Generally, smaller provinces rely on an annual fee that varies by sector and by the volume of water withdrawn. Annual fees usually are in the range of several hundred to several thousand dollars (for very large water withdrawals). On the other hand, larger provinces typically also impose a volumetric charge in addition to the annual fee. Examples of provincial volumetric charges are the following:

- British Columbia: Taking effect in 2016, withdrawals for Industrial and Commercial purposes will be charged $2.25 per million litres (British Columbia 2014).
- Ontario: Withdrawals for selected industrial sectors (food and beverage, concrete production, non-metallic minerals production) are currently charged $3.71 per million litres (Ontario Ministry of Energy and Environment n.d.).
- Manitoba: Industrial withdrawals rate ranges from $1.00 to $2.00 per million litres depending on volume of withdrawals (Manitoba 2015).

The final feature of the provincial permitting system concerns water quality. For the most part, provinces regulate water quality separately from the allocation of water resources. Thus, one set of laws and regulations establish limits to discharges of pollutants into water bodies and another set of laws and regulations govern which users may withdraw what volume of water. In addition, provinces have broadly eschewed the use of economic instruments to regulate water quality. In contrast to a number of other jurisdictions in the OECD which have followed the 'polluter pay' principle and adopted fees and charges to promote improved water quality, Canadian provinces have pursued water quality objectives by relying on quantitative limits on emissions (OECD 2009).

11.3.1 Assessment

The features of the provincial approaches to regulating water use and, in particular, the limited role for the pricing of either water withdrawals or discharges into water bodies have had a number of implications. First, there is little in most provincial water policies that promotes economic efficiency in either water use or choices related to discharges into water bodies. The fees levied for water permits are very low and are not directly tied to the opportunity costs (current or future) associated with proposed water uses. This failure of water permit fees to reflect the costs of water use is reinforced by the fact that provincial governments have chosen to not levy charges for the environmental damages associated with polluters' discharges into receiving water bodies. As a result, there is little evidence to suggest they encourage water conservation, efficient water use or socially optimal decision-making regarding discharges into water bodies. The lack of transferability in water use permits is also an important impediment to promoting efficient water use as water cannot be easily reallocated in the face of changing water demands and values. At the same time, however, in some provinces (particularly British Columbia) the permit fees are sufficient to raise several hundreds of millions of dollars in government revenue and these revenues are used to support the government's water-related efforts (British Columbia 2014).

11.3.2 Innovations

While the role of pricing in most provinces' water policies has been quite limited and, as a result, provided relatively weak incentives for water users to adopt efficient decision-making regarding water use and discharges into water bodies, there have been some developments in the last decade which suggest some provinces are revising their water allocation policies in the direction of a greater reliance on pricing and other market-based instruments (Kenny et al. 2011). For example, several provinces (most notably Ontario and British Columbia) have raised their volumetric fees for water permits. In the case of British Columbia, most application fees and volumetric water use charges were doubled in 2014 (British Columbia 2015). In addition, Alberta has provided permit holders with the limited ability to trade, on a permanent or temporary basis, the rights to withdraw water (Bjornlund 2010). In another example, the province of Ontario is currently investigating the viability of introducing a water quality trading program to reduce nutrient (phosphorous and nitrogen) loadings in Lake Simcoe, a popular lake north of Toronto (Lake Simcoe Region Conservation Authority 2014). Under the proposed program, agencies responsible for emissions of nutrients may achieve targeted reductions in nutrient loading by purchasing "offsets" that are generated by load reductions achieved elsewhere in the watershed. Total nutrient loadings in the lake will be achieved by requiring that agencies purchase offsets at a ratio of 2.5–1 relative to their emissions (Lake Simcoe Region Conservation Authority 2014)

11.4 Municipal Water Pricing

Canadian households obtain their potable water (and have their wastewater treated) in one of three ways. The majority of Canadians (approximately 75 %) are connected to a municipal system and receive their water and sewerage treatment services from their municipal government (Environment Canada 2011b). The remaining Canadian households, however, who live in rural or remote areas are not connected to a municipal system and must rely on water drawn directly from a surface or groundwater source or on a commercial water provider. Unlike many other OECD countries which have a mix of public and private supply agencies at the municipal level, Canada relies almost exclusively on publicly owned and government-run water supply and sewerage agencies. Table 11.1 provides information on water rates

Table 11.1 Residential water metering, rate types and prices across Canada (2009)

Province/territory	% residential metered[a]	Average annual residential water use (litres/day/per capita)[a]	% pop served flat rate	% pop served constant or decreasing block rate[b]	Average of marginal prices[c] (2009$)
Newfoundland & Labrador	0.02	395	84.2	7.0	NA
Prince Edward Island	1.5	189	31.6	68.4	0.75
Nova Scotia	92.6	292	0.2	99.8	1.56
New Brunswick	49.1	394	17.9	81.0	1.79
Quebec	16.5	386	67.9	14.4	0.85
Ontario	91.2	225	2.1	87.3	1.96
Manitoba	97.2	199	0.1	99.9	3.07
Saskatchewan	98.2	238	0.8	92.2	1.78
Alberta	84.8	209	0.2	62.6	1.78
British Columbia	32.6	353	20.3	70.1	1.27
Territories	8–98[b]	391	2.6	97.4	1.48

Notes

[a]Data are for 2009. Source:Environment Canada (2011a)

[b]Data are for 2006. Source: Environment Canada (2011a). Source only provides single estimate for all Territories (Yukon, Northwest Territories and Nunavut)

[c]Marginal price is calculated at a 25 m^3/month consumption level. Rate is for water supply and sewerage service combined. Averaged over those municipalities reporting volumetric rate structures

across the provinces using the most recently available survey of municipal rate-setting.

Canadian provincial governments all have the authority to regulate the actions of their municipal governments. Thus, decisions regarding price setting, investments in infrastructure, and other important aspects of water agencies' operations are, in principle, subject to provincial regulation. Historically, the most important and active area of provincial regulation has concerned water quality and the setting of standards for drinking water. In addition, since almost all water suppliers operate as departments of a municipal government (the water agency in Halifax, Nova Scotia, being a rare exception as it is a public utility run separately from the municipal government), provincial regulations concerning municipal government borrowing and investments indirectly impact water suppliers and sewerage treatment providers.

Interestingly, the area of municipal agency operations which has received the least regulatory direction from provincial governments concerns the setting of water and sewerage rates. Provincial governments have, for the most part, left decision-making regarding the level and structure of water and sewerage prices to municipalities. One exception to this general trend has been observed in Canada's largest province, Ontario, where legislative changes are slowly moving in the direction of requiring all water suppliers to adopt full-cost accounting and to set rates consistent with that accounting. These changes, critics argue, have been delayed and 'watered down' in the face of opposition from municipalities (Environmental Commissioner of Ontario 2014).

One direct result of this limited degree of regulatory oversight has been that guidelines regarding cost accounting and rate setting published by the AWWA (American Water Works Association 2012) have become *de facto* standards for most municipalities in Canada. While the AWWA guidelines provide a careful balancing of rate-setting objectives (equity, cost-recovery, promoting conservation), they do not have the force of law and have been criticized by researchers for not doing enough to promote efficient water prices (Renzetti and Dupont 2015)

Another result of this limited degree of regulatory oversight (which has stood in contrast to provincial regulation of electricity, natural gas and other public utilities' pricing) was that municipal water rates have evolved separately of each other and have shown a wide variety of formats and structures (Renzetti 2007). In 1999, for example, nearly half of all Canadian households paid a fixed fee for their water service which did not depend on the volume consumed (Environment Canada 2001). Thus, these households faced an effective price per cubic metre of water of zero. This national average masked wide regional differences, however, as approximately 80 % of households in some provinces (Ontario, Saskatchewan and Manitoba) faced some form of volumetric charges. The same report indicates that, for those households facing a volumetric rate structure, combined water and sewerage prices through the 1990s were in the range of $1.00 Canadian per cubic metre.

Over the last 10–15 years, the state of municipal water and sewerage pricing has advanced significantly (Environment Canada 2011a). By 2009, for example, more than two-thirds of Canadian households' water use was metered (Table 11.1). Water price increases have generally been in excess of the rate of inflation and sewerage prices have risen even more rapidly. By 2009, the average combined water and sewerage price was $1.85 per cubic metre. In addition, a growing number of municipalities are switching to increasing block rate price structures. In the past, these changes were documented through a series of surveys of municipal agency operations conducted by the federal government. Unfortunately, the federal government ceased implementing the survey in 2009 and, as a result, it is difficult to assess the most recent changes to water and sewerage prices.

11.4.1 Assessment

Municipal water rates have been quite low by international standards (Renzetti 2009; OECD 2010). This has had significant negative consequences for the long-run financial health of water and sewerage systems. Renzetti (2009) documents that, for the period 1988–2008, revenues for all Canadian municipal water supply agencies were equal to their expenditures in only 1 year. Not surprisingly, investments in maintaining supply networks have suffered and municipalities face a growing infrastructure 'deficit' due to inadequate capital investments. In a national study of the state of the nation's infrastructure, it was reported that:

> 15.4% of the<water>systems were ranked "fair" to "very poor" for the condition of their pipes. The figures were not much better for plants, reservoirs and pumping stations, where 14.4% ranked "fair" to "very poor"…. The replacement cost for the drinking-water infrastructure in "fair" to "very poor" condition is $25.9 billion, or $2,082 per Canadian household (Federation of Canadian Municipalities 2013, p 3).

A second concern regarding Canadian municipal water prices relates to the extent that they promote efficient decision-making by water users. For households whose water use is not metered, the answer is quite clear: since they face no volumetric price for water, they have little or no incentive to conserve on its use. In the cases where households and businesses face volumetric charges, the key question is whether the price they face properly informs them of the full social costs of their use of the water supply and sewerage treatment systems. This is a remarkably understudied area (Renzetti and Dupont 2015) but the bulk of the available evidence indicates that municipal volumetric water prices fail in several ways to inform consumers of the costs of their water use (Canada West Foundation 2011). The first piece of evidence was presented above: that revenues have failed to meet expenditures at the industry level for much of the last two decades. Thus, water consumption is being subsidized in Canada from general tax revenues. Second, water prices continue to reflect the past costs of infrastructure development rather than the costs of future investments that are dependent on current consumption decisions. Third, while municipalities are increasingly tying water prices to the amount consumed, they rarely have tied prices to other important drivers of cost such as time of use or distance from the water source.

Because municipal water prices are often designed to address multiple objectives, these efficiency-related criticisms may need to be weighed against other relevant goals. For example, one important implication of the water industry's shift over the last decade towards a greater reliance on the volumetric component of rate schedule may have been to introduce more instability of revenue. In the case of municipalities using flat fees (which are independent of the volume of water consumed), it is relatively straightforward to predict revenues. However, the introduction of volumetric charges or a shift towards a greater reliance on volumetric charges will likely increase the variability of revenues. This occurs in contrast to costs which, in the short-run, are largely invariant to the volume of water supplied because of the capital-intensive nature of water supply and sewerage treatment networks. A second issue that arises when considering the efficiency-related features of water prices stems from the potential for pricing reforms to adversely impact low income households. In recent research Renzetti et al. (2015a) demonstrate that water expenditures tend to decline as a share of Canadian households' income as their income rises. The implication of this is that price increases may impact low-income households more strongly as their water expenditures make up a larger share of their household's income. As a result, care must be taken in the design of water pricing reforms if they are to improve their signal regarding the costs of water use without having regressive impacts on low-income households.

11.4.2 Innovations

In the last decade, there have been some signs of individual municipalities adopting innovative approaches to water pricing. It is important to remember that because most provinces have not created a framework to regulate water pricing in the same way in which they have regulated other public utility pricing, the innovations seen have largely been the result of decisions made by individual municipalities rather than province-wide initiatives. In addition, there has relatively little coordination across provinces of water or sewage pricing at the municipal level.

An increasing number of municipalities are investing in the capacity to learn more about the nature of their customers' water demands. These efforts are important as they provide the foundation for more accurate water demand forecasts as well as anticipating the impacts of reforming water prices and introducing non-price demand management strategies. In the case of water demand forecasting, many water agencies have historically used a 'fixed coefficient' approach to forecasting (i.e. assuming that future water demands could be predicted by multiplying future population by an assumed constant per capita water use). This approach has been demonstrated to be inaccurate as it fails to account for changes in water use due to changes in prices, regulations, and household preferences (Renzetti et al. 2015b).

A number of water agencies are also adopting or investigating new forms of water prices. The increasing use of increasing block rate structures has already been discussed earlier in the chapter. Two other examples are seasonal surcharges and pricing stormwater management services. Seasonal surcharges are a form of 'time of use' pricing which requires the water agency to identify the period when demands and the cost of supply are highest. Other public utilities have exploited advances in metering technology to move to prices differentiated by the hour but water agencies in Canada have been constrained by past practices and technology to differentiate prices by season. Seasonal surcharges have the capability of performing several tasks. They can encourage conservation during periods when demand is close to system capacity (and, thus, avoid the need for other, less efficient, means of reducing demands such as outdoor watering restrictions). Another related goal of seasonal surcharges is to accurately inform consumers of the full costs of their water use. The public sector pricing literature has long advocated that users of services during 'peak periods' should pay higher prices that reflect the costs of providing system supply capacity to meet their peak demands (Wilson 1992). An example of a seasonal surcharge has recently been adopted in Vancouver, British Columbia. During the rainy season, when the city's water supply is at its least expensive (November through May), residents and commercial businesses enjoy a relatively low off-peak rate of $2.48/m^3. During the drier months, however, when demand on the water system is high, rates increase by about 25 % ($3.11/m^3), to reflect the added cost of supplying water to the city (City of Vancouver 2013).

Stormwater management pricing is also being explored by a number of Canadian water agencies. A number of factors have motivated this interest including the costs associated with coping with treating stormwater flows, the potential for storms becoming more frequent and severe in the future and the recognition of the role played by impervious surfaces (such as large parking lots) in creating heavy storm loads on water treatment systems. In many cases, the stormwater charges are based on size of property. In London, Ontario, for example, a property's total area determines the monthly charge for storm water. The fees collected go toward the construction, maintenance and operations of the storm sewers, ditches, watercourses, retention ponds, etc. For most residential properties, the charge in 2015 is $14.49/month while for larger commercial properties the monthly charge is $120.57 per hectare (City of London 2013).

11.5 Future Trends

The preceding discussion points to several potentially pressing issues and possible directions for provincial water use fees and municipal water and sewerage pricing.

It is likely that many provinces will continue to raise fees for water withdrawals in the future. Thus far, the fees charged have imposed an inconsequential financial burden on most industries and research suggests future increases are needed to motivate behavioural changes. Rivers and Groves (2013) for example conclude that the introduction of a uniform charge for water withdrawals across all provinces of $0.013/m^3 has the potential to reduce industrial water withdrawals nationally by approximately 25 %.

There is also evidence that moving towards a greater reliance on market-based instruments to address water quality issues may be a cost-effective strategy for provinces in the future. One of the areas in which these types of policy measures may assist governments in achieving water quality objectives in a cost-effective manner is the growing problem of massive algal blooms and related water quality threats in Lake Winnipeg, Lake Erie and elsewhere in Canada. Because the major sources of the nutrients fueling these blooms are a combination of nonpoint agricultural sources and overburdened municipal sewage treatment plants, it is likely a hybrid approach of markets and pricing may be required (Brandes et al. 2008).

At the municipal level, the significant backlog in infrastructure investments would seem to require further increases in water and sewerage prices in Canada in the future. An unresolved challenge, however, is that for many Canadian municipalities, residential water use is declining. According to a 2011 Environment Canada report (Environment Canada 2011b), residential water use per capita fell by almost 15 % over the period 2001–2009. The trend of declining residential water use suggests water agencies may need to explore alternative sources or forms of funding (e.g. seasonal surcharges, storm water charges). Another trend is the growing importance of sewerage treatment costs. The same report indicates that, for many Canadian municipalities, sewerage charges now make up half of the average household's water and sewerage expenditures. Since many Canadian governments appear to be increasing their efforts to protect water quality, it is likely that more stringent (and costly) regulations for sewerage treatment may be seen in the future.

References

American Water Works Association. (2012). *M1 principles of water rates, fees and charges* (6th ed.). Colorado: Denver.

Bakker, K. (2014). The business of water: Market environmentalism in the water sector. *Annual Review Environment and Resources, 39*, 469–494.

Bjornlund, H. (2010). *The competition for water: Striking a balance among social, environmental, and economic needs*. C.D. Howe Institute Commentary, no. 302. www.cdhowe.org/pdf/commentary_302.pdf. Accessed 10 Apr 2014.

Brandes, O., Nowlan, L., & Paris, K. (2008). *Going with the flow? Evolving water allocations and the potential and limits of water markets in Canada*. Report prepared for the Conference Board of Canada. http://www.conferenceboard.ca. Accessed 2 Aug 2014.

British Columbia. (2014). *Pricing B.C.'s water*. http://engage.gov.bc.ca/watersustainabilityact/. Accessed 18 Mar 2014.

British Columbia. (2015). *Overview of new fees and rentals*. http://www.env.gov.bc.ca/wsd/water_rights/water_rental_rates/. Accessed 20 Oct 2015.

Canada West Foundation. (2011). *Summary of recent Canadian literature on water pricing*. Canada Water Policy Backgrounder Report 5. http://cwf.ca/projects/water-pricing. Accessed 8 Apr 2014.

City of London. (2013). *Water, wastewater and stormwater rates*. https://www.london.ca/residents/Water/water-bill/Pages/Water-and-Wastewate-Rates.aspx. Accessed 3 July 2015.

City of Vancouver. (2013). *Metered utility rates for water, sewer, and energy*. http://vancouver.ca/home-property-development/metered-rates.aspx. Accessed 7 July 2015.

Coffin, V., Poulton, D. W., & Vander Ploeg, C. (2011). *Our water and NAFTA: Implications for the use of market-based instruments for water resources management*. Canada West Foundation Environment Research Series. http://cwf.ca/publications-1/our-water-and-nafta. Accessed 14 Aug 2015.

Conference Board of Canada. (2014). *How Canada performs: Water withdrawals*. http://www.conferenceboard.ca/hcp/details/environment/water-consumption.aspx. Accessed 3 Apr 2014.

Environment Canada. (2001). *Municipal water pricing, 1991–1999*. Ottawa: Public Works and Government Services.

Environment Canada. (2011a). *2011 Municipal water pricing report: Municipal water pricing 2009 statistics*. http://ec.gc.ca/Publications/default.asp?lang=En&xml=992156D4-2599-4026-9B4C-47855D26CCB8. Accessed 28 July 2015.

Environment Canada. (2011b). *2011 Municipal water use report: Municipal water use 2009 statistics*. https://ec.gc.ca/Publications/default.asp?lang=En&xml=B77CE4D0-80D4-4FEB-AFFA-0201BE6FB37B. Accessed 11 Sept 2015.

Environmental Commissioner of Ontario. (2014). *Keeping Ontario's drinking water safe*. http://www.eco.on.ca/blog/2014/11/06/keeping-ontarios-drinking-water-safe/. Accessed 20 July 2015.

Federation of Canadian Municipalities. (2013). *Canadian infrastructure report card volume 1: 2012 municipal roads and water systems*. www.canadainfrastructure.ca. Accessed 19 Aug 2014.

Horbulyk, T. (2016). Water policy in Canada. In S. Renzetti & D. Dupont (Eds.), *Water Policy and Governance in Canada*. Springer.

Kenny, A., Elgie, S., Sawyer, D., & Wichtendahl, C. (2011). *Advancing the economics of ecosystems and biodiversity in Canada: A survey of economic instruments for the conservation & protection of biodiversity*. Report prepared for Sustainable Prosperity Institute, Ottawa. http://www.sustainableprosperity.ca/article1431. Accessed 8 Apr 2014.

Lake Simcoe Region Conservation Authority. (2014). *Lake Simcoe Phosphorus Offset Program (LSPOP)*. http://www.lsrca.on.ca/programs/lspop.php. Accessed 9 July 2015.

Manitoba. (2015). *Water use licensing – Water use fees*. www.gov.mb.ca/conservation/waterstewardship/licencing/wlb/fees. Accessed 28 May 2015.

OECD. (2009). *Managing water for all: An OECD perspective on pricing and financing*. www.oecd.org/env/resources/1934075.pdf. Accessed 21 Aug 2015.

OECD. (2010). *Pricing water resources and water and sanitation services*. Paris: OECD Environment Directorate. Organization for Economics Co-operation and Development.

Ontario Ministry of Energy and Environment. (n.d.). *Permits to take water*. www.ontario.ca/environment-andenergy/water-taking-report-and-charges. Accessed 28 May 2015.

Renzetti, S. (2007). Are the prices right? Balancing efficiency, equity, and sustainability in water pricing. In K. Bakker (Ed.), *Eau Canada: The future of Canada's Water* (pp. 263–279). Vancouver: UBC Press.

Renzetti, S. (2009). *Wave of the future: The case for smarter water pricing* (C.D. Howe Institute Commentary, 281). Toronto: C.D. Howe Institute Commentary.

Renzetti, S., & Dupont, D. P. (2015). Water pricing in Canada: Recent developments. In A. Dinar, V. Pochat, & J. Albiac (Eds.), *Water pricing experiences and innovations* (pp. 63–83). Cham: Springer Publishing.

Renzetti, S., Dupont, D. P., & Chitsinde, T. (2015a). An empirical examination of the distributional impacts of water pricing reforms. *Utilities Policy*. doi:10.1016/j.jup.2014.12.004

Renzetti, S., Dupont, D., & Price, J. (2015b). *Challenges and opportunities for water demand forecasting when agency capacity is constrained*. Submitted to Canadian Water Resources Journal.

Rivers, N., & Groves, S. (2013). The welfare impact of self-supplied water pricing in Canada: A computable general equilibrium assessment. *Environmental and Resource Economics, 55*, 419–445.

Wilson, J. (1992). *Nonlinear pricing*. New York: Oxford University Press.

Part IV
Politics and Governance/Management

Chapter 12
The Politics of Water Policy Development in Canada

B. Timothy Heinmiller

Abstract The aim of this chapter is to take stock of Canada's water policy processes and to identify some general trends in Canadian water policy-making. To this end, it argues that water policy development in Canada has a pluralist character involving multiple organized interests, but is typically dominated by established, region-specific water user groups. The dominance of user groups is largely the product of economic, historical, and institutional factors which have given water users considerable political power and influence over Canada's water policy decision-makers. The other organized interests typically involved in water policy development – environmentalists and First Nations – have increasingly challenged and opposed the water user groups by using the courts, cultivating favourable public opinion, and using science-based arguments to promote their causes. Though environmentalists and First Nations have achieved some notable policy successes, the preponderance of power still lies with the water user groups. These arguments are explored by comparing water policy development in two distinctive Canadian regions: the southern Prairies where irrigation is the dominant water use and the Great Lakes-St. Lawrence where hydro-electric generation, shipping, industrial, and other water uses must be balanced.

12.1 Introduction

As shown throughout this volume, water policies are a crucial factor influencing the way that water is used (and sometimes abused) in Canada. Though the importance of water policies is universally recognized, the origins of water policies, and why governments chose some water policy designs over others, is less well-known. Part of the difficulty stems from the fact that the uses of water, the actors involved in water policy development, and the sovereigns making water policy decisions vary considerably across Canada. Thus, there is no single water policy development

B.T. Heinmiller (✉)
Department of Political Science, Environmental Sustainability Research Centre, Brock University, 1812 Sir Isaac Brock Way, St. Catharines, ON L2S 3A1, Canada
e-mail: theinmiller@brocku.ca

S. Renzetti, D.P. Dupont (eds.), *Water Policy and Governance in Canada*, Global Issues in Water Policy 17, DOI 10.1007/978-3-319-42806-2_12

process in Canada, but a plethora of water policy development processes varying by region and taking place at multiple levels of governance, from the local level to the international level.

The aim of this chapter is to take stock of Canada's water policy processes and to identify some general trends in Canadian water policy-making. To this end, it argues that water policy development in Canada has a pluralist character involving multiple organized interests, but is typically dominated by established, region-specific water user groups. The dominance of user groups is largely the product of economic, historical, and institutional factors which have given water users considerable political power and influence over Canada's water policy decision-makers. The other organized interests typically involved in water policy development – environmentalists and First Nations – have increasingly challenged and opposed the water user groups by using the courts, cultivating favourable public opinion, and using science-based arguments to promote their causes. Though environmentalists and First Nations have achieved some notable policy successes, the preponderance of power still lies with the water user groups.

This thesis is explored by comparing water policy development in two Canadian regions, the southern Prairies and the Great Lakes-St. Lawrence. These two cases illustrate the diversity of Canadian water policy-making, with the southern Prairies characterized by perennial water scarcity and the Great Lakes-St. Lawrence characterized by relative water abundance. Water uses in these regions are also quite different with irrigation dominating in the southern Prairies, and hydro-electric generation, shipping, industrial, and other uses competing in the Great Lakes-St. Lawrence. Analytically, these two very different governance situations also provide a means for capturing some of the diversity of water policy development in Canada and, thereby, identifying general trends in water policy-making.

12.2 Water Policy Development as Pluralism

Water policy development in Canada is an intensely political process generally characterized by pluralism. Pluralism is a longstanding concept in policy studies, useful for describing public policy development in many policy fields in most modern democracies. It refers to the prevalence of organized groups in policy-making and the tendency of these groups to compete with each other to influence policy-makers and thereby shape public policy (Smith 1990, p 302–303). From a pluralist perspective, two of the most important factors influencing policy outcomes are: (1) the ability or inability of interests – of varying kinds – to organize themselves and participate in policy development; and, (2) the balance of power between organized interests in terms of their ability to influence policy-makers. This pluralist emphasis on the relative power of organized interests is different from more conventional conceptions of democratic policy-making which place emphasis on public opinion and public participation as key factors shaping public policy. In this chapter, we will examine water policy development as a pluralist process, emphasizing both the constellations of organized interests involved in water policy issues and the balances of power between them.

Although the organized interests involved in water policy development vary considerably from region to region and issue to issue, there are three types of interests who participate most regularly: water user groups, environmentalists, and First Nations.

Water user groups are those actors who rely on access to water for various instrumental, usually economic, purposes. Because water can be used for so many different purposes, and because the uses of water vary widely across Canada, different regions typically feature different mixes of water user groups. For example, in the dry southern Prairies, the main user groups are irrigators, hydro-electric power generators, cities, and, to a lesser extent, the oil and gas industry, while the Great Lakes-St. Lawrence region features a different mix of user groups that includes cities, manufacturing industries, hydro-electric power generators, nuclear power generators, and the shipping industry. The main commonality amongst the user groups is their material self-interest in water use, and their desire to protect their water access and use in water policy processes. It is quite common, however, for various user groups to clash because of conflicting water uses. Prairie irrigators, for instance, are primarily interested in extracting water from rivers to grow crops, while hydroelectric generators would prefer to leave most water instream to increase river flows and improve generation capacity. At least part of the water policy development process is about working out these types of user group conflicts by establishing water use rules and practices in government policy (Johns et al. 2008).

The second type of organized interest involved in Canadian water policy-making is environmentalists. Environmentalists value water in its natural state and typically deplore any water use that disrupts or impairs natural aquatic ecosystems. This often puts them in direct conflict with water user groups, and it is quite common for water policy development in Canada to be dominated by conflicts between environmentalists and user groups. In the southern Prairies, there is a perennial policy conflict between irrigators who want to extract water from Prairie rivers and environmentalists who want to preserve natural flows, and in the Great Lakes-St. Lawrence there is ongoing disagreement between shippers who want to further deepen and channelize the St. Lawrence Seaway, and environmentalists who want to return the region's rivers to a more natural state (Heinmiller 2008, 2013). Though all environmentalists share a common goal of preserving natural aquatic environments, it is far from a monolithic movement. Some environmentalist groups, such as Friends of the Earth, are national and international in scope, while many others are regionally based or form around a single local issue. Similarly, some groups are quite radical in their approach to environmental protection, such as Greenpeace, while others, such as most fish and game associations, are more moderate and even have an instrumental interest in protecting water sources. So, it is not uncommon for policy disagreements to surface within the environmentalist camp.

First Nations are the third set of organized interests increasingly involved in Canadian water policy-making. Many First Nations have a spiritual and cultural connection to water that predates European settlement, and some continue to rely on water sources for fishing, trapping, and other livelihoods (Walkem 2007). Too many First Nations communities across the country also experience perennial water supply problems due to polluted water sources, inadequate water infrastructure, or both. On some water policy issues, First Nations align themselves with environmentalists since they often have similar water protection and conservation goals. Other times, First Nations are closer to a water user group, such as the First Nations in southern Alberta who have undertaken irrigation projects. What sets First Nations apart, however, are their unique claims to water rights based on longstanding treaties with the British and Canadian governments. Some of these rights claims have been settled, often through the courts, but many have not (Walkem 2007). For example, in southern Alberta, the Piikani First Nation signed a water rights settlement with the provincial and federal governments in 2010, but the Stoney Nakoda and the Tsuu T'ina and Samson Cree still have ongoing water rights disputes that are yet to be resolved (Phare 2009; Laidlaw and Passelac-Ross 2010). Many First Nations regard themselves as sovereign nations rather than as organized interests, so they often insist on dealing with the provincial and federal governments on a nation-to-nation basis, eschewing the regular policy consultation processes in which water user groups and environmentalists regularly participate.

Since many Canadian water resources are shared inter-provincially and/or internationally, the range of water user groups, environmentalists, and First Nations involved in shaping water policy is often quite wide. It is not just domestic organized interests affecting water policy, but interests from other jurisdictions, as well. In the southern Prairies, for example, Canada and the US, under the auspices of the International Joint Commission (IJC), have a longstanding water apportionment agreement on the shared St. Mary and Milk rivers (Boundary Waters Treaty 1909). The agreement divides the rivers' flows between the two countries and is designed to accommodate the demands of water user groups, particularly irrigators, on both sides of the international border, though irrigators in Montana complain periodically that they are short-changed by the apportionment (Halliday and Faveri 2007). Similarly, there are a number of IJC agreements and protocols in the Great Lakes-St. Lawrence region governing various aspects of water quality and water quantity. For instance, there is an IJC order of approval and Board of Control governing the regulation of water levels in Lake Ontario and the St. Lawrence River, which can be manipulated through dam operations. Plan 1958-D requires that water levels are maintained within an established range to ensure that hydro-electric generators and shippers have high enough water levels to be viable, but cities and lakeshore property owners are not subjected to damaging floods. These rules were established through a process of international negotiation that involved a wide range of organized interests from both sides of the border, and they are revisited regularly by the IJC (Macfarlane 2014, p 180–189).

12.3 Political Power in Policy Development

Pluralism implies that organized groups, such as water users, environmentalists, and First Nations, compete to influence government policy, but it does not imply that the contest between groups is an even one (Smith 1990, p 303). It is quite common for one group to have more power and political influence than the others, and this is certainly the case in Canadian water policy. For various economic, historical, and institutional reasons outlined below, water user groups have had the most political power and have enjoyed the most influence in shaping water policy in Canada. In some regions, such as the southern Prairies, a single user group has dominated the policy process, while in other regions, such as the Great Lakes-St. Lawrence, a plethora of user groups have found accommodations with each other, often at the exclusion and expense of environmentalists, First Nations, and the wider public.

12.4 The Economic Power of Water User Groups

A good deal of the political power enjoyed by water user groups comes from their important places in regional economies. Water is a key factor in many agricultural, industrial, and transportation processes, and many regional economies in Canada are, in one way or another, based on water use. Irrigation in the southern Prairies and manufacturing and shipping in the Great Lakes-St. Lawrence are good examples of this. Accordingly, water user groups often have a lot of economic clout behind them, both in terms of wealth production and job creation. This provides them with the financial resources needed to engage fully in policy development processes and it gives them economic structural power they can use to influence policy-makers. User groups often use the implicit (or explicit) threat of economic decline and job losses to resist water policy measures contrary to their interests, threats that can give pause to vote-conscious politicians. Environmentalists and First Nations have no such economic threats at their disposal and, in this way, do not have the same kind of leverage over policy-makers.

Agriculture has long been one of the most important economic sectors in the southern Prairies, and irrigation agriculture has established itself as the dominant water use in this region. Without irrigation, much of the agricultural land in the southern Prairies is suitable only for large-scale ranching or the growth of low value fodder crops. With irrigation, however, this land can be used for intensive farming and the growth of higher value cash crops. Southern Alberta is Canada's largest irrigation economy (by far) and it is estimated that 13 % of regional gross domestic product, 19 % of regional production, and 30 % of regional employment opportunities are directly or indirectly associated with irrigation. This translates into about $5 billion worth of production per year and about 13,000 full time jobs. It is also estimated that, without irrigation, the regional population could be reduced by as much as 75 %, as land is taken out of production, farms are consolidated, and irrigation spin-off jobs disappear (Alberta Agriculture Food, Rural Development 2004). Southern Alberta's irrigators have used some of their water-based wealth to form the Alberta Irrigation Projects Association (AIPA) which actively lobbies the provincial government to protect irrigators' interests. During the development of the 1996 *Water Act*, AIPA was very active in helping to craft the legislation, and repeatedly warned policy-makers of job losses and economic decline when irrigators' water entitlements were threatened. Alberta's politicians heeded these warnings and irrigators' water entitlements were nearly untouched by the new legislation (Heinmiller 2013).

Rather than a single dominant water user group, the Great Lakes-St. Lawrence region has a plethora of important user groups, each of which make important contributions to the regional economy. Hydro-electric power accounts for about one-third of Ontario's electric power generation and, according to Ontario Power Generation, is the province's lowest cost power source (Ontario Power Generation 2014). Manufacturing is still a key industry in the region, although it has gone into decline in the past decade, accounting for almost 13 % of Ontario's GDP in 2013 (Ontario Ministry of Agriculture, Food, and Rural Affairs 2014). Manufacturing processes such as steel and automotive production require large amounts of water and, historically, ready access to low cost water has been one of the region's most attractive competitive advantages. Another competitive advantage has been access to cost effective bulk shipping through the St. Lawrence Seaway, which has facilitated the movement of manufactured goods, agricultural commodities, petroleum, and all sorts of minerals to and from the North American interior. It is estimated that since it opened in 1959, 2.5 billion tonnes of cargo with an estimated value of $375 billion has been transported through the Seaway (St. Lawrence Seaway Management Corporation 2015). All of these user groups, and others, have formed organizations to protect and advance their respective interests with respect to water, and have been actively involved in local, provincial, federal, and international water policy processes. Environmentalists and First Nations have also been involved, but without the economic clout of these water user groups.

12.5 The Historical Advantages of Water User Groups

In contemporary water policy processes, water user groups also enjoy a first mover advantage that has placed them in the position of defending established policy status quos. Historically, Canadian governments encouraged and assisted in the development of water resources for economic purposes, often undertaking massive projects and spending vast sums of money to do so. The construction of dams and irrigation projects on the Prairies, and the construction of dams, canals, and hydro projects in the Great Lakes-St. Lawrence are good examples of this. These government investments, and the water use policies accompanying them, established the dominance of water user groups and created entrenched policy regimes long before environmentalists and First Nations became serious participants in water policy development.

The environmentalist movement did not emerge in Canada until the late 1960s, and First Nations were marginalized and excluded from water policy processes until at least the 1970s. By the time they became involved, they were confronted with entrenched policies in which governments had heavily invested and through which government agencies had formed close ties with user groups. Accordingly, exacting water policy change, on both the water quantity and water quality fronts, has been a daunting task of breaking down entrenched policy status quos.

On the southern Prairies, irrigation was firmly entrenched as the dominant water use in government policy by the time environmentalists and First Nations gained entry into water policy processes in the 1970s. Governments in the region saw irrigation as a path to prosperity in dry, sparsely populated rural areas, and both levels of government invested heavily in irrigation development, starting as early as the 1920s. Public spending on irrigation development peaked in the post-war period, with the construction of the massive Gardiner Dam project in Saskatchewan in 1967 and the development and expansion of 13 government-funded and government-administered irrigation districts in southern Alberta (Prairie Farm Rehabilitation Administration 1982). In all of these projects, irrigators formed close working relationships with regional politicians, the provincial agricultural departments, and the federal Prairie Farm Rehabilitation Administration (PFRA). Moreover, irrigators were able to secure entitlements to massive amounts of Prairie water at an early stage, leaving little water available to late coming users (Percy 2005). So, when environmentalists and First Nations eventually emerged as organized interests to challenge the irrigators and demand shares of the Prairie rivers for non-irrigation purposes, they confronted a policy regime in which governments had invested hundreds of millions of dollars, in which irrigators worked closely with agricultural departments to protect their interests, and in which the irrigators controlled the lion's share of the water resources. Breaking down such an entrenched status quo is an almost herculean task, greatly advantaging the irrigators in contemporary policy processes.

In the Great Lakes-St. Lawrence, the region's waterways were radically transformed to serve shipping and hydro power interests long before environmentalists or First Nations became serious players in water policy development. Construction of the first commercial shipping channels in the region dates to the 1840s, while the first hydro-electric power facilities were built in the 1880s. Through subsequent decades of international negotiation and cooperation, hydro power facilities were greatly expanded at Sault Ste. Marie, Niagara Falls, and Cornwall, and the modern St. Lawrence Seaway was completed in 1959. Altogether, government investment in water management infrastructure totalled in the billions of dollars by the middle of the twentieth century, and large, powerful government agencies were established to administer the hydro power and shipping systems. There also grew up a complex and entrenched set of water regulation and use rules, such as the IJC's Plan 1958-D, which established the dominance of shipping and hydro power interests in the region:

> …[I]t can be argued that the riparian, navigation and hydroelectric generation rights institutionalized in [Plan 1958-D] created a cartel over St. Lawrence water that has seriously deprived other uses not included at that time. In other words, the riparian, navigation, and hydro interests have succeeded in capturing water rights in the St. Lawrence to the exclusion of such uses as recreational boating, swimming, commercial fishing, and environmental preservation. To be fair, most of these potential users were excluded simply because they were not prevalent at the time… (Heinmiller 2008, p 253–254).

By the time environmentalists and First Nations became involved in water policy-making, the hydro and seaway facilities were long-accomplished facts, and the priority of these uses had been effectively institutionalized. Efforts to return the region's rivers to something closer to their natural state, and improve water quality for human and ecosystem uses, have meant challenging a firmly entrenched policy status quo, defended by the region's established user groups.

12.6 The Institutional Power of Water User Groups

A third factor that has conveyed a political power advantage to water user groups is the institutional context of Canadian water policy-making. Pluralist competitions between organized interests always occur within institutions which provide the rules of policy-making. Some of these are formal-legal rules outlined in a constitution or in pieces of legislation, while others are more informal rules, such as generally accepted norms or conventions. The prevailing institutional rules of policy development are important because they can – purposely or inadvertently – provide advantages to some organized groups and disadvantages to others. Sometimes the rules can create such a tilted political playing field that they virtually guarantee particular policy outcomes. In Canadian water policy development, the most important institution is the constitutional division of powers between the federal and provincial governments, but other institutions, such as the prevailing electoral system, can have a major impact, as well. Though the institutional context of Canadian water policy-making differs markedly in different water issues, the institutions generally tilt in the user groups' favour.

The Canadian constitution outlines a complex division of powers related to water. Under section 109 of the *Constitution Act, 1867*, the provinces were granted proprietary rights over most of the public lands and resources within their borders. This has effectively made the provinces the main regulators of water use, water extraction, and water allocation in Canada, with some exceptions. These exceptions are the water-related powers of the federal government, outlined in section 91 of the *Constitution Act, 1867*, which includes powers over fisheries, navigation and shipping, Indians and Indian lands, and international relations (Johns and Rasmussen 2008). Jurisdiction over natural resources is a sensitive topic in Canadian federal-provincial relations, and the two levels of government have been embroiled in protracted resource conflicts in the past, particularly over control of oil resources. Not wanting to provoke further conflict, Ottawa has, since the 1980s, interpreted its water-related powers quite narrowly, deferring to the provinces on most water issues touching provincial jurisdiction. As a result, many water issues are dealt with unilaterally by the provinces, some water issues are dealt with jointly by the federal and provincial governments, and only a few water issues are dealt with unilaterally by Ottawa (Heinmiller 2014).

In the southern Prairies, water policy-making is primarily a provincial concern. The federal government is involved in the administration of established international and interprovincial apportionment agreements for the Prairie rivers, and is sporadically involved in shipping, fisheries, and First Nations issues, but generally respects the provinces' powers to allocate their own waters (Heinmiller 2014). Accordingly, Alberta and Saskatchewan have their own distinctive water allocation and use policies, and the contest between water user groups, environmentalists, and First Nations has largely played out in provincial policy venues.

For two key reasons, this institutional context has worked much to the advantage of the dominant user group in the region, the irrigators. The first reason is the single member plurality electoral system that exists in each province. It is a well-established fact that single member plurality systems tend to over-reward geographically concentrated interests who hold sway over blocks of legislative seats, and under-reward geographically dispersed interests whose support is spread out across many legislative seats (Cairns 1968). While irrigators are a geographically-concentrated interest, environmentalists and First Nations are geographically dispersed. Thus, in southern Alberta, the "irrigation belt" constitutes about a dozen constituencies in which irrigation advocates are regularly elected to the legislative assembly, whereas environmentalists and First Nations are spread out across the province, hold little sway over any single constituency, and have little representation in the legislature. The Green Party, for example, has never won a seat in the Alberta legislature (Heinmiller Forthcoming). The second reason pertains to the simple fact that most water policy-making in the region is confined to the provincial level. This makes it difficult for environmentalists and First Nations working for water policy change in the Prairies to make end runs around their provincial policy venues by seeking policy change at the federal level. This sort of "venue shopping" has been used quite effectively by American environmentalists in their federal system, but is a strategy not readily available in Canada. Thus, advocates of water policy change are often confined to their provincial policy venue, where they are not able to match to electoral influence of the irrigators (Heinmiller Forthcoming).

A good example of this dynamic was the struggle over the Oldman Dam in southern Alberta. A promise to construct the dam was first made during the 1975 provincial election, a promise designed to lure the irrigation belt constituencies away from the Social Credit party and towards the governing Progressive Conservatives. The dam proposal proceeded through a series of studies and consultations, receiving mixed reviews, until, under pressure from drought-stricken irrigators, the government decided to begin construction in 1984. By 1987, environmentalists had formed a concerted opposition movement through an umbrella organization called Friends of the Oldman River, and the local Peigan First Nation had moved to blockade construction. Unable to make any progress with provincial policy-makers, Friends of the Oldman River went to court – all the way to the Supreme Court – to force the federal government to apply its environmental assessment procedures to the Oldman Dam project. This was, in effect, an environmentalist effort at venue shopping. Ottawa appointed an environmental assessment panel and the panel produced a report recommending that the dam be decommissioned, or that its environmental effects at least be mitigated, but the federal government refused to enforce the recommendations because it did not want to spark another resource conflict with Alberta. This left the decision in the hands of the Alberta government, which saw the dam through to completion and gave the irrigators an important policy victory (Glenn 1999).

In the Great Lakes-St. Lawrence region, the institutional context of water policy-making is more complex and not as blatantly tilted in favour of the water user groups. The federal presence is much larger here as four of the five Great Lakes and the St. Lawrence River are bisected by the Canada-US border and shipping is one of the region's major water uses. International policy-making is also more prevalent, in a number of forums. The IJC and its various boards regulate water levels in the boundary waters and administer the Great Lakes Water Quality Agreement, a treaty designed to clean-up pollution hotspots and address emerging pollution threats. In the Great Lakes-St. Lawrence River Basin Agreement, the states and provinces also have a pact to harmonize the regulation of water withdrawals in their respective jurisdictions, aiming to prevent unsustainable withdrawals within the basin and large diversions outside of the basin. Thus, water policy-making occurs in a multi-jurisdictional, multi-layered, and interconnected complex of institutions, and it is difficult for any water user groups to dominate all of these policy-making forums. Nor is there a geographical block of water users that can take advantage of the single member plurality electoral system in the same way that the irrigators do in the southern Prairies. For all of these reasons, the institutional playing field is relatively even in the Great Lakes-St. Lawrence, notwithstanding the economic and historical advantages enjoyed by established water user groups.

One illustrative example of pluralist policy-making in the Great Lakes-St. Lawrence is the struggle over Plan 1958-D. Environmentalists have lobbied both the federal and provincial governments to return the St. Lawrence River to a more natural state of flow, to reduce channelization, restore wetlands, and increase fish stocks. The chief barrier to these reforms, however, has been the IJC's Plan 1958-D which prioritized riparian, hydro, and shipping uses to the exclusion of environmental and recreational uses. Advocates for change formed the International Water Levels Coalition to push for change at the IJC level, and have been opposed by the hydro generators and the shipping industry. In fact, the shippers have been pushing for an expansion and deepening of the seaway to accommodate larger, ocean-going vessels. The conflict came to a head in late 2000 when the IJC established the Lake Ontario-St. Lawrence River Study Board "…to assess and evaluate the current criteria used for regulating water levels…" (International Lake Ontario-St. Lawrence River Study Board 2004). The Board issued a report in 2006 outlining three new regulation plans, none of which, according to the environmentalists, would return the river to anything approximating natural flows. In the end, Plan 1958-D was allowed to stand and continues as the water levels policy for the St. Lawrence and Lake Ontario even today (Heinmiller 2008).

12.7 The Political Power of Environmentalists and First Nations

Given the economic, historical, and institutional advantages enjoyed by water user groups, one might be tempted to conclude that environmentalists and First Nations are nearly powerless in Canadian water policy development, but this is not the case. Environmentalists and First Nations have cultivated other sources of political power that have allowed them to compete with, and occasionally overcome, established water user groups. First Nations, for example, have been adept at using the courts to establish their rights to natural resources, including water, effectively bypassing political processes that have frustrated them for years. Environmentalists, and to a lesser extent First Nations, have also had some success using public protest and public opinion to put pressure on policy-makers and push them toward enacting water policy reforms. As well, environmentalists frequently draw on science and scientific reports to buttress their arguments for policy change, seeking to persuade policy-makers of the inherent 'rightness' of their policy positions. So, environmentalists and First Nations are not without power in Canadian water policy-making, even though they are usually not as powerful as the water user groups with whom they compete for policy influence.

Since the early 1970s, First Nations have made considerable gains in water policy through the use of the courts. Beginning with the *Calder* decision of 1973, the courts have recognized that Canada's First Nations have legal rights to their traditional lands – and waters – based on their historic occupation and use of these lands (Monahan 2002, p 444). These rights were then given a considerable boost in section 35(1) of the *Constitution Act, 1982* which states: "The existing aboriginal and treaty rights of the aboriginal peoples of Canada are hereby recognized and affirmed" (Constitution Act 1982).

First Nations across the country have been able to use section 35 to force provincial and federal governments to honour existing treaty rights and to have their treaty rights specified by the courts. Many of these treaty rights involve rights to the use of water for fishing, trapping, transportation, domestic, and commercial purposes (Monahan 2002, p 442–478). First Nations have also been able to use the courts to establish non-treaty rights to lands and other resources, most recently in the 2014 *Tsilhqot'in* decision which established an expansive test for determining First Nations' rights on their traditional lands (Hildebrandt 2014). In effect, First Nations have used a venue shopping strategy that has moved some of their water concerns from the legislative arena to the courts, a venue that has been much more receptive to their policy demands. The downside of this strategy has been its incremental and fragmented nature, as each First Nation has had to fight individually through the courts to have its rights recognized. Environmentalists have also used the courts at various times, as we saw in the Oldman Dam example, mostly in an effort to force reluctant governments to implement their own environmental assessment and protection laws. However, there are no constitutionalized rights to a healthy environment, in contrast to the constitutionalized rights of the First Nations, so environmentalists have had little success in using the courts to overturn existing water policies.

One, perhaps surprising, political strength of environmentalists in water policy processes has been supportive public opinion. Support for environmentalism and environmentalist causes in Canada has ebbed and flowed, reaching peaks in the early 1970s, the early 1990s, and the mid-2000s (Paehlke 1992, 2000). Yet, there is a strong baseline of public environmental consciousness that environmentalist groups have carefully cultivated through public outreach and public education activities. The latest World Values Survey data, for instance, showed that just over 68 % of Canadians chose environmental protection over economic growth and job creation, when asked to prioritize one of these two conflicting social values (World Values Survey Wave 5 2005–2008 (2009). Favourable public opinion is a very important power resource for environmentalists because, if activated on a given water issue, it can place enormous pressure on vote-conscious policy-makers. In the case of the Oldman Dam, environmentalists succeeded in turning public opinion against the dam, forcing the Alberta government into a difficult choice between its irrigation supporters and the broader public (Glenn 1999). In Ontario, environmentalists were able to use public opinion to block water exports by tanker from Lake Superior. A company named Nova Group received a permit from the Ontario government in 1998 to commence water exports, only to have it cancelled a few months later after a severe public backlash (Heinmiller 2003, p 505). Public opinion can be fickle, and it can be difficult to capture the public's attention on a particular water issue, but it is often one of the most important political resources at the environmentalists' disposal.

Environmentalists also frequently draw on scientific reports conducted by environmental scientists in their efforts to influence policy decision-makers. Scientific reports are important to environmentalists because they often lend credence and legitimacy to their claims that water sources are degraded and in need of protection and restoration. The influence of scientific reports depends on a number of factors, including the perceived legitimacy of those producing them, the persuasiveness of the evidence, and policy-makers' receptivity to science-based argument. Sometimes, the reports are so compelling that environmentalists can even use them to persuade water user groups that policy reform is necessary. This happened in the early 2000s in southern Alberta when a series of reports on the state of the South Saskatchewan River Basin (SSRB) persuaded irrigators to accept a moratorium on the issuance of new water licenses in three of the SSRB's four sub-basins (Heinmiller Forthcoming). Similarly, in the Great Lakes-St. Lawrence, science-based reports by the IJC have been an important resource for environmentalists, helping them to push governments in the region toward the adoption of various water protection and conservation measures.

12.8 Conclusion

There is no singular water policy development process in Canada. Instead, there are many local, provincial, federal, inter-provincial, federal-provincial, and international water policy processes in which crucial decisions are made about the allocation, use, and protection of Canada's water resources. This chapter has attempted to capture some of this diversity by focusing on water policy development in two very different regions, the southern Prairies and the Great Lakes-St. Lawrence. Despite their many differences, both regions are characterized by pluralism in water policy development involving efforts by competing water user groups, environmentalists, and First Nations to influence policy-makers and shape water policy decisions. As Clancy (2014, p 35) observed in his study of Canadian water politics, "[k]ey policy decisions reflect the power relationships among the key interests," and this chapter has shown how established water user groups have economic, historical, and institutional power sources that make them the dominant interests in Canadian water policy-making. In response, First Nations have worked around the user groups' political power by shifting many of their water concerns to the courts, and environmentalists have attempted to counter user groups' power through the cultivation of favourable public opinion and the use of science-based arguments to try to pressure and persuade policy-makers. First Nations have achieved a string of successes in the courts, and environmentalists have gained some policy victories through the political process, but water user groups are still the main beneficiaries of most established water policies in Canada.

Looking forward, there is little reason to expect the dominant place of water user groups to change anytime soon. For this to happen, the fundamental economic and/or institutional power bases of user groups would have to be greatly diminished. It is conceivable, for instance, that water could someday play a much lesser economic role in emerging post-industrial economies, thereby diminishing the economic clout of user groups, or that constitutional or electoral reform could reduce the institutional advantages currently enjoyed by some water users. It is also conceivable that the progressive onset of climate change could someday trigger crises that realign the power resources of the organized groups involved in water policy-making, fundamentally shifting the balance of power between them. However, any such developments seem a long way in the future, if they happen at all. Instead, Canadian water policy development is likely to continue as it has for the past 30 years or so, with water user groups using their considerable power resources to defend established water policies, and environmentalists and First Nations taking opportunities to push for policy reforms where and when they can, chipping away at entrenched water policy status quos.

References

Alberta Agriculture Food and Rural Development. (2004). *Irrigation development in Alberta – Water use and regional development*. http://www.ijc.org/rel/pdf/83_stmary-milk_letter.pdf. Accessed 8 May 2014.

Boundary Waters Treaty. (1909). http://www.ijc.org/en_/BWT. Accessed 2 Apr 2013.

Cairns, A. C. (1968). The electoral system and the party system in Canada, 1921–65. *Canadian Journal of Political Science, 1*(1), 55–80.

Clancy, P. (2014). *Freshwater politics in Canada*. Toronto: University of Toronto Press. Constitution Act (1982) n.d.

Glenn, J. (1999). *Once upon an oldman: Special interest politics and the oldman river dam*. Vancouver: UBC Press.

Halliday, R., & Faveri, G. (2007). The St. Mary and Milk Rivers: The 1921 order revisited. *Canadian Water Resources Journal, 32*, 75–92.

Heinmiller, B.T. (2003). Harmonization through emulation: Canadian federalism and water export policy. *Canadian Public Administration, 46*(4), 495–513.

Heinmiller, B.T. (2008). The St. Lawrence: From River to Maritime Superhighway. In M. Sproule-Jones, C. Johns, & T. B. Heinmiller (Eds.), *Canadian water politics: Conflicts and institutions* (pp. 243–260). Montreal/Kingston: McGill-Queen's University Press.

Heinmiller, B.T. (2013). Advocacy coalitions and the Alberta Water Act. *Canadian Journal of Political Science, 46*(3), 525–547.

Heinmiller, B.T. (2014). Multilevel governance and the politics of environmental water recoveries. In I. Weibust & J. Meadowcroft (Eds.), *Multilevel environmental governance: Managing water and climate change in Europe and North America* (pp. 58–79). Cheltenham: Edward Elgar.

Heinmiller, B.T. (Forthcoming). *Water policy reform in Southern Alberta – An Advocacy Coalition Approach*. Toronto: University of Toronto Press.

Hildebrandt, A. (2014). Supreme Court's Tsilhqot'in first nation ruling a game-changer for all. http://www.cbc.ca/news/aboriginal/supreme-court-s-tsilhqot-in-first-nation-ruling-a-game-changer-for-all-1.2689140. Accessed 25 June 2015.

International Lake Ontario-St. Lawrence River Study Board. (2004). About us. http://www.losl.org/about/about-e.html. Accessed 15 July 2004.

Johns, C., & Rasmussen, K. (2008). Institutions for water resource management in Canada. In M. Sproule-Jones, C. Johns, & T. B. Heinmiller (Eds.), *Canadian water politics: Conflicts and institutions* (pp. 59–89). Monteral/Kingston: McGill-Queen's University Press.

Johns, C., Sproule-Jones, M., & Heinmiller, B.T. (2008). Water as a multiple-use resource and source of political conflict. In M. Sproule-Jones, C. Johns, & B.T. Heinmiller (Eds.), *Canadian water politics: Conflicts and institutions* (pp. 19–55). Montreal/Kingston: McGill-Queen's University Press.

Laidlaw, D., & Passelac-Ross, M. (2010). *Water rights and water stewardship: What about Aboriginal peoples?* http://ablawg.ca/2010/07/08/water-rights-and-water-stewardship-what-about-aboriginal-peoples/. Accessed 20 Oct 2013.

Macfarlane, D. (2014). *Negotiating a River – Canada, the US, and the creation of the St. Lawrence Seaway*. Vancouver: UBC Press.

Monahan, P. (2002). *Constitutional law* (2nd ed.). Toronto: Irwin Law Inc.

Ontario Ministry of Agriculture, Food, and Rural Affairs. (2014). Ontario Gross Domestic Product (GDP) for Ontario Economy, 2007 to 2013. http://www.omafra.gov.on.ca/english/stats/economy/gdp_all.htm. Accessed 19 June 2015.

Ontario Power Generation. (2014). *Hydroelectric Power.* http://www.opg.com/generating-power/hydro/Pages/hydro.aspx. Accessed 19 June 2015.

Paehlke, R. (1992). Eco-history: Two waves in the evolution of environmentalism. *Alternatives Journal, 19*, 18–23.

Paehlke, R. (2000). Environmentalism in one country: Canadian environmental policy in an era of globalization. *Policy Studies Journal, 28*(1), 160–175.

Percy, D. R. (2005). Responding to water scarcity in Western Canada. *Texas Law Review, 83*, 2091–2107.

Phare, M.-A. (2009). *Denying the source: The crisis of First Nations water rights*. Surrey: Rocky Mountain Books.

Prairie Farm Rehabilitation Administration. (1982). *History of irrigation in Western Canada*. Ottawa: Government of Canada.

Smith, M. J. (1990). Pluralism, reformed pluralism and neo-pluralism: The role of pressure groups in policy-making. *Political Studies, 28*(2), 302–322.

St. Lawrence Seaway Management Corporation. (2015). Great Lakes – St. Lawrence Seaway Facts. http://www.greatlakes-seaway.com/en/seaway/facts/. Accessed 19 June 2015.

Walkem, A. (2007). The land is dry: Indigenous peoples, water, and environmental justice. In K. Bakker (Ed.), *Eau Canada – The future of Canada's water* (pp. 303–319). Vancouver: UBC Press.

World Values Survey Wave 5 2005–2008. (2009). Official Aggregate v.20140429. www.worldvaluessurvey.org. Accessed 25 June 2015.

Chapter 13
Coordinating Water Policies: Necessary, But Not Sufficient

Rob de Loë

Abstract Responsibility for water is divided between the federal, provincial and territorial governments in Canada, and each government has pursued its own path. The resulting water policy landscape is highly fragmented. Issues exist that may warrant regional or national policy coordination. Examples relate to climate change adaptation, water exports and ecosystem protection, to name a few. Previous efforts to coordinate water policy in Canada have largely resulted in failure. This appears to leave Canada in a trailing position relative to numerous countries around the world that have pursued a more coordinated approach to water policy. Paradoxically, however, the failure of previous water policy coordination efforts creates an opportunity. A strong national policy that compartmentalizes water and isolates it from related social, environmental and economic concerns may be counterproductive in an increasingly inter-connected world where water has become material to a host of new actors. In this chapter, I argue that the most viable way to establish a more coordinated, strategic approach to water policy in Canada may be to shift attention to the ways in which water matters to other actors and sectors whose support is needed to achieve water policy objectives.

13.1 Introduction

Water has provided an important part of the foundation for Canada's social development and economic prosperity (Renzetti et al. 2011), and our water resources offer us a host of potential opportunities in a world in which water security increasingly will be a determinant of future prosperity. This foundation is not assured. Despite having access to considerable water resources, Canadians face significant water management challenges. These include providing clean, safe drinking water for all, adapting to increasing climatic variability, protecting and restoring aquatic ecosystems, and coping with floods, to name a few. Responsibility for water in

R. de Loë (✉)
School of Environment, Resources and Sustainability, University of Waterloo,
200 University Avenue West, Waterloo, ON N2L 3G1, Canada
e-mail: rdeloe@uwaterloo.ca

S. Renzetti, D.P. Dupont (eds.), *Water Policy and Governance in Canada*,
Global Issues in Water Policy 17, DOI 10.1007/978-3-319-42806-2_13

Canada is widely shared among public and private actors, including individual citizens. Hence, increased policy coordination among the many organizations and individuals involved in water governance in Canada has long been pursued (e.g., Lloyd 1969).

Despite the growing influence and importance of private actors such as corporations in the water domain, governments in Canada at the federal, provincial and territorial levels remain at the forefront of water governance (de Loë 2015). Canada's constitution divides responsibility for water among these senior levels of government, and each addresses its water-related responsibilities differently. At the provincial and territorial level, there is a further subdivision of responsibility, with key water management functions assigned to municipalities and special purpose bodies such as irrigation districts in Alberta, conservation districts in Manitoba, watershed organizations in Quebec, and conservation authorities in Ontario. In the territories, land claims and settlement agreements establish Aboriginal[1] governments as key actors, while in southern Canada, on First Nations reserves, Band Councils are responsible for provision of water to their members. The resulting water policy landscape is highly fragmented.

Shared and divided responsibility for water is the norm in countries around the world. This reflects considerations such as the presence of multiple levels of government in many countries (e.g., local, regional, state, Indigenous and national); historical decisions to assign responsibility for water to government agencies with mandates for areas such as agriculture, natural resources, and environment; and the existence of special purpose bodies such as watershed management agencies, irrigation districts, and soil and water conservation districts. In this chapter, I take the position that polycentric water governance – where there are many centres of decision making that share authority and are independent from each other (Ostrom 2010) – is normal, unavoidable, and potentially desirable if it increases accountability. Hence, I argue that the focus should be on achieving an appropriate amount of *coordination* among relevant actors.

The definition of coordination offered by Peters (1998, p 296) provides a starting point, coordination is:

> an end-state in which the policies and programs of government are characterized by minimal redundancy, incoherence and lacunae.

Whose policies and programs need to be coordinated, the amount of coordination that is needed, and how coordination will occur, all depend strongly on the

---

[1] Section 35 of Canada's Constitution Act identifies three distinct "aboriginal" peoples: Indians (First Nations), Métis and Inuit. Section 35 recognizes and affirms existing aboriginal and treaty rights of the aboriginal peoples of Canada. The relationship between First Nations and the federal Crown is strongly influenced by the *Indian Act*; a similar statute does not exist in the case of Métis and Inuit peoples. In this chapter, the term "Aboriginal" is used in contexts where it is based in law, or where it is considered locally appropriate (e.g., in the Northwest Territories). The term Indigenous has become the globally accepted way of referring to first peoples. The term First Nations is used specifically to refer to the Indigenous peoples described as "Indians" under the Constitution Act.

policy problem (Dinar 1998). Thus, coordination is both an outcome (policies are coordinated) and an ongoing, never-ending process. Numerous approaches to coordination exist, including memorandums of understanding, joint working groups and task teams, regulatory alignment initiatives and formal agreements. In this chapter, I focus on the role of "coordinating water policies" in the public realm. These typically take the form of formal, explicit, public statements of intention and action. They may be contained in legislation, in documents approved by Cabinets or legislatures, or they simply may be issued by governments.

The record of water policy coordination within and among Canada's senior governments is decidedly mixed. Some have been proactive in creating policies or strategies that articulate overall policies for water in their own jurisdictions, and define specific objectives and actions for implementing those policies. Examples include Alberta, Québec and the Northwest Territories. The federal government led the way in Canada with the creation of its *Federal Water Policy* in 1987 (Environment Canada 1987). However, its policy was allowed to languish, and the federal government has not made any significant efforts since the 1990s to revitalize water policy coordination within its own bailiwick. At the same time, the federal government has not demonstrated an appetite to coordinate water policy nationally. The provinces and territories have tried on their own to coordinate water policy nationally (The Council of the Federation 2010). Non-government organizations such as the Forum for Leadership on Water (Morris et al. 2007), Pollution Probe (2008) and the Canadian Water Resources Association (de Loë 2008) have tried to motivate governments to take action. These initiatives have not resulted in a more coordinated approach to water policy in Canada.

Should past failures to coordinate water policy across Canada motivate renewed efforts and stronger commitments? Or should they be viewed as an opportunity to take stock, and to rethink the narrative around water policy coordination in Canada? In this chapter, I will argue the latter position. Globally, the water community is starting to recognize that many – if not all – of the world's major water challenges cannot be resolved by water managers alone (e.g., United Nations World Water Assessment Programme 2009). From this perspective, focusing primarily on increasing coordination within the water sector may not be an effective way to deal with current and emerging problems that involve drivers at multiple scales, and are influenced by actors outside of the water sector. Increased coordination of water policy in Canada – whether within jurisdictions, or among them – will undoubtedly have the positive benefits that proponents point to, such as reduced waste, increased efficiency and effectiveness in program delivery, and stronger prospects for long-term sustainability of water resources (de Loë 2009). However, the potential also exists that a "water-centric" water policy for Canada could make matters worse if it draws attention away from, or fails to consider, the role played by critical external actors, drivers and institutions that are shaping the ability of the water community around the world to achieve its objectives.

In the next section, I briefly explore the Canadian water policy landscape, and contrast it with circumstances in other countries. I then examine the arguments that have been offered by the Canadian water community for increased coordination, and explore whether or not they hold up relative to the rapidly changing global waterscape. The chapter concludes with a case for a different way of thinking about coordination in Canadian water policy.

13.2 Canada's Approach to Water Policy Coordination Relative to the World

In the public realm, water policies can exist at all levels. They are being created by individual municipalities, sub-national governments, nation states, and in international organizations such as the United Nations. This chapter is concerned with water policies designed to draw together disparate policies within the public sector (both within and between or among jurisdictions). These "coordinating water policies" typically offer overarching visions and principles for water management and governance, along with implementation strategies. Surveying the global landscape, coordinating water policies exist in many jurisdictions, including the European Union, Australia, New Zealand, and Brazil, among others (de Loë 2008). Experiences in Australia and the European Union are briefly reviewed here to showcase different approaches to coordinating water policies. These experiences are then briefly contrasted to Canadian experiences.

Like Canada, Australia is a federation in which responsibility for water is shared between the national government (the Commonwealth) and state/territorial governments (Tisdell et al. 2002). Also like Canada, policies for water in each state and territorial government have developed on separate tracks, with the Commonwealth government supporting the implementation by states of national strategies using a host of mechanisms ranging from policies to funding (Government of Australia, Department of the Environment, Water, Heritage and the Arts 2007). The critical Murray-Darling Basin is one region in which the states and the Commonwealth have cooperated during the previous century and adopted a coordinated approach. Starting in 1994, an extremely high degree of policy coordination has taken place in Australia, beginning with the 1994 Council of Australian Governments (COAG) *Water Reform Framework* (WRF), and followed by the 2004 *National Water Initiative* (NWI), and the 2007 *National Plan for Water Security* (NPWS). These initiatives emerged in response to drivers such as growing environmental concern about the condition of Australia's rivers, pressure to restructure the water industry, the effects of multi-year droughts, and a neo-liberal reform agenda. Australia is an excellent example of a high degree of formal water policy coordination within a federation.

Experiences in the European Union illustrate a unique continental approach to coordinating the water policies of independent nation states. Water policy in the European Union's 28 member countries is guided by the overarching Water Framework Directive (WFD) (Lagacé 2010). The WFD, established in 2000 under Act 2000/60/EC, replaced a host of individual European water directives created in previous decades relating to drinking water, groundwater, swimming water, pesticides, and other concerns. It put in place a broad framework for water management, with detailed standards, targets and requirements contained in "daughter directives" (Watson and Howe 2006). Principles and objectives in the framework relate to community engagement and public involvement, the precautionary principle, sustainable development, pollution prevention, pricing and water use efficiency, and polluter pays. River basins provide a key spatial organizing framework for integrated water management. Member states are required to implement the WFD through their own laws, regulations and processes. European experiences under the Water Framework Directive show how water policy coordination can take place across national boundaries, at the continental scale.

Canada's experiences with policy coordination reveal considerably less ambition. As noted in the introduction, the federal government established an overarching water policy in 1987. The *Federal Water Policy* (Environment Canada 1987) was created following a comprehensive Inquiry on Federal Water Policy (Pearse et al. 1985). Inquiry commissioners travelled across the country, held hearings and met with numerous stakeholders. The policy provided an overall objective, two specific goals, five strategies, and 25 specific policy statements. It was grounded in the perspective that water had to be viewed "both as a key to environmental health and as a commodity that has real value…" (Environment Canada 1987, p 1). In this respect it was in line with thinking about the new idea of "sustainable development" contained in the World Commission on Environment and Development's report *Our Common Future* (1987).

The *Federal Water Policy* demonstrated that the federal government of the day saw a clear leadership role for itself, and for governments in general. It opened with the claim that:

> Governments must go beyond the symptoms to the causes of the problems and take new approaches with the changing realities in mind (Environment Canada 1987, p 1).

Hence, while it clarified the federal government's jurisdiction, it explicitly accounted for the constitutional roles of the provinces. In this context, the overall objective of the policy was:

> to encourage the use of freshwater in an efficient and equitable manner consistent with the social, economic and environmental needs of present and future generations (Environment Canada 1987, p 3).

The two overarching goals were "To protect and enhance the quality of the water resource" (Environment Canada 1987, p 3), and "To promote the wise and efficient management and use of water" (Environment Canada 1987, p 3). Strategies that would be used to achieve the objectives and goals included water pricing, science leadership, integrated planning, legislation and public awareness. Specific policy statements to support implementation related to concerns such as the management of toxic chemicals, inter-basin transfers, "native" water rights, flooding, and technological needs.

Environment Canada continues to portray the *Federal Water Policy* as a current and relevant coordinating policy (Environment Canada 2015). However, few outside of the federal government would agree that the policy has any relevance today

(Morris et al. 2007). Furthermore, since 1987, there has been little appetite within the Government of Canada for a renewed federal coordinating water policy, let alone a leadership role in the development of a national policy for the country as a whole (Clancy 2014). Instead, the federal government has relied on focused initiatives, such as the 2007 *Action Plan on Clean Water*, which provided funds to support actions including targeted pollution remediation efforts and drinking water on First Nations communities (de Loë 2008). Indeed, critics suggest that recent legislative changes instituted by the former Conservative government have significantly narrowed the scope of key federal statutes – notably the *Fisheries Act* and the *Navigable Waters Protection Act*. These changes were taken as a sign that Canada's federal government, under the Conservatives, had even less appetite for national leadership on water than it did a decade ago (e.g., Nikiforuk 2013; Barlow 2015).

In the absence of federal leadership on water policy, provincial and territorial efforts to coordinate water policies have taken on even more importance. Experiences in Alberta, Quebec, and the Northwest Territories showcase different approaches to water policy coordination.

- Alberta's *Water for Life* (Alberta Environment 2003) strategy was established in 2003 in response to growing concerns about the pace of industrial, agricultural and municipal development, pressure on water resources, drought, and other concerns. This policy, which was created by the provincial Cabinet, establishes Alberta's vision for water management. It is centered on three main goals: safe, secure drinking water; reliable, quality water supplies for a sustainable economy; and healthy aquatic ecosystems. Actions towards these goals are to be implemented through partnerships that operate at scales ranging from local to provincial.
- Quebec's *National Water Policy* was instituted in 2002 in response to a long-standing desire to coordinate the activities of a dozen provincial ministries with water-related responsibilities (Francoeur 2006). Public demands for action on issues ranging from flooding to water quality contamination added pressure (Gouvernement du Québec 2002; Gentès 2007). The policy aims to protect water quality and quantity in the province, and to promote a participatory approach to water governance (Gouvernement du Québec 2002; Luyet et al. 2005). It is also grounded in a suite of broad principles and specific objectives.
- Northwest Territories' coordinating water policy is called *Northern Voices, Northern Waters* (Government of the Northwest Territories 2010). The policy sets as an overall goal of ensuring that the waters of the territory are "substantially unaltered". Uniquely in Canada, this strategy was developed in partnership between the Government of the Northwest Territories and Aboriginal Governments.[2] Hence, it places considerable emphasis on the cultural, spiritual

[2] In the Northwest Territories in general, and in the water stewardship strategy in particular, the terms "Aboriginal People" and "Aboriginal Governments" are used. The term "Indigenous" does not appear in the strategy.

and historical importance of water to Aboriginal people and their involvement in governance for water. Like the other examples discussed here, the strategy also includes a combination of overall guiding values and principles, and more detailed objectives.

The provinces and territories have attempted, without the participation of the federal government, to coordinate water policy at a very high level. In 2010, the Council of the Federation endorsed a *Water Charter* (The Council of the Federation 2010). The Council represents the ten provinces and three territories. The Council's Charter recognizes common challenges and opportunities in Canada, and encourages its signatories to work together to reinforce water conservation, to make water use more efficient, to enhance monitoring, and to collaborate on shared issues. Unfortunately, the practical impacts of the charter in terms of policy coordination appear to be minimal (Sandford 2015).

There have also been attempts by Canadian non-government organizations to coordinate water policy across the country (e.g., Pollution Probe 2008; de Loë 2008). For example, the Canadian Water Resources Association and its partners aimed to provide a degree of policy coordination on nationally significant issues (de Loë 2008). This effort to develop a "pan-Canadian water strategy" was led from outside of governments, but depended on the federal, provincial and territorial governments taking up the cause. The initiative was abandoned due to a lack of interest among the governments.

13.3 Challenging Underlying Assumptions

The mixed record of success in creating coordinating water policies in Canada is an opportunity to take stock, and to reflect on the problems that coordinating water policies are supposed to address, and whether or not they really are the most appropriate way to address those problems. The discussion in this section focuses on some of the core assumptions and basic frames underlying typical efforts to develop coordinating water policies. The implications of these assumptions and frames are highlighted.

A basic assumption underlying the coordinating water policies discussed in this chapter, I argue, is that water problems have solutions that can and should be addressed by those responsible for water. This reflects a normal division of labour. The policy challenges that confront governments around the world are complex, and assignment of responsibilities based on major policy fields (e.g., transportation, environment, trade, health care) is commonplace. This division of labour is mirrored in higher education, where water expertise is often developed in specialized water focused programs such as hydrology and water resources engineering. The end result of this kind of division of responsibility is the formation of an epistemic community around water. This community is global, and comprises policy makers, resource managers, researchers, and advocates for water located in government agencies and nongovernment organizations. Members of this epistemic community often share common principles and beliefs about the significance of water in society (Kramer and Pahl-Wostl 2014; Mukhtarov and Gerlak 2014). As a result, they may make similar kinds of judgments about how water problem should be addressed, and about the boundaries that should be used to determine which issues and concerns are relevant in addressing those problems (Ison 2010; Edelenbos et al. 2013). These judgements have practical implications. For example, Blomquist and Schlager (2005, p 105) emphasize that:

> boundaries that define the reach of management activities determine who and what matters. Inside the boundaries, individuals and groups may participate in decision making and have their interests, values, and concerns addressed. Those who fall outside the boundaries have fewer and indirect ways of participating.

A common assumption held by many members of the world's water community is that watersheds (catchments) and river basins provide an appropriate and logical boundary for management, and even governance (Molle 2006). Watersheds are a powerful tool for understanding interconnections among flows of water, and between water and land. However, the relevance of watersheds and river basins in sectors beyond water should not be assumed. For example, land use planning is normally organized around municipalities, whose boundaries rarely if ever coincide with watersheds. Water management and land use planning are deeply interrelated. To illustrate, a decision to transform a forested area into a new residential subdivision has implications for water flows, storage, sediment transport, and water quality. Nonetheless, in the absence of organizations such as Ontario's conservation authorities, which explicitly seek to integrate economic, environmental and social concerns (Mitchell 2006), the domains of water management and land use planning frequently operate in relative isolation from each other (Gober et al. 2013). This speaks to a fundamental problem confronting the water sector: people outside of the water community may use different frames, and may neither understand nor accept the watershed as a relevant frame for their problems. Thus, it cannot be assumed that people outside of the water community will accept the water community's framing of problems, and the significance of water. In fact, it cannot be assumed that the interacting causes and effects associated with a water problem within a watershed are contained within the boundaries of that watershed (Mollinga et al. 2007; Davidson and de Loë 2014). This has significant implications for the viability of coordinating water policies, which tend to be grounded deeply in the water community's framing.

Building on these concerns, a second basic assumption underlying most efforts to coordinate water policy can be problematic. Coordinating water policies are typically developed or led by the water sector, which, in turn, is assumed to have within it the actors whose policies and activities are most in need of coordination. Put another way, efforts to coordinate water policies are often premised on the notion that the actors whose policies are being coordinated are the ones who have the most influence on desired outcomes. In some cases this assumption is entirely appropriate. For example, Canada's provinces and territories have had a good record of success in coordinating plumbing standards and drinking water quality regulations (de Loë 2009). These examples illustrate that coordination efforts can be highly effective in circumstances in which the problems are well-defined and water specialists have the necessary knowledge and, importantly, are viewed as the people who should address those problems. The case is considerably less clear cut in circumstances in which policy problems have been inappropriately framed as water problems, or the traditional water-centric framing fails to capture critical actors and drivers. Two examples are briefly discussed here to illustrate these circumstances: drinking water quality on First Nations communities, for the former, and contemporary algal blooms in Lake Erie, for the latter.

13.3.1 Drinking Water Provision on First Nations Communities

Drinking water provision on First Nations reserves in Canada is an archetypical example of a "water" problem that cannot and should not be addressed solely – or even primarily – from a water-centric perspective, by the water community (see Indigenizing Water Governance in Canada (Adeel 2016), this volume, for the larger context of water and Indigenous peoples in Canada, and Placing Canada's Water Policies in an International Context (Bradford et al. 2016), this volume, regarding the issue of water as a basic human right). Under the *Indian Act*, responsibility for drinking water provision on "Indian Reserves" is shared between the Band Council and the federal government, through Aboriginal Affairs and Northern Development Canada.

The knowledge and technology needed to provide clean, safe drinking water are robust and have existed for well over a century. New contaminants undeniably challenge existing treatment systems (Gavrilescu et al. 2015). However, in most cases the communities in Canada dealing with poor drinking water quality, and under drinking water advisories, are faced with more prosaic challenges such as coliform counts and breakdowns in chlorination equipment that can be addressed using existing technologies and improved practices. In 2008, 1766 communities were under drinking water advisories, and these included 93 First Nations communities (Eggertson 2008). Most of the 2675 First Nations communities in Canada are located in northern and remote parts of the country. Therefore, isolation and distance are often offered as reasons for poor drinking water quality in these communities (Swain et al. 2006). However, even First Nations communities in southern Ontario are dealing with drinking water quality problems that are not faced by neighbouring non-Indigenous communities (Plummer et al. 2013).

Properly bounding the question of poor quality of drinking water in Canada's First Nations communities requires accounting for interacting circumstances such as the legacy of colonialism; tensions on many reserves between leaders who derive their legitimacy from the Indian Act and those whose legitimacy originates with traditional sources; funding and financial management problems; the special and difficult relationship between First Nations and the federal Crown, and the fact that provincial laws, regulations and programs relating to drinking water do not apply on reserves; growing Indigenous aspirations for self-governance; a legal landscape that is continually shifting and being redefined by the courts; and the federal government's approach to dealing with its fiduciary responsibilities to First Nations (Mascarenhas 2007; Phare 2009). From this perspective, poor drinking water quality in First Nations communities is not simply a water problem. Instead, it is a symptom of vast and unresolved historical, cultural, legal, economic and social issues that define the lives of Canada's First Nations peoples. The water sector, including First Nations water managers who are dealing with water issues in their communities on a daily basis, are small players in this larger drama.

The federal government attempted to coordinate its own policies through the *Safe Drinking Water for First Nations Act* (S.C. 2013, c. 21). This statute establishes a federal regulatory framework for drinking water quality, training and certification, treatment, monitoring, and source water protection. On its face this is a laudable initiative. However, the response to the new framework has been mixed at best, with significant opposition coming from First Nations leaders and others, who characterize the Act as a takeover by the federal government that involves imposed solutions, raises but does not answer questions about the role of external parties, creates funding uncertainty rather than clarity, and contributes to the further derogation of rights of First Nations (Hewson 2014). It is fair to say that these criticisms could have been anticipated because the challenge of providing drinking water on First Nations communities was approached as a *water* problem, rather than as a response to the much larger issues that confront First Nations people in Canada.

13.3.2 Algal Blooms in Lake Erie

In August, 2014, the City of Toledo, Ohio, had to shut down its water supply because microcystin, a toxin produced by cyanobacteria (blue-green algae) growing in the western basin of Lake Erie, was discovered in the water supply. Lake Erie is the shallowest and most biologically productive of the Great Lakes shared by Canada and the United States, and nutrient pollution has long been a concern. During the 1960s nutrient pollution in Lake Erie was so severe that it was popularly described as "dead" (International Joint Commission [IJC] 2014). The Great Lakes Water Quality Agreement of 1972 was created by Canada and the United States under the framework of the Boundary Waters Treaty of 1909 to address issues such as this (Linton and Hall 2013). Thanks to an intensive effort on the part of the governments of Canada and the United States, and their provincial and state partners, the condition of Lake Erie improved dramatically. By the mid-1980s, due largely to regulations on phosphorous in laundry detergents, phosphorous loadings in Lake Erie were reduced by more than half compared to the 1970s (IJC 2014). However, algal blooms returned starting in the 2000s, with the worst ever bloom occurring in 2015 (National Oceanic and Atmospheric Administration 2015).

Bounding the contemporary algal bloom problem correctly requires understanding policy changes in other sectors not directly related to water. The dissolved reactive phosphorous causing the algal blooms in Lake Erie has many sources on both sides of the Canada-US boundary (Smith et al. 2015). Manufacturers of laundry detergents have not re-introduced phosphorous. Instead, more than half of the nutrients producing the contemporary algal blooms in the western basin of Lake Erie are attributed to monitored streams in rural areas; the Maumee River in Ohio is specifically identified as a significant source of dissolved reactive phosphorous (IJC 2014). Farmers in the Maumee watershed are using production practices (tillage and fertilizer application) designed to maximize corn production. They are focusing on intensive corn production in response to market signals that originate with policy decisions well outside the water domain.

In the early 2000s, the government of the United States put in place policies designed to enhance domestic energy security. The Energy Policy Act of 2005 and the Energy Independence and Security Act of 2007 (EISA) have caused a significant increase in the price farmers can receive for corn that can be used as feedstock in ethanol production (Fausti 2015). At the same time, the Federal Agriculture Improvement and Reform Act of 1996 removed acreage restrictions, and gave farmers more scope to decide which crops to plant (Fausti 2015). Other incentives at the federal and state levels also encouraged biofuel feedstock production. Corn production in the United States grew in response to these drivers: from 75.7 million acres in 2000 to 95.9 million acres by 2012 (30.63–38.81 million ha) (Gillon 2012; Avalos 2014). In the US portion of the Lake Erie basin, harvested corn increased by approximately 789,000 acres between 2002 and 2007 (United States Department of Agriculture 2007). Production increases were facilitated by intensive production practices, including fertilizer application rates and timing practices that maximized crop yields, but also resulted in water quality degradation due to increased fertilizer runoff (IJC 2014).

Primary responsibility for addressing this water quality problem falls to the relevant environmental and agricultural agencies in Canada and the United States. Lake Erie is shared by Canada and the United States. Hence, Canada and the United States have also been working together under the Great Lakes Water Quality Agreement. The International Joint Commission, the binational organization created under the Boundary Waters Treaty of 1909 that is responsible for the Great Lakes Water Quality Agreement, published a major study of the nutrient issue in 2014 (IJC 2014). Policy coordination within and among the governments of Canada, the United States, Ontario, and the eight Great Lakes states, is clearly needed to address the algal blooms in Lake Erie. However, in key respects the nutrient issue in the western basin of the lake is linked to decisions made by US energy policy makers who are not part of the water sector, and whose decisions are not encompassed by the boundaries of the Great Lakes Basin. Therefore, it is important to ask whether or not nutrient pollution in Lake Erie is a problem that the binational water sector can solve on its own.

13.4 Rethinking Water Policy Coordination

In 2007 a group of researchers, experts and citizens who came together as "FLOW" – the Forum for Leadership on Water – published *Changing the Flow: A Blueprint for Federal Action on Freshwater* (Morris et al. 2007). One of the members of FLOW, Ralph Pentland, had been the federal official responsible for drafting the 1987 *Federal Water Policy*. FLOW identified seven priority areas and 25 linked actions that, together, were designed to revitalize the federal government's water policy. Priority actions related to historical and emerging concerns were identified in FLOW's report, e.g., enhancing national capacity for freshwater protection, responding to the impacts of climate change and energy production, and protecting aquatic ecosystems and "Aboriginal" water rights (Morris et al. 2007). While it drew national attention to the need for federal leadership on water, FLOW's report and follow-up work did not produce the outcomes its members desired.

FLOW's initiative suggests that renewed federal leadership on water – even simply in terms of its own jurisdiction – would be welcomed by some members of the Canadian water community. However, in light of the issues raised in this chapter, it is instructive to ask whether or not a renewed federal water policy – or even the kind of national coordinating water policy sought by other non-government organizations – would be a sound basis for addressing contemporary water challenges. I argue that water policy coordination efforts such as the *Federal Water Policy*, Alberta's *Water for Life* strategy, Quebec's *Water Policy*, the Northwest Territories' *Northern Voices, Northern Waters* and even the pan-Canadian water policy concept advocated by the Canadian Water Resources Association, are valuable and can play critical roles. Specifically, they can focus on key water issues, and they can help to coordinate the actions of disparate water sector actors who have a role in creating or addressing those issues. However, on their own, these kinds of coordinating water policies may not be an effective way to address contemporary water problems that are created, or at least strongly influenced, by actors and drivers external to the water domain. As the examples of nutrient pollution in Lake Erie and drinking water quality on First Nations reserves demonstrate, there are circumstances under which a water-centric orientation can be a liability.

At the global level, recognition of the limitations of a water-centric perspective is growing. To illustrate, in its 2009 global overview, the United Nations World Water Assessment Programme [UNWWAP] (2009, ii) argued that:

> Many paths to sustainable development are linked to water, but the decisions that determine how water resources are used or abused are not made by water managers alone.

The report's authors suggested that the water sector had become trapped in a "water box", where the ability of water managers to address challenges such as provision of water for humans and ecosystems was constrained by external forces over which they had little or no control (e.g., political decisions about funding for water relative to other priorities). Escaping the water box, the UNWWAP (2009, ii) argued, requires that:

water managers inform the initial decision-making and participate in planning the appropriate responses, interacting with the principal actors and with the managers of other sectors.

This alone would be challenging, but perhaps feasible. More problematically, the UNWWAP suggested that "These [external] leaders must learn to recognize water's role in attaining their objectives and act accordingly" (UNWWAP 2009, p 1).

Implicit in this recommendation is the assumption that these external leaders will want to engage the water sector on *its* terms. Instead, these external leaders are likely to have different priorities. To illustrate, mining and food processing companies that have always used water, but viewed it as an operational concern, are now seeing it as a strategic priority (Barrett 2009; Lambooy 2011). Business interests that previously saw water as a peripheral concern – or did not think about it at all – are seeing water-related risks and opportunities (e.g., IBM 2009; Henderson and Parker 2012). These kinds of new actors are using their elite-level access and resources (MacDonald 2007; Clapp and Fuchs 2009) to shape how water is perceived by state actors whose concerns relate to economics and security rather than to water (e.g., see Office of the Director of National Intelligence 2012; World Economic Forum 2013). There is no reason to think that their interests necessarily align with those of the water community.

Moving beyond a water centric approach to water policy coordination may require different assumptions and strategies. As a starting point, it will be essential to know when it is necessary to use a different approach, and when a water centric approach will be sufficient. The transaction costs associated with policy coordination within the water sector are already significant. Policy coordination among water and other sectors – e.g., food, energy, health – is likely to have even higher transaction costs. Some water problems have relatively self-contained causes and effects, and well-defined interests; these may continue to be well-suited to water-centric approaches. For instance, in the case of drinking water supply and wastewater treatment, a host of operational decisions normally can and should be taken by water experts. In contrast, more systemic approaches may be needed in cases in which key policy sectors such as security, health, trade and energy intersect at multiple scales; where water-centric concepts such as the watershed fail to capture key actors; where water problems are symptomatic of larger issues; and where actors and institutions outside of the water sector are making decisions and taking actions that shape the ability of the water sector to achieve its goals.

In these circumstances, water policy coordination will be necessary to ensure that water issues are addressed coherently within the sector – but this will only be one of potentially many starting points. If we accept that that decision makers in sectors external to water may not necessarily want to take on board the basic assumptions of the water sector, it may be necessary for the water sector to rethink how it approaches water policy coordination. To illustrate, rather than starting from the assumption that watersheds and basins can be used to define the scope of governance, as is common in Canada, it may be necessary to focus on the actual scale and extent of problems and causes (Mollinga et al. 2007; Davidson and de Loë 2014; Muller 2015). From a practical perspective, strategies will be needed that allow policy makers and researchers to identify and assess critical links and relationships, and to evaluate capacity for addressing them. Promising insights and practical approaches already are being pursued to support the kind of diagnosis that will be necessary. For example, Mitchell (2005) argued the importance of thinking "comprehensively" at the normative and strategic levels, but shifting to an "integrated" perspective that involves fewer variables and relationships at the tactical and operational levels. Ostrom's (2009) social-ecological systems framework provides another useful starting point for considering interactions among critical variables, while systems thinking (e.g., Ulrich 1993; Ison 2010), the extensive literature of institutional analysis (e.g., Ingram et al. 1984; Young et al. 2008), and recent efforts to develop diagnostic approaches that reveal key social-ecological interactions pertinent to water (e.g., Garrick et al. 2013; Hinkel et al. 2015) provide appropriate problem framings and strategies.

13.5 Conclusion

The record of Canada's federal, provincial and territorial governments on water policy coordination has been mixed. Some provinces and territories – such as Alberta, Quebec and the Northwest Territories, discussed in this chapter – have developed coordinating water policies within their own jurisdictions. In contrast, the record of the federal government has been disappointing. From a leadership position in 1987, the Government of Canada under the Conservative Party retreated to a caretaker approach to water policy. Efforts by non-government organizations to create a national vision and strategy for water in Canada have failed (e.g., Pollution Probe, Canadian Water Resources Association). Even the Council of the Federation's Water Charter does not appear to have had practical impacts on water policy coordination.

In the face of growing water challenges and significant opportunities, one could argue that efforts to develop coordinating water policies in Canada should be redoubled. However, this chapter offers a more nuanced view. The water landscape is changing dramatically. New actors have emerged whose perspectives on water do not necessarily align with those of the water community. The traditional frames used by the water community in Canada, notably the watershed, may be a poor match with the actual boundaries of problems. Finally, as the examples of drinking water on First Nations reserves and algal blooms in Lake Erie demonstrate, it should not be assumed that the people and organizations formally tasked with water-related responsibilities are well placed to address current and emerging water challenges.

From this more nuanced perspective, coordinating water policies potentially may be *necessary*, but they almost certainly will not be *sufficient*. Hence, those who promote and develop coordinating water policies will need to ensure that they fit the problems they are designed to address. Different kinds of analyses may be needed to distinguish water issues that can be addressed from a traditional "water-centric" perspective from those that cannot. In the case of the latter, the water community should rethink how best it can achieve the outcomes it desires. Engaging with the political and economic actors who have the most influence over desired outcomes is likely to be unavoidable. In this light, coordinating water policies that offer a clear vision for water in Canada are going to be a starting point for discussions with people beyond water, rather than the final word on how Canada's water resources will be viewed and managed.

Acknowledgements I would like to thank Dr. James Patterson for his enormous contribution to helping me think through how we proceed after coming to the realization that water centric approaches no longer will be sufficient. Dr. Patterson was a post-doctoral researcher in the Water Policy and Governance Group in 2014–2015 and helped me to launch my SSHRC-funded project *Rethinking Water Governance: Towards a New Agenda for Research and Practice*. I would also like to thank Dr. Bruce Mitchell for his constructive comments on a draft of the chapter.

References

Adeel, Z. (2016). Placing Canada's water policies in an international context. In S. Renzetti, D. Dupont (Eds.), *Water policy and governance*. Dordrecht: Springer.

Alberta Environment. (2003). *Water for life: Alberta's strategy for sustainability* (Publication Number I/955). Edmonton: Alberta Environment.

Avalos, F. (2014). Do oil prices drive food prices? The tale of a structural break. *Journal of International Money and Finance, 42*, 253–271.

Barlow, M. (2015). *Blue betrayal: The Harper Government's assault on Canada's freshwater*. Ottawa: The Council of Canadians.

Barrett, D. J. (2009). Thinking outside the lease – Towards a strategic view of regional water management by the mining industry. *Mining Technology, 118*(3/4), 131–141.

Blomquist, W., & Schlager, E. (2005). Political pitfalls of integrated watershed management. *Society and Natural Resources, 18*(2), 101–117.

Bradford, L. E. A., Ovsenek, N., & Bharadwaj, L. (2016). Indigenizing water governance in Canada. In S. Renzetti, & D. Dupont (Eds.), *Water policy and governance*. Dordrecht: Springer.

Clancy, P. (2014). *Freshwater politics in Canada*. Toronto: University of Toronto.

Clapp, J., & Fuchs, D. (2009). Agrifood corporations, global governance, and sustainability: a framework for analysis. In J. Clapp & D. Fuchs (Eds.), *Chapter in corporate power in global agrifood governance* (pp. 1–26). Cambridge, MA: The MIT Press.

Davidson, S. L., & de Loë, R. C. (2014). Watershed governance: Transcending boundaries. *Water Alternatives, 7*(2), 367–387.

de Loë, R. (2008). *Toward a Canadian national water strategy* (Final report). Guelph: Rob de Loë Consulting Services.

de Loë, R. (2009). A Canadian vision and strategy for water in the 21st century. *Policy Options, 30*(7), 21–24.

de Loë, R. C. (2015). Water governance in Canada: Challenges and opportunities. In B. Mitchell (Ed.), *Resource and environmental management in Canada: Addressing conflict and uncertainty* (pp. 345–369). Toronto: Oxford University Press.

Dinar, A. (1998). Water policy reforms: Information needs and implementation obstacles. *Water Policy, 1*, 367–382.

du Québec, G. (2002). *L'Eau. La Vie. L'Avenir. Politique Nationale De L'Eau*. Québec: Gouvernement du Québec.

Edelenbos, J., Bressers, N., & Scholten, P. (Eds.). (2013). *Water governance as connective capacity*. Surrey: Ashgate Publishing Limited.

Eggertson, L. (2008). Investigative report: 1766 boil-water advisories now in place across Canada. *Canadian Medical Association Journal, 180*(2), 1261–1263.

Environment Canada. (1987). *Federal water policy*. Ottawa: Environment Canada.

Environment Canada. (2015). *Federal policy and legislation*. https://www.ec.gc.ca/eau-water/default.asp?lang=En&n=E05A7F81-1#Section3. Accessed 6 Oct 2015.

Fausti, S. W. (2015). The causes and unintended consequences of a paradigm shift in corn production practices. *Environmental Science and Engineering, 52*, 41–50.

Francoeur, L.-G. (2006). La politique nationale de l'eau du Québec de 2002: une œuvre inachevée. In A. Brun & F. Lasserre (Eds.), *Politiques de l'eau: Grand principes et réalités locales* (pp. 47–68). Québec: Les Presse de l'Université du Québec.

Garrick, D., De Stefano, L., Fung, F., Pittock, J., Schlager, E., New, M., & Connell, D. (2013). Managing hydroclimatic risks in federal rivers: A diagnostic assessment. *Philosophical Transactions of the Royal Society A: Mathematical, Physical & Engineering Sciences, 371*, 20120415.

Gavrilescu, M., Demnerová, K., Aamand, J., Agathos, S., & Fava, F. (2015). Emerging pollutants in the environment: Present and future challenges in biomonitoring, ecological risks and bioremediation. *New Biotechnology, 32*(1), 147–156.

Gentès, P. (2007, December 5). *Biologist*. Personal communication with L. Arsenault regarding Direction des politiques de l'eau. Unpublished.

Gillon, S. (2012). United States biofuel production as climate policy: Tensions between greenhouse gas reduction, agricultural economies, and agro-ecological practice. *Norteamérica, 7*(SPE), 129–164.

Gober, P., Larson, K. L., Quay, R., Polsky, C., Chang, H., & Shandas, V. (2013). Why land planners and water managers don't talk to one another and why they should! *Society and Natural Resources, 26*, 356–364.

Government of Australia, Department of the Environment, Water, Heritage and the Arts. (2007). *Water resources: Government action*. http://www.environment.gov.au/water/action/index.html. Accessed 4 Dec 2007. Unpublished.

Government of the Northwest Territories. (2010). *Northern voices, Northern waters: NWT water stewardship strategy*. Yellowknife: Environment and Natural Resources.

Henderson, D., & Parker, N. R. (2012). *The blue economy: Risks and opportunities in addressing the global water crisis*. Toronto: Blue Economy Initiative.

Hewson, S. (2014, October 23). Drinking water in First Nations communities: Has Canada breached its fiduciary duty? *McGill Journal of Law and Health* (MJLH Online). http://mjlh.mcgill.ca/blog.php?blog_id=126

Hinkel, J., Cox, M. E., Schlüter, M., Binder, C. R., & Falk, T. (2015). A diagnostic procedure for applying the social-ecological systems framework in diverse cases. *Ecology and Society, 20*(1), 32–44.

IBM. (2009). *Water: A global innovation outlook report*. Armonk: IBM.

Ingram, H. M., Mann, D. E., Weatherford, G. D., & Cortner, H. J. (1984). Guidelines for improved institutional analysis in water resources planning. *Water Resources Research, 20*(3), 323–334.

International Joint Commission. (2014). *A balanced diet for Lake Erie: Reducing phosphorous loadings and harmful algal blooms*. Ottawa: International Joint Commission.

Ison, R. (2010). *Systems practice: How to act in a climate-change world*. London: Springer.

Kramer, A., & Pahl-Wostl, C. (2014). The global policy network behind integrated water resources management: Is it an effective norm diffusor? *Ecology and Society, 19*(4), 11–34.

Lagacé, E. (2010). *Working together: What does Canada have to gain?* Toronto: Walter and Duncan Gordon Foundation.

Lambooy, T. (2011). Corporate social responsibility: Sustainable water use. *Journal of Cleaner Production, 19*, 852–866.

Linton, J., & Hall, N. (2013). The Great Lakes: A model of transboundary cooperation. In E. S. Norman, A. Cohen, & K. Bakker (Eds.), *Water without borders: Canada, the United States, and shared waters* (pp. 221–243). Toronto: University of Toronto Press.

Lloyd, T. (1969). A water resources policy for Canada. In M. J. Chambers & J. G. Nelson (Eds.), *Water* (Process and Method in Canadian Geography, pp. 278–293). Toronto: Methuen.

Luyet, V., Rousseau, A., Schlaepf, R., & Villeneuve, J.-P. (2005). Gouvernance Participative dans la gestion integrée de l'eau par bassin versant au Québec: Etat des lieux et réflexions. *VECTEUR Environnement, 38*(6), 36–49.

MacDonald, D. (2007). *Business and environmental politics in Canada*. Peterborough: Broadview Press.

Mascarenhas, M. J. (2007). Where the waters divide: First Nations, tainted water and environmental justice in Canada. *Local Environment, 12*(6), 565–577.

Mitchell, B. (2005). Integrated water resource management, institutional arrangements, and land-use planning. *Environment and Planning A, 37*(8), 1335–1352.

Mitchell, B. (2006). IWRM in practice: Lessons from Canadian experiences. *Journal of Contemporary Water Research and Education, 135*(1), 51–55.

Molle, F. (2006). *Planning and managing water resources at the river-basin level: Emergence and evolution of a concept* (Comprehensive Assessment Research Report). Colombo: International Water Management Institute.

Mollinga, P. P., Meinzen-Dick, R. S., & Merrey, D. J. (2007). Politics, plurality and problemsheds: A strategic approach for reform of agricultural water resources management. *Development Policy Review, 25*(6), 699–719.

Morris, T. J., Boyd, D. R., Brandes, O. M., Bruce, J. P., Hudon, M., Lucas, B., Maas, T., Nowlan, L., Pentland, R., & Phare, M. (2007). *Changing the flow: A blueprint for federal action on freshwater*. Toronto: The Gordon Water Group of Concerned Scientists and Citizens.

Mukhtarov, F., & Gerlak, A. K. (2014). Epistemic forms of integrated water resources management: Towards knowledge versatility. *Policy Sciences, 47*(2), 101–120.

Muller, M. (2015). The 'nexus' as a step back towards a more coherent water resource management paradigm. *Water Alternatives, 8*(1), 675–694.

National Oceanic and Atmospheric Administration. (2015, November 10). *Experimental Lake Erie Harmful Algal Bloom Bulletin – Bulletin 27*. Ann Arbor: National Centers for Coastal Ocean Science and Great Lakes Environmental Research Laboratory.

Nikiforuk, A. (2013). *Gutting of fisheries act a 'politically motivated abrogation': Biologists*. October 31, 2014. http://thetyee.ca/Blogs/TheHook/2013/10/31/Fisheries-Act-Gutting/. Unpublished.

Office of the Director of National Intelligence. (2012). *Global water security: Intelligence community assessment* (Vol. ICA 2012-08). Washington, DC: Office of the Director of National Intelligence.

Ostrom, E. (2009). A general framework for analyzing sustainability of social-ecological systems. *Science, 325*(4), 419–422.

Ostrom, E. (2010). Beyond markets and states: Polycentric governance of complex economic systems. *The American Economic Review, 100*(3), 641–672.

Pearse, P. H., Bertrand, F., & MacLaren, J. W. (1985). *Currents of change: Final report, inquiry on Federal Water Policy*. Ottawa: Inquiry on Federal Water Policy.

Peters, B. G. (1998). Managing horizontal government: The politics of co-ordination. *Public Administration, 76*, 295–311.

Phare, M. A. (2009). *Denying the source: The crisis of First Nations water rights*. Surrey: Rocky Mountain Books.

Plummer, R., de Grosbois, D., Armitage, D., & de Loë, R. C. (2013). An integrative assessment of water vulnerability in First Nation communities in Southern Ontario, Canada. *Global Environmental Change, 23*, 749–763.

Pollution Probe. (2008). *A new approach to water management in Canada: Vision and strategy*. Toronto: Pollution Probe.

Renzetti, S., Dupont, D., & Wood, C. (2011). *Running through our fingers: How Canada fails to capture the value of its top asset*. Toronto: The Blue Economy Initiative.

Sandford, R. W. (2015). *Storm warning*. Victoria: Rocky Mountain Books.

Smith, D. R., King, K. W., & Williams, M. R. (2015). What is causing the harmful algal blooms in Lake Erie? *Journal of Soil and Water Conservation, 70*(2), 27A–29A.

Swain, H., Louttit, S., & Hrudey, S. (2006). *Report of the expert panel on safe drinking water for First Nations*. Ottawa: Minister of Indian Affairs and Northern Development and Federal Interlocutor for Métis and Non-Status Indians.

The Council of the Federation. (2010, August). *Water Charter*. Council of the Federation, Ottawa.

Tisdell, J., Ward, J., & Grudzinski, T. (2002). *The development of water reform in Australia*. Cooperative Research Centre for Catchment Hydrology, Griffith University.

Ulrich, W. (1993). Some difficulties of ecological thinking, considered from a critical systems perspective: A plea for critical holism. *Systems Practice, 6*(6), 583–611.

United Nations World Water Assessment Programme. (2009). *The United Nations world water development report 3: Water in a changing world*. Paris/London: UNESCO/Earthscan.

United States Department of Agriculture. (2007). *Corn harvested for grain – Changes in acreage: 2002 to 2007*. Census of Agriculture 2002. http://www.agcensus.usda.gov/Publications/2007/Online_Highlights/Ag_Atlas_Maps/Crops_and_Plants/Field_Crops_Harvested/07-M166-RGBDot2-largetext.pdf. Accessed 17 Aug 2015.

Watson, N., & Howe, J. (2006). Implementing the EU water framework directive: Experiences of participatory planning in the Ribble Basin, north west England. *Water International, 31*(4), 472–487.

World Commission on Environment and Development. (1987). *Our common future*. London: Oxford University Press.

World Economic Forum. (2013). *Global risks 2013* (8th ed.). Geneva: World Economic Forum.

Young, O. R., King, L. A., & Schroeder, H. (Eds.). (2008). *Institutions and environmental change: Principal findings, applications, and research frontiers*. Cambridge: The MIT Press.

Chapter 14
Managing the Fraser River Basin

David Marshall, Steve Litke, and Theresa Fresco

Abstract The Fraser River Basin is British Columbia's largest watershed (240,000 km^2) boasting a wide diversity of plants, wildlife and natural resources, and home to 2.7 million people. From its headwaters to the Fraser River Estuary, the Fraser River Basin is a place where people of many cultures, languages and religions thrive. Yet, there are social, economic and environmental challenges that confront communities of the Basin. These include serious impacts from climate change, air and water pollution, multiple demands on natural resources, strains on ecosystems, downturns in local economies, changing demographics and a growing demand for public services. How well people manage these issues over time will depend on a common understanding of – and commitment to – sustainability, that being 'social well being supported by a vibrant economy and sustained by a healthy environment'. Drawing on over 20 years of experience working on these issues in the Fraser River system this chapter will focus on the evolution of river basin management in Canada ranging from traditional river basin management approaches to more comprehensive methodologies. The significance of recognizing that comprehensive river basin management extends well beyond water-related considerations to include other environmental media such as air shed and land use matters as well as economic, institutional, legal, social and political aspects will be revealed. Examples of this evolution and the translation of the theoretical goal of sustainability into tangible realities will be used to demonstrate the value and contributions to improved policy and decision-making. The importance of collaborative governance processes will be highlighted as well as the usefulness of measuring performance through the development and application of sustainability indicators. Links to recent court decisions will also demonstrate the emerging significance of the accommodation of Aboriginal interests in this new approach to river basin management. Lessons learned from the Fraser River Basin experience will be made to provide guidance to the management of river basins elsewhere.

D. Marshall (✉) • S. Litke • T. Fresco
Fraser Basin Council, 470 Granville Street, Vancouver, BC V6C 1 V5, Canada
e-mail: dmarshall@fraserbasin.bc.ca; slitke@fraserbasin.bc.ca; tfrescro@fraserbasin.bc.ca

S. Renzetti, D.P. Dupont (eds.), *Water Policy and Governance in Canada*,
Global Issues in Water Policy 17, DOI 10.1007/978-3-319-42806-2_14

14.1 Context and Background

Canada is blessed with a significant abundance of freshwater water resources. It has 7% of the world's renewable supply of freshwater and 20% of the world's total freshwater resources (including waters captured in glaciers and the polar ice cap).

It is important to remember that all of this water is interconnected. The area that drains all precipitation received as a runoff or base flow (groundwater sources) into a particular river or set of rivers is called a drainage basin or watershed. A number of small watersheds combine to make up regional watersheds or river basins.

Rivers are of immense importance to the geology, biology, history and culture of Canada. Although they contain only a small portion of the total amount of water in the country, at any given time rivers play a vital role in the hydrological cycle. They act as drainage channels for surface water; provide habitat, nourishment and means of transport to countless organisms; offer travel routes for exploration, commerce and recreation; leave valuable deposits of sediments; provide the power to produce electrical energy; and create the majestic scenery that is such an integral part of the Canadian landscape.

Flowing water has provided Canada with more than inspiration. For instance, long before settlers arrived in Canada, rivers were central to First Nations for food, medicines and transportation as well as in cultural and spiritual ceremonies. Rivers have always been central to the political and economic development of Canada. During the period of European colonization, rivers carried furs, trade goods, and explorers, heralding the influx of settlers into the wilderness. For example, the arrangement of streams and rivers flowing into Hudson Bay and into the Mackenzie and St. Lawrence Rivers permitted canoes to travel west and north across the length and breadth of the land that became Canada.

Nationally, Canada possesses 23 river basins of varying sizes and descriptions. These river basins exhibit varying degrees of development, experience a plethora of issues such as too little water to too much water, and/or pristine to polluted water quality. Due to the somewhat autonomous provincial responsibility for managing inland water resources represented by these river basins, a number of unique responses have developed. It is from these unique responses that the management of British Columbia's Fraser River Basin has emerged.

14.2 The Fraser River Basin

The Fraser River Basin is the fifth largest river basin in Canada, spanning an area of 238,000 km^2—an area comparable to the state of California. The Fraser River is 1399 km in length, flowing through diverse terrain, starting in the Rocky Mountains along the eastern border of British Columbia, traversing through vast forested areas, plateaus, canyons, rolling uplands, wetland and estuaries, finally discharging into the Strait of Georgia (Fig. 14.1). Aboriginal peoples have lived in the Basin for more

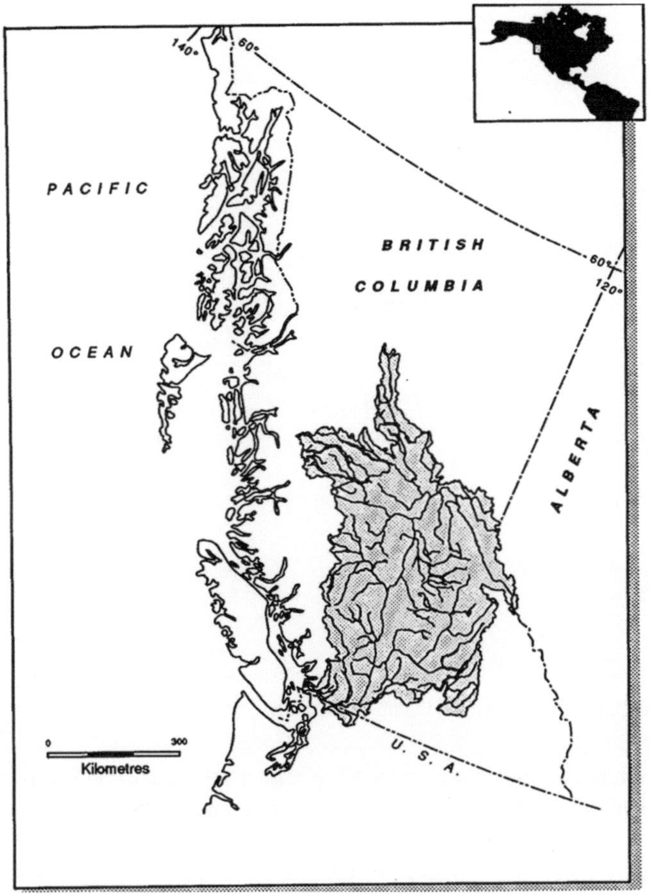

Fig. 14.1 The Fraser River Basin, British Columbia

than 10,000 years, and their populations likely never exceeded 50,000. Since Simon Fraser's historic trip down the Fraser in 1808, the population in the Basin increased significantly in the nineteenth century and exploded in the past 100 years, particularly after World War II. The population in the Basin today is over 2.7 million. The current use of the Fraser River varies along its length, with forestry, pulp and paper, fishing and recreational uses dominating the sparsely populated upstream region, and industrial, trade and municipal uses becoming more dominant in the more densely populated downstream region of Greater Vancouver (Fig. 14.2).

The Fraser Basin is one of the greatest salmon-producing river systems in the world, and contains a wide diversity of fish and waterfowl species. As Canada's largest producer of salmon, over 60 % of the sockeye, 60 % of the pink and 16 % of the Chinook salmon catches are from the Fraser. An average annual return of $300 million is realized from commercial, recreational and Aboriginal fisheries. Having

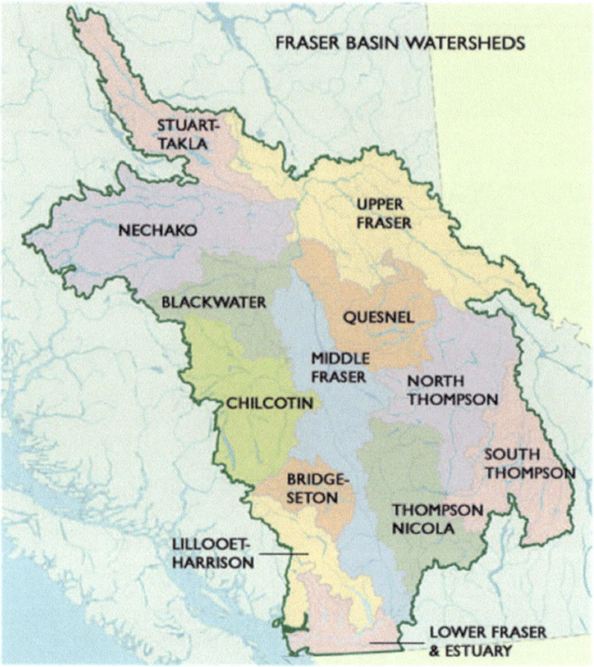

Fig. 14.2 The Fraser River Basin's 13 sub-watersheds

fished on the banks of the Fraser since time immemorial, salmon is also of particular cultural and spiritual significance to Aboriginal communities.

The Fraser River estuary also supports the highest density of wintering waterfowl, shorebirds and birds of prey in Canada. In a typical year, about 1.5 million birds (150 species) use the estuary and its adjacent upland, especially between October and December for birds migrating from eastern Siberia and western North America to South America.

It has been estimated that about 512 species of birds, mammals, fish, reptiles and amphibians reside within the Fraser River Basin. About one in ten vertebrate species that live in the Basin is "red-listed", meaning they have exhibited significant declines in population abundance. Of the 111 "red-listed" vertebrates in B.C., about 42 % of these are present in the Fraser Basin, illustrating the importance of this Basin in supporting biodiversity in British Columbia.

Over the past century, the Fraser River has been degraded by a wide variety of pollutants including toxic chemicals from sawmills, pulp and paper production, mining and other industries, municipal sewage, agricultural wastes and, increasingly, urban run-off. During the 1980s, the increasing pace of industrial developments, rapid urbanization and over-fishing throughout the Fraser Basin were raising environmental alarm bells. It was becoming clear to governments and the general public that the overall health of the River was in jeopardy.

Although serious water resource management problems have been created by development in the Fraser River Basin, it is important to see these in perspective of the broader context of the watershed. There are still many undeveloped headwaters with pristine water quality conditions. The large size of the Basin, the enormous main stem flows, and the concentration of development pressures close to the main stem and particularly in the downstream sub-basin have diluted the impacts. The main stem of the River has not yet been dammed, unlike the Columbia River. And overall, the main stem of the Fraser River has fairly good water quality conditions when compared with systems such as the St. Lawrence River and the Great Lakes, which have been subject to longer and more intensive development and have become much more seriously polluted.

14.3 Management of the Fraser River Basin

14.3.1 Water Governance

Because Canada is a federation, different orders of government have different jurisdictional roles related to water management, with many areas of shared commitment. Canadian provinces and territories have primary jurisdiction over most areas of water management and protection. Most provincial governments delegate certain authorities to municipalities, especially drinking water treatment and distribution, and wastewater treatment operations. Provinces may also delegate some water resource management functions to local authorities that may be responsible for a particular area or river basin. Most major uses of water in Canada are permitted or licensed under provincial water management authorities.

Water resource management in British Columbia (BC) involves numerous agencies and organizations and includes all orders of government. At the federal level, Environment Canada (EC) and the Department of Fisheries and Oceans (DFO) are the primary agencies responsible for water resource management. In relation to water management issues, EC manages water pollutants through such efforts as compiling water quality and quantity data, and by setting national water quality guidelines, while DFO is primarily concerned with the management of anadromous fish stocks, such as salmon, and their habitat through such instruments as stock enhancement plans and habitat protection programs.

14.3.2 River Basin Management

A complex web of legislation and regulatory tools guide the management of water resources in BC, and thus in the Fraser River Basin. For a number of years, and particularly in the last three decades, there have been continuing and substantial innovations in the governance systems for managing water and associated resources in British Columbia. From the outset, the Fraser River Basin has been a major focus of those innovations. The original water rights system was designed in response to the conflicts in its tributary streams that arose during the early gold rush. The Fraser River Estuary Management Program (FREMP), started in the 1970s, was at the forefront of attempts to manage water resources in an integrated and collaborative fashion. However, there was no comprehensive plan to focus on water resource management and broader sustainability issues in the entire Fraser River system.

The Fraser River Basin Action Plan (FRAP) was first announced as part of Canada's *Green Plan* in the early 1990s. This effort was the third time in the history of the Fraser that an attempt was made to develop a comprehensive plan for the Basin. The first attempt to develop a comprehensive plan for the Basin was begun by the Dominion-Provincial Board in the early 1950s and was a response to the disastrous flood of 1948. In many ways, it was ahead of its time. After encountering tremendous difficulties in generating information about the natural resources of the Basin and their development, a summary was published in an interim report and the comprehensive initiative then faded as attention narrowed to focus on flood control issues.

Thirty years later, a second attempt was made to develop a comprehensive plan for the Basin. In the late 1970s, in response to increasing conflicts in the Province's watersheds and the weaknesses of the earlier federal-provincial comprehensive river basin studies, such as the Okanagan Study, the province initiated a Strategic Planning program.

It began with a prototype investigation in the Nicola watershed and envisaged that initial Strategic Plans for each of 41 planning units in the Province would be completed at a rate of eight plans each year, and then progressively refined. Once again, an ambitious initiative that could have produced a set of plans for the entire Fraser Basin did not come to fruition. By 1989, only 16 plans had been completed and most of them fell short of the comprehensiveness that had been envisaged. Again, the task was found to be daunting, even when approached strategically and with a great deal more data available than when the Dominion-Provincial Board had struggled with its studies in the early 1950s.

Eventually, a comprehensive assessment of the Fraser River Basin's sustainability was completed in 1991 by UBC's Wastewater Research Centre (Dorcey 1991). The recommendations arrived at by this study formed the foundations of the Fraser Basin Management Board initiated in May 1992, which subsequently became the Fraser Basin Council Society in June 1997.

14.3.3 Fraser River Estuary Management Program

The first attempt by governments to take proper steps to address Basin concerns focused on the River's precious and ecological significant estuary. The Fraser River Estuary Management Program began in the late 1970s during an era of ambitious comprehensive planning schemes, and continued until 2012. Coastal zone management had joined river basin planning as a major preoccupation of both government and non-government organizations that advocated a break from narrow, compartmentalized resource management. The deterioration of water quality and the increasing incidence of multiple use conflict fixed public attention on this, the most heavily populated and industrialized estuarine region along Canada's west coast.

Concerned with a good deal more than salmon, FREMP comprised analytical and regulatory efforts to protect fish as well as wildlife habitat through an explicit process of multi-agency, multi-government planning. Thus, it represented a clear attempt to forge consensus on issues central to sustainable development.

Protection and enhancement of salmon stocks and associated habitat were among the leading public concerns motivating the creation of FREMP, and with DFO as a signatory to the formative agreement, such concerns remained high in program activities. Single agencies continued to look after their own mandate whether that was DFO's for fish stocks or the Harbour Commissions' for port management: FREMP played a role only where the regulated activities begin to affect one another.

14.3.4 Fraser River Action Plan

In 1990, the Government of Canada identified the Fraser River Basin along with the St Lawrence River and the Great Lakes as three major freshwater systems requiring priority action under its new *Green Plan (*Canada's first formal national environmental agenda). The Fraser River system was identified primarily due to recent declines in salmon stocks. Under the *Green Plan*, the Fraser River Action Plan (FRAP) was established. Jointly sponsored by Environment Canada and the Department of Fisheries and Oceans, FRAP was modeled on the successful Fraser River Estuary Management Program that focused on improving inter-agency cooperation on estuary management. FRAP set targets for restoring the environmental health of the watershed. The Program focused on ecosystems within the entire watershed, encouraged collective stewardship and cooperative partnerships, and involved the public so that British Columbians could better understand how their actions could harm or improve watershed health. FRAP had three distinct components: building partnerships with all levels of government, industry, communities, and Aboriginal groups in the Basin, cleaning up and preventing pollution in the Basin, and restoring the productivity of the natural environment.

14.3.5 Fraser Basin Management Program

In developing its *Green Plan*, the federal government had concluded that building partnerships would be critical to its success, thus funding from the *Green Plan* was combined with funds provided from the provincial and local governments to create the Fraser Basin Management Program (FBMP). An agreement was reached on May 26, 1992 to initiate the FBMP, to be led by a multi-governmental and non-governmental, multi-interest committee, called the Fraser Basin Management Board (FBMB). The FBMB included representation from the federal, provincial, and local governments, First Nations and seven members at large including an impartial chair.

This Board was required to make decisions by consensus and to encourage consensus-based decision making in all basin activities, as well as to facilitate the development of locally based decision-making processes. No formal basin wide organization existed prior to the FBMB. The Board was given five years to accomplish its goal. The FBMB set a path, with much public involvement, toward a very different kind of collaborative governance structure dedicated to advancing sustainability of the Fraser Basin. At the end of the five years in 1997, a new institution was established—the Fraser Basin Council.

One of the Board's priorities was to promote a healthy and sustainable river basin. It did this by introducing a project designed to check up on the Basin's health by monitoring its vital signs. Monitoring efforts included looking at specifics like groundwater quality, pulp mill effluents, employment, education levels and economic diversification. This monitoring required developing new ways of gathering information that would indicate progress—or lack thereof—towards sustainability of the Basin.

In 1995, the Board released its *State of the Fraser Basin* report, a compendium of eight reports that addressed key sustainability issues in the Basin. The report was a first attempt to answer the questions of basin residents about whether progress was being made in achieving sustainability.

In conjunction with the *State of the Fraser Basin* report, the Board released its first *FBMP Board Report Card* in 1995 and a second in 1996. These *Report Cards* graded overall progress on some of the more critical issues facing the Basin. The 1996 *Report Card* showed a slight increase in the Basin's progress towards a more sustainable future. The state of sustainability reports became a precursor to the highly visible and comprehensive State of the Fraser Basin reports produced by the Fraser Basin Council from 2000 to 2009.

One of the Board's key deliverables in its 5-year mandate was a strategic plan for sustainability of the Fraser Basin. This plan, named the *Charter for Sustainability* of the Fraser Basin, was published on 27 February 1997. The *Charter* is a prescription developed by, and for, the people who live, work and play in the Fraser Basin. It was conceived and confirmed as a "good faith" document and as a long-term plan to sustain the Fraser Basin. The *Charter for Sustainability* continues as the guiding set of principles for the Basin and all activities of the Fraser Basin Council.

These principles are based on an agreed upon vision: The Fraser Basin is a place where social well-being is supported by a vibrant economy and sustained by a healthy environment.

14.3.6 Fraser Basin Council Society

The Fraser Basin Management Board expected that a significant portion of the *Charter* would be self-implementing through individual and institutional actions. The Board strongly believed there was a need for an organization that was independent of government to oversee implementation of the *Charter*. It also recognized, however, that government, as well as the private sector, needed to be involved and, if possible, be part of the management structure of the organization. After examining a number of models, the Board recommended one similar in composition to the former board that would be funded by federal, provincial and local governments, as well as other sources of funding through not-for-profit status. On 3 June 1997, the Fraser Basin Council Society was established as a unique collaborative governance model that would include government, the private sector and civil society.

The mandate of the FBC is to bring people together to solve complex, multi-jurisdictional issues in the Fraser Basin, to take advantage of opportunities, and to strengthen the capacity of institutions and individuals to deal with emerging issues that threaten the overall sustainability of the Basin. Thus, the evolution of management efforts for the Fraser River Basin may be thought of as a journey from top down management towards fully inclusive, collaborative planning and decision-making processes.

14.4 Fraser Basin Council

The operational arm of the Fraser Basin Council Society is the Fraser Basin Council (FBC). As a custodian of the strategic plan for the Basin, the FBC has a mandate of *'educating on the need for sustainability'* (purpose statement in FBC's constitution) and thus, oversees implementation of the *Charter for Sustainability*. The Council is unique because it represents an innovative collaborative governance approach that cuts across traditional adversarial lines and differing perspectives to promote social well-being or quality of life. It is a non-governmental organization informed by governments, the private sector and civil society. While governments (i.e. federal, provincial, local and First Nations) are part of the Council's governance structure they also remain at "arms length" from the Council's management structure (Fraser Basin Council 2004).

The FBC was initially structured around 36 directors: three from each of the federal and provincial governments, one from each of the eight regional districts (i.e. local governments) in the Basin, representatives from each of the eight First Nations language groups in the Basin, ten non-government representatives from each of the five geographic regions of the Basin and four Basin wide Directors including the impartial Chair that provide environmental, social and economic perspectives. In 2009, a revision of the Council's by-laws enabled the Council to add two limited term directors, one that would represent a youth perspective (i.e. under

30 years of age) and one with significant connection's with BC's financial community.

From 1997 to 2009, FBC focused its work on the Fraser River system. During the lead up and development of the Council's second Five Year Strategic Plan (2011–2016) and the focus on three strategic programs, Climate Change and Air Quality, Water Resources and Watershed Management, and Sustainable and Resilient Communities and Regions, FBC expanded its mandate to be province wide with a focus on the Fraser Basin. The rationale for this decision was that activities and projects within each of these three strategic programs were relevant and beneficial to the Province as well as to the Fraser River Basin. As such, the work of the Council transformed and evolved to a more comprehensive approach to river basin management by moving beyond the conventional approach and extending well beyond water-related considerations to include other environmental media such as air shed and land use matters as well as economic, institutional, legal, social and political aspects of river basin management.

14.4.1 Fraser Basin Council Programs

Based on the priorities identified by the Board and its committees, the Fraser Basin Council becomes involved as a conflict resolution agent, catalyst and facilitator in specific issues of concern. It also provides secretariat, interim project management and coordinator services for many projects, and has initiated programs to educate the public, government, business and professional organizations on many aspects of sustainability. In all its activities, the FBC is first and foremost an advocate for sustainability. Basin wide and regional programs illustrate the breadth of issues that the FBC has undertaken and the degree of collaboration it has developed throughout the Basin.

Six of the more successful projects of the Council are as follows. Success constitutes effective problem solving, extent and breadth of communities and individuals influenced, local engagement, transparency and performance measurement techniques.

14.4.1.1 Britannia Mine Remediation

In 1998 the FBC helped bring public attention to a serious pollution problem at the old Britannia Mine site, located partway between Vancouver and Whistler.

An active copper mine from 1902 to 1974, Britannia Mine became the largest single point source metal contamination problem in North America. Acid Rock Drainage from the old mine had severely damaged marine life in Howe Sound. Contaminated water, soil and sediments at the site were also a significant obstacle to the long-term economic and social development in the community.

FBC invited all orders of government, private sector bodies and the community to come together on a solution. As a result of that early work, later negotiations among the parties, and new contaminated sites legislation, progress was made. A settlement agreement was reached between the Province and former mine owners.

A new water treatment facility was built in 2005, and the results have been substantial. Today, Britannia Beach is a vital, healthy community. And by 2011 Pink salmon were once again beginning to appear in the area — something hardly imaginable a decade ago!

14.4.1.2 Fraser Salmon and Watersheds

Nothing epitomizes the spirit of British Columbia quite like wild Pacific Salmon. Yet managing wild salmon and watersheds sustainably is a daunting challenge. Many salmon stocks are at risk, or in decline from such threats as habitat loss, overfishing and climate change impacts. To ensure wild salmon in our future, policies and practices need to change. No longer can a Fisheries and Oceans Canada manage this critical species alone. Their survival requires a collaborative effort of both fishers and regulators working in tandem.

The Fraser Salmon and Watersheds Program (FSWP) was a breakthrough initiative. Co-managed by the Pacific Salmon Foundation (PSF) and the Fraser Basin Council between 2006 and 2012, FSWP brought together British Columbians from different regions, sectors and areas of interest to improve the health and sustainability of wild Pacific salmon populations and the watersheds of the Fraser Basin.

Thanks to the BC Living Rivers Trust Fund, Fisheries and Oceans Canada and other funders, FSWP provided $13.6 million in funding for 300 projects across the Basin, in four focus areas:

- Watershed planning and governance
- Habitat restoration and stewardship
- Sustainable fisheries management
- Education and engagement.

14.4.1.3 Flood Hazard Management

Two devastating floods inundated the lower Fraser Valley and other parts of the Fraser Basin in 1894 and 1948. Scientists predict that there is greater than one-in-three chance that a flood of similar proportions will occur within the next 60 years. Such a catastrophic flood coupled with storm surges and anticipated sea level rise would result in severe social, economic, and environmental impacts. These include risk of injury and loss of life; billions of dollars in damage to private and public property; temporary loss of infrastructure and community services including communication and transportation; disruption of business and trade; and numerous impacts to water quality, as well as fish and wildlife habitat.

Over the years, FBC has worked with the province and local governments to develop a range of essential flood hazard management tools including digital flood hazard reference maps, a GPS survey of flood protection dykes and a hydraulic model to predict water elevations associated with the flood of record and other flood flows. These tools assist all orders of government and other agencies to better understand, plan for and mitigate potential flood hazards.

In 2014, FBC launched Phase 1 of the Lower Mainland Flood Management Strategy to prepare all lower mainland private and public sector interests for the next great lower mainland flood. This initiative has already secured unprecedented financial contributions from 44 public and private sector funding partners. Phase 1 involves a review of a range of Fraser River and coastal flood scenarios that may threaten the Lower Mainland as well as a regional analysis of flood vulnerabilities, protective works, policies and programs.

14.4.1.4 Sustainable Communities

In 2002, the FBC established "Strengthening Communities" as a priority program. Strong and economically diversified communities are essential for the sustainability of the Fraser Basin. In BC, many rural communities face challenging economic transitions as traditional forestry, mining and fishing industries undergo profound restructuring. The FBC and its partners such as the Canadian Rural Partnership (CRP) worked together to develop an action plan to address rural concerns; improve the delivery of services to rural communities and strengthen community capacity.

Emanating from the Strengthening Communities program came the Smart Planning for Communities (SPC) program, a collaborative initiative supported by a number of organizations to assist local and First Nations communities in addressing their long-term sustainability challenges by providing resources and tools for planning socially, culturally, economically and environmentally sustainable communities.

SPC offers practical information and advice on integrated community sustainability planning from expert sustainability facilitators. SPC helps to build mutually beneficial partnerships and develop an information sharing and resource network. The result is stronger, more vibrant and sustainable communities. SPC has been developed in response to the call for 'Integrated Community Sustainability Planning' (ICSP) in the 2005 Federal/Provincial Gas Tax Agreement. It recognizes that a flexible approach to integrated sustainability planning is needed in BC – an approach that allows communities to build on, enhance and integrate existing planning processes while also exploring innovative tools and frameworks.

14.4.1.5 Sustainability Indicators

Sustainability indicators are select data that help reflect social, economic and environmental health and that are tracked over time with a view to ascertaining trends and the state of sustainability.

The Fraser Basin Council began an indicators program in 2000, with a primary emphasis on sustainability of the Fraser Basin. The goals of the Program are to:

- increase public awareness and understanding about sustainability,
- identify critical issues and responses to improve progress,
- inform decisions and influence actions, and
- advance sustainability

FBC has published a series of Sustainability Snapshot reports on social, economic and environmental indicators — including Aboriginal/Non-Aboriginal Relations, Climate Change, Air Quality, Agriculture, Energy, Education, and more. Each report is supplemented by case examples and opportunities for action. The reports are intended to raise awareness about the state of sustainability for diverse audiences in the Fraser Basin and BC.

14.4.1.6 Aboriginal Engagement

The Fraser Basin Council's predecessor organization, the FBMB, cultivated an excellent working relationship with the eight First Nations language groups within the Fraser River Basin. This relationship was instrumental in the development of the Charter for Sustainability. In fact, the Charter's 11th principle of sustainability:

> Aboriginal Rights and Title: We recognize that aboriginal nations within the Fraser Basin assert rights and title. These rights and title now being defined must be acknowledged and reconciled in a just and fair manner

was strongly put forward by the three First Nations members of the FBMB who emphasized that First Nations assertion of their rights and title was a key element of a sustainable river basin. As such, the Charter became the first document in British Columbia, signed by two senior provincial government officials, to recognize the importance of 'Aboriginal rights and title' in decision-making processes. Principle 11 became vital in the development of the Council's policy on Aboriginal engagement. This policy states clearly that FBC must take into account aboriginal considerations in all of its programs and activities.

This acknowledgement of First Nations 'rights and title' by the FBMB in 1997 preceded their formal and legal basis by 17 years when on June 26, 2014, the Supreme Court of Canada unanimously declared that the Tsilhqot'in Nation have Aboriginal Title to approximately 1700 km^2 of land southwest of Williams Lake, BC. This court decision, the first in Canadian history to formally recognize Aboriginal Title, constituted a victory for the Tsilhqot'in people, who have been fighting for control over their traditional lands for 150 years.

The Supreme Court of Canada affirmed the test for Aboriginal title as set out in *Delgamuukw* v *British Columbia* that requires an Aboriginal group to show the land was exclusively occupied prior to sovereignty. This is an important change from the Court of Appeal decision that held that an Aboriginal group must demonstrate that its ancestors intensively used a definite tract of land with reasonably defined boundaries at the time of European sovereignty. The Supreme Court rejected the "postage-stamp" theory and held that Aboriginal Title is not confined to specific sites of settlement but also includes broad territorial tracts of land that were regularly used for hunting, fishing or otherwise exploiting resources.

As recognized holders of Aboriginal title, the Tsilhqot'in Nation now has the right to decide how Aboriginal title lands will be used; the right of enjoyment and occupancy of the land; the right to possess the land and the right to the economic benefits of the land. The unanimous Court specifically set out that this is not merely a right to first refusal but the right to proactively control and manage the land.

14.5 Collaborative Watershed Governance in BC

Since its inception the Council has been influential in advancing collaborative watershed governance (CWG) in the Basin and throughout BC. The Council also has a long history of facilitating, coordinating and supporting collaborative watershed governance activities on the ground. FBC has collated this experience and knowledge to add to the conversation of how collaborative watershed governance approaches can be implemented to fulfill the provisions for alternative governance arrangements, which may be enabled through British Columbia's new Water Sustainability Act.

The Council defines watershed governance as the range of political, organizational and administrative processes through which decision-makers are chosen, citizens and various groups articulate their interests, decisions are made and implemented, and decision-makers are held to account for their decisions. With the Royal Assent of the Water Sustainability Act in 2014 and its legislative proposal to enable 'alternative governance arrangements', the conversation about new and innovative forms of governance and shared decision-making is timely and marks an exciting opportunity.

Historic approaches to managing natural resources (i.e. provincial decision-making authority distributed across multiple, narrow organizational silos), has not been effective in protecting the health and sustainability of watersheds. In some cases, fragmented or uncoordinated decisions regarding land, water and other natural resources, along with intensified resource development, has led to adverse impacts on watershed ecosystems and the communities that depend on them. In some cases, community-led, collaborative, watershed initiatives have emerged in response to the limitations of this historic approach to governance. A rise in these types of initiatives has been observed since the 1990s (Brandes and O'Riordan 2014), demonstrating an increasing desire among watershed-based groups to be involved and take on a more significant role in governance and decision-making.

The growing numbers of watershed initiatives and organizations also raises the question of what constitutes good watershed governance and the effective performance of collaborative approaches. Though a collaborative and inclusive approach may help to reflect the myriad of issues within a watershed, bringing the right people together is no easy task. Done well, however, this process can lead to efficient, credible and cost effective means of developing integrated and robust solutions to complex problems.

Since the 1990s there have been many drivers of the shift to collaborative watershed governance throughout BC. These include a combination of the following:

- *The decreasing capacity of government* – With shrinking financial and human resources and increasing amount of regulations to manage, the capacity of the Province to fulfill their responsibilities related to water is lessening. To fill the gaps, there has been a rise of collaborative watershed entities across BC.
- *A changing climate* – Climate change is predicted to alter the hydrology of watersheds across the province, including the amount and timing of precipitation and stream flow.
- *Continuing and emerging role of First Nations* – British Columbia has a particularly unique context with respect to First Nations rights and title to their traditional lands and waters. For any collaborative initiative to be successful, First Nations need to be incorporated into the decision-making process.
- *Population growth and development pressures* – Increasing population and urban development also increases competition for water resources and can put stress on watershed ecosystems.
- *Cumulative effects on watersheds and communities* – In many parts of BC the cumulative effects of multiple resource development projects and other human activities are adversely affecting the health of watersheds and the communities that depend upon them.

Collaborative watershed governance is also complex due to a number of factors, including, but not limited to a myriad of legislation, regulations, and distributed jurisdictions; a diversity of different biophysical characteristics, resource uses and critical issues across different watersheds; a range of different levels of capacity and available resources across different governance organizations; and the complex web of human and organizational relationships in communities and across the landscape in BC ranging from trust to mistrust, from conflict to consensus, and from partners to adversaries. Collaborative approaches to governance are therefore largely shaped by their local ecological, socio-economic and political context in addition to overarching provincial, federal and/or global influences. No one size or shape fits all.

Several common success factors in collaborative watershed governance have emerged that reflect different sizes of watersheds, different regions of the province, and different issues within watersheds. These resolve around four key areas of success:

- *Enabling Policy Framework* – This includes the policy aspects of governance such as working within the current legislative, legal and regulatory framework as well as emergent, evolving and/or recommended aspects of policy such as new enabling legislation and evolving roles for existing jurisdictions.
- *Organizational Design* – This includes the formal structure, composition, and procedures of collaborative governance organizations such as their vision, man-

date, membership and terms of reference as agreed to by the members of the organization.
- *Capacity and Resources* – This includes the overall capacity of the governance organization relative to its mandate including the quality and quantity of human, financial and information resources.
- *Other Success Factors* – This includes the more qualitative human elements that are necessary for effective collaborative processes such as leadership, respect, trust and commitment.

14.5.1 Enabling Policy Framework

Enabling legislation for collaborative watershed governance needs to be in place with a clear regulatory framework that speaks to the roles, responsibilities and commitments of provincial governments and other jurisdictions and organizations involved. The clarity of an enabling policy/legislative framework helps to establish external legitimacy (i.e. credibility of those influencing decisions and credibility of outcomes).

Consensus decisions are often reached through a form of negotiation whereby participants enter into discussions articulating their views, interests, positions and preferences for the purpose of finding a mutually acceptable resolution or agreement. With successful collaborative watershed governance initiatives being inclusive of a diversity of perspectives, this participatory decision-making style enables different perspectives to be heard and accommodated. This more broad based agreement can increase legitimacy of decisions as well as support for implementation. In situations where conflict is high, the use of an impartial facilitator is a key to success to build trust between the members of the initiative.

14.5.2 Organizational Design

At the outset of a collaborative process it is important to establish a set of guiding principles and adopt those within a Terms of Reference document or Governance Manual. These principles are helpful and necessary to guide the activities and approaches of these initiatives. Guiding principles are helpful in establishing vision and mission statements and a related governance structure. Examples of guiding principles used by some initiatives include accountability and transparency with respect to communications, decision-making, reporting on performance and financial management. Other commonly used principles are inclusivity and equitable opportunity to participate.

Guiding principles should be based on values representative of the different interests at the table as well as reflective of the issues a watershed entity would like to address. For example, if there are groups in a given watershed that believe that past decisions have not properly consulted those affected, principles of legitimacy, transparency, consensus decision-making, and accountability may be high priority principles to adopt at that table. The principles that guide a governance body should be strong enough that they may be looked to for guidance and stability in times of conflict or debate; they should help steer the organization to a place of common understanding. The consistent application of these principles is likely to increase not only legitimacy but also the quality of decision-making (Brandes and O'Riordan 2014).

14.5.3 Capacity and Resources

Successful collaborative watershed governance mechanisms have established adequate and sustained human, financial and information resources to successfully deliver on their vision and mandates. In these cases, adequate human resources are established within their membership as well as in their technical and administrative support functions. In all cases, a critical foundation for governance is leadership from a community champion or organization. Other necessary human resources include a coordinator and technical and communications expertise. Securing the funds to hire a coordinator position in particular is crucial to advance the work and outcomes of these entities as they provide core administrative, communications and facilitation functions as well as help retain the institutional memory of the initiative. With respect to membership and representation, it is important to include all relevant interests and jurisdictions within the watershed at the outset of the collaborative process. In addition, certain groups that may have barriers to participation such as First Nations and NGO representatives may need the appropriate support such as travel assistance and/or honoraria.

Information resources are also critically important. River basin or watershed management decisions need to be based on best available data, information and traditional ecological knowledge. This includes monitoring data such as water quality, water quantity and use, local and traditional knowledge and citizen science. Though the inclusion of diverse interests can enhance access to these various forms of data and knowledge, it can still be a challenge given there is no comprehensive database on water and watersheds in BC (Brandes and O'Riordan 2014).

14.5.4 Other Success Factors

In addition to the more institutional and organizational keys to success, the human factors influencing collaborative watershed governance are also crucial.

Effective Leadership As referenced earlier, one of the critical foundations of effective collaborative watershed governance observed is the leadership of an individual champion or organizational champion especially through the 'storming and norming' periods of group development (Tuckman 1965). These champions are most effective in bringing relevant interests to the table, building trust between the different interests and driving the collaborative process. As a trusted 'steward of the process', these leaders help establish procedural legitimacy, mediate tensions and conflict and enable groups to explore mutual gains. In the cases where collaborative watershed governance is initiated as a result of an ecological crisis or conflicts between different users, trusted leaders and impartial facilitators can be especially crucial.

Mutual Respect and Interpersonal Trust Mutual respect and interpersonal trust among participants are also necessary for effective collaboration. Therefore, considerable time to build trust and strengthen working relationships is a necessary and productive investment.

Long-Term Commitment to Process and Outcomes It takes time to resolve challenges within watersheds because watersheds are complex ecological and socio-political systems. In many cases, even the best available data doesn't provide an immediate "right" answer and groups need to embrace a longer-term learning curve. This commitment to good process is better achieved when principles of transparency, inclusivity and accountability are incorporated into the activities of the watershed governance initiative.

14.6 A Final Thought…

Managing the Fraser River Basin is an enormous and challenging task. Yet, as a river basin that contributes about 80 % to the Province's economy and about 10 % to Canada's economy, it is essential that this be done in a comprehensive and collaborative manner. The Fraser Basin Council is a unique partnership of public and private interests, assuring broad representation from all sectors of society and emphasizing an integrated approach to realizing social, economic and environmental goals. It provides a non- confrontational, consensus-based forum in which no participant is first among equals, and in which the interests of all parties can be presented.

The FBC is a means of doing business together across insurmountable boundaries and it allows for inclusive, collaborative decision-making. All orders of government were made part of FBC's management structure in an effort to allow for greater influence on government policy and programs than is normally generated through traditional non-governmental organizations. The FBC is designed to achieve its goals by facilitating cooperative and collaborative action throughout the Basin. It has not been established to duplicate or compete with existing governmental and non-governmental mandates and activities. It has, however, and continues to act as a catalyst to avoid duplication and facilitate harmonization and collaboration among diverse basin interests. As long as the Fraser River Basin continues to be 'the heart and soul' of British Columbia, this form of collaborative governance will remain essential.

References

Brandes, O. M., & O'Riordan, J. (2014). A blueprint for watershed governance in British Columbia. In *POLIS Project on ecological governance.* University of Victoria. http://poliswaterproject.org/sites/default/files/POLIS-Blueprint-web.pdf. Accessed 24 Feb 2014.

Dorcey, H. J. (1991). *Perspectives on sustainable development in water management: Towards agreement in the Fraser River Basin* (Vol. 1 & 2). Vancouver: Westwater Research Centre.

Fraser Basin Council. (2004). *River basin research, Managing water resources at the lowest appropriate level.* The World Bank, The Fraser River Basin, British Columbia, Case Study Background Paper.

Tuckman, B. W. (1965). Developmental sequence in small groups. *Psychological Bulletin, 63*(6), 384–399.

Chapter 15
Indigenizing Water Governance in Canada

Lori E.A. Bradford, Nicholas Ovsenek, and Lalita A. Bharadwaj

Abstract Water-related governance challenges confronting Indigenous people are well documented and have significant implications for water governance in Canada. Indigenous people have traditionally had a lack of voice, and little participation or significant representation at higher political points of authority to influence or sanction laws of water protection, regulation and enforcement on traditional territories. Canadian water policy and management decision processes are dominated by Western scientific viewpoints and exclude Indigenous values, norms and conceptions of water governance. Indigenous people's relationship to water is often strongly connected to the spiritual world. Water is thought of as sacred, a sentient being, a gift from, and the life breath of the Creator, and, is fundamental for the wellbeing of the earth and all people. This worldview defines unique socio-cultural relationships with water and informs Indigenous water management and governance processes which are divergent from, and unrepresented within current Canadian water governance frameworks. Indigenous people are responding to water-related governance challenges by voicing their concerns, reclaiming their roles in water governance and calling for adaptation and realignment of current Canadian water policy regimes to include Indigenous water governance processes. This chapter will give historical, social and political context to the water-related governance challenges facing Indigenous peoples of Canada. Water governance challenges will be explored through discussions on Treaty Rights and jurisdictional fragmentation which at times, impedes those rights. We explore how community engagement, participation and empowerment and the duty to consult are challenging for issues of water governance. We find insight in the reclamation of sense of place through water in Indigenous communities. Indigenous socio-cultural relations to water, and traditional strategies of water resource use, monitoring, management and protection will be highlighted drawing on examples across Canada and globally. It will provide a

L.E.A. Bradford • L.A. Bharadwaj (✉)
School of Public Health, University of Saskatchewan, 104 Clinic Place, Saskatoon, SK S7N 5E5, Canada
e-mail: lori.bradford@usask.ca; lalita.bharadwaj@usask.ca

N. Ovsenek
College of Medicine, Department of Anatomy and Cell Biology, University of Saskatchewan, 104 Clinic Place, Saskatoon, SK S7N 5E5, Canada
e-mail: nick.ovsenek@usask.ca

S. Renzetti, D.P. Dupont (eds.), *Water Policy and Governance in Canada*,
Global Issues in Water Policy 17, DOI 10.1007/978-3-319-42806-2_15

review of literature on some principles of, and models for Indigenous governance of water. A critical discussion of the commonalities and diversities between Indigenous and Western Scientific approaches to water governance will be woven through the text and include debate on the creation of new governance and decision making frameworks that are truly inclusive, respect ancestral knowledge, and introduce culture and governance mechanisms so that Indigenous people can fully participate in the political, organizational, administrative and decision making processes and approaches to water governance in Canada.

15.1 Water Rich, Yet Governance Challenged

Canada, the second largest nation by area, possesses seven percent of the world's renewable water supply (Sullivan 2002; Environment Canada 2007). Water is arguably Canada's most significant and valued natural resource. It is a vital component of the energy, forestry, mining, tourism and agricultural sectors all of which contribute dynamically to the economic wealth, health and wellbeing of the Canadian population. Sixty percent (60 %) of the fresh surface water flows north, away from the majority of the population and the economic activities occurring in the south; therefore, regional water shortages are issues of concern for water users in various locations across the country. The Canadian population and its relatively advanced economy are predicted to rise by 25 % and 55 % respectively by the year 2050. Along with envisaged regional changes in precipitation, temperature and extreme weather events, pressures on the long-term sustainability of Canada's water resources are anticipated. In this chapter, we explore the problems with the current water governance regimes in Canada and their inclusion of good governance principles as laid out by the United Nations. We describe expert researchers' calls for enhancing water governance for the future, and outline points of convergence between current water governance systems, and those proffered by Indigenous[1] peoples in Canada. We use three examples of drinking water governance research to highlight ways of Indigenizing water governance in Canada as a suggested pathway to improving water governance globally. Our chapter is reflective; we return to the problems of jurisdictional fragmentation with some solutions derived from Indigenous governance models explored during our examples (Fig. 15.1).

[1] Indigenous herein means descendants of the original inhabitants of Canada which includes, First Nations people (referring to status and non-status), Inuit (the Indigenous people of the Arctic) and Métis (having a mixed-ancestry of Indian-European or Inuit-European people). First Nation also refers to a geographical place where a group of First Nations people live on land set aside for them specifically.

Fig. 15.1 Good governance principles: A reflective practice

15.2 Water Governance in Canada Today

In this section we will explore the emergence of water governance systems in Canada. We will highlight some key definitions of water governance terms, and describe good governance principles from across a variety of systems. We begin with conveying why the system of governance for water in Canada struggles to equitably distribute the vast resource most Canadians enjoy.

15.2.1 The Problems of Jurisdictional Fragmentation

In an age in which the role of the state in a globalized and interdependent world is under scrutiny, there is debate about the core concepts of good governance. In the past, the nation state was the omnipotent architect of *governance*; the processes of interaction and decision-making in a collective problem that lead to the creation, reinforcement, or reproduction of social norms and institutions (Hufty 2011). Modern flows of information among local and regional actors, and increasingly complex networks has meant that people can now create alliances of their own to provide new ways of organizing societies and managing services and resources in which personal stakes are claimed (Franks and Cleaver 2007). Hierarchical governance no longer suits public interest in many contexts. It is particularly so in the case of Canadian water resources.

Water governance is defined most commonly as:

governing water wisely … so that the involvement of the public and the interests of all stakeholders are included in the management of water resources (Global Water Partnership (GWP) 2000, p 15).

This definition makes reference to the range of political, social, economic and administrative systems in place to manage water resources, and the delivery of water services, at different levels of society (GWP 2000). It has been variously criticized as overly instrumental (considering water simply a 'resource' not a necessity), not measureable, contradictory, and too idealistic leading to artificial separation between water research and practice (Lautze et al. 2011; Castro 2007; Franks and Cleaver 2007). Others put forth the notion that water governance definitions need to include sustainability or they ignore depletion, justice and livelihood issues across the long-term (Wiek and Larson 2012). New definitions have emerged that attempt to guide specific aspects of water governance, such as requiring mechanisms for good governance of water (see for example, GWP-Med 2001 or Huitema et al. 2009), and requiring well-thought out processes and institutions (UNDP 2004). The confusion about what to include in definitions of water governance parallel the confusion of what is good governance of water in practice; it is difficult to affirm what to include in definitions because there is little evidence supporting what good governance practice for water resources looks like on the ground (Biswas and Tortajada 2010).

The sustainability of Canada's water resources relies on effective water governance (Bakker and Cook 2011). Canada is a decentralized federation and the governance of water resources is highly complex, multi-jurisdictional and involves all levels of government. The constitutional authority for water resources belongs primarily to the ten provinces (British Columbia, Alberta, Saskatchewan, Manitoba, Ontario, Quebec, New Brunswick, Nova Scotia, Prince Edward Island and Newfoundland and Labrador), that are situated south of the sixtieth parallel and largely strung along the United States border to the south, as well as the three territories north of the 60th parallel (Yukon Territory, Northwest Territories and Nunavut). These provincial/territorial jurisdictions and their subsequently delegated municipal and regional authorities exert direct control over the water resources within their respective boundaries. The constitutional authority of the federal government, seated in Ottawa, is limited to fisheries management, navigation, federal land management including Indigenous lands and international water governance through various legislative, regulatory and policy mechanisms. The provinces and territories have exclusive and explicit constitutional authority over water resources and water supply, within the constraints of provincial or national boundary agreements (i.e., the Master Agreement on Apportionment between the Prairie Provinces, and the Canada-U.S. Great Lakes Water Quality Agreement for the Great Lakes) (Environment and Climate Change Canada 2016). Responsibility for the management of municipal water supply is subsequently delegated by the provinces and territories to municipalities. Given this constitutional framework, jurisdictional fragmentation in Canadian water governance has and continues to present a number of challenges within the federation and also across the international border with the United States (Bakker and Cook 2011).

The various federal, provincial/territorial and municipal institutions/agencies (for example there are 20 federal agencies operating under 11 separate legislative frameworks) have competing and overlapping mandates and divisions of powers. This leads to a number of complex logistical problems, inter-ministerial conflicts, duplication of efforts, and impediments to information sharing. The constitutional framework, though widely considered to be well-formulated and purposeful in the overall Canadian context, presents a unique and interesting matrix of difficulties in coordination of water governance across jurisdictions. As a result, the formulation and implementation of decisions around water governance and management are often fragmented (Hrudey and Hrudey 2004). Further complexity is posed by the constitutional arrangement of the federal government's fiduciary responsibility to its Indigenous people (over 600 First Nations and additional Inuit and Métis Nations) as well as Indigenous efforts towards re-establishing self-determination (Slowey 2008; MacDonald 2011).

15.2.2 Fragmented Governance and Indigenous People in Canada

Canada's First Nation people, and their rights and freedoms, have been governed by the *Indian Act* since its enactment in 1867 by the Parliament of Canada (Indian Act 1985). The provisions of Section 91(24) of the Constitution Act, 1867; a major part of Canada's Constitution originally enacted as the British North America Act (BNA), provided Canada's federal government exclusive authority to legislate in relation to "Indians and Lands Reserved for Indians" and acknowledged that First Nations had special status (Slattery 1992). Parliament has amended the *Indian Act* frequently over the years. For example, in 1876 the *Act* was consolidated authorizing the Department of Indian Affairs (now known as Indigenous and Northern Affairs Canada) to intercede significantly in First Nation governance overseeing fundamental issues such as land management, resources, finances and defining the status of individuals. The fundamental premise of the *Act* is that the Crown has responsibility to care for and protect the interests of First Nation people. The overtly paternalistic nature of the act continues to be a divisive element woven into the fabric of Canadian society and the granting of special status to First Nation people under the *Act*, for example, is inherently divisive (White et al. 2011). Over the years, the legislation has become increasingly restrictive and controlling of the lives of First Nation people, and the management of their own lands and resources.

In the current state of affairs, the *Indian Act* devolves governing authority to Chief and Council in First Nation communities; however, the federal government retains ultimate authority. At the same time, First Nations are exerting increasing control over issues such as social services, health, education, policing, finances, infrastructure and resources. In the Canadian federation, since health education, social services and resources are usually the responsibilities of provincial governments, the direct relationship between the federal government and First Nation communities poses a logistical restriction for service provision in these spheres of wellbeing. Likewise, provision of safe drinking water has been advanced in most places through provincial and municipal governments. In the context of First Nation communities, however, a complex tri-departmental federal structure (Indian and Northern Affairs Canada (INAC), Health Canada, and Environment Canada) has shared responsibility for safe delivery of drinking water. In contrast to the provincial/municipal infrastructures, the Federal agencies responsible for water on reserve are not equipped with the appropriate human, financial, technical, management, monitoring and communications capacities to deliver all of the necessary components required for provision of safe drinking water. As a result, uncertainties, inconsistencies, and failed systems have perennially been the norm in First Nation communities. Ironically, the federal government has considerable financial capacity and has invested billions of dollars toward the issue, but the misalignment of governance is a continuing source of frustration. The allocation of funds to the various federal agencies and to the numerous First Nation communities is fragmented resulting in implementation of 'band-aid' solutions and the existence of opaque lines of responsibility (Patrick 2011; Basdeo and Bharadwaj 2013). These chasms have dire implications for First Nation water governance.

15.3 New Water Governance Research

In this section we explore how advances in water sciences like hydrology, modeling, climate science and political studies have contributed to new research needs for good water governance. We described some recent advances globally, and in the Canadian context, and present a table comparing water governance frameworks. We build the case for introducing Indigenous worldviews into these frameworks for holistic water governance in order to overcome recognized challenges.

15.3.1 Advances in Water governance

Water governance as a practice and research field has developed rapidly over the last two decades. Models of allocating water that were previously based on first-come, first-serve bases have been enhanced through advances in hydrological and social sciences, climate science, and political studies. Integrated Water Management (IWM) was introduced in 2011 as a way to blend water modeling, social science findings, and political studies (Viessman Jr 2011). Nazemi and Wheater (2014) make the case for better hydrological modeling to inform decision making, especially in headwater regions and locations facing increased uncertainty due to the effects of climate change. Wheater and Gober (2013) make it clear that the water problems of today are not those of the past; that because of non-stationarity, the models that are used to predict hydrological dynamics suffer from errors, and governance decisions should be cautious to compensate accordingly. They conclude that systems science and modeling needs to be combined with social and policy sciences to address issues of water security within a social learning framework. Sivapalan et al. (2014) make efforts to do this by returning humans to the water quantity and use equation for modeling and predicting global water availabilities; and by building a new science of water governance called socio-hydrology. Doing so integrates on-the-ground human water use into models that measure water resource change. Humans also compete for water across political boundaries. Indeed, Cascão and Zeitoun (2010) clarify how pillars of power transform water politics beyond the simple geographical struggles. There are not just physical, but also social, economic, military and cognitive battlegrounds between competing nations (Cascão and Zeitoun 2010). These dimensions alter water use and governance practice to maximize powerful societies' access to water. In the Canadian context, real divisions exist among viewpoints about water security and governance depending on geography and sustainability leanings (pragmatic versus idealistic) (Strickert et al. 2016). In short, water experts are calling for a unified, and interconnected drive to help create more sustainable, evidence-based, locally- and socially-responsive water governance systems aimed at securing, measuring, allocating and predicting water use for human use. Therein lies a dilemma; is water simply a resource for human use decided upon using measurable scientific evidence? If so, why haven't our advances in integrating the sciences of water use and modeling resulted in equitable sharing of water resources? This, to us, speaks of opportunity for convergence with Indigenous worldviews and knowledge holders who have long-lived experience with sustainable water management. We will thus make a case for the indigenizing of water governance on a global level. Now, we review the current state of water governance in Canada.

15.3.2 Governments and Good Governance

The current suite of Canadian water governance practices range from Indigenous spiritually-guided principles in self-governed locations, to more concrete and measureable systems being applied at the regional, provincial, watershed, and transboundary levels. Each system follows their own principles of good governance which were cultivated from experience through cultural evolution, and cycles of engagement with stakeholders. Table 15.1 presents a range of systems espousing good governance practices from very local, place-based sources to the more globally-relevant United Nations (UN)-supported systems signed onto by a host of countries around the world. These principles can be broadly categorized into four types: rationally-based and measureable, spiritually-based, social and normative, and stewardship-based. Each of the models presented here encompasses some, but not all of those categories. When we explore the three projects later in this chapter, we will reveal how the use of all four principles may help overcome the water governance struggles encountered today.

Localized government systems such as those on reserve and in northern Canadian regions where devolution has occurred are contested for their abilities to manage resources (see for example, Cameron and Campbell 2009; Hill et al. 2008). There are, however, recognized benefits to having more locally-based governance systems; place-based cultural values legitimize social norms and the legitimization means that norms are respected (Licht et al. 2007; Feitelson and Fischhendler 2009). When a locally-relevant and legitimate rule about shared resources is imposed, there is less deviance from that norm because of the social cost (Bejaminsen and Lund

Table 15.1 A spectrum of good governance principles

Good governance model	First nations governance models		Canadian government governance characteristics		UN and other international body principles of good governance		
Examplar → Category of governance principle ↓	Cree Nation GovernanceWorking Group[1]	Métis[2, a]	Saskatchewan river	Env. Canada[b] basin survey[4]	Institute on governance (IOG) – based on UNDP[6]	UNc[c,d,e]	World Bank[10, f, g]
Spiritually-guided principles	*Taapwaayaayihtimuwin* (Spiritual Beliefs) The land named by their forefathers and the stories the land is named after – the "Spirituality' and "Place" are sacred						
Stewardship-based principles	*Respect. Eeyou Uchimaaun* (The homeland of yesterday, today and tomorrow) *Eeyou Pimaatisiiun* (caretaking) "*Power with*" (partnership with the land).	*Honouring and Rituals Seasonality* and migration as a means of preserving resources and life		Recognition of the values of water to environment, economy and society			

Social and normative principles	*Eeyou Kischihuwin* (The inherent right of government – right to govern oneself) *Cree Identity/Being Iiyiyuu* (meaning being Cree honestly and with pride – the Eeyou way of doing) *Eeyou Miyupimaatisiiun*, Health and Social Wellbeing. *Youth and Education. Integrity* (Building a strong foundation)	*Autonomy of Nation* (developed through sovereignty attitudes) *Adaptive and Community based Law* (eg. Rules of the Buffalo Hunt[h, i]) *Being their own boss* (individual autonomy; being freemen)	*Equitable and inclusive* (1): all affected by decisionmaking have the right to speak and be heard *Transparent* (2): the justification for decisions are communicated clearly to all concerned *Accountable* (5): it is clear who is responsible for the decisions and consequences	*Stakeholder representation*, support, and involvement *Inclusive and transparent governance* and coordination mechanisms (boards, authorities) *Partners and stakeholders collaborate* at basin and sub-basin levels, and decisionmaking is integrated, timely and adaptive	*Accountability* (accountability and transparency) *Legitimacy and voice* (participation and consensus orientation)	*Accountable*: it is clear who is responsible for the decisions and consequences *Consensus oriented*: all groups are heard and a diverse interests are represented in decisions *Participatory:* decision making includes the input of a broad group *Responsive*: there is timely two-way communication among stakeholders *Equitable and inclusive*: all affected by decision making have the right to speak and be heard *Transparent:* the justification for decisions are communicated clearly to all concerned	*Predictable, open and enlightened policy making* *Strong civil society participating in public affairs* *Executive arm accountable for its actions*
			Responsive (4): there is timely two-way communication among stakeholders *Participatory* (6): decision making includes the input of a broad group *Consensus oriented* (7): all groups are heard and a diverse interests are represented in decisions				

(continued)

Table 15.1 (continued)

Good governance model	First nations governance models		Canadian government governance characteristics		UN and other international body principles of good governance		
Rationalized principles (bureaucracies, legalities, measureable progress indicators, and economic aspects)		*Fairness Mediation over formal Adjudication and Punitive Action*	*Effective and efficient* (3): decisions meet the needs of society while making best use of resources *Follows the rule of the law* (8): decisions are made fairly, and according to law	*Market and non-market values for aquatic resources* IWM must be *grounded in a strong legal and regulatory framework* and carried out by designing and using an appropriate mix of instruments and tools Clear focus and orientation towards *results and evidence based decision-making*; and *basis in scientific principles, sustainable management, and precautionary approaches*	*Fairness* (equity and the Rule of Law) *Direction* (Strategic Vision) *Performance* (responsiveness, effectiveness and efficiency)	*Follows the rule of the law*: decisions are made fairly, and according to law *Effective and efficient*: decisions meet the needs of society while making best use of resources	*Following the rule of law Bureaucracy imbedded with a professional ethos*

[a]Royal Commission on Aboriginal Peoples (RCAP): Report of the Royal Commission on Aboriginal Peoples: Restructuring the Relationship, Vol. 2, Part 2 (Ottawa: Canada Communications Group, 1996) (Co-chairs: R. Dussault & G. Erasmus) [4]Strickert and Bradford (2015) Stakeholder buy-in to the Principles of Good Governance across a flood. (in preparation)

[b]Environment Canada (2010). Integrated Watershed Management. Accessed June 2015 from https://www.ec.gc.ca/eau-water/default.asp?lang=En&n=13D23813-1#governance [6]Madden, J., Graham, J., & Wilson, J. (2005). *Exploring options for Métis governance in the 21st century* (p. 21). Ottawa: Institute on Governance

[c]United Nations (2015) Governance. Accessed June 22, 2015: http://www.un.org/en/globalissues/governance/

[d]Graham, J., Amos, B., & Plumptre, T. (2003). Principles for good governance in the 21st century. *Policy brief, 15*, 1–6

[e]Dobriansky, P. (2003). "Principles of Good Governance": Economic Perspectives; http://usinfo.state.gov/journals [10]World Bank (1994). Governance: The World Bank's experience. Washington, DC: The World Bank

[f]Williams, D., & Young, T. (1994). Governance, the World Bank and liberal theory. *Political Studies, 42*(1), 84–100

[g]One party leads for the duration of the hunt and has supreme authority: Democratic one-person one vote to elect ten captains, elected captains choose leader, rules of hunt are amended from previous years if needed, then put in place by this leader and group of captains

[h]Cree Nation Governance Working Group (2011). Cree Nation Governance Working Group Annual Report. Accessed June 21, 2014 from http://creetrappers.ca/wp-content/uploads/2014/02/CTAActivityReport2010.pdf

[i]Chartrand, L. (2008). 'We Rise Again': Metis Traditional Governance and the Claim to Metis Self-Government. *We Rise Again': Metis Traditional Governance and the Claim to Metis Self-Government" in Yale Belanger, Aboriginal Self-Government in Canada (3rd ed.) (Saskatoon: Purich Publishing, 2008)*, 145–157

[1]The Cree Nation represents the most widely distributed group of the identified First Nation people ranging geographically from Alberta in the west to Québec in the east and across subarctic to plains regions.

[2]The Métis are people of mixed European and Indigenous ancestry, and one of the three recognized indigenous peoples in Canada. They are geographically spread across all the provinces and territories

2003; McLeod et al. 2015). More specifically, clear understandings of watershed dynamics and boundaries formed through living and using water in those boundaries create buy-in for water stewardship across a watershed (Cohen and Davidson 2011). Reiterating local historical narratives about water practices act as intergenerational lessons which reinforce the value of water resources to people (Anderson 2010). In Table 15.1 we include two examples of spiritual and pragmatic governance principles from the Cree Nation Governance Working Group and from Métis traditional governance working groups.

In the Cree principles, the value of *Taapwaayaayihtimuwin* (meaning spiritual value) is the first of 11 core guiding values agreed on by Nations from across central Canada for good governance going into the twenty-first century (Cree National Working Group 2011). Operationally, it means that affording Nature and her elements autonomy is necessary. Further, the concept teaches people to conceive of Nature as a spiritual being which may be respectfully approached to give both the individual and others her bounty in order for life to be sustained (Cree National Working Group 2011; Gillis 2014).

The principle of *Seasonality* emerging from Métis conceptions of good governance provides another point of divergence from other models of good governance (Chartrand 2008). The principle emphasizes respect for the natural rhythms of water flow, food availability, and patterns of consumption of natural resources that emerge from natural seasonal variations. This principle is pragmatic and stewardship-centered; protecting the seasonality of resources, and sharing knowledge of that variation ensures the consistent return of resources each year. This is in direct opposition to principles of effective and efficient use of resources driven by consumer demand, rather than natural supply.

Given the call for a truly integrated approach for water governance, and the gaps identified above in current principles guiding good governance, we will now explore how some of Canada's Indigenous worldviews differ, and how they provide points of convergence for development of better water governance systems to be practiced around the globe.

15.3.3 Indigenous Peoples of Canada in Profile

The constitutional term applied collectively to the three groups of Indigenous people in Canada (Indian, Inuit, and Métis) is "Aboriginal" (Health Canada 2003). Use of the term "Indigenous" is the preference of Indigenous groups with whom we work, and is gaining transnational acceptance (First Nations Studies Program 2009). In total, there are 1,836,035 people who report having Indigenous ancestry from one of the three groups. This represents 5.6% of the Canadian population (AANDC 2014). In 2011, 851,560 people identified as First Nations, 451,795 people identified as Métis and 59,445 identified as Inuit (Statistics Canada 2011). The Indigenous population is growing at a faster rate than the general Canadian population, partly due to reporting; that is, there may be a growing comfort in self-identifying when, historically, it was formerly associated with oppression (Canadian Ministry of

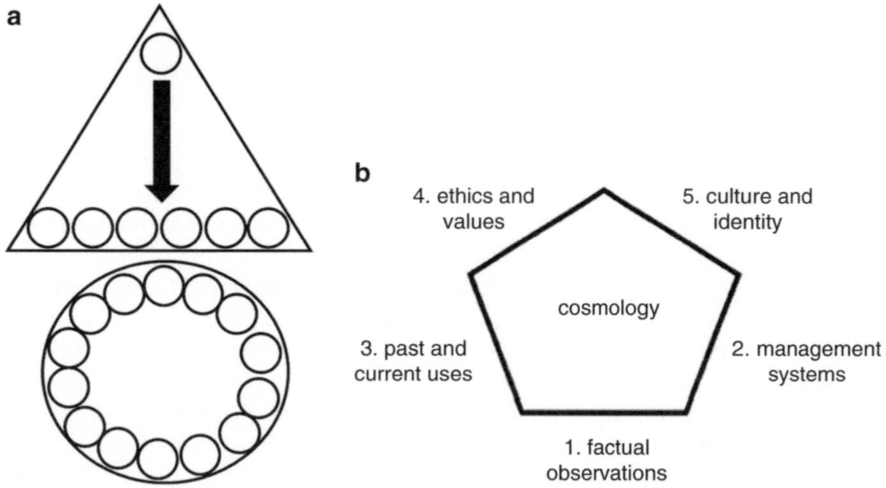

Fig. 15.2 Two examples of indigenous worldviews (**a**) 'Power over' (*top*) and 'Power with' (**b**) Houde's (2007) pentagram of traditional ecological knowledge

Supply and Services 1996; Indigenous Corporate Training Inc. 2015). The Indigenous population increased by 232,385 people, or 20.1 % between 2006 and 2011, compared with 5.2 % for the non-Indigenous population (Statistics Canada 2011). "First Nations", is the common title used in Canada to describe the various societies of Indigenous peoples who are accorded status as "Indians" under the Indian Act (1985), and, who are not of Inuit or Métis descent. First Nation communities or reserves[2] in Canada are located widely across the country. As of March 2014, according to the Indian Register, an official record identifying all registered Indians, there are 901,053 Registered Indians living in Canada (AANDC 2014). The First Nation population comprises the largest proportion (60 %) of the Indigenous population in Canada. Approximately 47.4 % of First Nations people live off reserves, largely in urban settings. The other 52.6 % of First Nations live on reserves and in communities on Crown land (AANDC 2014).

There are 617 First Nations governments/bands in Canada and 3100 reserves across Canada, not all of which are permanently occupied (AANDC 2014). First Nations in Canada have a combined reserve land base that totals approximately 3.8 million hectares. It is expected that the total reserve area will increase by 1.1 million hectares as a result of treaty land entitlement and specific land claims (AANDC 2014). A defining characteristic of Indigenous peoples in Canada is the connection of their individual, and cultural wellbeing to the land and to the water of their home (Wilson 2003; Richmond and Ross 2009).

[2] In Canada an "Indian Reserve" is specified by the Indian Act as a "tract of land, the legal title to which is vested in Her Majesty, that has been set apart by Her Majesty for the use and benefit of a First Nations government/band.

15.3.4 Indigenous Worldviews and Decision Drivers

The rise of Indigenous cultures, languages and social systems has significantly shaped the development of Canada. Many Indigenous epistemologies speak of 'pondering great mysteries that lie no further than the self' (Ermine 1995). Some Indigenous worldviews encompass self-reflection and introspection for the purpose of finding meanings and knowledge through prayer, fasting, dream interpretation, ceremonies and silence (Baskin 2006). In these worldviews, ways of discovery are often passed generationally through Elder teachings and by way of blood memories. Within many Indigenous worldviews, individuals have the inherent ability for introspection; this self-reflection is often conducted alone, is unique for each individual, and provides an individual with purpose and guidance on what they can offer to a whole (Ermine 1995; Baskin 2006; Hampton 1995). Knowledge, for many Indigenous peoples, is thus based on experience. Experiences through inward journeys provide both individual teachings, and collective ones through each individual's contribution to the community's culture (Ermine 1995; Baskin 2006).

While Indigenous cultures and worldviews come from different ontologies and epistemologies, five common elements are generally integral; these include having a holistic perspective, understanding interconnectedness, understanding the dynamics of life, being connected to the land and community, and believing in the strength of 'power with' instead of 'power over.' (McKenzie and Morrissette 2003; Kainai Board of Education 2005; Simpson 2000; Graham 2002a, b; Weaver 2001). These five elements can be found in Indigenous storytelling, art, models of knowledge and cultural expression. They can guide the governance norms of each community in specific and combined ways; for instance, stressing that all things are interconnected including one's family, community, land, animals, water, and the spirit world reminds individuals that respecting a norm shows one's respect for everyone, everything, and every place associated with that norm (Simpson 2000; Graham 2002a, b; De La Torre 2004). To divide these realities into separate categories is to dishonour Indigenous worldviews.

Indigenous worldviews apply at both the individual and social level, and focus on both individual and collective responsibility, knowledge sharing, and reinforcement of values for all members of one's community (Graham 2002a, b; Maurial 1999; Baskin 2006; Rice 2005). Through emphasizing "power with" (symbolized by a circle) rather than "power over," (a triangle) there is a levelling of power among all members of the community and the environment where they live (Fig. 15.2a) (Alberta Education n.d.).

Humans are neither above, nor below others within the circle, and all are dedicated to the rhythm of that system. On the spiritual level, personal and community health, and that of the environment are understood to be interrelated. Thus, to Indigenous people the wellbeing of the earth is essential for survival. The concept of "power with" supports governance systems that are based on respect for environments and resources because of the recognition that without the environmental resource counter-part, the culture cannot continue. No mastery of 'management' through identifying ways of having power over a resource is used to gain advantages over other people (McKenzie and Morrissette 2003). The concept of 'Power With' equates to equity, fairness, inclusivity and voice, while 'Power Over' reflects hierarchy, bureaucracy, and power imbalances. Figure 15.2b presents a different conception of an Indigenous worldview. Houde's (2007) pentagram of traditional ecological knowledge is derived from his work with First Nation communities in Canada. In the pentagram, six 'faces' of traditional knowledge are presented in deepening level, with one's cosmology (beliefs and spiritual relationship with the environment) acting as the foundation for the other faces (Houde 2007). The pentagram's interconnectedness contrasts with governance principles that separate observable and measurable principles (i.e., efficiency) from spiritual principles.

Indigenous worldviews encompass a fluid sense of time and the cyclical nature of change across a variety of scales; day and night, the seasons, life and death (McKenzie and Morrissette 2003; Cajete 2000; Rice 2005). The implicit connection to the Earth emphasizes the importance of place and belonging, and imbues members with strong stewardship principles; the land and resources shape their way of life, and to maintain that, they must respect the provider. Knowledge is often embedded within the language of the community and the languages themselves evolve as environments change. Information tends to be framed around relationships such as the interconnectedness of humans, animals, plants, the all-inclusive environment and the creator. Information is gathered individually, and shared holistically (Maurial 1999; Castellano 2000; Whitt et al. 2001). The oral traditions are used to ensure knowledge is shared from generation to generation (De La Torre 2004). Social learning, in this way, is inherent to governance in some Indigenous populations.

15.3.5 What Does Water Mean to Indigenous Peoples?

Water for Indigenous peoples is the basis of all life. Many Indigenous groups relay that at the beginning of time, the creator gave them instructions to respect water, air, and the earth by keeping it pure. These original instructions are reflected in some Indigenous cultural beliefs and values today. Indigenous place names reflect the waters near to which they are located: on a height of land in northern Manitoba, there is a pond from which water flows east into the Hayes River, and west into the Nelson. This is the *Echimamish*, "the-river-that-flows-both-ways." In tribute to the almost mystical reversal of flow that made portaging between the rivers so easy, Indigenous people leave offerings each time they crossed this divide. The Inuit of Canada's North have a special relationship with water and ice. They derive a great deal of their sustenance from it – seals and fish still remain the mainstay of many traditional Inuit diets. In the long dark winter, villages were established on or near the ice to provide ready access to this food source. In the spring and summer, they fished for trout in inland lakes and rivers. Many Inuit now live in permanent villages, and no longer move from camp to camp, however, the perils of living in the

North remain, as does the necessity of understanding and respecting the role of water. Knowing how water can be used for nourishment and transportation is a life-saving skill. Many Inuit families are teaching the traditional ways to their children on the land and water through hunting and fishing (Tagalik 2009; Nunavut Tunngavik Incorporated 2011; Inuit Tapiriit Kanatami 2015). Both pragmatic and spiritual values for water are reported by Indigenous groups.

Grand Chief B.G. Cheechoo, chief of the Nishnawbe-Aski Nation, explains one example of how the relationship is conceived:

> Our history is tied to these waters. Our continued reliance on fishing, trapping and hunting and our desire to do so is dependent on these waters. Our future is based on these waters ... Any threat to such waters poses a direct threat to our survival. (Environment Canada 2010).

15.3.6 The Links from Water to Worldviews

Indigenous worldviews are thus deeply rooted in identification with the land and water. One of the most important and most common cultural teachings is that people live in harmony with the natural world. Common to the oral stories and legends passed on by Indigenous Elders from one generation to another is that people are a part of the whole of creation and all objects in the environment—whether animate or inanimate—and are treated with the utmost respect. In that way, Indigenous peoples often emphasize the replacement of principles of rational maximization for one's benefit with a communal respect and stewardship principle based on the values of wisdom, love, honesty, humility, and truth. Good governance can enable people to live in a way that promotes harmony and balance with everyone and everything in creation. Indigenous and "mainstream Canada" sometimes differ by their commitment to thankfulness versus economic maximization rooted in Eurocentric worldviews and Western scientific principles. Whereas Indigenous people do not regularly demand proof before beliefs change through introspection and sharing, Westernized cultures tend to resist change that is not evidence-based. Because of this dependence on empiricism, western cultures judge their wellbeing through measureable indicators. Wellbeing in Indigenous cultures is described in both tangible and intangible ways; it is tied to both the holistic environment and to one's introspective journey and the relating of that journey to others; as such, it is less measureable using the western scientific tradition.

15.4 Challenges for Good Water Governance: Three Examples

In this section we describe water governance challenges being faced by communities in Canada.

15.4.1 Water Challenges for Indigenous Communities in Canada

There exists a significant inequity in the provision of adequate water resources for Indigenous peoples in comparison to the generally excellent standard experienced by the majority of the Canadian population. Until recently, problems with water quality in Indigenous communities have not received the same degree of public or political attention in comparison to those occurring in other communities such as Walkerton Ontario (Graham 2002a, b; Godlewska et al. 2010; Morrison et al. 2015). The occurrence of waterborne infections in reserve communities is an alarming 26 times the national average, and approximately 30 % of community water systems themselves are deemed to pose a high risk to water quality (Health Canada 2016). Boil water advisories have been in effect for more than a decade in a number of communities. For example the Neskantaga First Nation (of approximately 300 people), in the James Bay lowlands, has had a boil water advisory in effect since 1995. As of July 2015, 133 First Nations communities were under drinking water advisories (Health Canada 2016). Given the history and worldview of Indigenous people and their strong cultural, physical, mental and spiritual ties to the land and water, the current state of water supply and security is especially devastating to the population.

Though many issues of water management and governance are relevant to the discussion, here we will consider drinking water supply and quality as an example to illustrate the issues surrounding water governance for Indigenous communities in Canada. The problems associated with provision of safe drinking water in these communities stem largely from constitutional/institutional arrangements described above, but also relate to different worldviews driving water governance. Three additional considerations that are worthy of discussion include limitations of capacity (human, financial and technical) in individual communities, geographical factors and divergence of governance principles (Bradford et al. 2016).

15.4.2 Capacity and Indigenous Water Governance

The issue of capacity is directly related to the current centralized and hierarchical water governance structure. In Canada, First Nations communities range in population from as little as a few hundred people to 13,000. Inherent limitations of capacity of individual communities due to scale is a key challenge to the provision of safe drinking water, as is true for small and rural water operators in Canada and around the globe (see, for example, Kot et al. 2015). Chief and Council are responsible for the design and construction of water and wastewater systems on each reserve, must assume 20 % of the costs, and are tasked with operation and maintenance including monitoring water safety and ensuring the presence of trained operators. Indigenous water operators typically lack sufficient financial and human resources and the technical knowledge to design, build and maintain effective community water systems.

Good communications with the generally well-functioning provincial/municipal infrastructures is lacking because of the primacy of relationships with the federal government. In addition, First Nation communities receive block funding from the federal government but there is significant competition for the allocation of these resources within communities from other immediate areas of concern such as social services, emergency response, health, education and community infrastructure amongst others. By virtue of these imperatives, and other factors, funding is often siphoned from long-term water provision solutions and infrastructure. Many communities have become dependent on expensive and sometimes inappropriate technology to fix immediate water quality problems. Thus the issue of limited capacity is complex and compounded by a number of interrelated factors within and outside First Nation communities.

How does the example of capacity relate to how practitioners conceive water governance in the Canadian context? The objectification of water as a 'resource' places it in direct competition with other resource needs on reserve; social support, emergency response, education, and heath are direct competitors with water for funding and action. Indigenous worldviews would conceive of water differently; viewing water as a pragmatic and spiritual necessity not solely for human use removes its provision from the rationalized principles of good governance (Table 15.1) to spiritually- and stewardship-based principles. In short, water would be prioritized for action and resourcing, not simply for the sake of human use, but because of its spiritual and pragmatic value.

15.4.3 Example 1: Six Nations of the Grand Reserve

A poignant and somewhat ironic example is the water provision for the Six Nations of the Grand River reserve, located in Ontario, the most populous First Nations in Canada, with a registered population of 24,000 people with 13,000 people living on reserve. In January 2014, Six Nations opened its new $41-million water treatment plant. The new plant is expected to provide safe drinking water to new and existing users reaching an equivalent of 7370 people over a 10-year period. However, of the 2674 households in Six Nations, only 460 are connected to the community water system (Dupont et al. 2014). As many as four out of five homes in this community are not connected to water lines, driving most families to rely on wells or cisterns for their water. It is anticipated that the plant won't improve access for the roughly 2200 homes, schools and businesses that are not currently on water lines and the extension of water lines to every home on the reserve will cost at least $120 million. A source of financial and infrastructure resources required to rectify this situation has not been identified. Viewing water as a consumable resource and the addition of infrastructure as an act of economic development failed water governance in this community through a lack of pragmatic understanding on the local human social environment. Indeed, the result of the disconnect (both physically and metaphorically) will serve to reinforce existing dependence on bottled water as a pragmatic practice, and spiritually-entrenched belief due to the increasing mistrust of modern water interventions on reserve (Dupont et al. 2014).

15.4.4 Geography and Water Governance in Canada

Geographical considerations and the remote location of many First Nations communities are additional factors germane to the discussion of water governance in relation to drinking water. In some cases it relates to the access, supply and quality of source water (Bradford et al. 2016).

How does the example of geography relate to how practitioners conceive water governance in the Canadian context? The technical ways in which water governance has been operationalized has created barriers for public participation in water decisions and thereby, removed human geographies from contributing to water policy (Agnew 2011). In many ways, knowledge of the belief systems for, and patterns of use of local waters, is a precursor to developing drinking water systems that communities respect and use (Failing et al. 2007; Scodanibbio 2011; VanTol 2011).

15.4.5 Example 2: Yellow Quill First Nation Water Treatment Plant

For example, the Yellow Quill First Nation, a community of 1000, located in eastern Saskatchewan, was on a boil water advisory for more than 8 years before 2004 when an alternate and treatable water source was identified for the community. Community members became dependent on bottled water. Although the reserve had access to ample groundwater supplies, the quality of the water was deemed so poor it was ruled out as a sustainable drinking water source. The nearest source of high quality raw water was 80 km away, a distance beyond practical and economic delivery of supply. As a result of their location, the Yellow Quill reserve had no other option than to rely on the groundwater source previously deemed as inadequate due to its extremely high content of dissolved salts, organics, iron, manganese, ammonium and arsenic. The Yellow Quill First Nation installed costly water treatment plant facilities in 2004 for purifying the groundwater source to bring the water in the aquifer up to acceptable standards. The maintenance costs alone for this system were not sustainable for the community. In 2010, the treatment plant burned down and a new plant was built in its place. Many community members continue to use bottled water even though the new plant is highly effective. This burdens the community with unnecessary costs. Residents report not trusting the source water due to the smell and taste, and a long history of boil water advisories (Dupont et al. 2014). They also report not trusting the 'high-tech' plant to restore the water to an acceptable standard, not liking the smell and taste of treated water, and not having a spiritual connection with the water itself in their region because of the degradation of their local environments, wildlife, and waterways. Similar situations are common to many First Nations communities that rely on insufficient or inadequate raw water sources to sustain their community drinking water supplies, and where costly advanced technological solutions are required (Patrick 2011; Plummer et al. 2013; White and Leblanc 2015). Researchers now emphasize the importance of learning the history and human/physical geographies of a community prior to engaging in research to inform decision making with Indigenous communities (Castleden et al. 2012; Leeuw et al. 2012), particularly when it comes to water management (Woodward and Marrfurra McTaggart 2015; Daley et al. 2015).

15.4.6 Geography, Community Dynamics and Voice

In other cases, geographic location and community dynamics present challenges with participation in decision making processes (Kearney et al. 2007). The remote location of many First Nations communities is a physical barrier restricting opportunities for First Nations engagement and participation in policy forums (Tarlier et al. 2007). First Nations have voiced concerns regarding their ability to provide input, and exercise authority to make decisions and implement solutions related to water usage, protection and management (Wotherspoon and Hansen 2013). Among North American Indigenous people, there is growing concern over the rigidity of imposed political boundaries that have little relation to accepted Indigenous geographic knowledge (Nadasdy 2012; Norman 2014; Thom 2014). The reliance on established colonial political boundaries means duty to consult in many cases, does not reach all potentially affected people in water governance systems.

15.4.7 Example 3: W.A.C. Bennett Dam

An example is the 50 year-old decision by the Provincial government of British Columbia to build the W.A.C. Bennett Dam, a dam that controls and alters water flow of the Peace River for the purpose of generating power. Environmental changes caused by the damming of the Peace River meant dependence, isolation, alienation, and illness for many Indigenous communities downstream over a vast area in the watershed covering at least two additional provincial jurisdictions and numerous devolved autonomous governance entities in the Northwest Territories (Loo 2007; Loo and Stanley 2011). Environmental impacts of the dam rendered some downstream land unlivable, and has led to the destruction of hunting and fishing grounds. Water quality and quantity issues affecting the livelihoods of communities downstream extend as far as the Slave River and Delta in the Northwest Territories (Lenaghan 2010; Das et al. 2015).

This example emphasizes the differences between water governance that spawns from rationalized principles of effectiveness (i.e., providing water but also harnessing energy to serve local populations and export to consumers) and efficiency (doing so at acceptable costs); and governance that emerges from social and normative principles (i.e., consensus orientation, participatory, derived from community-based law, and focused on health and social wellbeing) (refer to Table 15.1). With the Safe Drinking Water for First Nations Act in place in Canada, the legal responsibility for safe water provision on reserve is now in the hands of local leadership. Thus, Chiefs and Council can be penalized (financially and criminally) for water-related illnesses in their communities, even though they inherit the pre-existing infrastructure and previous decisions of external authorities on what types of water services the reserve operates. The strong regulatory and legal framework and drive for efficient resource utilization clashes with values and belief systems in remote Indigenous communities who have little say in water governance in general, and had few opportunities for meaningful engagement in development of the Bennett Dam, and at a federal level, review and development of Bill S-8 (Brody 1999; Loo 2007; Baird and Plummer 2013). In the case of the Bennett Dam, downstream community leaders can now be held responsible for waterborne illnesses, even when the source water is not trusted by members of their own communities. Not surprisingly, there is growing dissent and advocacy among Indigenous groups and allies in Canada over water rights and input into water governance (Gilio-Whitaker 2015; Hnidan 2015).

15.4.8 Additional Challenges Face by Remote Indigenous Communities

Isolation of communities poses obvious problems related to the building and maintenance of infrastructure, access to training, consultation, and other aspects of water governance-related capacity. In some instances the terrain is not conducive for infrastructure related to household water distribution systems. Additionally, individuals within communities are often distributed widely across vast landscapes creating situations where communities are left to rely on bottled, trucked-in, or trucked-to-cistern drinking water supplies. Water governance and management frameworks for these relatively inefficient water supply systems are in their infancy, often leading to failed water service for individuals and communities. And this is just one dimension of several water-focused governance challenges; that is, researchers have yet to explore the inequities in wastewater services, hydroelectric projects, water-related natural hazards, water valuation and water system maintenance over the long-term. We will now summarize the barriers to good water governance in Canada with respect to the examples we explored above, and future needs.

15.5 Barriers to Good Water Governance in Canada

There are several challenges facing good water governance in Canada; in this chapter we explored three; first, we describe disconnections between pragmatic practices, spiritual beliefs, and treating water as a consumer product. Second, we described some challenges of geography and decisions on water infrastructure. Third, we explained one example, of many, where a focus on regulatory and legal frameworks for policy and governance of water devolved blame, rather than provision of effective solutions for water services for Indigenous people. The three examples point to larger ideological differences in water governance models across Canada. In this section of the chapter, we look to bridging those differences by finding common ground in the models and providing pathways for blending governance systems in the future.

15.5.1 Convergence and Divergence: Where Are the Overlaps in Good Governance Principles?

Water experts (e.g., Zeitoun and Mirumachi 2008; Bakker and Cook 2011; Wheater and Gober 2013) have called for adaptive, holistic, value-driven, evidence-based systems of water governance that include social learning and are informed by integrated water models. In the examples described above on drinking water in First Nations reserves, we see that few of the spiritually- or stewardship-based governance principles espoused by Indigenous peoples are in place. The social and normative beliefs central to some Indigenous conceptions of good water governance are infrequently applied. Instead, decisions for drinking water on reserves are made in very operationalized, hierarchical, context-independent, and ineffective ways (see for example, Patrick 2011; Morrison et al. 2015). In Canada, new governance systems for First Nations drinking water management could be developed through combining aspects of both westernized and Indigenous good-governance principles and practices. The relatively small sizes of reserve populations give additional merit for using locally-relevant, pragmatic and spiritually-based principles to help guide drinking water decision making. For example, a recent project with Elders from Yellow Quill provided evidence of past examples of using spiritual guidance for drinking water management. Elders described ceremonies they performed regularly around a local lake that provided them with drinking water. The ceremonies taught water stewardship through respect for that lake and the waters within, as well as the proper ways of taking the water so they did not damage the resource or suffer water-borne illness. Elders said that at times when these ceremonies were not performed (they were outlawed during the residential school era), respect for the lake, knowledge of safe drinking water practices, and any deeply-held beliefs about protecting the lake were abandoned. As a result, the community feels that they cannot even trust treated water accessed from that raw supply.

On a larger scale, misguided attempts at maximizing water use could benefit from incorporating more Indigenized principles into management regimes. In the case of the South Saskatchewan River's interprovincial Alberta-Saskatchewan-Manitoba 1969 Master Agreement on Water Apportionment, Alberta must pass on 50 % of the natural flows in the river to downstream Saskatchewan. In the Alberta allocation and water sharing system, the over-allocated system and subsequent poor quality water coming downstream could be improved by emphasizing a belief in the concept of "Power With" and in the social interconnectedness of the people who are dependent on the river. The current system is based on upstream users having 'power over' those downstream. Operationally, it means that upstream users maximizing their water ration through licenses and trading schemes. Operationally, it means that the license holders take out everything they can from the system (Alberta Water Council 2009); there is no inherent belief in the seasonality or spirituality of the river and its natural flow patterns. In recent research, the anchoring to the first-in-time, first-in-right principle in the Alberta water allocation systems remains prevalent (see for example, Strickert et al. 2015). With pressures including climate warming and its effects on glaciers, snowpacks, and evaporation, cyclic flood and drought, and rapidly increasing human activity, a crisis in water quantity and quality is on the horizon for the Prairie water systems (Schindler and Donahue 2006). Replacing the maximization-based system with one that rewards water quality improvement as a sign of respect for the downstream users, the environment and the Creator would provide greater benefits to the Prairie water system as a whole. The turn to a unified, "Power With" shared respect for the resource could promote adaptability in that during flood years, the upstream reaches could take more than their operationalized limit and assist those downstream from disastrous consequences. The mutuality and communications that would be a part of this kind of system would make the river users better prepared for times of need (like during drought and flood) and help Albertans realize their connections to Saskatchewan.

15.5.2 Summary and Path Forward

We have now seen how western and Indigenous good governance principles are applied in the context of water governance in some cases across Canada. Globally leading organizations lean on measurable, and maximization-based systems that sometimes offer tokenistic social and spiritual components. Indigenous systems balk at measurability, but instead depend on reflection and sharing to gauge the wellbeing of water systems. Systems straddling those boundaries often focus on the social principles of good governance at the cost of including spiritually-based and operationalized principles. Hence, problems exist: there is inertia to overcome and stigmas to conquer in including more spiritually- and stewardship-based approaches to water governance, just as there is resistance to measureable, hierarchical, and strategic principles of governance at the other end of the spectrum. But, how do we provide a path for stakeholders in water governance to use relevant principles in the

Fig. 15.3 The path forward through indigenized water governance in Canada

right contexts within legitimate social and geographical boundaries? How do we shift from rationally-based governance systems that exist in the global sphere (i.e., UN, UNEP, World Bank) to ones that are based on interconnectedness, autonomy, and spiritually-driven for small scale First Nations reserve settings? How do we encourage operationalization in systems that are ineffectual? The balance may lie in using the social principles that appear universal in models of good governance; By giving people their inherent right to self-govern, and by providing the means for accountable, participatory, collaborative, and transparent groups to construct their water governance principles for their location in an integral way, we can achieve that goal (Fig. 15.3). This is already occurring in some ways through bureaucratic means with water governance decisions undergoing consultation with local stewardship groups in the Prairie Provinces, however, truly co-created and self-governed water governance systems are slow to emerge in Canada's Indigenous communities and other non-Indigenous jurisdictions.

15.6 Inoculating Water Governance with Indigenous Beliefs

Like most paradigm shifts, incremental changes across long time spans result in improvements for complex problems. In the case of water governance, noticing real change is difficult, but key to our learning is critical reflection and analysis of current processes (Pahl-Wostl et al. 2011). Even in their seminal evaluation of paradigm changes for water management, Paul-Wostl et al. (2011) miss Indigenous water management as potential path in overcoming the prediction and control (rationalized principles), and integrative and adaptive (i.e., reactive incorporation of some social and normative principles) paradigms dominating water governance today.

If today's water experts are themselves resistant to examining how Indigenous systems can contribute to water governance, we need more efforts in sharing Indigenous water beliefs, governance practices and success stories in research literature and indeed, more broadly across society. We begin right now. In early 2016, the Province of British Columbia's new Water Sustainability Act came into effect (Bill 18—2014 Water Sustainability Act 2014). The new Act is accompanied by seven policy directions, of which the seventh is to 'enable a range of governance approaches' (British Columbia 2016). The recent support of tribal parks and self-governance of natural resources (see Murray and King 2012; and Borrini-Feyerabend and Hill 2015) lends hope to the incorporation and parallel existence of Indigenous governance principles alongside other water governance systems. The relationship is also reciprocal, the Tribal Parks' governance principles includes practicing the traditions of ancestors, having transparent communication with all stakeholders, and *also* working in close cooperation with neighbors (Tla-o-qui-aht First Nations n.d.).

In closing, we leave you with some reflection on the structure of this chapter. The current state of water governance in Canada is indicative of the processes in place around the world, but unique to our society is the drive for reconciliation with Indigenous people. We seek to contribute to that worthy goal by making a call for sharing research and practices of good water governance by Indigenous people in Canada. We seek to point out inequities and challenge the 'Power Over' rhetoric inherent to rationalized principles driving water governance. We seek to inoculate water governance with a spiritually-guided, pride-inducing and stewardship-focused approach that emphasizes the health and wellbeing of both people and the environment. If we, like some Indigenous conceptions, regard water itself being deserving of dignity and mutual respect "Water will agree to help you with anything you ask of it" (Ellen White, as quoted in Anderson 2010, p 21).

References

Aboriginal Affairs and Northern Development Canada (AANDC). (2014). *Aboriginal peoples and communities*. https://www.aadnc-aandc.gc.ca/eng/1100100013785/1304467449155. Accessed July 2016.

Agnew, J. (2011). Waterpower: Politics and the geography of water provision. *Annals of the Association of American Geographers, 101*(3), 463–476.

Alberta Education. (n.d.). *Worldviews and aboriginal cultures: Where hearts are rooted.* Alberta, Canada. https://education.alberta.ca/media/307113/o02.pdf. Accessed Sept 2015.

Alberta Water Council. (2009). *Recommendations for improving Alberta's water allocation transfer system.* Alberta, Minister of the Environment. http://www.awchome.ca/Portals/0/pdfs/WATSUP_web_FINAL.pdf. Accessed Sept 2015.

Anderson, K. (2010). *Aboriginal Women, Water and Health: Reflections from Eleven First Nations, Inuit, and Métis Grandmothers*. Winnipeg: Prairie Women's Health Centre of Excellence.

Baird, J., & Plummer, R. (2013). Exploring the governance landscape of indigenous peoples and water in Canada–An introduction to the special issue. *Indigenous Policy Journal, 23*(4).

Bakker, K., & Cook, C. (2011). Water governance in Canada: Innovation and fragmentation. *Water Resources Development, 27*(2), 275–289.
Basdeo, M., & Bharadwaj, L. (2013). Beyond physical: Social dimensions of the water crisis on Canada's First Nations and considerations for governance. *Indigenous Policy Journal, 23*(4).
Baskin, C. (2006). Aboriginal world views as challenges and possibilities in social work education. *Critical Social Work, 7*(2), 1–9.
Bejaminsen, T. A., & Lund, C. (2003). Formalisation and Informalisation of Land and Water Rights in Africa: An Introduction. *Securing Land Rights in Africa, 14*, 1.
Bill 18 – 2014 Water Sustainability Act. (2014). *2014 legislative session: 2nd session, 40th parliament*. Third Reading. Legislative Assembly of British Columbia. https://www.leg.bc.ca/Pages/BCLASS-Legacy.aspx#%2Fcontent%2Flegacy%2Fweb%2F40th2nd%2F3rd_read%2Fgov18-3.htm. Accessed 16 Mar 2016.
Biswas, A. K., & Tortajada, C. (2010). Future water governance: Problems and perspectives. *Water Resources Development, 26*(2), 129–139.
Borrini-Feyerabend, G., & Hill, R. (2015). Governance for the conservation of nature. In G. L. Worboys, M. Lockwood, A. Kothari, S. Feary, & I. Pulsford (Eds.), *Governance and Management of Protected Areas* (pp. 169–206). Canberra: ANU Press.
Bradford, L. E. A., Okpalauwaekwe, U., Waldner, C. L., & Bharadwaj, L. A. (2016). Drinking water quality in Indigenous communities in Canada and health outcomes: A scoping review. *International Journal of Circumpolar Health, 75*, 32336. http://dx.doi.org/10.3402/ijch.v75.32336.
British Columbia. (2016). *Water sustainability act*. https://engage.gov.bc.ca/watersustainabilityact/. Accessed July 2016.
Brody, H. (1999). *Assessing the project: Social impacts and large dams*. Contributing Paper for WCD Thematic I.1.
Cajete, G. (2000). *Native science: Nature laws of interdependence*. Santa Fe: Clear Light Publishers.
Cameron, K., & Campbell, A. (2009). The devolution of natural resources and Nunavut's constitutional status. *Journal of Canadian Studies/Revue d'études canadiennes, 43*(2), 198–219.
Canadian Ministry of Supply and Services. (1996). People to people, Nation to nation: Highlights from the report of the Royal Commission on Aboriginal Peoples. MSS publication No. C96-980366-4. Ottawa.
Cascão, A. E., & Zeitoun, M. (2010). Power, hegemony and critical hydropolitics. In *Transboundary water management: Principles and practice* (pp. 27–42). Washington, DC: Earthscan.
Castellano, M. B. (2000). Updating Aboriginal traditions of knowledge. In G. J. S. Dei, B. L. Hall, & D. G. Rosenburg (Eds.), *Indigenous knowledges in global contexts*. Toronto: University of Toronto Press.
Castleden, H., Morgan, V. S., & Lamb, C. (2012). I spent the first year drinking tea: Exploring Canadian university researchers' perspectives on community-based participatory research involving Indigenous peoples. *The Canadian Geographer/Le Géographe canadien, 56*(2), 160–179.
Castro, J. E. (2007). Water governance in the twentieth-first century. *Ambiente & sociedade, 10*(2), 97–118.
Chartrand, L. (2008). We rise again: Metis traditional governance and the claim to Metis self-government. In B. Yale (Ed.), *Aboriginal self-government in Canada* (3rd ed., pp. 145–157). Saskatoon: Purich Publishing.
Cohen, A., & Davidson, S. (2011). The watershed approach: Challenges, antecedents, and the transition from technical tool to governance unit. *Water Alternatives, 4*(1), 1.
Cree Nation Governance Working Group. (2011). Cree Nation Governance Working Group Annual Report. http://creetrappers.ca/wp-content/uploads/2014/02/CTAActivityReport2010.pdf. Accessed 21 June 2014.

Daley, K., Castleden, H., Jamieson, R., Furgal, C., & Ell, L. (2015). Water systems, sanitation, and public health risks in remote communities: Inuit resident perspectives from the Canadian Arctic. *Social Science & Medicine, 135*, 124–132.

Das, A., Sagin, J., Van der Sanden, J., Evans, E., McKay, H., & Lindenschmidt, K. E. (2015). Monitoring the freeze-up and ice cover progression of the Slave River 1. *Canadian Journal of Civil Engineering, 42*(999), 1–13.

De La Torre, J. (2004). In the trenches: A critical look at the isolation of American Indian political practices in the nonempirical social science of political science. In D. A. Mihesuah & A. C. Wilson (Eds.), *Indigenizing the academy: Transforming scholarship and transforming communities* (pp. 174–190). Lincoln: University of Nebraska Press.

Dupont, D., Waldner, C., Bharadwaj, L., Plummer, R., Carter, B., Cave, K., et al. (2014). Drinking water management: Health risk perceptions and choices in first nations and non-first nations communities in Canada. *International Journal of Environmental Research and Public Health, 11*(6), 5889–5903.

Environment and Climate Change Canada (2016). *Agreements*. http://www.ec.gc.ca/default.asp?lang=En&n=241E833B-1. Accessed July 2016.

Environment Canada. (2007). *A primer on fresh water: Questions and answers*. Cat. no. En37-90/2006E-PDF. ISBN 0-662-43776-4. Ottawa, Canada. http://publications.gc.ca/collections/collection_2007/ec/En37-90-2006E.pdf. Accessed Sept 2015.

Ermine, W. (1995). Aboriginal epistemology. In M. Battiste & J. Barman (Eds.), *First Nations education in Canada: The circle unfolds* (p. 108). Vancouver: University of British Columbia Press.

Failing, L., Gregory, R., & Harstone, M. (2007). Integrating science and local knowledge in environmental risk management: A decision-focused approach. *Ecological Economics, 64*(1), 47–60.

Feitelson, E., & Fischhendler, I. (2009). Spaces of water governance: The case of Israel and its neighbors. *Annals of the Association of American Geographers, 99*(4), 728–745.

First Nations Study Program. (2009). *Terminology*. http://indigenousfoundations.arts.ubc.ca/home/identity/terminology.html. Accessed 10 Mar 2016.

Franks, T., & Cleaver, F. (2007). Water governance and poverty a framework for analysis. *Progress in Development Studies, 7*(4), 291–306.

Gilio-Whitaker, D. (2015). Idle no more and fourth world social movements in the new millennium. *South Atlantic Quarterly, 114*(4), 866–877.

Gillis, J. (2014). Indigenous cosmology in global contexts: A remediation of the paradigm of sustainable development in natural resource extraction policies. *Journal of the International Relations and Affairs Group, 4*(1), 197.

Global Water Partnership. (2000). *Towards water security: A framework for action* (p. 48). The Hague: World Water Forum.

Godlewska, A., Moore, J., & Bednasek, C. D. (2010). Cultivating ignorance of Aboriginal realities. *The Canadian Geographer/Le Géographe canadien, 54*(4), 417–440.

Graham, J. (2002a). *Safe water for First Nation communities: Learning the lessons from Walkerton*. Ottawa: Institute on Governance.

Graham, T. C. (2002b). Using reason for living to connect to American Indian healing traditions. *Journal of Sociology and Social Welfare, 29*(1), 55–75.

GWP-Med. (2001, December 20–21). *Presentation on effective water governance*. Workshop of the GWP-Med sub-regional working groups of North Africa and Middle East on Effective Water Governance. Marriott Hotel, Cairo.

Hampton, E. L. (1995). Memory comes before knowledge: Research may improve if researchers remember their motives. *Canadian Journal of Native Education, 21*(Supp), 46–54.

Health Canada. (2003). *A statistical profile on the health of First Nations in Canada: Determinants of health, 1999 to 2003*. http://www.hc-sc.gc.ca/fniah-spnia/pubs/aborig-autoch/2009-stats-profil/indexeng.php. Accessed July 2016.

Health Canada. (2016). Drinking water advisories in First Nations communities. http://www.hc-sc.gc.ca/fniah-spnia/promotion/public-publique/water-dwa-eau-aqep-eng.php. Accessed 10 Mar 2016.

Hill, C., Furlong, K., Bakker, K., & Cohen, A. (2008). Harmonization versus subsidiarity in water governance: A review of water governance and legislation in the Canadian provinces and territories. *Canadian Water Resources Journal, 33*(4), 315–332.

Hnidan, T. (2015). Treating water: Engineering and the denial of indigenous water rights. *International Journal of Engineering, Social Justice, and Peace, 4*(1–2), 1–16.

Houde, N. (2007). The six faces of traditional ecological knowledge: Challenges and opportunities for Canadian co-management arrangements. *Ecology & Society, 12*(2), 34.

Hrudey, S. E., & Hrudey, E. J. (2004). *Safe drinking water: Lessons from recent outbreaks in affluent nations* (pp. 103–108). London: IWA Publisher.

Hufty, M. (2011). Investigating policy processes: The Governance Analytical Framework (GAF). In U. Wiesmann & H. Hurni (Eds.), *Research for sustainable development* (p. 403). Bern: Geographica Bernensia.

Huitema, D., Mostert, E., Egas, W., Moellenkamp, S., Pahl-Wostl, C., & Yalcin, R. (2009). Adaptive water governance: assessing the institutional prescriptions of adaptive (co-) management from a governance perspective and defining a research agenda. *Ecology & Society, 14*(1), 26.

Indigenous Corporate Training Inc. (2015). Indian Act and the Pass System. http://www.ictinc.ca/blog/indian-act-and-the-pass-system. Accessed 16 Mar 2016.

Inuit Tapiriit Kanatami. (2015). *Inuit Today*. https://www.itk.ca/about-inuit/inuit-today. Accessed 16 Mar 2016.

Kainai Board of Education, et al. (2005). *Aboriginal perspectives: Aboriginal studies 10* (pp. 4–5). Edmonton: Duval House Publishing Inc.

Kearney, J., Berkes, F., Charles, A., Pinkerton, E., & Wiber, M. (2007). The role of participatory governance and community-based management in integrated coastal and ocean management in Canada. *Coastal Management, 35*, 79–104.

Kot, M., Gagnon, G., & Castleden, H. (2015). Drinking water compliance challenges: How do Canadian small water systems respond? *Water Policy, 17*, 349–369.

Lautze, J., De Silva, S., Giordano, M., & Sanford, L. (2011). Putting the cart before the horse: Water governance and IWRM. In: *Natural resources forum*. Blackwell Publishing Ltd., *35*(1), 1–8.

Leeuw, S. D., Cameron, E. S., & Greenwood, M. L. (2012). Participatory and community-based research, Indigenous geographies, and the spaces of friendship: A critical engagement. *The Canadian Geographer/Le Géographe canadien, 56*(2), 180–194.

Lenaghan, M. (2010). Reflection on the reservoir: The dark side of 20th century hydro development in BC. *The Clearihue Review, 43*(2), 35.

Licht, A. N., Goldschmidt, C., & Schwartz, S. H. (2007). Culture rules: The foundations of the rule of law and other norms of governance. *Journal of Comparative Economics, 35*(4), 659–688.

Loo, T. (2007). Disturbing the peace: Environmental change and the scales of justice on a northern river. *Environmental History, 12*(4), 895–919.

Loo, T., & Stanley, M. (2011). An environmental history of progress: Damming the peace and Columbia Rivers. *Canadian Historical Review, 92*(3), 399–427.

MacDonald, F. (2011). Indigenous peoples and neoliberal "privatization" in Canada: Opportunities, cautions and constraints. *Canadian Journal of Political Science, 44*(2), 257–273.

Maurial, M. (1999). Indigenous knowledge and schooling: A continuum between conflict and dialogue. In L. M. Semali & J. L. Kincheloe (Eds.), *What is indigenous knowledge: Voices from the academy* (pp. 59–77). New York: Falmer Press.

McKenzie, B., & Morrissette, V. (2003). Social work practice with Canadians of Aboriginal background: Guidelines for respectful social work. In A. Al-Krenawi & J. R. Graham (Eds.), *Multicultural social work in Canada: Working with diverse ethno-racial communities* (pp. 251–282). Don Mills: Oxford University Press.

McLeod, F., Viswanathan, L., Whitelaw, G S., Macbeth, J., King, C., McCarthy, D. D., & Alexiuk, E. (2015). Finding common ground: A critical review of land use and resource management policies in Ontario, Canada and their intersection with First Nations. *International Indigenous Policy Journal, 6*(1), 1–23.
Morrison, A., Bradford, L., & Bharadwaj, L. (2015). Quantifiable progress of the first nations water management strategy, 2001–2013: Ready for regulation? *Canadian Water Resources Journal/Revue Canadienne des Ressources Hydriques, 40*(4), 352–372.
Murray, G., & King, L. (2012). First Nations values in protected area governance: Tla-o-qui-aht tribal parks and Pacific Rim National Park Reserve. *Human Ecology, 40*(3), 385–395.
Nadasdy, P. (2012). Boundaries among kin: Sovereignty, the modern treaty process, and the rise of ethno-territorial nationalism among Yukon First Nations. *Comparative Studies in Society and History, 54*(3), 499–532.
Nazemi, A., & Wheater, H. S. (2014). How can the uncertainty in the natural inflow regime propagate into the assessment of water resource systems? *Advances in Water Resources, 63*, 131–142.
Norman, E. S. (2014). *Governing transboundary waters: Canada, the United States, and Indigenous communities*. London: Routledge.
Nunavut Tunngavik Incorporated. (2011). *The Status of Inuit Children and Youth in Nunavut.* ISBN 978-0-9784035-8-4. https://www.tunngavik.com/files/2012/11/2010-11-SICS-Annual-Report-Eng.pdf. Accessed 16 Mar 2016.
Pahl-Wostl, C., Jeffrey, P., Isendahl, N., & Brugnach, M. (2011). Maturing the new water management paradigm: Progressing from aspiration to practice. *Water Resources Management, 25*(3), 837–856.
Patrick, R. J. (2011). Uneven access to safe drinking water for First Nations in Canada: Connecting health and place through source water protection. *Health & Place, 17*(1), 386–389.
Plummer, R., de Grosbois, D., Armitage, D., & de Loë, R. C. (2013). An integrative assessment of water vulnerability in First Nation communities in Southern Ontario, Canada. *Global Environmental Change, 23*(4), 749–763.
Rice, B. (2005). *Seeing the world with Aboriginal eyes: A four dimensional perspective on human and non-human values, cultures and relationships on Turtle Island.* Winnipeg: Aboriginal Issues Press.
Richmond, C. A., & Ross, N. A. (2009). The determinants of First Nation and Inuit health: A critical population health approach. *Health & Place, 15*(2), 403–411.
Schindler, D. W., & Donahue, W. F. (2006). An impending water crisis in Canada's Western prairie provinces. *Proceedings of the National Academy of Sciences, 103*(19), 7210–7216.
Scodanibbio, L. (2011). Opening a policy window for organisational change and full-cost accounting: The creation of BC Hydro's water use planning program. *Ecological Economics, 70*(5), 1006–1015.
Simpson, L. (2000). Anishinaabe ways of knowing. In J. Oakes, R. Riew, S. Koolage, L. Simpson, & N. Schuster (Eds.), *Aboriginal health, identity and resources* (pp. 165–185). Winnipeg: Native Studies Press.
Sivapalan, M., Konar, M., Srinivasan, V., Chhatre, A., Wutich, A., Scott, C. A., Westcoat, J. L., & Rodríguez-Iturbe, I. (2014). Socio-hydrology: Use-inspired water sustainability science for the Anthropocene. *Earth's Future, 2*(4), 225–230.
Slattery, B. (1992). First Nations and the Constitution: A question of trust. *Canadian Bar Review, 71*, 261–293.
Slowey, G. (2008). *Navigating neoliberalism: self-determination and the Mikisew Cree First Nation*. Vancouver: UBC Press.
Statistics Canada. (2011). *Table 1. Aboriginal identity population. Canada, 2011.* https://www12.statcan.gc.ca/nhs-enm/2011/as-sa/99-011-x/2011001/tbl/tbl01-eng.cfm. Accessed July 2016.
Strickert, G., Chun, K. P., Bradford, L., Clark, D., Gober, P., Reed, M. G., & Payton, D. (2015). Unpacking viewpoints on water security: Lessons from the South Saskatchewan River Basin. Water Policy wp2015195.

Strickert, G., Chun, K. P., Bradford, L., Clark, D., Gober, P., Reed, M. G., et al. (2016). Unpacking viewpoints on water security: Lessons from the South Saskatchewan river Basin. *Water Policy, 18*(1), 50–72.

Sullivan, C. (2002). Calculating a water poverty index. *World Development, 30*(7), 1195–1210.

Tagalik, S. (2009). *Inunnguiniq: Caring for children the Inuit way*. National Collaborating Centre for Aboriginal Health, University of Northern British Columbia. http://www.nccah-ccnsa.ca/docs/fact%20sheets/child%20and%20youth/Inuit%20caring%20EN%20web.pdf. Accessed 16 Mar 2016.

Tarlier, D. S., Browne, A. J., & Johnson, J. (2007). The influence of geographical and social distance on nursing practice and continuity of care in a remote First Nations community. *Canadian Journal of Nursing Research (CJNR), 39*(3), 126–148.

Thom, B. (2014). Reframing indigenous territories: Private property, human rights and overlapping claims. *American Indian Culture and Research Journal, 38*(4), 3–28.

Tla-o-qui-aht First Nations. (n.d.) *'Hishuk ish ts'awalk'* Everything is one. Tla-o-qui-aht Tribal Park. http://www.tribalparks.ca/?ref=nf. Accessed 16 Mar 2016.

United National Development Programme (UNDP). (2004). *Water governance for poverty reduction. Key issues and the UNDP response to the Millennium Development Goals*. New York: Water Governance Programme Bureau for Development Policy, UNDP.

VanTol, K. (2011). *Knowledge engagement in collaborative water governance: A New Brunswick example*. https://uwspace.uwaterloo.ca/bitstream/handle/10012/6270/VanTol_Katherine.pdf?sequence=1&isAllowed=y. Accessed 16 Mar 2016.

Viessman, W., Jr. (2011). Integrated water management. *Journal of Contemporary Water Research and Education, 106*(1), 2.

Weaver, H. N. (2001). Indigenous identity: What is it, and who really has it? *American Indian Quarterly, 25*(2), 240–255.

Wheater, H., & Gober, P. (2013). Water security in the Canadian Prairies: Science and management challenges. *Philosophical Transactions of the Royal Society of London A: Mathematical, Physical and Engineering Sciences, 371*(2002), 20120409.

White, R., & Leblanc, I. (2015). Collaborating with Pikangikum First Nation toward resilience through technologies for water and sanitation. In: Humanitarian Technology Conference (IHTC2015). IEEE Canada International, pp. 1–5.

White, J., Peters, J., Dinsdale, P., & Beavon, D. (2011). *Aboriginal Policy Research: Voting, Governance and Research*. Toronto: Thompson Educational Publishing.

Whitt, L. A., Roberts, M., Norman, W., & Grieves, V. (2001). Belonging to land: Indigenous knowledge systems and the natural world. *Oklahoma City University Law Review, 26*, 701.

Wiek, A., & Larson, K. L. (2012). Water, people, and sustainability—A systems framework for analyzing and assessing water governance regimes. *Water Resources Management, 26*(11), 3153–3171.

Wilson, K. (2003). Therapeutic landscapes and First Nations peoples: an exploration of culture, health and place. *Health & Place, 9*(2), 83–93.

Woodward, E., & Marrfurra McTaggart, P. (2015). Transforming cross-cultural water research through trust, participation and place. *Geographical Research, 54*(2), 129–142.

Wotherspoon, T., & Hansen, J. (2013). The 'Idle No More' Movement: Paradoxes of First Nations Inclusion in the Canadian Context. *Social Inclusion, 1*(1), 21.

Zeitoun, M., & Mirumachi, N. (2008). Transboundary water interaction I: Reconsidering conflict and cooperation. *International Environmental Agreements: Politics, Law and Economics, 8*(4), 297–316.

Part V
Drinking Water Policy

Chapter 16
Canadian Drinking Water Policy: Jurisdictional Variation in the Context of Decentralized Water Governance

Gemma Dunn, Leila Harris, and Karen Bakker

Abstract This chapter reviews Canada's approach to drinking water governance, focusing on the regulations, policies, practices and institutions related to the management and provision of drinking water. This review is significant given Canada's highly decentralized approach to water governance. We critically evaluate the implications of decentralization for drinking water safety, examining both the uptake of voluntary national guidelines across Canadian jurisdictions, as well as application of day-to-day microbial risk assessment and management practices in various agencies in two provinces (Ontario and BC). Learning from these analyses, we identify a high degree of variability, specifically: (1) variation in the uptake of national Drinking Water Quality Guidelines across provinces and territories; and (2) considerable variability in microbial risk assessment and management practices across provinces and between agencies. We discuss the implications of these findings in light of ongoing harmonization and subsidiarity debates, as well as discussions as to whether compliance should be voluntary or legally binding. Our analysis indicates that the Canadian approach has contributed to data gaps and urban-rural disparities, and reduced capacity for integrated decision-making and effective oversight.

G. Dunn (✉)
Program on Water Governance, University of British Columbia,
439-2202 Main Mall, Vancouver BC V6T 1Z4, Canada
e-mail: gdunn@uisce.ca

L. Harris
Institute for Resources, Environment, and Sustainability (IRES), Center for Women's and Gender Studies (CWAGS), University of British Columbia, 2202 Main Mall,
Vancouver BC V6T 1Z4, Canada
e-mail: lharris@ires.ubc.ca

K. Bakker
Department of Geography, Program on Water Governance, University of British Columbia,
1984 West Mall, Vancouver BC V6T 1Z2, Canada
e-mail: bakker@geog.ubc.ca

S. Renzetti, D.P. Dupont (eds.), *Water Policy and Governance in Canada*,
Global Issues in Water Policy 17, DOI 10.1007/978-3-319-42806-2_16

16.1 Introduction

Threats to drinking water across Canada and internationally have drawn attention to the importance of governance in safeguarding drinking water and public health (Eggerston 2008; Hrudey 2011; Peterson and Torchia 2008). A number of Canadian researchers have argued that previous failures to provide safe drinking water can be attributed to institutional and organizational issues such as inadequate financial resources, weak standards, failure to implement or enforce standards, lack of skilled staff, rivalries among agencies and insufficient political will (see e.g. de Loë and Kruetzwiser 2005; Ivey et al. 2006; Plummer et al. 2011). These factors are often referred to by the general term "governance"—the laws, rules, regulations, policies, practices and institutions related to the use and management of water (Pahl-Wostl 2009; OECD 2011; Pahl-Wostl et al. 2012; Gupta et al. 2013).

Within these broader debates on drinking water safety governance, there are two important perspectives of interest to Canadians. The first perspective focuses on tap water quality regulations, particularly the allocation of responsibility for setting and implementation of drinking water standards. Focusing solely on tap water regulations and testing (also referred to as 'end-product testing') however, can be regarded as a post-hoc, reactive approach to managing drinking water—as it can only highlight a potential health problem after the water has been consumed (Hrudey 2003). The second (more recent) perspective focuses on broader and more comprehensive understandings of the risks to water quality through evaluation of the entire water supply system. This concept is known as the 'source-to-tap' framework, whereby protection of a high quality raw water source, in combination with effective treatment and safe distribution, supported by legislation and water quality testing, will yield water that is safe for human consumption, thus improving drinking water safety (Dunn et al. 2014a). The evaluation of the entire water supply system is regarded as a proactive approach. Increasingly, preventative approaches are considered to be more reliable and cost effective to protect public health (Byleveld et al. 2008; Dominguez-Chicas and Scrimshaw 2010; Hamilton et al. 2006).

This chapter reviews some of the evidence from our analysis of these two perspectives within the Canadian context (see for example, Dunn et al. 2015, 2014a, b, c). Our goal is to situate the Canadian experience in relation to the literature (including recommended best practices), in order to shed light on the effectiveness of current decentralized governance strategies for drinking water safety. In the first section, we provide an overview of the Canadian approach to water governance. In the second section we highlight results from analysis of the two dimensions of drinking water safety. First, focusing on the compliance testing perspective, we analyze the uptake of drinking water quality guidelines across all of the provinces and territories. Second, we offer an assessment of the day-to-day implementation of microbial risk assessment and management practices (part of the recent emphasis on a more 'proactive' whole systems approach) in two Canadian provinces (British Columbia [BC] and Ontario [ON]). In both cases we find highly variable approaches. Together, these two lines of evidence provide a basis to consider the central question of our chapter: what are the implications of decentralized governance for drinking water safety? We consider this evidence in light of ongoing debates as to whether oversight should be harmonized or delegated and linked to this, whether compliance should be voluntary or legally binding. In the conclusion we offer suggestions as to ways in which the effectiveness of current decentralized drinking water governance approaches could be strengthened.

16.2 Decentralized Drinking Water Governance: The Canadian Context

Canada is the only Group of Eight (G8) country and (along with Australia) only one of two Organization for Economic Cooperation and Development (OECD) member states that does not have legally enforceable drinking water quality standards at the national level. It is also unique amongst OECD countries with respect to the relatively passive role taken by the federal government in matters of drinking water regulation. Canada has no national water policy, and no legally enforceable drinking water quality standards at the national level. Some aspects of microbial testing for water quality are mandated across all jurisdictions (*Escherichia coli* and total coliform monitoring of treated drinking water). However, beyond general guidelines recommending the adoption of a multi-barrier approach [by the Canadian Council of Ministers of the Environment Water Quality Task Group and the Federal–Provincial–Territorial Committee (CCME FPTC) on Drinking Water], there is no overarching Canadian framework specific to drinking water governance, or microbial risk assessment and management, anywhere along the source-to-tap framework (CCME 2004). The absence of a mandated national framework has resulted in considerable diversity across the country (from province to province, and from municipality to municipality) in both drinking water regulation as well as risk assessment and management practices. This diversity raises some important questions related to the potential consequences of decentralization from a drinking water quality management perspective, namely the implications of (i) the decentralization of authority and (ii) the high degree of voluntarism.

Decentralization is often defined as a shift of decision-making and administrative authority away from central government, often associated with "rescaling", whereby decision-making is shifted away from nation-state, towards provincial or other subnational (e.g. municipal or watershed) scales of governance. The term could also refer to a situation of decentralized governance more broadly, and need not always refer to a shift away from centralized authority (the situation could be established as highly decentralized from the outset). Over the past several decades, decentralization has gained popularity in policy and academic literatures, particularly in relation to environmental governance (Cohen and McCarthy 2014). Among other purported benefits, decentralization is often promoted as key to increasing efficiency, and improving policy performance and the effectiveness of service delivery (Rogers and

Hall 2003; Brandes and Kriwoken 2006; Swanson et al. 2010). For water governance in particular, there has been considerable attention to both the positive and negative outcomes associated with decentralized approaches (Cohen and Davidson 2011; Conca 2006). Some experts argue that greater harmonization and formal regulation of drinking water quality (and thereby a lower degree of voluntarism and subsidiarity) is necessary (Bakker 2007; Harrison 1997; Hill et al. 2008; Saunders and Wenig 2007; Weibust 2009). Others argue in favor of decentralization, advocate subsidiarity, and assert that public health protection will be improved by greater 'system knowledge' and contextually specific approaches, rather than formal and standardized regulations (Hrudey 2004, 2011; Hrudey et al. 2006). Among other claims, some argue that movements towards watershed or ecosystem scale governance serve to mobilize local interest, align ecosystem needs and governance practices, as well as increase citizen involvement and promote reciprocity and consensus in ways that are thought to be best suited to sustainable water management (Rogers and Hall 2003; Brandes and Kriwoken 2006). Although decentralized approaches are increasingly promoted, they also often remain challenging to implement successfully in practice (Swanson and Bhadwal 2009).

In Canada, decentralization has arisen through the constitutional division of powers between scales of government (federal, provincial and municipal), with both federal and provincial governments having specific roles with respect to water. The federal government's water responsibilities include navigable waters, fisheries and transboundary waters (but no direct role in water quality or drinking water). Canada's ten provinces, which derive their authority from the 1867 Constitution Act, have constitutional responsibility for fresh water, natural resources, and health (including drinking water). The three territories (Northwest Territories, Nunavut, and Yukon) differ from the provinces in that their power and mandates do not derive from the constitution, but rather are delegated under the authority of Canada's Parliament—although the Government of Canada is increasingly transferring province-like powers to territorial governments. Put simply, this means that each jurisdiction has the ultimate responsibility for formulating (an implementing) their own drinking water standards, and risk management practices, to ensure that public health and related goals are met. Whilst responsibility for drinking water quality in Canada may be shared between provincial health authorities and municipalities, it is largely the municipalities that have been delegated the responsibility for the direct provision of safe drinking water (Kot et al. 2011; Hill and Harrison 2006). The result is that each province is developing its own unique approach to water governance for drinking water quality, and beyond that, practices further vary from municipality to municipality, producing considerable variability in regulatory approaches, guidelines and management practices. Our research documents that variability occurs not just between provinces, but also across different agencies engaged in water management (e.g., water utilities, health authorities and watershed authorities) even within the same watershed. The result is considerably more complexity and variability than is often appreciated.

For the large and geographically varied context of Canada, the benefits of subsidiarity[1] have long been argued, with the suggestion that decentralization of natural resource management has the potential to facilitate more appropriate local responses (e.g. adaptation of regulations) in line with context specific biophysical conditions and changes (Swanson et al. 2010). The decentralized approach to water governance also finds resonance in the federated approach to governance as adopted in the constitutional division of powers—the very idea of a federation is based on a similar logic related to consolidating certain federal responsibilities while each jurisdiction maintains authority for its own internal affairs. However, a growing number of Canadian scholars are expressing concern about the impacts of decentralized governance for drinking water quality, in particular. Critics argue that Canada's pursuit of subsidiarity may be detrimental to drinking water management (see e.g. Hill et al. 2008; Weibust 2009; Bakker and Cook 2011), contributing to urban–rural disparities (Hrudey 2011), as well as regulatory disparities (Cook et al. 2013; Dunn et al. 2014a, b), in ways that foster and entrench inequities (Boyd 2011), poor data collection and quality, and inadequate information sharing (Dunn and Bakker 2011). Another concern is that decentralization may occur without necessary and sufficient capacity; for example, local actors might lack the power, capacity or financial resources to fulfill expected mandates (Brown and Purcell 2005). Although the full consequences of a highly decentralized approach are difficult to disentangle, there is value in evaluating and learning from the current situation, particularly as Canada is in many ways a paradigmatic example of decentralized water governance. It is noteworthy that other federations (such as the United States [US]) have highly decentralized approaches on a number of issues (e.g., education), yet for drinking water, have chosen more harmonized and centralized 'minimum standards' and requirements—likely due to its considerable importance, as well as due to concerns for inequities which can plague federated governance approaches (Anderson 2010). Canada thus offers a relatively unique case in which to consider the question of the impacts of decentralized drinking water governance—the focus of the remainder of the chapter.

16.3 What's Happening on the Ground?

This section summarizes key findings from two related studies. The first analysis examines the uptake of the Canadian Drinking Water Quality guidelines across Canada's 13 provinces and territories ("jurisdictions"). The second analysis involved an inquiry into microbial risk assessment and management practices in BC and

[1] Subsidiarity refers to the delegation of decision-making and policy implementation to the lowest-appropriate scale (Bakker and Cook 2011).

Ontario.[2] Both research efforts were undertaken between 2012–2014, with the results published in four journal articles (Dunn et al. 2015, 2014a, b, c). In this section, we provide details, related to analysis of water quality guidelines (with greater focus on regulatory dimensions), followed by the microbial risk study to highlight on-the-ground practices beyond the regulatory realm.

16.3.1 Canadian Drinking Water Quality Guidelines

One of the few attempts at harmonization of drinking water quality in the Canadian context has been made through a non-binding federal process: the Federal-Provincial-Technical Committee on Drinking Water (FPTCDW). This federal agency, with membership from the provinces and territories, develops the Canadian Drinking Water Quality Guidelines (CDWQG), which are currently comprised of 94 chemical, physical, microbial, and radiological parameters. For this investigation we compared every province and territory's drinking water guideline/standard against the 2012 Canadian Drinking Water Quality Guidelines. These national guidelines are voluntary and non-enforceable. Each Canadian province and territory is able to choose: (i) whether or not to adopt the CDWQG; (ii) to what extent they apply the guidelines, and (iii) whether to make them (or aspects of them) legally enforceable.

16.3.1.1 Uneven Uptake of Canada's Drinking Water Quality Across Territories and Provinces

In total, there are 127 unique drinking water parameters used by the 13 Canadian provinces and territories (94 of which comprise the 2012 Canadian Drinking Water Quality Guidelines). Only four provinces use the CDWQG in their entirety. There are only 16 (of the 94) Canadian Drinking Water Guidelines that all provincial and territorial jurisdictions apply uniformly (i.e., using the same value). Effectively, these comprise the current "common baseline" for drinking water quality, which all provinces and territories share. There are 18 additional parameters that only a few jurisdictions use (none of which are included in the CDWQG).

[2] For the first component of the research, each province and territory drinking water guideline/standard was compared against the 2012 Canadian Drinking Water Quality Guidelines (CDWQG), which are comprised of 94 chemical, physical, microbial, and radiological parameters. These parameters include health considerations (maximum acceptable concentrations (MAC)), aesthetic considerations (aesthetic objectives (AO)), and operational guideline values (operational guidance (OG)) (FPT CDW 2012). The data was compiled into a database, and analyzed for variation across jurisdictions. The second part of this research employed a case study approach examining 'on-the-ground' microbial risk assessment and management practices along the source-to-tap spectrum in three case study watersheds from two Canadian provinces (BC and Ontario).

16.3.1.2 There Is Substantial Variation in the Strength of Governance Approaches

Most notable is the distinction between mandatory and voluntary guidelines. For example, 8 of the 13 provinces and territories have mandatory (i.e. legally binding) drinking water standards. Of these 8, only the Northwest Territories, has adopted all 94 CDWQG into provincial regulation (Nova Scotia and Manitoba have adopted the majority of the CDWQG into provincial regulation). Five jurisdictions have non-legally binding approaches (British Columbia, New Brunswick, Prince Edward Island, Newfoundland and Labrador and Nunavut). Of these five, British Columbia follows the CDWQG in its entirety but only the microbial indicators are legally binding.

Most Canadian provinces and territories have, at a minimum, legally binding standards for microbiological parameters, usually two bacteriological indicators—total coliforms and *E. coli*.[3] The 2012 CDWQG now include guidelines for *Giardia, Cryptosporidium*, and enteric viruses, but a handful of provinces and territories have yet to formally embrace these newer standards. By way of comparison, the United States has more extensive microbiological guidelines than Canada, which include Maximum Contaminant Level Goals (MCLGs) of zero for emerging contaminants *Cryptosporidium*, *Giardia lamblia*, *Leguinella* and enteric viruses. As Hrudey (2011) notes, whilst the Canadian Drinking Water Quality guidelines carefully stress that microbial pathogens pose the greatest risk to human health from drinking water, there remains an overriding emphasis on chemical hazards.

In sum, our research found that the CDWQG are used in a variety of ways across Canadian jurisdictions: either in their entirety or partially, as guidelines or legally binding standards. In short, the voluntary system of drinking water guidelines at the federal level has produced considerable heterogeneity across jurisdictions, both in terms of the application of drinking water quality parameters as well as the extent of legal enforcement.

16.3.2 Microbial Risk Assessment and Management

For this research we examined the day-to-day microbial risk assessment and management practices undertaken in six different watersheds across two of Canada's most populous provinces, BC and Ontario. This investigation provided 'on-the-ground' understanding of the degree to which microbial risk assessment and management approaches are occurring in different agencies involved in water management (watershed agencies, water utilities, and public health authorities). We

[3]There are limitations to using *E. coli* as a surrogate indicator for all pathogens, since strong empirical evidence indicates that *E. coli* is not always predictive of pathogens occurrence, particularly for viruses and protozoa. I.e. the absence of *E. coli* is not a guarantee that a water sample is pathogen free (Harwood et al. 2005).

looked for key similarities and differences, and situated them in relation to international best practice derived from literatures related to microbial risk assessment and management. As an example, best practice advocates transitioning from implicit (implied but not formally expressed) to explicit (formalized) and routine application of risk assessment and management procedures that are applied system-wide.

The terms risk assessment and risk management are intrinsically linked and often conflated; however, their meanings are quite distinct. Risk assessment is a scientifically based process involving four key steps: (i) hazard identification/measurement; (ii) hazard characterization (e.g. dose–response analysis); (iii) exposure assessment; and (iv) risk characterization (Hunter et al. 2003). Risk assessment undertaken within the context of a source-to-tap framework should offer an integrated understanding of the various components of the water supply system, their strengths and weaknesses, and the existing and potential threats to water quality so that informed decisions can be made. Risk management refers to the control options, the legal considerations and risk management decisions (including economic and social factors) to reduce or mitigate risk. This includes the task of managing the assessed risks in the face of uncertainty, balancing consideration of potential hazards with available treatment and mitigation strategies as well as resources (Hamilton et al. 2006). Examples of formalized risk assessment and management procedures include: Hazard Analysis and Critical Control Points (HACCP), Qualitative Microbial Risk Assessment (QMRA) and Water Safety Plans (WSP). In the following chapter, Kot et al. discuss the most notable of all the risk management procedures in the water realm, specifically examining the uptake of Water Safety Plans (WSP)[4] in Alberta.

In parallel to our work on variable drinking water guidelines, our research on microbial risk assessment and management also found that Canadian approaches to risk management and assessment practices vary considerably across jurisdictions. Potentially more important, we again found that these practices also diverge considerably from best practices in the literature (Reimann and Banks 2004). Adding to the literature through our focus on everyday management and on-the-ground practices, we find that the variability occurs not just between provinces but also within provinces and even between agencies. In general, risk assessment practices across the 18 agencies interviewed are limited and variable in terms of types of methods used, scope of application and frequency of use. There is limited uptake of the established formalized methods identified in the literature[5] (e.g. Water Safety Plans) with different agencies using different risk tools, even within the same watershed. For example, interviewees were familiar with (and sometimes using) industry standard tools such as Hazard Analysis and Critical Control Points (HACCP) and

[4] WSP's are a risk-based preventative approach to managing drinking water safety from catchment to consumer (source-to-tap). WSP is a comprehensive risk assessment and management plan to identify and prioritize potential threats to water quality at each step in a specific system's water supply chain (from source to tap) implementing best practices to mitigate threats to drinking water (Bartram et al. 2009; Gelting 2009; Hrudey 2011)

[5] In the original article (Dunn et al. 2014a) Table 1 lists key industry risk assessment and management tools. The introduction to this paper highlights how risk assessment practices have evolved.

Critical Control Points (CCP). However, full implementation of these assessment frameworks was limited. Typically, these tools were used informally; drawing on elements of these tools to develop their own tailor made risk assessment suited to specific needs or resources (both financial and staff capacity). As mentioned earlier, best practice in the literature recommends that risk assessments be formalized, routine (or continuous) and applied system-wide (i.e. the entire water supply system — source-to-tap). However, our findings indicate that this approach has not been adopted widely. Rather, the most common applications of risk assessments are short-term or one-off assessments, for example, prior to the design and construction of a water treatment plant.

Although few agencies among the 18 interviewed have microbial risk management plans, all water utilities and health authorities have an Emergency Response Procedure (ERP), which is a regulatory requirement in BC and Ontario (and certainly part of a risk based approach to water management). An ERP is activated once an actual (e.g. *E.coli* is present in treated drinking water) or perceived (e.g. elevated total coliform count or turbidity measurement) microbial contamination event occurs and typically includes a communication strategy. In line with this finding, several interviewees described their agency's approach to risk assessment and management as reactive rather than proactive. This was a general theme that was apparent from the analysis—risk assessment and management is variable, tends to be partial and limited in focus, and is generally reactive and one-off, rather than part of an ongoing monitoring and governance approach. Moreover, given that each agency engages with these questions and concerns differently, using different data and approaches, we find that this represents a barrier to more integrated and comparative understandings, including potential sharing of data between agencies, or ability to use risk based assessment to understand broader trends and shifts.

16.3.3 Variability in Drinking Water Governance

Taken together, these studies show that there is as high degree of variation between Canadian jurisdictions in their approach to drinking water governance. We found considerable variability in the number of drinking water parameters being applied from jurisdiction to jurisdiction as well as differing enforcement approaches. The CDWQG are used in a variety of different ways: either in their entirety or partially, as guidelines or legally binding standards. In addition, on-the-ground microbial risk assessment and management practices also vary between provinces as well as vary within provinces and between similar types of agencies. Even in the same watershed, there is often inconsistency and lack of compatibility between approaches used by different agencies, limiting the potential for integration of management approaches, or data sharing. Further the use of risk assessment tools is limited and variable, with limited use of risk management plans, and other significant differences in on-the-ground risk assessment practices across the two study provinces.

16.4 Discussion

What are the implications of our findings with respect to the long-term effort to secure safe drinking water? Whilst there has been no research to date to provide evidence to show whether a jurisdiction with tighter regulation does in fact have higher water quality than another jurisdiction with lower standards, it does nonetheless raise some relevant concerns: Which arrangement brings better outcomes: harmonized versus decentralized governance? Should we have voluntary guidelines or mandatory standards? Are there particular risks of highly decentralized approaches that deserve further consideration, particularly to mitigate against such risks if this approach is adopted? We explore these debates below.

16.4.1 Harmonized vs. Decentralized Governance

In Canada, regulatory and institutional arrangements in most provinces and territories tend to be highly complex, with a myriad of statutes, regulations and agencies governing water resources. For example, in BC, there are 13 government agencies with responsibilities relating to drinking water and source protection, and 29 statutes that deal with matters of relevance to drinking- and source-water protection (Dunn et al. 2015; Fraser Basin Council 2005; British Columbia Inter-Agency Memorandum of Understanding British Columbia Inter-Agency Memorandum of Understanding 2006). In Ontario, there are 14 ministries engaged in water policy and 56 statutes related to land-use planning and water management (Cook 2011, 113). Only two jurisdictions (Manitoba and Saskatchewan) have a single agency tasked with responsibility for all aspects of water in that province.

During the course of our research, practitioners (in BC and Ontario) identified a number of features in the decentralized governance framework that impede their capacity to implement risk assessment and management strategies to reduce microbial risk in recreational and drinking water. These include: multiple regulations, agencies and stakeholders; limited coordination and clarity on roles and responsibilities of different agencies; lack of clarity in terms of what institution or regulatory requirement takes precedence. One of the most significant challenges, identified by practitioners, relates to the fragmentation of land and water jurisdiction, and of regulation and mandates (including difficulties prioritizing between land-use needs and conflicts). It is increasingly understood that land use and water quality are inextricably linked. The protection of source waters (a critical strategy to reduce microbial risk in recreational and drinking water) is critically dependent on the integration of land-use planning and watershed management (Peckenham et al. 2005; Plummer et al. 2011). However, in practice, these issues are often addressed in isolation (including problem framing, compartmentalization of infrastructure, institutional organization and mandates, regulatory and governance approaches). While decentralization does not rule out coordination, our interviews and case study analysis highlighted that the lack of cohesion and coordination (both institutional and regulatory) in multi-level governance contexts is a considerable impediment in terms of the capacity of these agencies and practitioners to mitigate risks to drinking water quality or to fully operationalize the multi-barrier approach.

Overall, there are several specific concerns as a result of variability, decentralization, and delegation of responsibilities, which we detail below: (i) data collection challenges, (ii) urban-rural disparities and, (iii) poor water quality in First Nations communities, and other inequities. Together, these concerns speak to the problems inherent when a uniform approach is not applied—leading to a patchwork of approaches that risk further entrenching these difficulties, rather than working to overcome them.

16.4.2 Data Collection

There is currently no national drinking water surveillance system, or standardized approach to collect drinking water quality information in Canada[6] (Christensen et al. 2010). For example, there is no systematic collection of water-borne disease outbreaks (Schuster et al. 2005). Moreover, data collection and methods vary from province to province and this heterogeneity renders data collection and data sharing extremely difficult and prone to inconsistencies and data gaps (see for example, Dunn and Bakker 2009, 2011; Norman et al. 2012). Limited data has impeded our ability to fully understand the impacts of burden of disease (including economic impacts); compare water quality between jurisdictions; and monitor national trends. At a minimum, data needs to be comparable and integrated, and resources and expertise need to be shared in order to overcome some of the issues identified by practitioners. Our findings from this research affirm, overall, that the highly decentralized approach in Canada, and the variable approaches to drinking water parameters and microbial risk aggravate these data collection challenges. In light of this, if decentralized governance is pursued, it likely needs to be coupled with specific measures in parallel that will serve to promote data integration and sharing.

16.4.3 Urban-Rural Disparities

The variable approach we have documented at the provincial scale is exacerbated when local systems are considered. Indeed, the literature suggests that there is growing disparity between larger and smaller (mostly rural) systems in Canada (Christensen et al. 2010; Hrudey 2008; Charrois 2010; de Loë and Kruetzwiser

[6] Whilst there are many "pockets" of data (collected by various federal, provincial, municipal agencies and local interest groups), there is neither an overarching central surveillance or information retrieval system, nor consistency (or guidelines) in collection methods.

2005). Larger municipalities typically test for more drinking water parameters than smaller systems:

> Larger systems are held to more rigorous standards as those systems are generally able to purchase better technology and hire specialized personnel (Hamilton et al. 2006).

In contrast, smaller systems typically have fewer resources and less capacity. For example, smaller systems may lack the financial capacity to build extensive water treatment facilities (Davies and Mazumder 2003); have fewer highly trained personnel (Kot et al. 2011; Hrudey 2008; Charrois 2010; de Loë and Kruetzwiser 2005); and limited capacity to meet sampling requirements (Kot et al. 2011). Shortfalls in the technical, managerial and financial capacities have been identified as contributing factors in major waterborne disease outbreaks in Canada, including the one that occurred in Walkerton in 2000[7] (de Loë and Kruetzwiser 2005; Jalba et al. 2010). Each province has responded differently to the challenge of managing water quality in smaller systems. For microbial water quality, for example, Ontario has created a separate regulatory framework for small drinking water systems; British Columbia has exemptions for smaller systems, as has Québec for systems serving less than 21 persons (Cook et al. 2013).

Some commentators have argued that as such, smaller communities are at higher risk and are more likely to experience an outbreak of infectious disease than people receiving their water from a large public utility (Kot et al. 2011; Peterson and Torchia 2008; Hrudey 2008). To illustrate this point, consider the use of Boil Water Advisories (BWAs) across the country. BWAs are public notifications of drinking water quality concerns, and meant to be used as temporary, precautionary measures to protect the public from possible waterborne illnesses (Grover et al. 2012). A nationwide tally conducted in 2008 documented 1766 Boil Water Advisories across Canada in small towns, cities and townships (Eggerston 2008). Many of these BWAs have been in effect for over a year, and in extreme cases, some last decades; the majority of which were on small systems (BC Ombudsman 2008; Grover 2011).[8] The situation of poor water quality, or higher risk, for these communities speaks to the question of whether or not more uniformity, or federally mandated minimum standards would be helpful. Of course, the argument against requiring higher standards, or specific treatment options, in small communities is always at once a question of resources, and financial investments—BWAs have in effect become a default response to the inability to ensure safe drinking water in all communities.

---

[7] In May 2000, contaminated drinking water supplies in Walkerton, Ontario resulted in 2300 illnesses (almost half the town's population at that time), 65 hospitalizations, 27 people suffering acute kidney failure and seven deaths (Hrudey and Hrudey 2004; Hamilton et al. 2006; Dupont and Jahan 2012).

[8] This 2008 report identified British Columbia as having the highest number of BWAs per capita in the country—530 (Eggerston 2008). We conducted a more recent tally of publicly posted BWAs in BC (in November 2013) and found 533 BWAs in effect (Dunn et al. 2014c).

16.4.4 Voluntary Guidelines vs. Mandatory Standards

Canada's voluntary system of drinking water guidelines and microbial risk assessment and management practices has produced considerable heterogeneity at the provincial level in terms of application of drinking water quality parameters, the extent of legal enforcement, and application of risk assessment and management tools. This research is supported by findings in Cook et al. (2013), which also highlighted significant provincial variation in regulatory approaches including uneven compliance and enforcement of microbial drinking water quality (drawing on comparisons between BC, Ontario and Quebec).

Our research found that only 8 provinces have mandatory standards; 5 use voluntary approaches. This finding is aligned with other studies that have found that voluntary approach results in even more fragmentation, variability, and ad hoc approach across provinces and territories. This variability in terms of degrees of adoption creates patchwork systems, which are especially prone to gaps and overlaps (Hrudey 2011; Boyd 2011; Hill et al. 2008; Weibust 2009; Boyd 2006; Bakker and Cook 2011). There are obvious practicalities that would make testing of all 94 parameters a significant challenge if the entire suite of the guidelines were made mandatory (particularly the cost along with staff and time implications). However, it is important to note that in the judicial inquiry that followed the Walkerton *E. coli* outbreak, Justice O'Connor's report recommended that Canada move towards greater regulation and uniformity in drinking water standards. This included the specific recommendation that such standards also be applied to First Nations reserves (Boyd 2011; O'Connor 2002). The Walkerton Inquiry report also recommended that drinking water quality standards have the force of law, moving away from the more voluntary structure at present (O'Connor 2002). Prominent Canadian non-governmental organizations echoing the Walkerton Inquiry Report have similarly called for legally binding regulations at the national level arguing that the current approach is of serious concern (Boyd 2006; Christensen 2011; Christensen et al. 2010). While the suggestion of pursuing more uniform, or minimal guidelines and standards is a recommendation that is not necessarily straightforward, working towards this goal might contribute a great deal to ensuring all Canadians are able to access safe water, equitably, regardless of where they live.

By way of comparison, the US and the European Union, both of which share some similarities with Canada (i.e., geographically varied federation and confederation respectively), have adopted uniform and harmonized federal regulations (or, in the case of the EU, supra-federal regulations) to ensure consistent standards across their respective territories. The US introduced national drinking water standards in 1974, through the Safe Drinking Water Act (SDWA) (McKay and Moeller 2000). Prior to this, individual states had responsibility for public water supplies. The US introduced national standards for a number of reasons, including: reporting and surveillance deficiencies, poor compliance and the increasing prevalence of waterborne disease (US Congressional Budget Office 1995). The European Community implemented mandatory drinking water standards in 1980. Australia is the only other geographically varied federation (in the industrialized world) like Canada that does not have mandatory water quality standards at the national level. Similarly, Australian researchers have expressed concerns that their non-statutory approach is inadequate with data gaps, urban-rural disparities and lack of national surveillance being identified as specific concerns (see e.g. McKay and Moeller 2000).

Of course, strict regulations also pose a problem: without adequate resources, this risks becoming an "unfunded mandate" or a requirement that goes above and beyond what might be necessary, creating significant problems—especially for small communities. Rural communities are substantial in number, fragmented and dispersed across Canada, in addition to being geographically remote in some instances, making these systems harder to monitor (Hrudey 2008; Charrois 2010). Some commentators suggest that these disparities have effectively created "two-tier system of supply roughly split along the urban-rural divide" (Christensen et al. 2010; Hrudey 2008), while others have frequently referenced a two-tier system for First Nation reserves and non-reserve communities (Christensen et al. 2010). Cook et al. (2013) suggest unique regulatory approaches for systems of different sizes may be necessary to resolve the specific challenges facing small water systems. However, our analysis suggests that the absence of a baseline, minimal, uniform and regulated approach (such as the type used in the US or the EU), means that different populations are effectively exposed to different levels of risk—a classic environmental justice concern, as often it is lower income or vulnerable populations that may not enjoy adequate levels of protection. Again, the issue of equity is one that is often a key concern for federations, as summarized by Anderson (2010, p 11), such an approach:

> may result in citizens across the country getting very different treatment … Thus for reasons of equity and program integrity (in such instances) the central government should have a role.

Another, related argument is that greater emphasis should be placed on proactive, preventative multi-barrier approaches and developing institutional "system knowledge" (i.e., in-depth understanding of the water supply system from the source to treatment and distribution) and capacity (institutional, financial, technical, human resources, social). Hrudey, for example, emphasizes the importance of well-trained and competent operators with in-depth knowledge and understanding of the water supply system (source, treatment and distribution). He argues that these skills should be supported by regional capacity, an important factor given economies of scale and pooling of resources. He suggests:

> …throwing money at a problem without having the money guided by those who understand what is required will achieve little, or may even be counter-productive (Hrudey 2011, p 17).

In this vein, some commentators argue in favour of "light handed" regulations that facilitate regional variation and allow greater emphasis to be focused on achieving and demonstrating competence relative to the specific risks facing a system (Hrudey 2011; Kot et al. 2011; Hrudey et al. 2006; Bakker and Cook 2011).

The considerations regarding effective use of scarce resources are key: monitoring and testing can be expensive, and can often be an ineffective use of limited funds. Moreover, it is neither physically nor economically feasible to test for all drinking-water quality parameters in all systems and locations. Accordingly, a key factor in drinking water safety is competent, well-trained operators at the local scale. Our analysis also implies that an overarching structure of drinking water quality standards/guidelines may be also important, as it creates a higher or lower threshold of acceptable drinking water quality for which local operators are held accountable. Greater uniformity in application of risk assessment and management practices could facilitate more informed, context-specific, evidence-based approaches to variability in drinking water treatment and compliance—in this case, the variability would be justified based on consideration of risk in a way that could inform difficult decisions about balancing trade-offs with respect to investments, resources, and capacity, for different sites and systems. In sum, a balance needs to be struck between adequately protecting drinking water safety through regulation, whilst accommodating spatial variation and considering capacity, particularly for smaller systems. Our results suggest that more regulation and uniformity, at least in some aspects (e.g. minimal regulated quality standards, or consistent risk assessment and management approaches), can be meaningful aspects of an approach to guide progress towards this balanced approach.

A risk management perspective may complement this argument. Compliance testing alone (relative to specific guidelines) is sub-optimal as it is inherently reactive—problems may only be identified after they have occurred. Hence, we are seeing growing interest in risk management via a source-to-tap framework, which is thought to be critical, particularly given the recognition that hazards are innumerable and resources to deal with them are limited (Dominguez-Chicas and Scrimshaw 2010; Gelting 2009; Hamilton et al. 2006; Hrudey 2004, 2009). Risk based approaches offer considerable promise in this regard. In other sectors, risk assessment and management are central to operations and protocols. Such practices have been utilized for decades in the energy utility sector, in industries such as automotive and food, and among High Reliability Organizations (HROs) such as aviation and nuclear power plants. Comparatively speaking, formalized, explicit and routine risk assessment practices in the water sector are relatively newer and less widespread (Egerton 1996; Pollard et al. 2004). Nevertheless, they are quickly gaining popularity around the world and are already widely used across Europe, Africa and the Americas (Gelting 2009). Interestingly, a number of jurisdictions (such as Iceland, The Netherlands and parts of Australia) have started to introduce risk assessment and management into legally binding requirements. In Canada, the 2002 Walkerton Inquiry report recommended that source-to-tap approaches for securing drinking water should include risk assessment and management (Ivey et al. 2006; Hrudey 2011; O'Connor 2002). However, our research shows that water quality risk assessment and management have yet to become mainstream practice across Canada (Kot and colleagues share their insights on the Albertan experience in the following chapter).

16.5 Conclusions

This chapter reviewed Canada's approach to drinking water governance, with particular attention to the implications of decentralization for drinking water safety, with reference both to the variable guidelines and regulations across Canadian jurisdictions, as well as day-to-day microbial risk assessment and management practices. Our analysis suggests that fragmented governance has created significant gaps and reduced capacity for integrated decision-making, or effective oversight. While we recognize some possible benefits of decentralized approaches (such as flexibility in drinking water provision and management, and accommodation for the limited financial capacity of small and rural communities), adequate oversight in order to maintain equity and public health goals remains a key concern.

Recognizing some of the potential benefits of subsidiarity, but also cognizant of the risks outlined above, we suggest that decentralization could be more effective if: (1) A coordinating framework or "master plan" was put in place to cohesively tie the disparate efforts together including risk assessment, or sharing of data and expertise (OECD 2011; Gupta et al. 2013). This should include clear planning and reporting to ensure broad management goals are being met (Swanson and Bhadwal 2009). (2) Policy development were to move from a piece-meal to an integrated approach—particularly to be better able to link human and ecosystem health perspectives for water supply, as well as wastewater, stormwater and land-use planning (Jacobs and Pulwarty 2003; Pahl-Wostl 2009; Bakker 2012; Pahl-Wostl et al. 2012; Gupta et al. 2013). (3) Institutional rules and relationships were reshaped with an eye to systemic incentives to ensure that water challenges are addressed proactively (Dovers 2001; Ison 2010). (4) Stronger support and direction from senior authorities were made available to guide or incentivize these processes. To this last point, although decentralization often means delegation of authority, the authors concur with other commentators who suggest that centralized or higher levels of government likely still has a crucial role to play (Swanson and Bhadwal 2009). Finally, (5) increased financial resources and staffing investments are made towards this end, particularly to enable more contact, communication and trust building across key entities who play a role in monitoring, management and provision of drinking water (Plummer et al. 2011). In terms of where the investments may be needed, stronger attention to communication, cohesion, coordination and integration could help to address some of these challenges to better build on experiences, learn across sectors, issues and jurisdictions, and to better identify workable solutions (Dunn et al. 2015).

References

Anderson, G. (2010). *Fiscal federalism: A comparative introduction*. Oxford: Oxford University Press.

Bakker, K. (2007). *Eau Canada: The future of Canada's water*. Vancouver: UBC Press.

Bakker, K. (2012). Water security: Research and opportunities. *Science, 337*(6097), 914–915. doi:10.1126/science.122633710.1126/science.1226337.

Bakker, K., & Cook, C. (2011). Water governance in Canada: Innovation and fragmentation. *International Journal of Water Resources Development, 27*, 275–289.

Bartram, J., Corrales, L., Davison, A., Deere, D., Drury, D., Gordon, B., et al. (2009). *Water safety plan manual: Step-by-step risk management for drinking-water suppliers*. Geneva: World Health Organization.

Boyd, D. (2006). *The water we drink: An international comparison of drinking water quality standards and guidelines*. Vancouver: David Suzuki Foundation.

Boyd, D. (2011). No taps, no toilets: First Nations and the constitutional right to water in Canada. *McGill Law Journal, 57*, 81–133.

Brandes, O. M., & Kriwoken, L. (2006). Changing perspectives-changing paradigms: Taking the "soft path" to water sustainability in the Okanagan Basin. *Canadian Water Resources Journal, 31*, 75–90.

British Columbia Inter-Agency Memorandum of Understanding. (2006). Inter-agency accountability and coordination on drinking water protection. 16 Oct 2006, Version 7.

Brown, J. C., & Purcell, M. (2005). There's nothing inherent about scale: Political ecology, the local trap, and the politics of development in the Brazilian Amazon. *Geoforum, 36*, 607–624.

Byleveld, P. M., Deere, D., & Davison, A. (2008). Water safety plans: Planning for adverse events and communicating with others. *Journal of Water and Health, 6*(1), 1–9.

Canadian Council of Ministers of the Environment (CCME). (2004). *From source to tap: Guidance on the multi-barrier approach to safe drinking water*. Ottawa: Canadian Council of Ministers of the Environment Water Quality Task Group and the Federal–Provincial–Territorial Committee on Drinking Water.

Charrois, J. W. A. (2010). Private drinking water supplies: Challenges for public health. *Canadian Medical Association Journal, 182*, 1061–1064.

Christensen, R. (2011). *Waterproof III: Canada's drinking water report card*. Vancouver: Ecojustice.

Christensen, R., Goucher, N., & Phare, M. A. (2010). *Seeking water justice: Strengthening legal protection for Canada's drinking water*. Vancouver: Ecojustice.

Cohen, A., & Davidson, S. (2011). The watershed approach: Challenges, antecedents, and the transition from technical tool to governance unit. *Water Alternatives, 4*, 1–14.

Cohen, A., & McCarthy, J. (2014). Reviewing rescaling strengthening the case for environmental considerations. *Progress in Human Geography, 39*(1), 3–2.

Conca, K. (2006). *Governing water: Contentious transnational politics and global institution building*. Cambridge: MIT Press.

Cook, C. (2011). *Putting the pieces together: Tracing jurisdictional fragmentation in Ontario water governance*. Dissertation, University of British Columbia. http://hdl.handle.net/2429/39930. Accessed May 2014.

Cook, C., Prystajecky, N., Ngueng Feze, I., Joly, Y., Dunn, G., Kirby, E., et al. (2013). A comparative analysis of microbial drinking water quality testing in three Canadian provinces. *Canadian Water Resources Journal, 38*, 185–195.

Davies, J. M., & Mazumder, A. (2003). Health and environmental policy issues in Canada: The role of watershed management in sustaining clean drinking water quality at surface sources. *Journal of Environmental Management, 68*, 273–286.

De Loë, R., & Kruetzwiser, R. D. (2005). Closing the groundwater implementation gap. *Geoforum, 36*, 241–256.

Dominguez-Chicas, A., & Scrimshaw, M. D. (2010). Hazard and risk assessment for indirect potable reuse schemes: An approach for use in developing water safety plans. *Water Research, 44*(20), 6115–6123.

Dovers, S. (2001). Institutional barriers and opportunities: Processes and arrangements for natural resource management in Australia. *Water Science & Technology, 43*(9), 215–226.

Dunn, G., & Bakker, K. (2009). *Canadian approaches to assessing water security: An inventory of indicators* (Policy report). Vancouver: Program on Water Governance.

Dunn, G., & Bakker, K. (2011). Fresh water related indicators: An inventory and analysis. *Canadian Water Resources Journal, 36*(2), 135–148. doi:10.4296/cwrj3602815.
Dunn, G., Harris, L., Cook, C., & Prystajecky, N. (2014a). A comparative analysis of current microbial water quality risk assessment and management practices in British Columbia and Ontario Canada. Science of the Total Environment, 468–469, 544–552.
Dunn, G., Bakker, K., & Harris, L. (2014b). Drinking water quality guidelines across Canadian provinces and territories: Jurisdictional variation in the context of decentralized water governance. *International Journal of Environmental Research and Public Health, 11*, 4634–4651.
Dunn, G., Henrich, N., Holmes, B., & Harris, L. (2014c). Microbial water quality communication: Public and practitioner insights from British Columbia, Canada. *Journal of Water and Health, 12*(3), 584–595.
Dunn, G., Harris, L., & Bakker, K. (2015). Microbial risk governance: Challenges and opportunities in freshwater management in Canada. *Canadian Water Resources Journal, 40*(3), 237–249. doi:10.1080/07011784.2015.1043648.
Dupont, D., & Jahan, N. (2012). Defensive spending on tap water substitutes: The value of reducing perceived health risks. *Journal of Water and Health, 10*(1), 56–68.
Egerton, A. J. (1996). Achieving reliable and cost effective water treatment. *Water Science and Technology, 33*(2), 143–149.
Eggerston, L. (2008). Investigative report: 1766 boil-water advisories now in place across Canada. *Canadian Medical Association Journal, 178*, 1261–1263.
Federal-Provincial-Territorial Committee on Drinking Water (FPT CDW). (2012). *Guidelines for Canadian drinking water quality*. Ottawa: Health Canada.
Fraser Basin Council. (2005). *Authorities affecting source water protection in British Columbia*. Vancouver: Fraser Basin Council.
Gelting, R. (2009). Water safety plans: CDC's role. *Journal of Environmental Health, 72*(4), 44–45.
Grover, R. (2011). *Boil, boil, toil and trouble: The trouble with boil water advisories in British Columbia*. Master's thesis. University of British Columbia. https://circle.ubc.ca/handle/2429/33790. Accessed 19 July 2013.
Grover, R., Copes, R., Mavinic, D., & Teschke, K. (2012). *Boil water advisory protocol. Water security guidance document, Part 3 Section 7*. Vancouver: Program on Water Governance.
Gupta, J., Akhmouch, A., Cosgrove, W., Hurwitz, Z., Maestu, J., & Ünver, O. (2013). Policymakers' reflections on water governance issues. *Ecology and Society, 18*(1), 35.
Hamilton, P. D., Gale, P., & Pollard, S. J. T. (2006). A commentary on recent water safety initiatives in the context of water utility risk management. *Environment International, 32*, 958–966.
Harrison, K. (1997). *Passing the buck: Federalism and Canadian environmental policy*. Vancouver: UBC Press.
Harwood, V. J., Levine, A. D., Scott, T. M., Chivukula, V., Lukasik, J., Farrah, S. R., et al. (2005). Validity of the indicator organism paradigm for pathogen reduction in reclaimed water and public health protection. *Applied and Environmental Microbiology, 71*, 3163–3170.
Hill, C., & Harrison, K. (2006). Intergovernmental regulation and municipal drinking water. In G. B. Doern & R. Johnson (Eds.), *Rules, rules, rules, rules: Multilevel regulatory governance* (pp. 234–258). Toronto: University of Toronto Press.
Hill, C., Furlong, K., Bakker, K., & Cohen, A. (2008). Harmonization vs. subsidiarity in water governance: A review of water governance and legislation in the Canadian provinces and territories. *Canadian Water Resources Journal, 33*, 1–18.
Hrudey, S. E. (2003). Drinking water risk management principles for a total quality management framework. In S. E. Hrudey (Ed.), *Drinking water safety—A total quality management approach* (pp. 3–14). Ottawa: Institute for Risk Research, University of Waterloo.
Hrudey, S. E. (2004). Drinking-water risk management principles for a total quality management framework. *Journal of Toxicology and Environmental Health Part A, 67*, 1555–1566.

Hrudey, S. E. (2008). Safe water? Depends on where you live! *Canadian Medical Association Journal, 178*(8), 975. doi:10.1503/cmaj.080374.
Hrudey, S. E. (2009). Chlorination disinfection by-products, public health risk tradeoffs and me. *Water Research, 43*(8), 2057–2092.
Hrudey, S. E. (2011). *Safe drinking water policy for Canada—Turning insight into foresight*. Toronto: C.D. Howe Institute Commentary.
Hrudey, S. E., & Hrudey, E. J. (2004). *Safe drinking water: Lessons from recent outbreaks in affluent nations*. London: IWA Publishing.
Hrudey, S. E., Hrudey, E. J., & Pollard, S. J. T. (2006). Risk management for assuring safe drinking water. *Environment International, 32*, 948–957.
Hunter, P. R., Payment, P., Ashbolt, N., & Bartram, J. (2003). Chapter 3—*Assessment of risk. Assessing mi- crobial safety of drinking water: improving approaches and methods* (pp. 79–109). Published on behalf of World Health Organization and organization for economic co-operation and development. London: IWA Publishing.
Ison, R. L. (2010). *Systems practice: How to act in a climate change world*. London: Springer.
Ivey, J. L., de Loë, R. C., & Kreutzwiser, R. D. (2006). Planning for source water protection in Ontario. *Applied Geography, 26*(3), 192–209.
Jacobs, K., & Pulwarty, R. (2003). Water resource management: Science, planning and decision-making. In R. G. Lawford, D. D. Fort, H. C. Hartmann, & S. Eden (Eds.), *Water: Science, policy, and management—Challenges and opportunities* (pp. 117–204). Washington, DC: American Geophysical Union.
Jalba, D. I., Cromar, N. J., Pollard, S. J. T., Charrois, J. W., Bradshaw, R., & Hrudey, S. E. (2010). Safe drinking water: Critical components of effective inter-agency relationships. *Environment International, 36*, 51–59.
Kot, M., Castleden, H., & Gagnon, G. A. (2011). Unintended consequences of regulating drinking water in rural Canadian communities: Examples from Atlantic Canada. *Health and Place, 17*(5), 1030–1037.
McKay, J., & Moeller, A. (2000). Statutory regulation of water quality in modern Australia: Has it been forgotten by the regulators? *Water International, 25*(4), 595–609.
Norman, E. S., Dunn, G., Bakker, K., Allen, D., & Cavalcanti de Albuquerque, R. (2012). Water security assessment: Integrating governance and freshwater indicators. *Water Resources Management, 27*, 535–551. doi:10.1007/s11269-012-0200-4.
O'Connor, D. R. (2002). Chapter 2: Government oversight of the delivery of drinking water: Introduction. In *Report of the Walkerton inquiry: A strategy for safe drinking water, Part Two*. Toronto: Ministry of the Attorney General. http://www.attorneygeneral.jus.gov.on.ca/english/about/pubs/walkerton/part2/Chapter_2.pdf. Accessed 23 Apr 2014.
Ombudsman, B. C. (2008, June). *Fit to drink: Challenges in providing safe drinking water in British Columbia* (Special Report No. 32). Legislative Assembly of British Columbia. Victoria: Ombudsman.
Organisation for Economic Cooperation and Development. (2011). *Water governance in OECD countries: A multi-level approach*. Paris: OECD.
Pahl-Wostl, C. (2009). A conceptual framework for analyzing adaptive capacity and multi-level learning processes in resource governance regimes. *Global Environmental Change, 19*, 354–365.
Pahl-Wostl, C., Lebel, L., Knieper, C., & Nikitina, E. (2012). From applying panaceas to mastering complexity: Toward adaptive water governance in river basins. *Environmental Science & Policy, 23*, 23–34.
Peckenham, J. M., Schmitt, C., McNelly, J., & Tolman, A. (2005). Linking water quality to the watershed: Developing tools for source water protection. *Journal of American Water Works Association, 97*(9), 62–69.
Peterson, H., & Torchia, M. (2008). Safe drinking water for rural Canadians. *Canadian Medical Association Journal, 179*(1), 55. doi:10.1503/cmaj.1080061.

Plummer, R., de Grosbois, D., de Loë, R., & Velaniskis, J. (2011). Probing the integration of land use and watershed planning in a shifting governance regime. *Water Resources Research, 47*(9). doi:10.1029/2010WR010213.

Pollard, S. J. T., Strutt, J. E., MacGillivary, B. H., Hamilton, P. D., & Hrudey, S. E. (2004). Risk analysis in the water utility sector: A review of drivers, tools and techniques. *Process Safety and Environmental Protection, 82*(6), 453–462.

Reimann, C., & Banks, D. (2004). Setting action levels for drinking water: Are we protecting our health or our economy (or our backs!)? *Science of the Total Environment, 332*, 13–21.

Rogers, P., & Hall, A. W. (2003). *Effective water governance* (Global Water Partnership Technical Committee (TEC) Background Papers No. 7). Stockholm: Global Water Partnership.

Saunders, J. O., & Wenig, M. M. (2007). Whose water? Canadian water management and the challenges of jurisdictional fragmentation. In K. Bakker (Ed.), *Eau Canada: The future of Canada's water* (pp. 119–141). Vancouver: UBC Press.

Schuster, C., Aramini, J., Ellis, A., Marshall, B., Robertson, W., Medeiross, D. T., & Charronm, D. F. (2005). Infectious disease outbreaks related to drinking water in Canada. 1974–2001. *Canadian Journal of Public Health, 96*, 254–258.

Swanson, D., & Bhadwal, S. (Eds.). (2009). *Creating adaptive policies: A guide for policy-making in an uncertain world.* New Delhi: Sage Publications Inc, International Development Research Centre (IDRC).

Swanson, D., Barg, S., Tyler, S., Venema, H., Tomar, S., Bhadwal, S., & Drexhage, J. (2010). Seven tools for creating adaptive policies. *Technological Forecasting and Social Change, 77*, 924–939.

US Congressional Budget Office. (1995). *Report to Congress on the cost and benefits of federal regulations.* US GPO: 4, Washington, DC.

Weibust, I. (2009). *Green leviathan: The case for a federal role in environmental policy.* Burlington: Ashgate Publishing Limited.

Chapter 17
Preparing for Success – Drinking Water Safety Plans and Lessons Learned from Alberta: Policy Considerations Contextualized for Small Systems

Megan Kot, Heather Castleden, and Graham A. Gagnon

Abstract Safe drinking water is best ensured through a proactive, multiple barrier water management approach. A water safety plan is an internationally-recognized framework for managing drinking water resources, one that emphasizes multiple barriers between the water source and the consumer's tap, and facilitates competence-building among those responsible for ensuring safe drinking water. In 2011 the province of Alberta required all public water utilities to implement Drinking Water Safety Plans (DWSPs), a first in Canada. With examples from international water safety plan implementation, and from the province of Alberta, we raise questions around the ability of communities, in particular those that are small, to effectively implement and utilize water safety plan policy. Drawing from the health literature, we examine the water policy uptake challenge through a lens of community readiness, and propose readiness for change as a first step in preparation for new water policy. By raising the issue of readiness, we facilitate a discussion around water policy implementation and common reasons for failure, while offering an approach for making 'good ideas' a reality on the ground. An emphasis on readiness challenges both communities and regulators to first examine whether the foundational groundwork and preparedness to support a particular policy is in place, and removes the expectation of a positive policy outcome away from implementation alone.

M. Kot (✉) • G.A. Gagnon
Centre for Water Resource Studies, Dalhousie University, 1360 Barrington Street, Halifax, NS B3H 4R2, Canada
e-mail: megan.kot@dal.ca; graham.gagnon@dal.ca

H. Castleden
Health, Environment, and Communities Research Lab, Department of Geography and Planning, Queen's University, 99 University Ave, Kingston, ON K7L 3N6, Canada
e-mail: heather.castleden@queensu.ca

S. Renzetti, D.P. Dupont (eds.), *Water Policy and Governance in Canada*,
Global Issues in Water Policy 17, DOI 10.1007/978-3-319-42806-2_17

17.1 Introduction

Safe drinking water is important to the health and wellbeing of all Canadians, yet outbreaks of waterborne illness can and do still occur in public water systems. Ensuring that treated water from a public water utility is safe to drink demands considerable diligence. However, many models of water management are reactive, meaning contamination is usually addressed only after it has entered a water supply (Dunn et al. 2014). In some cases, this means contaminated water may be well on its way to consumers before contaminants are detected. As an alternative, a proactive approach to water management emphasizes prevention, often by relying on multiple barriers throughout a water supply, each of which provides a 'check point' where water quality is assured (Health Canada 2003; Hrudey et al. 2006; Plummer et al. 2010; Summerscales and McBean 2011). A 'water safety plan' approach reflects these proactive principles, while also serving to develop competence among those responsible for ensuring safe drinking water (World Health Organization 2004; Hrudey 2011). Although a water safety plan is considered "the most effective means of consistently ensuring the safety of a drinking-water supply" (Bartram et al. 2009, p. 1), its uptake is limited, with many municipalities, regions, and nations worldwide still relying on reactive approaches.

The province of Alberta is the first in Canada to require all public water utilities to have a water safety plan in place. Introduced in 2011 as Drinking Water Safety Plans (DWSPs), the approach is comprehensive, guiding water providers through a source-to-tap assessment of existing and potential risks, accounting for present and potential threats to water quality, and establishing protective barriers at key points throughout a system with the goal of minimizing impact, should contamination occur (Reid et al. 2014; Gagnon et al. 2015). In principle, by following a DWSP approach, anyone responsible for a public water supply should be well positioned to provide safer water quality, with fewer risks of contamination, than with a more traditional 'response management' approach. In practice however, the experience in Alberta shows that some communities, especially those that are small,[1] have had challenges with implementing DWSPs (Perrier et al. 2014; C2H 2014). There is further concern that as a result of these challenges, not all communities have or will benefit equally from a DWSP approach (Perrier et al. 2014).

In this chapter, we provide an overview of the water safety plan approach as a tool to: improve competence among water suppliers, ensure better risk assessment and risk comprehension, streamline recordkeeping, and facilitate not only best practices but also better protection of public health. With examples from international water safety plan implementation, and from the province of Alberta's recently-implemented DWSP approach, we raise questions around the ability of communities, in particular those that are small, to effectively implement these types of

[1] Health Canada defines a small community water supply system as one serving fewer than 5000 customers, however definitions of a small community vary significantly across Canada and internationally (NCCPH 2011).

'blanket' or 'one size fits all' policies. Drawing from the health literature, we examine the water policy uptake challenge through the conceptual lens of 'community readiness', and propose readiness for change as a first step when introducing new water policies. In this way, we facilitate a constructive discussion around water policy implementation and common reasons for failure, while offering a pathway forward to make 'good ideas' a reality on the ground. Doing so, we believe, challenges both communities and regulators to first examine whether the foundational groundwork and preparedness to support a particular policy is in place, and removes the expectation of a positive outcome away from policy implementation alone.

17.2 Global WSP Uptake

A hallmark of any water safety plan is its flexibility, which – at least theoretically – makes the approach suitable for a wide range of settings, including large cities, small and remote communities (Rinehold et al. 2011; Perrier et al. 2014; Kot et al. 2014). While many examples of water safety plan implementation exist, these can vary from location to location. In general, however, these plans follow three key principles, which together facilitate consistency in drinking water safety (Bartram et al. 2009; WHO 2011). These are:

1. A system assessment of the entire water supply from the source to the consumer's tap: The purpose of the assessment is to ensure the current system can provide drinking water of a quality that meets required standards and regulations.
2. Identification of control measures in the drinking water supply system: Each control measure must be matched with a means of operational monitoring to ensure timely action should water quality deviate from the norm.
3. A management plan is in place that describes both actions for normal activity, and actions should an incident occur: The purpose of the management plan is to guide any action in a water supply system, and should include plans for monitoring and communication.

While some regions have had water safety plans in place for over a decade, others are only just starting out. As a result, the information available on this approach, including benefits and challenges, can be best presented as a snapshot, with no one experience providing the 'true' water safety plan experience (Kot et al. 2014). For example, Iceland legislated water safety plans as early as 1995, and a recent study favourably linked this decades-old approach with improvements to water system operations, water quality and public health (Gunnarsdóttier et al. 2012). Other regions, including Australia, Europe and the Americas, have reported water safety plans as contributing to improved regulatory compliance (Metge et al. 2003; Dyck et al. 2007), better risk awareness among operators (Mullenger et al. 2002; Rinehold et al. 2011; Gelting et al. 2012), a reduction in customer complaints about water quality (Mullenger et al. 2002), and in general, improved efficiency around how water resources are being managed (Medema et al. 2001; Davison and Deere 2007; Jayaratne 2008).

A water safety plan approach also facilitates goal setting, and can provide utilities with guidance for short-term, incremental improvements that result in progressively better public health protection. By helping to identify and prioritise goals, a water safety plan approach may be useful in reducing a strain on resources by focusing attention on eliminating or alleviating immediate risks first, while helping to organize community-level planning around achieving longer-term goals (Davison et al. 2006; Seghezzo et al. 2013). Achieving a series of short-term goals can also help a utility maintain motivation in carrying out a water safety plan approach over the long term, creating a constructive feedback loop for policymakers and stakeholders (Rinehold et al. 2011; Mudaliar 2012). Ultimately, any measures taken to prevent an outbreak of waterborne illness will be less costly than allowing an outbreak to occur. A case in point is the total cost of the outbreak of waterborne illness in Walkerton, Ontario, which resulted in seven deaths and over 2300 people falling ill, many of whom were hospitalized, was estimated at $64.5 million, not including human suffering which would have pushed the total well past $150 million (Livernois 2002).

In some regions, small communities have completed 'simplified' water safety plans or received additional training and support. This was the case in Iceland, where small communities now complete an abbreviated version of the water safety plan, yet still report benefiting positively as a result (Gunnarsdóttir and Gissurarson 2008). In a similar effort, small communities in Bangladesh used pictorial tools to assist with hazard identification where local expertise was lacking (Mahmud et al. 2007). Finally, local facilitators in the Pacific Islands were trained to deliver information on water safety plan principles to small and remote communities in a culturally appropriate manner and in the local dialect (Hasan et al. 2011). The World Health Organization, anticipating that some small communities face greater challenges in implementing water safety plans, has published specific guidelines that emphasize risk identification and management while making accommodations for limited capacity (WHO 2012). Still, the lack of uptake suggests further challenges remain.

17.3 Water Safety Plan Uptake in Canada

Despite advances in water treatment technology, substantial water operator training, and stringent water quality regulations, many communities in Canada continue to face challenges associated with maintaining the ultimate trifecta: *clear*, *unscented*, and *safe* drinking water (Hrudey 2011; NCCPH 2011; Dunn et al. 2014). Unlike other developed countries, Canada does not have a system in place for federal jurisdiction over drinking water quality (see Placing Canada's Water Policies in an International Context). The federal government has national guidelines for the design, operation, maintenance and monitoring of water systems, but there is no national water policy to enforce compliance with these recommendations (Dunn et al. 2014). These responsibilities are generally devolved to the provinces and territories, which in turn regulate drinking water according to their own standards, with much of the responsibility pass down to regional municipalities (Bakker 2007; de Loë 2009; Bakker and Cook 2011). Matters are further complicated with respect to on-reserve First Nations communities, where jurisdictional complexity regarding drinking water responsibilities exists due to Section 35 and Sub-Section 91(24) of our Constitution, which grants the federal government exclusive jurisdiction for 'Indians and lands reserved for Indians' (see Indigenizing Water Governance in Canada and Changing Currents: A Case Study in the Evolution of Water Law in Western Canada).

Major public water contamination events have helped to shape how water is managed in Canada (Plummer et al. 2010). Following the May 2000 outbreak of waterborne illness in Walkerton, mentioned above, a public inquiry made numerous recommendations, including that a multiple barrier approach be adopted in the province to help prevent future outbreaks from occurring, and that ensuring operational competence, more so than compliance with regulations alone, would be made a priority (O'Connor 2002). Across the country, while significant changes resulted from these recommendations, water management in general remains focused on compliance, and only one province, Alberta, has adopted a water safety plan approach. Introduced in 2011, 'drinking water safety plans' (DWSPs) were made a requirement for all of the province's regulated community water systems. Although modelled after international water safety plans, the Alberta framework is uniquely tailored to include a checklist of common water quality challenges known to occur in that province (AEP 2014). The Alberta DWSP policy presents a new step for utilities in addition to existing water quality standards and procedures. For Canada, the Alberta example also provides a first look at water safety plan implementation in this country, and provides an opportunity to examine benefits and challenges.

17.4 Alberta's DWSP Approach

Alberta's DWSP is similar to international models and comprises four stages (AEP 2014; Reid et al. 2014): 1. Collecting detailed information about a water supply; 2. Identifying and understanding existing and potential risks; 3. Identify how these risks are or should be addressed to minimize their impact; and 4. Ensure resources are available and a management plan is in place to reduce these risks to an acceptable level. Risks are assessed in terms of *likelihood* (the probability of something occurring that will impact water safety) and *consequences* (the outcome of any hazardous event, based on the control measures in place). The DWSP tool itself is a form-fillable Microsoft Excel workbook that is pre-populated with more than 200 common risks, with additional space for other unique risks. Once all risks are identified and evaluated based on the above-described likelihood and consequence, a matrix is used to determine an overall risk value and the highest scoring risks are prioritized for action. As new risks emerge (and old risks are addressed) the workbook must be updated. In this way, Alberta regulators expect utilities to treat DWSPs as a 'living' document (AEP 2014).

With a focus on risk prevention, a DWSP approach guides water providers to consider real and potential risks while matching these with appropriate preventative measures (barriers) (AEP 2014). The detail afforded in a DWSP approach encourages water providers to look beyond known risks and consider what might or could occur under certain conditions. In anticipation that some of the province's smaller communities would be less prepared to take on the DWSP requirement, a series of training workshops and on-the-ground support opportunities were offered through both the Government of Alberta and through a provincial capacity building program called the Closer to Home Initiative (AWWOA 2014; C2H 2014). While some training efforts are ongoing, the majority occurred during the DWSP's 2-year implementation period, which ran from 2011 to 2013 (C2H 2014).

17.5 Adopting DWSPs: Experiences of Small Water Systems in Alberta

Taking on any new water policy, in particular one that seeks to change how resources are managed at the community level, can be both challenging and time consuming. Those responsible for drinking water may be required to make significant adjustments not only to the types of decisions being made within a water supply, but also the processes used to make these decisions. In resource-constrained communities, responding to newly identified risks may require a re-prioritization of services provided by the community, with water taking greater (and unexpected) precedence (Perrier et al. 2014). Alberta's DWSP approach is only just underway, with long-term impacts likely to remain unknown for some time. What follows is an overview of the known experiences, challenges and benefits emerging as a result the DWSP requirement, drawing from two recent projects and their subsequent reports, both of which examined the challenge of DWSP implementation specific to small communities. The first is a qualitative study involving semi-structured, open-ended interviews with water operators and community decision-makers (Perrier et al. 2014). The second is from a report on the DWSP sessions offered through the Closer to Home Initiative (C2H 2014), and which supports many of the findings made by Perrier and colleagues.

Completing, implementing and maintaining a DWSP approach is time-intensive, particularly at the outset. Water operators in small communities often work alone with water system management being only one of many responsibilities in their job description, such as maintaining roads in the wintertime or servicing a number of public spaces year-round. This can make it difficult to devote more of their time to water-related duties. Such time constraints are compounded by the absence of detailed, long-term record keeping in many small communities, meaning those responsible for completing a DWSP may need to map out a water system for the first time before being able to conduct the initial risk assessment (Perrier et al. 2014; C2H 2014). When new tasks are introduced, allocating time away from these other responsibilities may be negatively perceived by those in positions of authority, and by those in the community (Perrier et al. 2014). Ideally, operators should to be provided with “protected time” (C2H 2014, p 5), the sole purpose of which is to complete the DWSP, and resolve DWSP-identified issues either alone or in collaboration with stakeholders.

A DWSP approach tasks operators with the burden of responsibility for assessment, risk identification, and risk prioritization. However, action to mitigate risks often relies on a broader range of individuals in a community, including elected officials who make financial decisions. This means operators may, in some cases, be required to ‘make the case’ for a specific improvement or early preventative action that may appear to be unnecessary to the untrained eye. Reporting from experience, some operators expressed concern that not all identified risks (and associated preventative measures) would or could resonate with elected officials as requiring immediate action. The rationale for this challenge may be two-fold: elected officials often lack the specific background knowledge required to understand some water-related issues, while operators may lack the skills (communication or otherwise) required to persuade elected officials to take action (Perrier et al. 2014; C2H 2014). Furthermore, raising awareness around risks present in a water supply may have a negative effect on operators, and perceptions around the quality of their work ‘on the job’, by those who are not familiar with DWSP principles, and who expect operators to provide safe drinking water that is (unrealistically) risk-free (Perrier et al. 2014).

Implicit in a DWSP approach is that action will follow as a result of the risks identified, and in many communities these actions are anticipated to have financial consequences (Reid et al. 2014). The Province of Alberta has in place a voluntary Full Cost Accounting Program that requires municipalities to ensure that costs associated with providing safe and secure drinking water may be met (AEP 2015); however, it appears that many smaller communities have yet to adopt such measures (Dore 2015). This leaves water-related improvements, including those identified through a DWSP approach, in direct competition with other maintenance needs in the community (Perrier et al. 2014). The dependence on external funding for water-related improvements that persists among many small communities limits the ability to respond accordingly to a DWSP approach, even when risks are clearly outlined (C2H 2014).

Costs associated with implementing preventative actions outlined in a DWSP are likely to be the highest at the outset of plan implementation, as the most significant risks are being addressed (Gregor and Winstanley 2005; Chang et al. 2013). As a result, early support from decision-makers is critical, as the financial consequences and personnel time required to respond to such risks can be substantial (C2H 2014; Perrier et al. 2014). This is particularly the case when decision-makers are required to invest proactively in a water supply, without the pressure of a ‘real-time’ disaster such as contamination to provide motivation (Perrier et al. 2014).

Alberta's progress with a DWSP approach is positive, however, some of the shortcomings identified in smaller communities during these early stages of implementation could be addressed collectively using the foundational elements present in internationally-applied water safety plan approaches. In particular is the World Health Organization's emphasis on the pre-implementation stage, during which an inclusive team of stakeholders are brought together and tasked with carrying out the water safety plan process. Here, stakeholders may include decision makers, water operators, and landowners, with each playing a role in supporting the goals set out for reducing risk in a water supply (Davison et al. 2006; Schmoll et al. 2011). As an inclusive approach, this brings together people (and resources) to accomplish the mutually beneficial goal of safe drinking water (Davison et al. 2006), while also helping to remove the burden of responsibility placed on operators as is currently the case in the Alberta model.

17.6 Adopting DWSPs: Are Small Communities Ready?

To be effective, a DWSP approach requires that communities work continuously to identify risks, and to respond to these risks in a timely manner. Yet for many small communities, inadequate financial resources, ageing infrastructure and limited human capacity undermines their capacity to respond (Kot et al. 2011; Perrier et al. 2014; C2H 2014). The result is an implementation gap, whereby an otherwise good idea (such as DWSPs) become difficult to enact given local circumstances. The province of Alberta has made some efforts to address this gap by providing additional training for operators working in small communities, yet there are indications that small communities' DWSPs may face more significant challenges on the ground at the community level (C2H 2014; Perrier et al. 2014; Kot 2015).

To help identify and address these challenges, we suggest that a community's readiness for a change be considered prior to initiating new water policy. A lens of community readiness helps illuminate the process through which a community can prepare for, and then implement, a specific change (Donnermeyer et al. 1997; Chilenski et al. 2007). Models of community readiness have been successfully applied in health-based prevention and awareness building programs (see, for example, Kelly et al. 2003; York et al. 2008). We suggest that a readiness approach is complementary to water quality improvement goals, and would assist communities to prepare for DWSPs, even under resource-constrained conditions (Kot 2015).

In general, readiness describes the motivation to act that a group of individuals must possess in order to take action and make a positive change (Goodman et al. 1998; Foster-Fishman et al. 2007). Without readiness, any change, including effective solutions to pressing problems, have the potential to fail or result in negative outcomes or consequences. The community readiness model developed by Oetting and colleagues (1995) provides a framework for identifying strengths and weaknesses related to a desired change, and offers an approach for improving readiness for that change. Readiness is conceptualized in the model as six categories or

Fig. 17.1 The six categories or 'dimensions' of community readiness (From Oetting et al. 1995)

'dimensions': community efforts, community knowledge of efforts, leadership, community climate, community knowledge about the issue, and resources (Fig. 17.1). These dimensions form the basis for assessment and determining where a community is at in terms of readiness. The model also identifies nine 'stages' of readiness: no awareness, denial/resistance, vague awareness, preplanning, preparation, initiation, stabilization, confirmation/expansion, and professionalization. In this way, communities can range from having 'no awareness' that a problem exists, to taking ownership of a problem (professionalization) and serving as a mentor to other communities (Oetting et al. 1995; Edwards et al. 2000). Once aware of their own readiness gaps, communities can address these gaps using approaches outlined in the model. Failing to understand readiness gaps prior to implementing a change can undermine the success of an otherwise 'good idea' or promising practice (Oetting et al. 1995; Chilenski et al. 2007). A community readiness model approach thus seeks to prepare communities for change, and works strategically towards ensuring the change is implemented appropriately and effectively.

While the original community readiness model includes three steps: assessment, stage-identification, and readiness building, only the assessment portion has been applied in the Alberta DWSP context and is discussed below (Kot 2015). Further study is needed to determine the utility of the complete model in building readiness for DWSP implementation in other jurisdictions, including the ease at which regulators or communities themselves can use the model and administer its findings. A preliminary study examining the uptake of DWSPs among a cohort of eight small communities in Alberta found low to medium readiness scores across all six readiness dimensions. This suggests that the full potential of a DWSP approach is not being realized in all Albertan communities, despite being a requirement across the province (Kot 2015).

Understanding community readiness for change is achieved through a comprehensive understanding of related strengths and weaknesses in a particular community, and an assessment questionnaire is used to gather this information. The questionnaire presents a number of questions related to each of the six readiness dimensions (Table 17.1). Each question is tailored to reflect the desired change (in

Table 17.1 Dimensions of community readiness for DWSP policy uptake with suggested questions for Key informants

Six Dimensions of Community Readiness for DWSP Policy Uptake
Dimension 1. Community efforts: What is being done to make water safe?
How much of a concern/priority is tap water in the community?
What policies and practices related to tap water quality are in place in the community?
What are the primary concerns related to tap water in the community?
Dimension 2. Community knowledge of efforts: How much is known about safe water?
Do people in the community talk formally (or informally) about tap water quality?
How aware is the general public about policies and practices in place for safe drinking water?
How are people in the community informed of these policies and practices?
Dimension 3. Leadership: How supportive are leaders?
How much of a concern is tap water quality to the leadership in the community?
What opportunities are there for leaders to become engaged or involved?
Would the leadership agree to devote additional resources for water improvement?
Dimension 4. Community climate: What is the prevailing attitude?
What is the overall feeling among community members regarding tap water?
Do community members provide any support for ensuring tap water quality?
Do people in the community drink tap water?
Dimension 5. Community knowledge: What do people know about the issue?
How knowledgeable are community members about tap water quality or quantity?
What information is available in your community regarding the tap water?
Who can community members contact to obtain information regarding their tap water?
Dimension 6. Resources: What resources are available to support efforts?
Is there a volunteer base in the community with an interest in water issues?
How are current efforts related to tap water quality funded?
Are there plans to seek out additional funding?

Kot (2015, p. 101)

this case, implementing and maintaining a DWSP approach), and input is (ideally) gathered from several knowledgeable individuals in the community (Kot 2015; Plested et al. 2006).

The first dimension, *community efforts*, describes programs or applicable water policies already in place in the community that relate to DWSPs. This can include compliance with water quality regulations, ongoing source water protection planning, and community engagement activities. Communities who currently struggle to achieve regulatory compliance will likely be less ready for DWSPs, as given existing water quality challenges the capacity to adopt new practices are likely limited. The second dimension, *community knowledge of efforts*, seeks to understand the existing relationship between water customers and their water supply. For example, are customers involved in source water protection efforts? Are customers aware of the drinking water source, and is there an understanding of how (and why) water is treated to ensure safety? A customer base that is aware and involved in local water issues may help balance customer expectations for water quality and their willingness to pay for these services. Questions around the third dimension of *leadership* focus on understanding the extent to which elected or appointed community leaders and other influential community members care about and play an active role in support of the local water system. When community leadership understands the added value of a DWSP approach, the greater the chances are that adequate measures to support necessary improvements will be taken, even if this means that limited budget will be reworked in support of preventative maintenance of a water supply (Perrier et al. 2014).

The *community climate*, the fourth dimension, refers to predominant attitudes and awareness in the community about local water supplies and efforts to produce safe drinking water. Community interest in engaging in water-related issues can range from one of avoidance ('it is what it is') to one reflecting a desire to ensure vigilance ('we cannot let our guard down'). For example, a preference for bottled water can be interpreted as a form of issue-avoidance, as customers are choosing an alternative source of water instead of lobbying decision makers to make improvements to local tap water. The fifth dimension seeks to understand *community knowledge* around improvement options. Questions posed regarding this dimension seek to gauge awareness about the need for continuous professional attention to water treatment and management. Without public understanding about the water operator's role in protecting public health, the importance of treatment in preventing waterborne illness, or what is gained through appropriate water pricing, it may be challenging to find adequate support for new water policies such as DWSPs.

The sixth and final dimension of community readiness examines available *resources*. Here, communities are encouraged to take a broad view and include both the human and financial resources necessary to support a DWSP approach. For example, is there a volunteer base that can be relied upon to assist with source water protection, to develop and carry out education and awareness programs, or to support operators in non-critical (i.e. general maintenance) tasks? Regarding financial resources, can additional financial capacity be made available or be sought from external sources, if required to support the risk prevention measures outlined in a DWSP?

The assessment described above is only the first step in understanding community readiness for change. Subsequent steps, including the scoring process and the identification of a stage of readiness, are described in full elsewhere (see Kot 2015 for a DWSP-specific example; Plested et al. 2006 for a general example). The model acknowledges that some communities may be 'more ready' (that is, earn a higher readiness score) in some dimensions than in others. In this case, communities would first target the lower-scoring dimensions. Doing so helps ensure the community can support the desired change drawing equally from all six dimensions. For example, if weakness in the leadership dimension is identified, one readiness building strategy would be to develop awareness among this critical cohort, first through one-on-one discussion to build recognition and understanding about local water issues, the rationale for a DWSP approach, and the benefits to the community over the long term (Kot 2015). In part, the municipal leader training on DWSPs offered through the Closer to Home Initiative would serve to increased readiness in this dimension (C2H 2014; AWWOA 2014), however this was not formalized as part of a community readiness building approach.

17.7 Concluding Comments

The introduction of DWSPs in the province of Alberta provides a first look at water safety plans in Canada, and raises a number of issues around the unintended consequences of 'universal' policy uptake in small communities. As with other regions worldwide, the outcome of DWSPs will be best observed over time, drawing from both short and long term outcomes. Some of the challenges observed in early studies within the province's smaller communities provide insights into the types of challenges that are likely to emerge elsewhere in Canada. To address these challenges, we put forward community readiness as an essential determinant in the success or failure of a new water policy like DWSPs. Drawing from an existing community readiness model, we suggest there is an opportunity to improve both how water policies are implemented, and how well the benefits of these policies play out in the community setting.

References

Adeel, Z. (2016). Placing Canada's water policies in an international context. In S. Renzetti, & D. Dupont (Eds.), *Water policy and governance*. Dordrecht: Springer.

Alberta Environment and Parks (AEP). (2014). *Drinking water safety plan*. http://environment.alberta.ca/apps/regulateddwq/DWSP.aspx. Accessed 15 Jan 2014.

Alberta Environment and Parks (AEP). (2015). *Full cost accounting for municipal drinking water facilities*. http://aep.alberta.ca/water/programs-and-services/drinking-water/knowledge/full-cost-accounting-for-municipal-drinking-water-facilities.aspx. Accessed 15 Feb 2016.

Alberta Water and Wastewater Operator's Association (AWWOA). (2014). *Water and wastewater resource materials*. https://awwoa.ab.ca/resources/water-and-wastewater-resource-materials. Accessed 15 Feb 2016.

Bakker, K. (2007). *Eau Canada: The future of Canada's water*. Vancouver: UBC Press.

Bakker, K., & Cook, C. (2011). Water governance in Canada: Innovation and fragmentation. *International Journal of Water Resources Development, 27*(2), 275–289.

Bartram, J., Corrales, L., Davison, A., Deere, D., Drury, D., Gordon, B., Howard, G., Rinehold, A., & Stevens, M. (2009). *Water safety plan manual: Step-by-step risk management for drinking-water suppliers*. Geneva: World Health Organization. http://www.who.int/water_sanitation_health/publication_9789241562638/en/. Accessed 19 Apr 2013.

Bradford, L. E. A., Ovsenek, N., & Bharadwaj, L. (2016). Indigenizing water governance in Canada. In S. Renzetti, & D. Dupont (Eds.), *Water policy and governance*. Dordrecht: Springer.

Brandes, O., & Curran, D. (2016). Changing currents: A case study in the evolution of water law in Western Canada. In S. Renzetti, & D. Dupont (Eds.), *Water policy and governance*. Dordrecht: Springer.

Chang, Z., Chong, M., & Bartram, J. (2013). Analysis of water safety plan costs from 380 case studies in the Western Pacific region. Water Science & Technology: Water 381.

Chilenski, S., Greenberg, M., & Feinberg, M. (2007). Community readiness as a multidimensional construct. *Journal of Community Psychology, 35*(3), 347–365.

Closer to Home (C2H). (2014). *Area Cohort Drinking Water Safety Plan (DWSP) completion sessions: Evaluation summary report*. http://awwoa.ab.ca/public/download/documents/13309. Accessed 15 Feb 2016.

Davison, A., & Deere, D. (2007). *Water safety plan workbook for drinking water: Materials for training of trainer*. Singapore: World Health Organization Western Pacific Regional Office. http://www.waterlinks.org/system/resources/W1siZiIsIjIwMTUvMDUvMDYvNXc2ejI3MTg2YV9XYXRlcl9TYWZldHlfUGxhbl9Xb3JrYm9va19mb3JfRHJpbmtpbmdfV2F0ZXIucGRmIl1d/Water%20Safety%20Plan%20Workbook%20for%20Drinking%20Water.pdf?sha=30a10e0d2c8f7062. Accessed Mar 2013.

Davison, A., Deere, D., Stevens, M., Howard, G., & Bartram, J. (2006). *Water safety plan manual*. http://www.who.int/water_sanitation_health/dwq/manual.pdf

de Loë, R. (2009, July–August 21–25). A Canadian vision and strategy for water in the 21st century. Policy Options.

Donnermeyer, J., Plested, B., Edwards, R., Oetting, G., & Littlethunder, L. (1997). Community readiness and prevention programs. *Journal of the Community Development Society, 28*, 65–83.

Dore, M. (2015). *Water policy in Canada: Problems and possible solutions*. Cham: Springer International Publishing.

Dunn, G., Bakker, K., & Harris, L. (2014). Drinking Water Quality Guidelines across Canadian provinces and territories: Jurisdictional variation in the context of decentralized water governance. *International Journal of Environmental Research Public Health, 11*, 4634–4651.

Dyck, A., Exner, M., & Kramer, A. (2007). Experimental based experiences with the introduction of a water safety plan for a multi-located university clinic and its efficacy according to WHO recommendations. *BMC Public Health, 7*, 34.

Edwards, R., Jumper-Thurman, P., Plested, B., Oetting, E., & Swanson, L. (2000). Community readiness: Research to practice. *Journal of Community Psychology, 28*(3), 291–307.

Foster-Fishman, P. G., Cantillon, D., Pierce, S. J., & Van Egeren, L. A. (2007). Building an active citizenry: The role of neighborhood problems, readiness, and capacity for change. *American Journal of Community Psychology, 39*(1–2), 91–106.

Gagnon, G., Castleden, H., Kot, M., Perrier, E., & Chandler, K. (2015). *Drinking water safety plans*. Canadian Water Network. http://www.cwn-rce.ca/assets/Uploads/Drinking-Water-Safety-Plan-Report-2015.pdf. Accessed July 2015.

Gelting, R. J., Delea, K., & Medlin, E. (2012). A conceptual framework to evaluate the outcomes and impacts of water safety plans. *Journal of Water, Sanitation and Hygiene for Development, 2*(2), 103.

Goodman, R., Speers, M., McLeroy, K., Fawcett, S., Kegler, M., Parker, E., & Wallerstein, N. (1998). Identifying and defining the dimensions of community capacity to provide a basis for measurement. *Health Education & Behaviour, 25*(3), 258–278.

Gregor, J., & Winstanley, A. (2005). Considering policy implementation alongside policy formulation in drinking water management in New Zealand and for Pacific Islands. *Proceedings of the Water Environment Federation, WEFTEC 2006: Session 11 through Session, 20*(14), 1404–1417.

Gunnarsdóttir, M. J., & Gissurarson, L. R. (2008). HACCP and water safety plans in Icelandic water supply: Preliminary evaluation of experience. *Journal of Water and Health, 3*(6), 377–382.

Gunnarsdóttir, M. J., Gardarsson, S. M., Elliott, M., Sigmundsdottir, G., & Bartram, J. (2012). Benefits of Water Safety Plans: Microbiology, compliance, and public health. *Environmental Science & Technology, 46*(14), 7782–7789.

Hasan, T. J., Hicking, A., & David, J. (2011). Empowering rural communities: Simple water safety plans. *Water Science & Technology: Water Supply, 11*(3), 309–318.

Health Canada. (2003). Public health initiatives related to drinking water quality in Canada. http://www.hc-sc.gc.ca/ewh-semt/alt_formats/hecs-sesc/pdf/pubs/water-eau/phi-isp/phi-isp-eng.pdf. Accessed 18 Jan 2016.

Hrudey, S. E. (2011). *Commentary 323 – Safe drinking water policy for Canada – Turning hindsight into foresight*. C.D. Howe Institute. http://www.cdhowe.org/safe-drinking-water-policy-for-canada-%E2%80%93-turning-hindsight-into-foresight/8585. Accessed 4 June 2013.

Hrudey, S. E., Hrudey, E. J., & Pollard, S. (2006). Risk management for assuring safe drinking water. *Environment International, 32*(8), 948–957.

Jayaratne, A. (2008). Application of a risk management system to improve drinking water safety. *Journal of Water and Health, 6*(4), 547–557.

Kelly, K. J., Edwards, R. W., Comello, M. L. G., Plested, B. A., Thurman, P. J., & Slater, M. D. (2003). The community readiness model: A complementary approach to social marketing. *Marketing Theory, 3*(4), 411–426.

Kot, M. (2015). *Readiness for ensuring safe drinking water in small community systems*. Dissertation, Dalhousie University.

Kot, M., Castleden, H., & Gagnon, G. A. (2011). Unintended consequences of regulating drinking water in rural Canadian communities: Examples from Atlantic Canada. *International Journal of Health and Place, 17*(5), 1030–1037.

Kot, M., Castleden, H., & Gagnon, G. A. (2014). The human dimension of water safety plans: A critical review of literature and information gaps. *Environmental Reviews, 23*(1), 24–29.

Livernois, J. (2002). *The economic costs of the Walkerton water crisis* (The Walkerton Inquiry, Commissioned Paper 14). http://www.archives.gov.on.ca/en/e_records/walkerton/part2info/commissuepapers/13livernois/14-Livernois1.pdf

Mahmud, S. G., Shamsuddin, S. A. J., Ahmed, M. F., Davison, A., Deere, D., & Howard, G. (2007). Development and implementation of water safety plans for small water supplies in Bangladesh: Benefits and lessons learned. *Journal of Water and Health, 5*(4), 585–597.

Medema, G. J., Payment, P., Dufour, A., Robertson, W., Waite, M., Hunter, P., & Andersson, Y. (2001). *Safe drinking water: An ongoing challenge. Assessing microbial safety of drinking water*. London: WHO/OECD.

Metge, S., Brodard, E., & Conan, M. (2003). Lyonnaise des Eaux: Application of HACCP principles for in drinking water. *Proceedings of water safety—Risk management strategies for drinking water. Umweltbundesamt*, (Federal Environmental Agency), Berlin. http://www.umweltbundesamt.de/en/publikationen/water-safety-berlin-28-30-april-2003-conference. Accessed 3 Oct 2013.

Mudaliar, M. M. (2012). Success or failure: Demonstrating the effectiveness of a water safety plan. *Water Science and Technology: Water Supply, 12*(1), 109–116.

Mullenger, J., Ryan, G., & Hearn, J. (2002). A water authority's experience with HACCP. *Water Science and Technology, 2*(5–6), 149–155.

National Collaborating Centres for Public Health (NCCPH). (2011). *Small drinking water systems*. http://www.nccph.ca/173/Definition_-_Small_Drinking_Water_System.ccnsp. Accessed 15 Feb 2016.

O'Connor, D. (2002). *Report of the Walkerton inquiry Part 2: A strategy for safe drinking water*. http://www.attorneygeneral.jus.gov.on.ca/english/about/pubs/walkerton/part2. Accessed 20 Apr 2013.

Oetting, E. R., Donnermeyer, J. F., Plested, B. A., Edwards, R. W., Kelly, K., & Beauvais, F. (1995). Assessing community readiness for prevention. *The International Journal of the Addictions, 30*, 659–683.

Perrier, E., Kot, M., Castleden, H., & Gagnon, G. (2014). Drinking water safety plans: Barriers and bridges for small systems in Alberta, Canada. *Water Policy, 16*(6), 1140–1154.

Plested, B., Edwards, R., & Jumper-Thurman, P. (2006). *Community readiness: A handbook for successful change*. The Tri-Ethnic Centre for Prevention Research. http://triethniccenter.colostate.edu/CRhandbookcopy.htm. Accessed Jan 2012.

Plummer, R., Velaniškis, J., de Grosbois, D., Kreutzwiser, R. D., & de Loë, R. (2010). The development of new environmental policies and processes in response to a crisis: The case of the multiple barrier approach for safe drinking water. *Environmental Science & Policy, 13*(6), 535–548.

Reid, D., Abramowski, K., Beier, A., Janzen, A., Lok, D., Mack, H., Radhakrishman, H., Rahman, M., Schroth, R., & Vatcher, R. (2014). Implementation of Alberta's drinking water safety plans. *Water Quality Research Journal of Canada, 49*(1), 5–9.

Rinehold, A., Corrales, L., Medlin, E., & Gelting, R. (2011). Water safety plan demonstration projects in Latin American and the Caribbean: Lessons from the field. *Water Science & Technology, 11*(3), 297–308.

Schmoll, O., Castellexner, C., & Chorus, I. (2011). From international developments to local practice: Germany's evaluation and dialogue process towards water safety plan implementation. *Water Science and Technology: Water Supply, 11*, 379–388.

Seghezzo, L., Gatto D'Andrea, M. L., Iribarnegaray, M. A., Liberal, V. I., Fleitas, A., & Bonifacio, J. L. (2013). Improved risk assessment and risk reduction strategies in the Water Safety Plan (WSP) of Salta, Argentina. *Water Science & Technology: Water Supply, 13*(4), 1080.

Summerscales, I. M., & McBean, E. (2011). Incorporation of the Multiple Barrier Approach in drinking water risk assessment tools. *Journal of Water and Health, 9*(2), 349–360.

World Health Organization. (2004). *Guidelines for drinking water quality*. Geneva: World Health Organisation.

World Health Organization. (2011). *Guidelines for drinking water quality. Chapter 4*. http://www.who.int/water_sanitation_health/water-quality/guidelines/dwq-guidelines-4/en/. Accessed 15 Feb 2016.

World Health Organization. (2012). *Water safety planning for small community water supplies: Step-by-step risk management guidance for drinking-water supplies in small communities*. http://www.who.int/water_sanitation_health/publications/small-comm-water_supplies/en/. Accessed 15 Nov 2015.

York, N., Hahn, E., Rayens, M., & Talbert, J. (2008). Community readiness for local smoke-free policy change. *American Journal of Health Promotion, 23*(2), 112–120.

Chapter 18
Public Health at the Watershed Scale

Karen Morrison, Martin J. Bunch, and Lars Hallström

Abstract Public health is usefully defined and managed at a variety of spatial spaces. This place-based, healthy settings approach has led to the formation of a number of successful programs and policies, such as the healthy cities, healthy neighborhoods, healthy schools and healthy homes movements. The application of a healthy settings approach to watersheds creates a powerful frame for public policy that enhances both natural and social systems, and is particularly relevant to discussions around climate change. Health and well-being are impacted by the governance and management of watersheds, at multiple scales, in ways that go well beyond the traditional focus on flooding or drinking water. Drawing on examples from watersheds, this chapter outlines the policy implications of watersheds as a setting for health and well-being. It highlights the need for a more strategic approach to watershed governance that actively seeks linkages with public health institutions in order to more effectively leverage scarce resources to meet common goals. It fills a gap in our understanding of the links between watershed-level programming and public health.

18.1 Introduction

The traditional disconnect between water and public health policy is being bridged by responses to global climate change and associated implications for human and environmental health (McMichael et al. 2006, 2012; Vineis 2009). At the heart of the debate is the role that public health agencies can and should play in environmental policy. In this chapter we argue that active engagement of public health professionals in water policy discussions is imperative if we are to confront the challenge

K. Morrison (✉) • M.J. Bunch
Faculty of Environmental Studies, York University, 4700 Keele St, Toronto, ON, Canada, M3J 1P3
e-mail: kmorriso@yorku.ca; bunchmj@yorku.ca

L. Hallström
Resource Economics and Environmental Sociology, University of Alberta, 116 St & 85 Ave, Edmonton, AB, Canada, T4V 2R3
e-mail: lars.hallstrom@ualberta.ca

S. Renzetti, D.P. Dupont (eds.), *Water Policy and Governance in Canada*, Global Issues in Water Policy 17, DOI 10.1007/978-3-319-42806-2_18

of water governance in the twenty-first Century (Huntington 2006; Trenberth 2011). Through its concern for equity and its increasing acceptance of the importance of ecological determinants of health, the public health community is slowly making inroads into this area. Water policy has traditionally been the purview of politicians, engineers, and more recently, ecologists. As pressure on our water resources increases, there is a need for an inclusive social-ecological systems perspective to water governance grounded in public health principles.

Approaches to public health based on physical places or settings are a well-established practice. There are many examples of such initiatives, including the World Health Organization's Healthy Cities campaign (Ashton et al. 1986; Northridge et al. 2003), and a variety of local Healthy Schools (Frumkin and Fox 2011; Warwick et al. 2005), Healthy Neighbourhoods (El-Askari et al. 1998), and Healthy Homes programs (Turcotte et al. 2014). Yet the settings concept has only recently been applied to ecosystem boundaries, such as watersheds (Bunch et al. 2011; Morrison et al. 2012; Parkes et al. 2010) and wetlands (Horwitz et al. 2012). There are many advantages to framing watersheds as health issues using the settings rhetoric: it is easy for public health professionals to understand (once they learn what a 'watershed' is); it links to a familiar intersectoral program of action (WHO 2010, 2015); and it expands existing health programming to include natural ecosystems so as to manage ecological determinants of health.

Watershed boundaries are particularly well-suited to these discussions. Watershed governance was recognized by the Millennium Ecosystem Assessment (Cork et al. 2005) as a cornerstone of the "adaptive mosaic" scenario – a future that appears feasible and (to us) more desirable than the other scenarios presented. The scenario foresees regional watershed-scale ecosystems as the focus of political and economic activity, with the development of networks among communities, regions, and nations to better manage the global commons (Cork et al. 2005). Watershed-based organization with an intersectoral focus has strong justification in the water management community, and forms a key element of the 'ecosystem approach' implemented by the International Joint Commission for management of the Great Lakes Water Quality Agreement (Allen et al. 1994; International Joint Commission 2012). There is a wide range of watershed governance systems currently operating in Canada, with a notable gap in such programming in Northern Canada (Robins 2007; Roy et al. 2008; Senecal and Madramootoo 2005). Several provinces also have widespread systems of quasi-governmental or voluntary watershed planning (e.g. Conservation Authorities in Ontario, the Fraser Basin Council in BC and Save Our Seine in Manitoba). For the most part, the role of public health has been limited to concerns about drinking water and flooding. As one watershed manager told us:

> When watershed organizations work with local health organizations, it is generally on the same projects, from different perspectives, not from a singular perspective that connects health and the environment (Morrison et al. 2012).

This disconnect persists, despite long-standing identification. In fact, the need for such a social-ecological perspective was recognized by the public health community in the Ottawa Charter for Health Promotion (WHO 1986):

> Our societies are complex and interrelated. Health cannot be separated from other goals. The inextricable links between people and their environment constitute the basis for a socioecological approach to health.

> The overall guiding principle for the world, nations, regions and communities alike is the need to encourage *reciprocal maintenance* – to take care of each other, our communities and our natural environment.

More recently, this notion was restated by the Canadian Public Health Agency in its discussion paper on the ecological determinants of health (Hancock et al. 2015). There are ample opportunities, and indeed an urgent need, for this public health role to be expanded to include upstream/downstream issues around water sovereignty and equity, social justice, peace building and ecosystem services protection so as to ensure that watershed ecosystems positively support the health of current and future Canadians.

The health dimensions of watershed ecosystems will need particular attention in the face of a changing climate that will influence not only water supply and quality, but also the vulnerability many populations. Climate change will have dramatic effects on all ecosystems, with water resources particularly affected. Falkenmark refers to the ideas of hydrosolidarity and of a hydrological imperative to remind us that our ground and surface water resources (including the so-called 'green' water found in our soils), is ultimately determined by precipitation patterns which we know are changing to include more frequent and severe weather events such as storms and drought (Falkenmark 2003; Falkenmark and Folke 2002). Similarly, the impacts of climate change on health are well-documented in the Intergovernmental Panel on Climate Change reports (Confalonieri et al. 2007). The health impacts specifically associated to water management are summarized in Table 18.1.

18.1.1 Nested Hierarchy of Watersheds

The threats to population health that can be linked to water resources management are social, economic and ecological. As a result, adopting a multi-level nested hierarchy of watersheds is well-suited to address these concerns. This concept illustrates how the scale of a watershed relates to how quickly it reacts in terms of space and time to change. All watersheds (also referred to as catchments or river basins) share the same hydrological feature, that of draining precipitation from higher to lower elevations. Large systems (for example, river basins of 25,000 km^2) react more slowly to changing weather and climate than do smaller watersheds (e.g., 500 km^2).

A critical feature of the watershed construct is its ability to link local, regional and transnational concerns. The nested hierarchy of watersheds enables watershed governance to occur at multiple spatial scales, from the urban and rural creek to the large river or lake basin, and to a transboundary water system that may encompass one or more provinces or nation-states. In Canada, the Fraser, Lake Winnipeg and

Table 18.1 Health impacts, climate-related causes and potential health effects related to water management

Category	Climate-related causes	Projected/possible health effects
Extreme weather events	More frequent and violent thunderstorms, more severe hurricanes and other types of severe weather	Death, injury and illness from violent storms, floods, heavy snow, landslides, ice, etc.
		Road accidents caused by extreme precipitation;
	Heavy rains causing landslides and floods	Social and emotional injury and long-term mental harm from loss of loved ones, property and livelihoods
	Canal breaching	
		Health impacts due to food or water shortages
	Rising sea levels and coastal instability	Hypothermia and frostbite from disruption of heating systems
		Illnesses related to drinking water contamination
	Severe ice storms	Effects of displacement of populations and crowding in emergency shelters
	Extreme snowfall events	
	Social and economic changes	Indirect impacts from ecological changes, infrastructure damages and interruptions to health services
		Psychological health effects, including mental health and stress-related illnesses
Contamination of food and water	Contamination of drinking and recreational water by run-off from heavy rainfall and rapid snowmelt	Outbreaks of strains of micro-organisms, such as *E. coli, Crytosporidium, Giardia, S. typhi*, amoebas and other water borne pathogens
	Cross-contamination of water mains by sewage	Food-borne illnesses
		Intoxication from the ingestion, dermal exposure and/or inhalation of algal toxins (several of which are also suspected carcinogens)
	Changes in marine and freshwater environments, including run-off patterns and turbidity, that result in algal blooms and higher levels of toxins in drinking and irrigation water, fish and shellfish	
		Other diarrheal and intestinal diseases
Vector-borne diseases	Changes in the biology and ecology of various disease-carrying insects, ticks and rodents (including geographical distribution)	Increased incidence of vector-borne infectious diseases, such as hantavirus and plague, including post-disaster exposure (e.g. to insects, mice and rats)
	Faster maturation for pathogens within insect and tick vectors	Introduction of infectious diseases new to Canada
	Longer disease transmission seasons	Possible emergence of new diseases, and of those previously eradicated in Canada

Adapted from Health Canada (2001) and Segun (2008)

Table 18.2 Selected links between public health and watershed scale

Watershed scale	Links to public health
Transboundary	Peace building/water wars; equity; justice; transparency;
Large regional	Upstream-downstream justice and equity; downstream eutrophication and cyanotoxins; securing baseline water flows; transparency over water permitting; cultural and spiritual values;
Regional	Pathogen and contaminant transportation; mental health; social capital; public access to information about land uses, water quality and quantity; education; equity and access to resources; consumption of fish and game
Sub-watershed	Social capital; community building; cultural uses; values; recreational water use;

Great Lakes-St. Lawrence Basins are three of a number of major systems that require focused resources and attention. Upstream and downstream tensions are critical to navigate. As a result of changes in precipitation patterns, increasing demand for water resources, and intense pressure on land and water resources, these tensions are anticipated to increase in their severity in the coming decades (see Table 18.2).

18.2 Watershed Governance and Public Health

Socio-ecological watershed systems are highly complex. The watershed context may be local or regional, it may be rural or urban, heavily or lightly managed. Unique combinations of topography, geology and biology react to, and store, water from precipitation events in different ways. Added to this are the socio-economic and cultural dynamics at play in different parts of the country. The intersection between watersheds and public health will play out differently in different places, and can incorporate concerns over such diverse topics as: shellfish harvesting, mercury contamination, acid rain, recreational water use and access, and industrial and agricultural permits to take water. As a result, the values placed on these topics by various parties may vary significantly, making the institutional, organizational and procedural methods adopted and adapted very important. In other words, how water is governed, and the policies, programs and tools that result, are also important.

Public policy is perhaps best defined as "anything a government chooses to do or not to do" (Dye 1972, p 2). Given the impacts and complexity noted above, we contend that the policy discourse about governments and other stakeholders operating on a watershed basis, and the suite of choices and options available to them, needs to be extended to conceptualize watersheds as settings for human health and well-being. Indeed, the failure to integrate health and well-being into policy-making at the watershed scale is a missed opportunity to address the social and ecological determinants of health that manifest within and at the scale of watersheds.

A watersheds-as-settings-for-health approach to public policy is perhaps best clarified using the traditional policy process heuristic. As an approach, conceptualizing watersheds as settings for health can, and should, influence all stages of the policy process, from problem identification, planning, design, and implementation through evaluation and adaptation. To do so has the additional benefit of potentially transcending traditional political and sectoral boundaries systems by emphasizing watershed-based scope, spaces and populations. This approach is consistent with a broader "policy design" (Bobrow and Dryzek 1987) perspective that sits "above" specific policy issues such as scarcity or quality, and that instead acknowledges the context, audience and values that are specific to watersheds and watershed governance.

As an example of this, in a 2012 workshop in Ontario, we brought together watershed agencies, academics, non-governmental organizations and public health representatives to discuss the benefits of thinking about watersheds as settings for population health. The group were asked to describe a suite of common values shared by public health and watershed agencies that have been useful in their experience to articulate the value of collective and collaborative action (Morrison 2015). These shared values are listed in Table 18.3 below. The values that emerge when framing watersheds-as-settings-for-health are quite different than those focused on more traditional approaches to water governance. Thus, the perspectives represented in such an exercise are a critical factor. Watershed boundaries unite local and regional governments intimately. For example, one community's effluvia may flow into the drinking water intakes of another community downstream. The analogy that commonly emerges is one of water as the lifeblood of a natural system, flowing through rivers, lakes and aquifers as the veins and arteries of a watershed ecosystem. It is this lifeblood that creates a unique space for watershed-scale conversations and engages governmental agencies, the private sector and civil society in a search for solutions.

Table 18.3 Shared values for healthy community and healthy watersheds (Ecohealth Ontario 2012)

List of shared values between watershed and public health
Concern with population and community health;
Recognition that well-being is a holistic concept that requires new and non-traditional thinking about health and social programming;
Commitment to promoting wellness and preventing illness and harm;
Understanding that natural or green infrastructure (i.e. forest cover) is as fundamental as built infrastructure for sustaining human and animal health and well-being;
Acknowledgement that citizens expect scarce resources to be effectively leveraged to achieve common societal goals;
Appreciation of the need to address social and environmental determinants of health at multiple scales, including the watershed (e.g. to improve equity, social capital, education, water quality and quality, etc.), and
Recognition that climate change is a watershed and a health issue; it is unavoidable and of serious concern.

Table 18.4 Millennium Ecosystem Assessment (MEA) definitions of five domains of human health and well-being (Corvalan et al. 2005)

Health domain	Millennium ecosystem assessment definitions
Security	"Safety of person and possessions, secure access to necessary resources, and security from natural and human-made disasters"
Basic material for a good life	"The ability to have a secure and adequate livelihood, including income and assets, enough food and water at all times, shelter, ability to have energy to keep warm and cool, and access to goods"
Health	"The ability of an individual to feel well and be strong, or in other words to be adequately nourished and free from disease, to have access to adequate and clean drinking water and clean air, and to have the ability to have energy to keep warm and cool"
Good social relations	"The presence of social cohesion, mutual respect, and the ability to help others and provide for children"
Freedom of choice and action	"The ability of individuals to control what happens to them and to be able to achieve what they value doing or being"

Building from this analogy, the Millennium Ecosystem Assessment (Corvalan et al. 2005) describes five domains of health and well-being that provide a useful framework to discuss the public health benefits of watershed management: security, basic material for a good life, health, good social relations, and freedom of choice and action (see Table 18.4). We elaborate on these domains with respect to watersheds as settings for human health and well-being, in the sections below.

18.2.1 Security

Effective water and land management provides a safe and secure living environment for human populations through protection from flood and drought as well as land subsidence (including landslides). Changing precipitation patterns and the loss of ice cover are symptoms of potentially massive changes to hydrology across the country linked to global climate change (Warren and Lemmen 2004; Kundzewicz et al. 2013; Corvalan et al. 2005). As previously mentioned, and as put forward by the MEA, much of our planned adaptation for climate change will by necessity be based within a watershed context. Water management and land management are essentially synonymous, as actions taken on land affect water resources and vice versa. For example, soil quality is affected by the so-called 'green water' that resides in plants and soil resources, and food security is threatened by inadequate access to both irrigated (blue) and green water resources (Postel 1996; Rosegrant et al. 2009). In this way the watershed construct is critical to both land and water resource management.

Secure resource access is a critical component of human health and well-being. Increasingly, access to public lakes and rivers is being limited through privatization, excluding some segments of society from the benefits (including physical, social, emotional and spiritual) of being out on, or in proximity to, the water. Water resources are also over-allocated in many major river basins in Canada. The over allocation of water threatens river baseflow (and thus the biota that depend on them). It also threatens the public's access to critical water resources in times of scarcity and drought. This is particularly important to consider as we enter an era of highly variable resource conditions. There are already rivers in Canada, for example, the Netchako River in British Columbia, that in times of drought do not reach the sea. This has dire consequences for salmon and other fish populations, and for the human populations that depend on these communities for physical, cultural and spiritual sustenance.

Tensions between nation states are only one manifestation of the broader tension between upstream and downstream communities in a watershed (Meyers 2004). In larger basins, upstream communities tend to be more rural. These areas have critical influence over land uses that affect water quality and quantity downstream. Downstream populations tend to be more urban with larger populations and demands on the water resources running through their communities. One of the clearest examples of this connectivity is that the sewage outfalls of upstream communities go into the very rivers that provide the drinking water intake for downstream communities. In some countries, such as Costa Rica, systems of Payments for Ecological Services are being used whereby the wealthier and more powerful urban communities subsidize upstream watershed protection and restoration efforts aimed at improving downstream water quantity and quality (Pagiola 2008). Another example of this is the Catskill Mountains that provide drinking water to New York City and are subject to a number of restrictions of land uses in return for compensation from the City (Pires 2004).

The importance and urgency of the need to address water/human health issues in particular watersheds is exemplified by the concept of 'water wars', whereby those who have the power and influence to acquire water resources undermine the access to water of others. The result may be a scenario in which a war is the inevitable outcome. In 1995, Ismail Serageldin, former Vice President for Special Programs of the World Bank stated that "If the wars of this century were fought over oil, the wars of the next century will be fought over water."

Heeding a warning such as this one increases in importance given the rising tensions in transboundary basins around the world (Annin 2006; Barlow 2009; Dinar 2009; Solomon 2011); many of these are captured in the Transboundary Freshwater Dispute Database (Oregon State University and Northwest Alliance for Computational Science and Engineering 2015).

18.2.2 Basic Material for a Good Life

That water provides the basic material for a good life is well-understood, although perhaps not adequately valued. Freshwater, moving through the hydrological cycle is the basis for all life on Earth. We ourselves are comprised of approximately 80 %

water. We have already noted the link between land and water management; yet the links between land-water systems, economics, and Canadian livelihoods are often less well-documented. For example, key domestic industries including food, lumber and aggregate extraction can be directly affected by changes in water quality and water supply. In Ontario, aggregate mining has emerged as a source of tension and public protest due to its impact on groundwater tables. In cases where the water is polluted, for example with acid rain, the lake flora and fauna and the rate of growth of forests are directly affected (Likens et al. 1996). According to the Canadian Acid Raid Deposition Science Assessment:

> The emerging science linking acid deposition to reductions in the productivity of forests suggests that reducing acid deposition could have significant economic benefits to the forestry sector (Environment Canada 2004).

In arid and drought-prone regions like Palliser's triangle in Southern Alberta and Saskatchewan, the impacts of climate change include the rapid melting of the Rocky Mountain glaciers that feed the region's rivers. Short term impacts on water resources are already being identified (Barnett et al. 2005; Schindler and Donahue 2006). Long term, and adverse consequences are anticipated for hydro-electric energy and food security in the region. As indicated in a Natural Resources Canada report on climate change:

> For much of western Canada, snowmelt and glacier runoff from mountainous areas are primary sources of water supply for downstream regions. With warmer conditions, the seasonal and long-term storage capacity of alpine areas may decrease, due to thinner snowpacks, more rapid spring runoff, and decreased snow and ice coverage. This, in turn, would result in lower summer river flows and therefore greater water shortages during the period of peak demand. Recent trends observed on the eastern slopes of the Canadian Rocky Mountains suggest that the impacts of diminishing glacier cover on downstream flows are already being felt (Lemmen and Warren 2004, p 38–39).

Maintaining "swimmable, fishable and drinkable" water resources was identified as a core guiding principle for the International Joint Commission for the Great Lakes (International Joint Commission 2012). Water security depends on resilient source water protection and the judicious use and protection of water by agricultural and industrial sources. Unsustainable practices leading to industrial and agricultural pollution, combined with natural contamination, such as some sources of mercury, already affect our ability to safely consume wild fish and game harvested from Canadian waters. The Ontario Sport Fish Consumption guidelines outline the limits on the amount of fish that can be eaten from Ontario lakes due to concerns over persistent organic pollutants and metals. For communities that rely on wild fish for subsistence, these contaminants pose serious health risks, particularly for children and women of child bearing age. In many Canadian cities, new immigrants are unaware of the potentially negative impacts of eating local fish and game and culturally appropriate messaging and outreach is required. This includes urban invertebrate populations, such as crawfish, which have not traditionally been harvested for food in urban areas, but are increasingly being identified as food resources by new Canadians.

The provision of a baseline quantity and quality of accessible (and affordable) water is considered a basic human right. In July 2010, the UN General Assembly and Human Rights Council recognized the human right to water and sanitation as "essential for the full enjoyment of life and all human rights" (Heller 2015).

This right had been heavily contested in the international arena due to the cost and difficulty inherent in providing universal access to effective water and sanitation services (Baer and Gerlak 2015; Alberta Centre for Sustainable Rural Communities 2012). The consequences of – and policy response to – poor water governance systems was dramatically illustrated in Walkerton, Ontario where seven people died and thousands were made ill due to an inadequate drinking water quality system and a "complex interplay and intertwining of social and ecological processes" (Ali 2004; see also O'Connor 2002) that were apparent at the scale of the watershed. This event was a wakeup call to Canadians and stimulated responses to protect people's health – for example, an overhaul of Ontario's drinking water legislation. In Canada, failures to provide this basic human right are particularly noticeable in rural, remote, Northern and First Nations communities, where safe and affordable drinking water cannot be taken for granted.

18.2.3 Health

The links between human health and wellbeing and watersheds include all aspects of health: physical, social, emotional, spiritual, cultural and mental. Many of the links to physical health are well-established (e.g. from flooding), but others, such as community-acquired infections (infections where the cause is unknown) are increasingly being linked to elements of watershed management. For example, Pintar et al. (2010) highlight the risk of acquiring cryptosporidium from contaminated recreational water due to agricultural waste in a Southern Ontario community. Similarly, Jiang et al. (2015) note the links between salmonellosis and extreme temperature and precipitation events in the Chesapeake Bay (Maryland and Virginia, USA). Thinking through the implications of, and the interventions necessary to contain contaminant and pathogen transport on a watershed scale is vital to the protection of human health, particularly as antibiotic resistance escalates and increasing pressure is put on finite and shared water resources.

The links between mental health and contact with nature have been the focus of a myriad of studies. Exposure to variegated natural landscapes, and in particular to water features (rivers, lakes, streams, oceans) and green spaces are critical features of this literature (Louv 2008; Townsend and Weerisuriya 2010). Theories such as affective responses to nature, nature deficit disorder, attention restoration and extinction of experience are all founded on the link between mental and emotional health and natural landscapes. The mental health implications of natural disasters, such as flooding, are also significant and are exacerbated by a lack of understanding of local watershed and floodplain dynamics (Goldmann and Galea 2014; Morrison 2015; Stanke et al. 2012). Although there are many potential health impacts, mental health is especially concerning post-flooding in many areas. The psycho-social impacts of flooding are well-recognized. For example, following the 2013 Southern Alberta flood, the province of Alberta created a new position, that of a Chief Mental Health Officer, to help guide the response to this regional disaster. In Manitoba, teams of psycho-social workers were dispatched to flood-affected communities following the 2013 floods (Morrison 2015).

The link between watersheds and cultural and spiritual health is well-illustrated by the role that salmon play in both indigenous and non-indigenous Canadian cultures. Salmon test our ability to manage sustainable watersheds through their intimate connection between the oceans and the headwaters of rivers and streams. Threats to salmon populations are key motivators for watershed protection measures on both coasts. Brown trout play a similar role in the Great Lakes region of Southern Ontario. Whereas sub-watershed units may be more appropriate for some issues, the protection of salmon populations requires nothing less than a full watershed approach.

18.2.4 Good Social Relations

Good social relations related to water is a factor relevant to watershed management that can be addressed at multiple scales, ranging from international transboundary water management committees to local water stewardship activities. Access to, and use of, the so-called "blue parks" is critical for a wide range of community associations, clubs and teams. Community groups interested in water protection, use and restoration have the benefit of engaging people from a wide cross-section of the community and provide significant opportunities for inter-generational exchange and learning. They increasingly foster inter-cultural exchanges related to the potential uses for these resources in a community. These follow the full spectrum of stewardship activities, from direct action and intervention, to advocacy, to research and education.

Good social relations are a critical feature of the 'peace building' role that water management agencies can play (Woodrow 2006) – in direct contrast to the water war/water security rhetoric. Indeed, studies have shown that resolving water tensions are more likely to lead to peace building solutions, which is a recognition of its vital role in supporting human life (Phillips et al. 2006; Wolf 2004).

Similar perspectives resonate within the broader community sustainability literature. While sustainable development under Agenda 21 has tended to emphasize economic and environmental sustainability, more recent perspectives such as the community capitals framework (Flora and Flora 2012) and community capacity assessments (Beckley et al. 2009) present a more holistic and integrative vision of the conditions required for positive, integrated sustainability action. Rather than emphasizing the pillars of sustainability (economic, social, environmental, cultural and institutional), these approaches focus upon the wide range of characteristics, resources and assets of the broader community itself. Thus, from a watershed and public health standpoint (and consistent with perspectives such as Kickbusch's Health Society (2007)), it is useful to recognize that human communities, and thus the watersheds of which they are part, have varying levels of capital (human, social, natural, built, financial, political and cultural (Flora and Flora 2012)) that contribute to the reduction of inequities.

Social and environmental justice concerns are clearly evident at the watershed scale; the aforementioned link to payments for ecological services is an indication of our emerging understanding of that feedback loop. Access to and allocation of water resources are contentious issues where a concern for justice is warranted. Decisions made about water, and in particular, decisions that have long-term ramifications such as water pollution, over-allocation, biodiversity loss and salinization are of intergenerational concern. A multi-generational outlook that respects the water requirements of future Canadians is needed to make sound and sustainable water management decisions under difficult circumstances.

18.2.5 Freedom of Choice and Action

The links between water resources and freedom of choice and action are predominantly those of access to information and education. These are components of empowerment which also require accessible and responsive watershed governance organizations. Public participatory access to these governance processes is necessary. There is also an urgent need for better public access to information about water resource quality, quantity and the threats facing these resources at multiple temporal and spatial scales. Building up our collective "Watershed IQ" is a vital part of this equation. Empowerment and access to information are well-served by responsive watershed governance agencies that involve multiple stakeholders. Even where such organizations exist, however, access to information is currently still an issue. There is a need to pull back the curtain on decisions related to permits to take water, certificates of approval, and data regarding contaminants and pathogens, among others. Canadians have the right to know about the current state of our water resources. The Access Initiative of the World Resources Institute targets these rights directly and provides a model to initiate discussions about these critical issues in Canada:

> The Access Initiative is the largest network in the world dedicated to ensuring that citizens have the right and ability to influence decisions about the natural resources that sustain their communities.
>
> Access to information, access to public participation, and access to justice (the three "access rights") are practical means of ensuring that decisions by governments consider sustainable development concerns and the interests of the poor.
>
> An informed and empowered public monitors government and corporate performance, is alert to problems, challenges the conventional wisdom of government or corporate decision-makers, discusses the issues, organizes social and political change, and demands improve-

ments. Where independent courts supply remedy and redress without political interference, the public can better hold decision makers accountable.

The Council for Canadians water campaign highlights the lack of a federal water policy framework, as well as the potentially deleterious impact of recent changes to federal regulations pertaining to water resources (e.g. the Fisheries Act; Navigation Protection Act (formerly the Navigable Waters Protection Act)) and widespread concern over the impact of fracking on groundwater resources (Lui 2014). The critical feature of these regulatory changes is lack of transparency in the process that is causing deep consternation among water management stakeholders. By embedding significant legislative and regulatory changes in omnibus budget bills (e.g. Bill C38 and C45), the legislative process and the ability of Canadians to respond to radical change has been short-circuited. In addition, trade agreements such as the Trans-Pacific Partnership have the potential to erode our ability to make our own choices about protection and access to water by transferring agency from Canadian governments to a panel of lawyers when Canadian environmental laws are challenged by foreign corporations (Gleeson and Friel 2013; Ku and Yoo 2013).

Some watershed organizations recognize the freedom of choice and action aspect of health in their work. For example in Ontario, the Credit Valley Conservation Authority, in partnership with researchers from York University, is working to identify the best ways to communicate links among human health and well-being and ecosystem services on a watershed basis. The 'Well-being and Your Watershed' project utilizes indictors and narratives to create a more fulsome representation of the benefits that watershed management has for human population health and well-being, in part to inform and empower watershed residents to be stewards of their communities. Other watershed agencies are also making these links, including the newly formed Nechako Watershed Roundtable in British Columbia, which is developing a similar watershed portal to engage governance stakeholders and citizens in learning, monitoring, sharing information about, and planning for watershed management. Ecohealth Ontario, a new organization that is jointly chaired by a watershed management agency and a public health agency, has as its goal to develop new collaborations among the human health, social and environmental sectors and to build a common agenda around research, communication and policy. The intent of these collaborations is to foster improved health and well-being outcomes for Ontarians through the provision of better ecosystem quality, increased green space and enhanced access to nature.

18.3 Policy Implications

Given the complexities and challenges of linking social, health and environmental questions of equity, access and governance across conventional geo-political jurisdictions, the watersheds-based approach promoted here leads to a number of implications or conclusions for public policy, watershed governance, public health and natural resource management.

18.3.1 Multi-Scalar Watershed Management is Needed Across Canada

The nested hierarchy of watersheds provides an important foundation to work through the governance and management of ecosystem-based settings for human health and well-being. The challenges facing human populations and the social-ecological systems in which they are embedded manifest differently at different spatial and temporal scales. For air-land-water-society relationships, watershed boundaries serve to coalesce these issues and give them a place-based focus. Issues around access to information are as important as those related to access to our physical resources, including clean water, soil and air. Defining watersheds at the macro-scale, such as in the Lake Winnipeg Basin, the Great Lakes-St. Lawrence Basin and the Fraser Basin is vital to confront looming issues of water justice, security and access. Upstream-downstream tensions can be mitigated through proactive planning, dialogue and the provision of resources at the large river and lake basin scale. Transboundary waters require particular attention. At smaller scales, the issues involved and the stakeholders engaged in the work become more and more local, with opportunities for education, dialogue and the building of social and natural capital. Framing this work in the context of public and population health encourages shared responsibility and a leveraging of resources to gain the multiple co-benefits of restoration and adaptation efforts. In many regions of Canada, watershed planning is not fully conceived as a setting for human health and well-being. Such a dialogue could prove very productive to inspire new and fulsome efforts to tackle systemic social and environmental issues in the new millennium.

18.3.2 Public Health Should Play a Proactive Role in Watershed Management Committees

As indicated above, it is important that public health professionals have a strong seat at the table and actively participate in issues around resource distribution, contamination, equity, access and protection in order to safeguard the health and well-being of current and future generations of Canadians. With the uncertainty caused by projections around hydrology, weather patterns and climate change, water and food security are going to be of interest to Canadians in the coming years. Securing public access to, and ownership of, these resources is a real concern. As the late Anthony McMichael (2008, on-line pagination) wrote, environmental sustainability is a "prerequisite for attaining human security, well-being, health and survival. Why else do we seek sustainability?"

Landscape and ecosystem level protection of the ecological determinants of health is another concern. These links were recently recognized by the United Nations (2015) in the Sustainable Development Goals, which include goals related to: good health and well-being; reduced inequalities; sustainable cities and communities; climate action; life below water; life on land; and, partnerships for the goals. Watershed management at multiple scales integrates all of these concerns.

18.3.3 Watersheds Should Frame Public Health Studies

Public health professionals are just beginning to recognize the importance of watershed-scale studies to better understand complex socio-ecological issues that exist in this space. This includes studies of waterborne pathogens, where much is being learned about the link between land uses (including the treatment of agricultural effluent), local hydrology and the prevalence of community acquired infections, such as cryptosporidium. The need to look beyond individual facilities and industries and take stock of the local, contextual watershed scale requires new approaches, tools and assumptions and is increasingly requiring inter- and transdisciplinary approaches in fields such as epidemiology. Building up a cadre of professionals with the skills and attitudes needed to approach this work is urgently needed. This includes having natural resource management professionals able to engage in a meaningful way with their public health colleagues and vice versa.

18.3.4 Natural Capital and Ecological Services Are a Source of Public Health Benefits

The need for investments in public health infrastructure was well-recognized after the global outbreak of severe acute respiratory syndrome (SARS) in 2002–2003 and Canada's Walkerton water contamination crises in 2000. Less recognized has been the need to invest in natural capital and ecological service provision to protect and enhance public health. The multiple co-benefits of these actions for both human and ecological health are sufficiently well-documented to inspire real change. Having strong public health voices on watershed-scale governance committees will help ensure that the connections between watershed-level actions and the ecological determinants of health are recognized and effectively leveraged. The potential co-benefits and synergies can make better use of limited financial resources in both the public and environmental sectors.

18.3.5 Adaptive Policy and Public Participation Is Critical

Given the complexity of the social, ecological, economic, ethical and health related issues that exist in our watersheds, we need resilient policies that can adapt to the tensions inherent in networked, multi-scalar governance arrangements. We also need interventions that take a long-term perspective on the value of our land and water resources to current and future generations of Canadians. By linking the social determinants of health (Mikkonen and Raphael 2010) with ecological determinants we have the ability to create a governance model for adaptive policy making that realizes the potential of watersheds-as-settings-for-health. The adaptive policy framework put forward by the International Institute for Sustainable

Development provides an excellent basis for this discussion. This framework makes use of monitoring of key system indicators and formalized and scheduled evaluation that may trigger "no-regrets" alternatives for anticipated change and automatic adjustments based on revised understanding of social-ecological conditions in situations where change is unanticipated. Policies informed by the framework are based on "principles of effective intervention in complex adaptive systems" (International Institute for Sustainable Development 2006).

Consideration of the context, values and audiences engaged in watershed governance at multiple scales will help shape a new policy platform that is more robust in its treatment of watershed-scale issues and their effect on public and population health.

18.4 Conclusions

Thinking about watersheds as settings for health and well-being is underdeveloped in Canada, despite our leadership in health promotion and advocating for a strong link between social and ecological systems (as per the Ottawa Charter for Health Promotion). Operationalizing these ideas at the watershed scale has the potential to reinvigorate policies and programs designed to protect the environment and promote human health and wellbeing. The issues at hand are not simply related to ecosystem protection and restoration. They are also social issues around equity, justice, access, transparency and the empowerment of Canadians. Seen in this light, it is evident that the public health voice needs to be heard in debates over landscape and watershed-level issues in order to protect the health and well-being of current and future generations of Canadians. As Sneddon et al. (2002) wrote:

> The problem is not so much that the ecological impacts of human societies are being ignored; in fact, there is increasing recognition that we are dealing with a human-dominated ecosphere. Rather, it is a failure within many of the natural sciences to engage more deeply with questions regarding how the organization of human societies, and the power relations within and between different social groups, might contribute to ecological degradation and conflicts over water.

Working with existing stakeholders, the public health sector has the potential to contribute to identifying and mitigating these critical issues in Canada.

References

Alberta Centre for Sustainable Rural Communities. (2012). *Scoping review of successful practices in water and watershed management in rural Albertan communities* (ACSRC report series, Vol. 15). Camrose: Alberta Centre for Rural Communities, University of Alberta.

Ali, S. H. (2004). A socio-ecological autopsy of the E. coli O157:H7 outbreak in Walkerton, Ontario, Canada. *Social Science & Medicine, 58*(12), 2601–2612. http://doi.org/10.1016/j.socscimed.2003.09.013

Allen, T. F. H., Bandurski, B. L., & King, A. W. (1994). *The ecosystem approach: Theory and ecosystem integrity* (Special report of the ecological committee to the Great Lakes Science Advisory Board, p. 64). Windsor: Great Lakes Science Advisory Board.

Annin, P. (2006). *The great lakes water wars*. Washington, DC: Island Press.

Ashton, J., Grey, P., & Barnard, K. (1986). Healthy cities—WHO's new public health initiative. *Health Promotion International, 1*(3):319–324. http://doi.org/10.1093/heapro/1.3.319

Baer, M., & Gerlak, A. (2015). Implementing the human right to water and sanitation: A study of global and local discourses. *Third World Quarterly, 36*(8):1527–1545. http://doi.org/10.1080/01436597.2015.1043993

Barlow, M. (2009). *Blue covenant: The global water crisis and the coming battle for the right to water* (Reprint edition). New York: New Press.

Barnett, T. P., Adam, J. C., & Lettenmaier, D. P. (2005). Potential impacts of a warming climate on water availability in snow-dominated regions. *Nature, 438*(7066):303–309. http://doi.org/10.1038/nature04141

Beckley, T. M., Martz, D., Nadeau, S., Wall, E., & Reimer, B. (2009). Multiple capacities, multiple outcomes: Delving deeper into the meaning of community capacity. *Journal of Rural and Community Development, 3*(3). http://journals.brandonu.ca/jrcd/article/view/217

Bobrow, D. B., & Dryzek, J. S. (1987). *Policy analysis by design*. Pittsburgh: University of Pittsburgh Press.

Bunch, M. J., Morrison, K. E., Parkes, M. W., & Venema, H. D. (2011). Promoting health and well-being by managing for social-ecological resilience: The potential of integrating ecohealth and water resources management approaches. *Ecology and Society, 16*(1):6. http://www.ecologyandsociety.org/vol16/iss1/art6/

Canada, H. (2001). *Climate change and health: A policy primer*. Ottawa: Health Canada.

Confalonieri, U. B., Menne, R. A., Ebi, K. L., Hauengue, M., Kovats, R. S., Revich, B., & Woodward, A. (2007). Human health. In M. L. Parry, O. F. Canziani, J. P. Palutikof, P. J. van der Linden, & C. E. Hanson (Eds.), *Climate change 2007: Impacts, adaptation and vulnerability. Contribution of working Group II to the fourth assessment report of the intergovernmental panel on climate change* (pp. 391–431). Cambridge: Cambridge University Press.

Cork, S., Peterson, G., & Petschel-Held, G. (2005). Chapter 8: Four scenarios. In *Millenium ecosystem assessment*. New York: United Nations.

Corvalan, C., Hales, S., & McMichael, A. (2005). *Ecosystems and human well-being: Health synthesis*. Geneva: World Health Organization.

Dinar, S. (2009). Scarcity and cooperation along international rivers. *Global Environmental Politics, 9*(1):109–135. http://doi.org/10.1162/glep.2009.9.1.109

Dye, T. R. (1972). *Understanding public policy*. Englewood Cliffs: Prentice Hall.

Ecohealth Ontario. (2012). Healthy communities depend on healthy watersheds. Position Paper, November. Toronto, Canada. http://www.ecohealth-ontario.ca/files/Watershed_Health_FINAL.pdf. Accessed 6 Apr 2016.

El-Askari, G., Freestone, J., Irizarry, C., Kraut, K. L., Mashiyama, S. T., Morgan, M. A., & Walton, S. (1998). The healthy neighborhoods project: A local health department's role in catalyzing community development. *Health Education & Behavior, 25*(2):146–159. http://doi.org/10.1177/109019819802500204

Environment Canada. (2004). *Canadian acid raid deposition science assessment: Summary of key results*. Queen's Printer's Press for Canada, Ottawa. http://s3.amazonaws.com/zanran_storage/www.mining.ca/ContentPages/758093769.pdf

Falkenmark, M. (2003). Water management and ecosystems: Living with change, background Paper No. 9. Global Water Partnership Technical Advisory Committee, Stockholm

Falkenmark, M., & Folke, C. (2002). The ethics of socio-ecohydrological catchment management: Towards hydrosolodarity. *Hydrology and Earth System Sciences, 6*, 1–9.

Flora, C. B., & Flora, J. L. (2012). *Rural communities: Legacy and change* (4th ed.). Boulder: Westview Press.

Frumkin, H., & Fox, J. (2011). Healthy schools. In A. L. Dannenberg, H. Frumkin, & R. J Jackson (Eds.). *Making healthy places: Designing and building for health, well-being, and sustainability* (pp. 216–228). Washington, DC: Island Press/Center for Resource Economics. http://dx.doi.org/10.5822/978-1-61091-036-1_14

Gleeson, D., & Friel, S. (2013). Emerging threats to public health from regional trade agreements. *The Lancet, 381*(9876):1507–1509. http://doi.org/10.1016/S0140-6736(13)60312-8

Goldmann, E., & Galea, S. (2014). Mental health consequences of disasters. *Annual Review of Public Health, 35*(1):169–183. http://doi.org/10.1146/annurev-publhealth-032013-182435

Hancock, T., Spady, D. W., & Soskolne, C. L. (Eds.). (2015). *Global change and public health: Addressing the ecological determinants of health: The report in brief.* Ottawa: Working Group on the Ecological Determinants of Health. Canadian Public Health Agency.

Heller, L. (2015). The crisis in water supply: How different it can look through the lens of the human right to water? *Cadernos de Saúde Pública, 31*(3), 447–449.

Horwitz, P., Finlayson, M., & Weinstein, P. (2012). *Healthy wetlands, healthy people: A review of wetlands and human health interactions* (Ramsar Technical Report No. 6). Geneva: Secretariat of the Ramsar Convention on Wetlands, Gland, Switzerland, & The World Health Organization.

Huntington, T. G. (2006). Evidence for intensification of the global water cycle: Review and synthesis. *Journal of Hydrology, 319*(1–4):83–95. http://doi.org/10.1016/j.jhydrol.2005.07.003

International Institute for Sustainabile Development. (2006). *Designing policies in a world of uncertainty, change, and surprise: Adaptive policy-making for agriculture and water resources in the face of climate change.* Winnipeg/New Delhi: International Institute for Sustainable Development and The Energy and Resources Institute.

International Joint Commission. (2012). Great lakes water quality agreement – 2012: Protocol amending the agreement between Canada and the United States of America on great lakes water quality, 1978, as amended on October 16, 1983, and on November 18, 1987. http://www.ec.gc.ca/grandslacs-greatlakes/A1C62826-72BE-40DB-A545-65AD6FCEAE92/1094_Canada-USA%20GLWQA%20_e.pdf

Jiang, C., Shaw, K. S., Upperman, C. R., Blythe, D., Mitchell, C., Murtugudde, R., & Sapkota, A. (2015). Climate change, extreme events and increased risk of salmonellosis in Maryland, USA: Evidence for coastal vulnerability. *Environment International, 83*:58–62. http://doi.org/10.1016/j.envint.2015.06.006

Kickbusch, I. (2007). Health governance: The health society. In *Health and Modernity* (pp. 144–161). New York: Springer. http://link.springer.com/chapter/10.1007/978-0-387-37759-9_9

Ku, J., & Yoo, J. (2013). Globalization and sovereignty. *Berkeley Journal of International Law, 31*, 210.

Kundzewicz, Z. W., et al. (2013). Flood risk and climate change: Global and regional perspectives. *Hydrological Sciences Journal, 59*(1), 1–28.

Lemmen, D. S., & Warren, F. J. (Eds.). (2004). *Climate change impacts and adaptation: A Canadian perspective*. Ottawa: Climate Change Impacts and Adaptation Directorate, Natural Resources Canada.

Likens, G. E., Driscoll, C. T., & Buso, D. C. (1996). Long-term effects of acid rain: Response and recovery of a forest ecosystem. *Science-AAAS-Weekly Paper Edition, 272*(5259), 244–245.

Louv, R. (2008). *Last child in the woods: Saving our children from nature-deficit disorder* (Updated and Expanded edition). Chapel Hill: Algonquin Books.

Lui, E. (2014). What Harper's 2013 budget bills are doing to watersheds today. Canadian Perspectives, Autumn, pp. 19–20.

McMichael, A. J. (2008). Population, human resources, health, and the environment: Getting the balance right. *Environment*. Jan/Feb. on-line pagination. http://www.environmentmagazine.org/archives/back%20issues/january-february%202008/mcmichael-full.html

McMichael, A. J., Woodruff, R. E., & Hales, S. (2006). Climate change and human health: Present and future risks. *The Lancet, 367*(9513):859–869. http://doi.org/10.1016/S0140-6736(06)68079-3

McMichael, A. J., Montgomery, H., & Costello, A. (2012). Health risks, present and future, from global climate change. *BMJ, 344*, e1359. http://doi.org/10.1136/bmj.e1359

Meyers, N. (2004). *Environmental security: What's new and different?* (Background paper for The Hague Conference on Environment, Security and Sustainable Environment). The Hague: The Hague Conference on Environment, Security and Sustainable Environment.

Mikkonen, J., & Raphael, D. (2010). *Social determinants of health: The canadian facts*. Toronto: York University School of Health Policy and Management.

Morrison, K. (2015). Application of the multiple exposures, multiple effects model to four flood events in Canada: Lessons learned for public health adaptation to extreme precipitation and flooding in the context of climate change (Report prepared for the Peterborough County-City Health Unit). Peterborough.

Morrison, K. E., Parkes, M. W., Hallstrom, L. H., Neudoerffer, R. C., Bunch, M. J., & Venema, H. D. (2012). *Ecohealth and watersheds: Watersheds as settings for health and well-being in Canada* (Network for Ecosystem Sustainability and Health Publication Series, No. 3. No. 978-1-894784-52-8) (p 37). Winnipeg: Network for Ecosystem Sustainability and Health and the International Institute for Sustainable Development. http://www.iisd.org/pdf/2012/eco-health_watersheds_canada.pdf

Northridge, D. M. E., Sclar, D. E. D., & Biswas, M. P. (2003). Sorting out the connections between the built environment and health: A conceptual framework for navigating pathways and planning healthy cities. *Journal of Urban Health, 80*(4):556–568. http://doi.org/10.1093/jurban/jtg064

O'Connor, D. R. (2002). *Report of the Walkerton inquiry: The events of May 2000 and related issues: A summary*. Toronto: Ontario Ministry of the Attorney General.

Oregon State University and Northwest Alliance for Computational Science and Engineering. (2015). Transboundary freshwater dispute database 2002: Atlas of international freshwater agreements. http://www.transboundarywaters.orst.edu/database/. Accessed 3 Nov 2015.

Pagiola, S. (2008). Payments for environmental services in Costa Rica. *Ecological Economics, 65*(4):712–724. http://doi.org/10.1016/j.ecolecon.2007.07.033

Parkes, M. W., Morrison, K. E., Bunch, M. J., & Venema, H. D. (2010). Towards integrated governance for water, health and social-ecological systems: The watershed governance prism. *Global Environmental Change, 20*:693–704. http://doi.org/10.1016/j.gloenvcha.2010.06.001

Phillips, D., Daoudy, M., McCaffrey, S., Öjendal, J., & Turton, A. (2006). *Trans-boundary water co-operation as a tool for conflict prevention and broader benefit sharing*. Ministry of Foreign Affairs Sweden. http://www.egdi.gov.se/publications62.htm

Pintar, K. D. M., Fazil, A., Pollari, F., Charron, D. F., Waltner-Toews, D., & McEwen, S. A. (2010). A risk assessment model to evaluate the role of fecal contamination in recreational water on the incidence of cryptosporidiosis at the community level in Ontario. *Risk Analysis, 30*(1):49–64. http://doi.org/10.1111/j.1539-6924.2009.01321.x

Pires, M. (2004). Watershed protection for a world city: the case of New York. *Land Use Policy, 21*(2):161–175. http://doi.org/10.1016/j.landusepol.2003.08.001

Postel, S. (1996). *Dividing the waters: Food security, ecosystem health, and the new politics of scarcity* (Worldwatch Paper 132). Washington, DC: Worldwatch Institute. http://www.popline.org/node/310100#sthash.LbdaCMJS.dpuf

Robins, L. (2007). Nation-wide decentralized governance arrangements and capacities for integrated watershed management: Issues and insights from Canada. *Environments, 35*(2), 1–47.

Rosegrant, M. W., Ringler, C., & Zhu, T. (2009). Water for agriculture: Maintaining food security under growing scarcity. *Annual Review of Environment and Resources, 34*(1):205–222. http://doi.org/10.1146/annurev.environ.030308.090351

Roy. A., Wenger, S., Fletcher, T., Walsh, C., Ladson, A., Shuster, W., & Brown, R. (2008). Impediments and solutions to sustainable, watershed-scale urban stormwater management: Lessons from Australia and the United States. *Environmental Management, 42*(2):344–359. http://doi.org/10.1007/s00267-008-9119-1

Schindler, D. W., & Donahue, W. F. (2006). An impending water crisis in Canada's western prairie provinces. *Proceedings of the National Academy of Sciences, 103*(19):7210–7216. http://doi.org/10.1073/pnas.0601568103

Segun, J. (Ed.). (2008). *Human health in a changing climate: A Canadian assessment of vulnerabilities and adaptive capacity*. Ottawa: Health Canada.

Senecal, C., & Madramootoo, C. (2005). Watershed management: Review of Canadian diversity. http://www.iwaponline.com/wp/00705/wp007050509.htm. Accessed 3 Nov 2015.

Sneddon, C., Harris, L., Dimitrov, R., & Özesmi, U. (2002). Contested waters: Conflict, scale, and sustainability in aquatic socioecological systems. *Society & Natural Resources, 15*(8):663–675. http://doi.org/10.1080/08941920290069272

Solomon, S. (2011). Water: The epic struggle for wealth, power, and civilization (Reprint edition). New York: Harper Perennial.

Stanke, C., Murray, V., Amlôt, R., Nurse, J., & Williams, R. (2012). The effects of flooding on mental health: Outcomes and recommendations from a review of the literature. *PLoS Currents, 4*. http://doi.org/10.1371/4f9f1fa9c3cae

Townsend, M., & Weerisuriya, R. (2010). *Beyond blue to green: The benefits of contact with nature for mental health and well-being*. Melbourne: Blue Beyond Ltd. http://www.hphpcentral.com/wp-content/uploads/2010/09/beyondblue_togreen.pdf

Trenberth, K. (2011). Changes in precipitation with climate change. *Climate Research, 47*(1):123–138. http://doi.org/10.3354/cr00953

Turcotte, D. A., Alker, H., Chaves, E., Gore, R., & Woskie, S. (2014). Healthy homes: In-home environmental asthma intervention in a diverse urban community. *American Journal of Public Health, 104*(4):665–671. http://doi.org/10.2105/AJPH.2013.301695

Vineis. P. (2009). Climate change and the diversity of its health effects. *International Journal of Public Health, 55*(2):81–82. http://doi.org/10.1007/s00038-009-0092-0

Warren, F. J., & Lemmen, D. S. (Eds.). (2004). *Canada in a changing climate: Sector perspectives on impacts and adaptation* (pp. 191–232). Ottawa: Government of Canada.

Warwick, I., Aggleton, P., Chase, E., Schagen, S., Blenkinsop, S., Schagen, I., & Eggers, M. (2005). Evaluating healthy schools: Perceptions of impact among school-based respondents. *Health Education Research, 20*(6):697–708. http://doi.org/10.1093/her/cyh024

WHO. (1986). *Ottawa charter for health promotion*. Geneva: World Health Organization.

WHO. (2010). *Adelaide statement on health in all policies: Moving towards a shared governance for health and well-being*. World Health Organization. http://www.who.int/social_determinants/hiap_statement_who_sa_final.pdf

WHO. (2015). Intersectoral action for health. http://www.who.int/kobe_centre/interventions/intersectorial_action/en/. Accessed 3 Nov 2015.

Wolf, A. T. (2004). *Regional water cooperation as confidence building: Water management as a strategy for peace*. Berlin: Adelphi Research. http://citeseerx.ist.psu.edu/viewdoc/download?doi=10.1.1.132.2717&rep=rep1&type=pdf

Woodrow, P. (2006). *Water can be a pathway to peace, not war*. Washington, DC: Woodrow Wilson Center, Environmental Change and Security Program. http://www.wilsoncenter.org/sites/default/files/ECSPReport13_NavigatingPeace.pdf

Chapter 19
Chlorination of Drinking Water – Scientific Evidence and Policy Implications

Madjid Mohseni, Edward A. McBean, and Manuel J. Rodriguez

Abstract Identification of the value of chlorine as an efficient oxidant and disinfectant resulted in its subsequent worldwide application for disinfecting water, and is considered one of most important discoveries of the early twentieth century, improving public health. The use of chlorine in various forms continued at worldwide scale, for subsequent decades, until the 1970s when it was identified that chlorine reacts with naturally occurring organic compounds, producing potentially harmful disinfection byproducts (DBPs). Regardless, despite widespread concerns around DBPs and the inclusion in drinking water guidelines and regulations for some DBPs, many communities (especially small and rural communities with limited resources) rely solely on chlorine as primary disinfectant. Also, the use of chlorine as a secondary disinfectant to ensure residual disinfectant in the distribution system is required in many parts of the world, including in Canada. Public resentment to having chlorine in their water is a challenge associated with the use of this disinfectant in drinking water supplies, especially in aboriginal communities, with many disliking the taste and odour of chlorine, making the policy around chlorination of water very difficult and complicated. While science supports the benefits of maintaining residual chlorine in water supplies, it warns of DBP formation and its potential impact to public health. Hence, resistance to implementation of policy due to public perception and resentment continues. The work presented will describe the role of science in shaping the policy and public opinion around the use of chlorine

---

M. Mohseni (✉)
Department of Chemical & Biological Engineering, University of British Columbia, 2360 East Mall, V6T 1Z3 Vancouver, BC, Canada
e-mail: madjid.mohseni@ubc.ca

E.A. McBean
School of Engineering, University of Guelph, 50 Stone Rd E, N1G 2 W1 Guelph, ON, Canada
e-mail: emcbean@uoguelph.ca

M.J. Rodriguez
École supérieure d'aménagement du territoire et de développement régional (ÉSAD), Laval University, 2325 Rue de l'Université, G1V 0A6 Ville de Québec, QC, Canada
e-mail: manuel.rodriguez@esad.ulaval.ca

S. Renzetti, D.P. Dupont (eds.), *Water Policy and Governance in Canada*, Global Issues in Water Policy 17, DOI 10.1007/978-3-319-42806-2_19

in drinking water supplies, with particular emphasis on small and rural communities.

19.1 Introduction

Elimination of the risk of microbiological contamination of drinking water is achieved through a combination of three complementary strategies: (i) ensuring protected or pathogen-free source water, (ii) applying appropriate purification or disinfection to eliminate/inactivate pathogens in the water, and (iii) ensuring a secure distribution system that delivers safe water to consumers. A combination of multiple stressors, such as population increases and natural resource extraction, has led to reduced availability of protected source water, and hence, the greater need and emphasis on providing effective water treatment to reduce microbiological contaminants. As well, there is an emphasis on improving distribution system integrity to ensure that contamination does not occur during drinking water transport from the point of treatment to individual consumers.

Treatment of water to reduce the risk of microbiological contamination is achieved via a process generally referred to as "disinfection". While there are a number of disinfection technologies (e.g., ultraviolet (UV) light, ozone, membrane filtration, etc.), chemical disinfection with chlorine is the most widely used disinfection process. The merit of disinfection of water using chlorine was one of the most important discoveries related to human health in the history of mankind. Since the late 1800s, when first applied in Europe and then in the United States in the early 1900s, chlorine has had major impacts on reducing the incidence of waterborne disease worldwide (Morris 2007).

Chlorine, in the form of a gas or solution, is the most commonly used disinfectant throughout the world. According to a 1995 EPA survey, approximately 64 % of community groundwater and surface water systems disinfect their water with chlorine (US EPA 2000). However, the importance of alternatives to chlorine should not be overlooked, including options such as UV light and ozone which are gaining widespread acceptance as alternatives to chlorine as the primary disinfectant during the treatment of water. Another key and critical function of chlorine is its relatively great stability which provides some level of residual chlorine to protect the water from contamination in the distribution system. "Residual chlorine" is a vital indicator to show that sufficient chlorine is present to prevent the regrowth of pathogens and/or introduction of opportunistic pathogens in the system, thereby protecting the water within the distribution system. Indeed, the need to maintain residual chlorine in the distribution system is currently a regulatory requirement in many parts of the world, including Canada. Definitely, a part of this identified policy exists to ensure a barrier sufficient to provide water protection beyond the last point of the treatment system.

It is appropriate to emphasize that "safe water" does not mean zero risk, but that it is safe enough for drinking and food preparation (Wikipedia 2015). It also means that in terms of policy, treatment and chlorination are adopted as a strategy to protect water consumers. Indeed, the use of chlorine as primary or secondary disinfectant provides a strategy to reduce as most as possible the microbiological risk associated with drinking water. As a general policy, disinfection of water with chlorine or with other disinfectants must not been compromised at any moment even considering the side-effects (e.g. generation of by-products and taste and odours).

The approach for use of chlorine as a primary disinfectant and also to leave some residual in the distribution system (secondary disinfection) was the primary strategy in all developed countries until the formation of disinfection by-products (DBPs) was identified in the 1970s. Given the potential negative health impacts of DBPs formed during chlorination, there have been changes to disinfection policy and approaches resulting in other drinking water disinfection mechanisms being employed (alternatives to chlorine for primary or secondary disinfection).

In response to the various aspects referred to above, the policy now exists that disinfection (of some variety) is adopted at all locations in North America. However, it is important to note that this policy is not universally accepted by much of the population living in small and remote communities in Canada, especially First Nations. In the paragraphs that follow, we review the science behind the use of chlorine and the various aspects of policy requirements in Canada, and indicate how the policy is stipulated.

In this Chapter, the authors review the objectives of water chlorination, the benefits of chlorine for microbial inactivation in primary and secondary disinfection, the problems associated with the formation of DBPs, the advantages and drawbacks of alternative disinfectants, public perceptions associated with chlorine in drinking in particular in aboriginal communities and the challenges for the future.

19.2 Functions of Water Chlorination

19.2.1 Primary Disinfection with Chlorine

Chlorination has been used in the water industry for many decades to inactivate pathogens present in source waters. This disinfectant was used for the first time in the late nineteenth century to inactivate pathogenic microorganisms present in water (US EPA 2000). In addition to inactivating pathogens, chlorination has been used for oxidation purposes in order to remove taste and odours, as well as iron and manganese. Chlorination has been also used to improve coagulation and filtration processes, reduce algal growth in sedimentation basins and filters, and prevent biological regrowth (in particular biofilm within water distribution systems (US EPA 1999)). In addition, unlike other disinfectants, chlorine is designed to provide a minimum level of residual throughout the distribution system and protect against microbial contamination of the distributed water. Also, residual chlorine contributes to the control of microorganism presence associated with contaminant intrusion in the distribution system (Islam 2015; Islam et al. 2015).

For many years, water utilities have faced the significant challenge of balancing the benefits (microbial risks) and the drawbacks (chemical risks) of disinfecting water with chlorine as the primary or secondary disinfection agent. Chlorine is widely used as a disinfectant because of its relatively good efficacy in inactivating bacteria and viruses, its stability and ease of use, and its relatively low cost (Golfinopoulos et al. 2007). It is also the most common disinfectant in small systems (Dore et al. 2013), used as the primary disinfectant (to contribute to microorganism inactivation within the treatment plant) and as a secondary disinfectant (to contribute to the control of microbiological contamination in the distribution system). Chlorine disinfection involves the destruction of microorganism cell structures, interference with energy yielding metabolisms, biosynthesis and growth of microbes (Montgomery 1985). Generally speaking, chlorine is effective at inactivating bacteria and viruses occurring in source waters. However, it is not effective at inactivating some pathogenic protozoa, such as *Cryptosporidium oocysts*, and *Giardia cysts*.

Disinfection with chlorine is affected by physical, chemical and biological factors. The capacity of chlorine to inactivate pathogens depends mainly on the following features: the type(s) of microorganism to be inactivated, water temperature, water pH, disinfectant dose and its contact time in water. Higher chlorine doses and contact times generally yield better microorganism inactivation. For the most part, disinfection with chlorine is more effective at a lower pH and higher temperatures. In fact, when chlorine gas is added to water, rapid hydrolysis occurs to hypochlorous acid and hypochlorite ion. Lower pH favours the presence of hypochlorous acid (a better disinfectant than the hypochlorite ion). Finally, water temperature influences the reaction kinetics of chlorine in water (Cornell and Connell 1997).

When disinfecting water with chlorine as part of the treatment process, the disinfection efficiency concept must be considered. Disinfection efficiency (*Ct*) is the product of the residual disinfectant concentration (typically expressed in mg/L) and the contact time of chlorine in water (typically expressed in minutes). The *Ct* product is used as a design parameter, in particular, to design the primary disinfectant contact chamber of the treatment facility. This product is also used to evaluate disinfection efficiency (Log removals) according to the type of microorganisms (bacteria, viruses, etc.). For example, 2-Log removal is equivalent to 99 % and 3-Log removal is equivalent to 99.9 % inactivation of microorganisms. In the routine operation of a water treatment plant, *Ct* can be calculated based on data from free chlorine concentration measurements and water flowrates. The US EPA (2015) provides *Ct* values according to the type of microorganisms and levels of log removals. Generally speaking, ozone and chlorine dioxide (sometimes used as alternatives to chlorine for primary disinfection) are stronger oxidants and hence, have higher disinfection efficiencies (lower required *Ct*) than chlorine.

Many water treatment plants use chlorine for both primary disinfection (inactivation) and secondary disinfection (distribution system protection). Since chlorine is not effective for inactivating pathogenic protozoa (i.e., *Cryptosporidium* and *Giardia*), such plants must use pre-treatment strategies to reach the required Log removals for such microorganisms (some type of filtration is frequently used for those purposes). For example, for treatment of surface water, 2- or 3-Log removal is required for protozoa *Cryptosporidium* and *Giardia* that is 99.0 % or 99.9 % of removal, respectively (Crittenden et al. 2005a; Scheili 2015). Potential reductions of THMs by 63 % and HAAs reductions by 39 % could be obtained by shifting the emphasis from pre-chlorination to post-chlorination (McBean et al. 2008). For secondary disinfection, the main difficulty consists of maintaining adequate levels of free residual chlorine within the distribution system. The benefits of maintaining minimum residual levels of free chlorine are well known. Free residual chlorine reduces risks of microbiological contamination due to biofilm or contaminant intrusion. For example, Francisque et al. (2009) used robust historical heterotrophic plate counts (HPC) data and other water quality parameters to quantify the positive impact of residual chlorine levels on HPC, a common sentinel of microbiological water quality in distribution systems (Allen et al. 2004). They developed a statistical model that predicts the probability of relatively high HPC as a function of water temperature and distribution system geographical locations (Francisque et al. 2009).

19.2.2 Importance of Residual Chlorine

Maintaining residual disinfectant in water distribution systems, from the treatment plant to the extremities, is essential to ensure microbiological water quality and control biofilm growth. Many jurisdictions have set policies stipulating a disinfectant (primarily chlorine) residual in the distribution system. For example, US EPA requirements specify that the chlorine residual shall not be less than 0.2 mg/L for any period greater than 4 h at the entrance to the distribution system, and cannot be undetectable in more than 5 % of the samples taken each month in the distribution system. In Canada, different provinces have different requirements with respect to maintaining a disinfectant residual entering, or in, the distribution system. These range from a measurable level in Newfoundland to a minimum level of 0.5 mg/L in Manitoba (CWWA personal communication).

In fact, residual chlorine concentrations decrease, and in many cases become very low (and sometimes undetected), in the extremities of distribution systems (Chambers et al. 1995). Ensuring adequate levels of residual chlorine is a major task because of the number of water quality and operational parameters that affect the residual depletion. In addition, the spatial and temporal variability of the source and treated water, generally associated with meteorological events (source water) and changing residence times of water within pipes (treated water), makes chlorine concentrations highly variable in time and space. In fact, chlorine decay is affected significantly by water temperature and water quality characteristics (pH, organics, etc.) that may change seasonally, monthly or daily. In addition, free residual chlorine depletion is greatly affected by hydraulics (in pipes and reservoirs). Thus, the challenge of maintaining minimum levels of free residual chlorine is greater when the quality of water that leaves the plant contains higher levels of organics (e.g., total or dissolved organic carbon) during warmer periods and in the network extremities.

19.2.3 Strategies to Maintain Residual Chlorine

Some strategies can be used to counterbalance residual chlorine decay and increase the levels of residual chlorine at distribution system extremities. Examples include reducing chlorine demand, increasing the chlorine doses, using booster chlorination, and improving hydraulic management to reduce residence time of water in the distribution system. It is also possible to consider an alternative disinfectant, such as chloramines, for secondary disinfection.

Reducing chlorine demand in water is an effective strategy to increase residual chlorine levels in the distribution system. However, this approach requires the implementation of physicochemical treatment procedures to increase the removal of organics before the water is chlorinated. This approach may require large investments. Increasing the chlorine dose will contribute not only to increasing the residual concentrations, but also to greater occurrence of DBPs (for example, trihalomethanes-THMs and haloacetic acids-HAAs) and could generate taste and odours, and citizen complaints. Indeed, some researchers have reported that the taste of chlorine might be perceptible when residual concentrations are higher than 0.2 mg/l (Crittenden et al. 2005b; de França Doria 2010; Nagata et al. 2011).

Use of booster stations for rechlorinating water in the distribution system is an interesting strategy to ensure higher levels of residual chlorine at system extremities (Islam et al. 2015). It also allows reductions, at the treatment plant, of the chlorine dose required for secondary disinfection. However, implementing chlorination booster stations increases the costs for infrastructure and will require further operator monitoring and management tasks. In recent years, there has been increasing interest in implementing strategies to ameliorate secondary disinfection based on booster rechlorination optimization (including minimal investment costs as optimization criteria). Islam (2015) reviewed different applications that have been reported to optimize booster chlorination. Such applications typically include optimization of booster station numbers, optimization of booster geographical distribution, and optimization of chlorine dosage. The aim of these applications for booster optimization is twofold: to ensure minimum levels of residual chlorine in every location of the distribution system to avoid microbiological contamination; and, (simultaneously) to reduce the formation of DBPs and the generation of chlorine-related taste and odours (and thus citizen complaints).

The use of alternatives to chlorine for secondary disinfection is also a valuable strategy (as it is frequently practised for primary disinfection, for instance with ozone or chlorine dioxide). For example, chloramines have been used in the United States for several years as an alternative to chlorine during secondary disinfection, mainly for the purpose of reducing the occurrence of chlorinated DBPs (Guay et al. 2005). Chloramines for secondary disinfection have other advantages: they are relatively stable in the distribution system, efficient for taste and odour control, and more effective than chlorine for biofilm control (Wilczak et al. 2003; Norton and Lechevallier 1997). However, they may generate nitrogenous DBPs, such as nitrosamines, that have higher toxicological relevance than THMs and HAAs (Brisson et al. 2013).

It is also possible to improve the hydraulic management and design in the distribution system to reduce residence times of water in pipes and storage tanks. Lower water ages favour higher levels of residual chlorine, but also reduce the formation of DBPs, such as THMs and HAAs. Identifying scenarios to reduce water residence times has been applied recently with emerging modelling techniques (Delisle et al. 2015).

Strategies to improve the management of residual chlorine in distribution systems require a good understanding of the factors affecting chlorine decay. For many years, chlorine decay in distribution systems has been modelled for decision-making purposes. Models for residual chlorine decay and evolution are useful to identify strategies to improve water quality within the distribution system. The most common strategy to describe residual chlorine has been the use of first- or second-order kinetic models (Jadas-Hécart et al. 1992; Dharmarajah et al. 1991; Rodriguez et al. 2002; Chambers et al. 1995; Rossman et al. 1994).

19.3 Challenges of Chlorine and Alternative Disinfectants

19.3.1 Disinfection By-Product Formation and Potential Health Impacts

The application and use of chlorine for drinking water disinfection has always been the subject of much debate, with many of its opponents arguing for the use of protected sources, rather than disinfection or the use of non-chemical disinfection technologies. This argument gained credential, in the mid-1970s, when it was demonstrated that free chlorine reacts with natural organic matter (NOM) in water to produce chlorinated organics, specifically the four trihalomethanes (THMs) (Rook 1974; Bellar et al. 1974): Chloroform – $CHCl_3$, Bromodichloromethane (BDCM) – $CHCl_2Br$, Dibromochloromethane (DBCM) – $CHClBr_2$, and Bromoform – $CHBr_3$. Shortly thereafter, regulations were introduced and limits set on allowable THM concentrations in potable water. Since then, and with the advances made in analytical methods and instruments, more byproducts (e.g., haloacetic acids) resulting from chlorination have been identified and limits established for many of these byproducts (US EPA 1998). Currently, there are over 600 DBPs that have been identified (Richardson 2002), with the majority of known DBPs being halogenated and associated with the use of chlorine as the primary or secondary disinfectant. Hence, THMs, and to a lesser extent HAAs, are currently used as indicator chemicals for all potentially harmful compounds formed by the addition of chlorine to water. Indeed, in many countries, the levels of THMs and HAAs in chlorinated water supplies are regulated based on this assumption. For example, THMs and HAAs are regulated in the United States by the EPA with maximum allowable concentrations of 80 μg/L and 60 μg/L, respectively, in drinking water. In Canada, there are guidelines solely for maximum concentrations of THMs and HAAs, at 100 μg/L and 80 μg/L, respectively (Health Canada 2013).

With stringent regulatory limits on the maximum allowable concentration of THMs and HAAs in drinking water and the concerns associated with the potential health impacts of DBPs, significant research and much effort has focused, in recent decades, on determining the true human health effects of these DBPs. However, there have been relatively few toxicological and epidemiological studies carried out on the effects of DBPs on health outcomes (Nieuwenhuijsen et al. 2000). So far, some of the main concerns have been reproductive health impacts such as low birth weight, pre-term delivery, spontaneous abortions, stillbirth, and birth defects. Although not conclusive, a number of toxicological and epidemiological studies point towards an association between THMs and low birth weight (Ruddick et al. 1983; Kanitz et al. 1996; Savitz et al. 1995; Levallois et al. 2012). However, it is important to note that administered doses in toxicological studies have been high and although epidemiological studies have mostly shown excess risks, these were often not significant and the assessment of exposure was often limited (Nieuwenhuijsen et al. 2000). Evidence of associations with other outcomes such as spontaneous abortions, stillbirth, and birth defects from epidemiological studies is even weaker, but gaining ground, although there is no evidence of an association between THMs and pre-term delivery (Dodds et al. 1999; Nieuwenhuijsen et al. 2000).

As noted, one of the main limitations of the epidemiological studies to date has been the assessment of exposure. There are very real difficulties in making any accurate assessment of DBP exposure and uptake because of the potential for variation in concentrations of DBPs in different parts of the distribution system, at different times of the year, and as a consequence of differences in individual consumer behaviour. The problem is further complicated by the lack of analytical data on most DBPs, other than THMs and HAAs, and the fact that exposure to volatile DBPs will increase through inhalation while showering, bathing or swimming.

19.3.2 Disinfection of Protozoa

Further to the challenge of DBP formation, the effectiveness of chlorine in controlling certain waterborne illnesses caused by protozoa is an issue. In the 1980s, the *Giardia lamblia* cyst was identified as an important waterborne pathogen. Because

G. lamblia is more resistant to chlorine than bacteria and viruses, more stringent standards for reduction of pathogens were established (US EPA 1991). Also, the *Cryptosporidium parvum oocyst*, was later identified as an important source of waterborne illness and is even more resistant to chlorine than *G. lamblia*. In fact, chlorination is ineffective against *C. parvum*, because it would take a significant amount of chlorine to achieve sufficient inactivation of this pathogen. It has been reported that the total loss of excystation ability of a human-derived strain of *Cryptosporidium* requires exposure times of 24 h at *Ct* value [concentration of free chlorine in milligrams per litre (or parts per million) multiplied by time in minutes] of 8,000–16,000 mg.min/l free chlorine (Smith et al. 1990). Studies on chlorine and chloramine reported by Sterling et al. (1989) also use in vitro excystation that showed impracticably high *Ct* values. Note that the aforementioned *Ct* values are orders-of-magnitude higher than those for inactivation of viruses and bacteria, thereby making chlorine a non-viable primary disinfectant for parasites.

19.3.3 Alternative Disinfectants

The aforementioned challenges, associated with the application of free chlorine, have led the scientific community and regulatory authorities to seek alternative disinfection technologies, e.g., ozone, chlorine dioxide and ultraviolet light, as primary disinfectants. These commonly used alternative disinfectants are known to produce lower levels of the four regulated THMs and most of the HAAs. Hence, many utilities have switched, or are in the process of switching, to one of the alternative techniques for primary disinfection. For example, ozone became the preferred primary disinfectant in some parts of Europe in the late 1960s and 1970s, and UV is fast replacing chlorine for the disinfection of surface water, especially in medium to large municipalities.

Although they address some of the problems posed by chlorine (e.g., no THMs and HAAs formation during disinfection with ozone), alternative chemical disinfectants come with their own unique challenges with respect to other priority DBPs. For example, the use of ozone can result in the formation of bromate, especially when elevated levels of bromide are present in the source waters. Bromate is a concern, because it has been shown to be a potential carcinogen in laboratory animals (WHO 2005). As another example, iodo-THM levels were highest in water disinfected with chloramines; dichloroacetaldehyde was highest in water disinfected with chloramines and ozone (with no biological filtration); and selected halonitromethanes were formed at higher levels during post-disinfection when pre-ozonation was used (Krasner et al. 2006). Differences in source-water conditions, including concentrations of bromide or iodide, concentrations of natural organic matter (precursors to DBPs), and pH, can also have a dramatic effect on the DBP species (chlorine-, bromine-, or iodine-containing) and the concentrations produced.

19.4 Chlorine and Public Perception

An equally important, and somewhat understated, challenge to the use of chlorine has been public resentment to its application in water. A significant portion of the population has always had an aversion to the use of chlorine, highlighting its impact on the aesthetic quality of water and seeking to avoid exposure to a chemical with such toxic properties, even at low concentrations. Opposition to the use of chlorine as a disinfectant is particularly strong in smaller and Aboriginal communities due to their stronger and closer connection to the environment and nature. Two studies conducted in 1994 and 2001 in Quebec, Canada, showed that consumers living nearest to a water treatment plant, where residual chlorine levels are highest, perceived the most risk and least satisfaction with the quality of their drinking water (Turgeon et al. 2004). Piriou et al. (2004) tested consumer sensitivity to free and combined chlorine in seven different demographic and geographic locations across the United States. Unlike the findings in the Quebec study, they found that although individual sensitivity varied widely, very few participants were able to recognize the chlorine flavour, even at concentrations close to the US maximum contaminant level (Piriou et al. 2004). Furthermore, there was no statistically significant difference in the sensitivity threshold to chlorine taste among tap, bottled, and filtered water drinkers. This study concluded that consumers did not switch to alternative tap water solutions based solely on their detection of free chlorine in the water (Piriou et al. 2004). Research in this field must focus on better understanding the factors that explain the rejection of chlorinated drinking water in small and Aboriginal communities and identify the alternative strategies that ensure safe water with use of chlorine or different treatment and management technologies and method.

19.5 Challenges of Aboriginal Traditional Knowledge and General Public Attitudes/Constraints

As an example "Water is what sustains us. Water is what brings us into this world and water is what keeps us in this physical world" as stated by Jan Longboard, as reported by Anderson (2010). Hence, issues of Aboriginal Traditional Knowledge (ATK), as it interplays with chlorination, are very important concerns with regard to the willingness to accept chlorination as a water treatment technology. While water is absolutely critical to human life, there exist issues of ATK that influence the acceptability of certain practices in relation to water. For example, in ATK the use of slow sand filters as a water treatment technology is generally considered as acceptable, since a slow sand filter replicates movement through soil as a natural purification effect of filtration, for "cleaning" water. Similarly, UV radiation is generally considered acceptable in ATK as a water treatment technology, as UV radiation is a part of the influence of sunlight as a natural disinfection process. However, reliance upon chlorine disinfection is not a natural process, therefore there is resistance.

Some segments of the population accept the idea of chlorination. The degree of acceptance of chlorine is evident in that just over half of First Nations residents and four in ten members of the general public state that they feel safe as a result of chlorine additions to tap water (EKOS 2011). However, there are many who do not trust chlorine.

As indicated in previous sections of this chapter, maintaining the chlorine residual at a measurable level throughout the water distribution system is widely practised as a policy, because it is a valuable indicator that the water is protected from bacterial contamination. In fact, if a communal water supply does not have this disinfection residual, a precautionary Drinking Water Advisory (DWA) could be issued. It is noteworthy that the number of DWAs issued to First Nations communities is substantial. To demonstrate the magnitude, in 2013, 65 of 158 in First Nations communities in Ontario were under a DWA (Health Canada 2013). Many of these were precautionary DWAs and relate to chlorine residual deficiency (LeBlanc 2015 personal communication), reflecting that for many, the water supply quality is not necessarily compromised, but is not protected. The numbers of DWAs change over time but many DWAs are consistently evident. As of May 31, 2014, Health Canada reported 84 DWAs in effect in First Nations communities in the Province of Ontario (Harvey et al. 2015).

To provide some context, it is important to note a number of situations where DWAs have been issued. In three of these communities, the advisories occurred as a result of low-pressure events and watermain breaks which lasted between two and 10 days i.e., there is a need to keep a residual when there are failures of the distribution system, as described by Murphy et al. (2015a, b). In many First Nations communities, there is definite acknowledgement that power outages have occurred, and the expectation of future power outages is real. In part, power outages arise owing to the fact that the communities are at the end of the electrical distribution system.

Although the evidence indicated in the preceding sections of this chapter clearly demonstrates that disinfectants have beneficial effects in terms of the safety of water as a result of chlorine as a residual, there is considerable resistance to this policy, resulting in resistance to apply/implement chlorination and/or turn their system off. ATK and the attitudes of people in First Nations communities do influence the implementation of the chlorine policy. While there is considerable resistance to chlorine being added to the water, the reasons for this resistance vary. Some of the resistance is the result of smell, odour, or a combination of both. This underscores the importance of obtaining new data about the factors that explain the rejection of chlorine and focus in research to alternatives to chlorine in these communities.

As specific examples, EKOS (2009 and 2011) completed a series of surveys specifically on the topic. The findings are very informative as they indicate people's attitudes toward chlorination. Both the taste and smell of chlorine in water are cited as elements that First Nations and residents from other small communities notice. The majority of both respondent groups say they notice a difference in the taste of water that has chlorine in it, compared to water without the chlorine and more than half of each group state that they do not like the taste.

With respect to perceptions of safety, there are regional differences across Canada. Support for including chlorine in water in BC is much lower than all other First Nations regions in Canada. Just one-third (33 %) of First Nations residents in BC support the use of chlorine, while slightly more than one-third (36 %) oppose it. There is also a slightly higher level of support from the smallest communities (41 %) as determined by interviews (EKOS 2011). Those who indicated that they opposed the inclusion of chlorine in their drinking water were asked to indicate their reason(s) why. Responses indicate that health concerns predominate among First Nations residents. Taken together, just over one-half of First Nations residents who oppose chlorine in their water cite a concern for chlorine's effect on health (53 %). Specifically, the concerns include the lack of knowledge of the effects of chlorine on health (19 %), the belief that chlorine is detrimental to health (17 %) and that chlorine is poisonous (17 %). A dislike of the taste of chlorine (17 %) also registers as a reason for opposition among those who oppose chlorine (EKOS 2011). Murphy et al. (2015b) reported that a water treatment operator in one community was not allowed to purchase chlorine.

Seven in ten First Nations and the general public have been reported as noticing the difference in smell of chlorinated water. In fact, roughly six in ten First Nations and half of other community residents have looked into alternate sources of water to avoid using water that has been chlorinated. First Nations residents are less likely to seek out alternate sources of water owing to its smell rather than to its taste, although just under half have, as have four in ten residents of the general public (EKOS 2011). The "killing of life in the water" is one expressed attitude, which indicates refusal to use the chlorinated water and instead, rely upon the use of an alternate water source or bottled water, or untreated water.

As a specific example of a community, many Anishnaabe have indicated they would prefer to drink cool, clear, surface water collected from a pristine natural environment or, in other words, water without treatment by "western" methods. Further, as Anderson (2010) reports from her interviews, First Nations people had once been able to drink directly from natural waters in their area, and this relationship has been hard to give up. When asked why people do not like to drink the chlorinated town water which is piped to their homes, Anderson (2010) reports that people do not like to drink the chlorinated town water:

> It's the taste. You can taste the chlorine. You can smell it. Even when you're running the tap and filling a glass, filling your kettle, whatever, you can smell the chlorine.

By imposing the DWA, there is no real choice as to whether or not the community (First Nations or general public) wants the DWA. For First Nations, this goes against what they believe in and foregoes consultation altogether. As a result of new legislation, the government does indeed have the power to impose DWAs on First Nations communities regardless of how they feel. This has been elaborated upon by Corston-Pine (2015, personal communication):

> It's like your parents feeding you vegetables when you are a kid. They don't always taste good but mom and dad know best. Does the Government truly know what's best for a FN community? Well according to the Indian Act, the Government does have the obligation, but the government always seems to forget to 'talk' with the communities.

Use of chlorine and other disinfectants to treat drinking water undermines First Nations historical and relational understandings of water. ATK indicates that nature should not be an issue, so why is there a need to treat? Using chlorine to treat poor source water is equivalent to putting a bandage on a broken leg. ATK indicates that if you see life in the water, it is okay to use and drink (fish and crustaceans, etc. things you can see). Adding chlorine to the water is not a choice and although they recognize that chlorine creates water that is safe to drink, they would prefer to have the choice of whether or not to add chlorine to their water; chlorine kills the spirit of the water, which is why many of the community members choose to drink the "pure," "life-giving" water from the lake or the pond.

For the general population, health concerns are less pronounced among those in smaller communities. Those in the general population are more likely to cite their belief that nothing needs to be added to water (31 %) or that they do not like the taste (27 %). In some communities in Newfoundland-Labrador, chlorine is not added because the community members cannot afford it.

19.6 Moving Forward

Clearly, we face a situation where to provide "safe" water, we need a disinfectant residual. How to move forward? As apparent from the text of this chapter, policy stipulates a disinfectant (which is primarily chlorine) residual in the distribution system. Hence, there is a problem in terms of knowledge/science and trust. So how do we better implement the policy? One of the clear dimensions needed is education and information, which are in addition to setting and implementing regulatory policies.

When prompted specifically about the types of information that would help to reassure First Nations community members about their tap water quality, roughly three-quarters of First Nations residents living on-reserve each said that more information about water quality testing procedures/frequency of testing or the acceptable levels in tap water, or information about the quality of tap water on their reserve, would make them feel safer (EKOS 2009). In interviews, Black and McBean (2016) were told that:

> water has spirit. Spirit is life. When chlorine is added to water, while it does have benefit, it kills the spirit of the water and anything else that is in it. If spirit dies, so too does the water. Giving us the idea that all that remains is 'dead water' (Black and McBean 2016, personal communication).

One of the features of adherence to ATK, evident from the preceding discussion, is that physical filtration is generally acceptable by First Nations people. The rationale is that physical filtration is much like Mother Earth in the way that nature cleans water. The net finding is that a combination of ATK and western knowledge, associated with information, will better help policy implementation.

19.7 Conclusions

Chlorination has been used in the water industry for many decades as a primary and secondary disinfection technique. This is largely because of its effectiveness at eliminating pathogens, in particular bacteria and viruses, ease of use and stability, and relatively low cost. The main drawbacks of chlorination are the formation of chlorination DBPs, generation of chlorine-related tastes (and associated citizen complaints), and limited efficacy to eliminate protozoa. Because of these disadvantages, the use of disinfectant alternatives to chlorine has increased in the last several years. Despite favourable policy, as well as regulatory requirements across North America towards maintaining adequate levels of residual disinfectant in the distribution systems to reduce the probabilities of microbiological contamination, many small and First Nations communities generally oppose the use of chlorine, largely because they reject the use of a chemical agent in water that will be distributed to the community.

Further research will need to: (1) update the state-of-the-art of chlorination practices in Canadian water systems; (2) increase the understanding of factors that explain rejection of chlorine as a disinfectant in some communities, especially First Nations; (3) develop strategies to ensure that each Canadian water supply system satisfactorily applies both primary and secondary disinfection; and (4) identify the socio-technical challenges of providing safe tap water to all Canadians. Research with these purposes will require the involvement of multidisciplinary teams of water engineers, social science scientists and health experts. This will eventually ensure a safer and cleaner drinking water supply for the public and help reduce the risks associated with the presence of harmful pathogens in drinking water.

References

Allen, M. J., Edberg, C. S., & Reasoner, D. J. (2004). Heterotrophic plate count bacteria – What is their significance in drinking water? *International Journal of Food Microbiology, 92*(3), 265–274.

Anderson, K. (2010, October). *Aboriginal women, water and health: Reflections from eleven first nations, Inuit and Metis Grandmothers* (p. 32). Atlantic Centre of Excellence for Women's Health, Prairie Women's Health Centre of Excellence.

Bellar, T. A., Lichtenberg, J. J., & Kroner, R. C. (1974). The occurrence of organohalides in chlorinated drinking water. *Journal of the American Water Works Association, 66*(22), 703–706.

Black, K., & McBean, E. (2016). Increased indigenous participation in decision-making: A Policy framework for the improvement of indigenous health. *The International Indigenous Policy Journal*, under review.

Brisson, I., Levallois, P., Tremblay, H., Sérodes, J.-B., DeBlois, C., Charrois, J., Taguchi, V., Boyd, J., Li, X. F., & Rodriguez, M. J. (2013). Spatial and temporal occurrence of N-Nitrosamines in seven drinking water supply systems. *Environmental Monitoring and Assessment, 185*(9), 7693–7708.

Chambers, V. K., Creasey, J. D., & Joy, J. S. (1995). Modelling free and total chlorine decay in potable water distribution systems. *Journal of Water Supply: Research & Technology – AQUA, 44*(2), 60–69.

Cornell, G., & Connell, G. F. (1997). *Chlorination and chloramination handbook* (Water disinfection series 1st ed., p. 174). Denver: AWWWA.

Corston-Pine, E. (2015). *Facilitator of First Nations*. Guelph: University of Guelph, personal communication.

Crittenden, J. C., Trussell, R., Hand, D., Howe, K., & Tchobanoglous, G. (2005a). Kinetics of disinfection. In *Water treatment: Principles and design, MWH* (2nd ed., pp. 1042–1067). Hoboken: Wiley.

Crittenden, J. C., Trussell, R., Hand, D., Howe, K., & Tchobanoglous, G. (2005b). Taste and odor. In *Water treatment: Principles and design, MWH* (2nd ed., pp. 100–110). Hoboken: Wiley.

Delisle, F.-J., Rochette, S., Pelletier, G., Asadzadehd, M., Tolson, B., & Rodriguez, M. J. (2015). Minimizing water residence time in Quebec City's main distribution network using hybrid discrete dynamically dimensioned search (HD-DDS): Part II. *Journal of Water Supply: Research & Technology – AQUA, 64*(3), 378–390.

Dharmarajah, H., Patania, N., Jacangelo, J., & Aieta, E. (1991, June 23–27). Empirical modeling of chlorine and chloramine residual decay. In*: AWWA annual conference proceedings*, Philadelphia.

Dodds, L., King, W., Woolcott, C., & Pole, J. (1999). Trihalomethanes in public water supplies and adverse birth outcomes. *Epidemiology, 10*(3), 233–237.

Dore, M., Singh, R., Acharib, G., & Khaleghi-Moghadam, A. (2013). Cost scenarios for small drinking water treatment technologies. *Desalination and Water Treatment, 51*(16–18), 3628–3638.

Doria, M. F. (2010). Factors influencing public perception of drinking water quality. *Water Policy, 12*(1), 1–19.

Ekos Research Associates Inc. (2009, February 27). *Water quality on-reserve quantitative research*. Report to Health Canada, Ottawa, Canada. PWGSC contract number: H1011-08009/001/CY, POR registration number: 095-08.

Ekos Research Associates Inc. (2011, March 25). *Perceptions of drinking water quality in First Nations communities and general population*. Report to Health Canada, Ottawa, Canada. PWGSC contract number: HT372-101292/001/CY, POR registration number: 087-10.

Francisque, A., Rodriguez, M. J., Miranda-Moreno, L., Sadiq, R., & Proulx, F. (2009). Modeling of heterotrophic bacteria counts in a water distribution system. *Water Research, 43*(4), 1075–1087.

Golfinopoulos, S., Nikolaou, A., Uyak, V., Toroz, I., Rodriguez, M. J., & Sérodes, J.-B. (2007). Occurrence, fate and transport modeling of DBPs in drinking water. In A. Nikolaou, L. Rizzo, & H. Selcuk (Eds.), *Advances in control of disinfection by-products in drinking water systems* (pp. 129–162). Hauppauge: Nova.

Guay, C., Rodriguez, M. J., & Sérodes, J.-B. (2005). Using ozonation and chloramination to reduce the formation of trihalomethanes and haloacetic acids in drinking water. *Desalination, 176*(1–3), 229–240.

Harvey, R., Murphy, H., McBean, E., & Gharabaghi, B. (2015). Using data mining to understand drinking water advisories in small water systems: A case study of Ontario First Nations drinking water supplies. *Water Resources Management, 29*(14), 5129–5139.

Health Canada. (2013). *Guidance for providing safe drinking water in areas of federal jurisdiction*, Version 2. Water and Air Quality Bureau, Healthy Environments and Consumer Safety Branch, Health Canada, Ottawa, Catalogue No. H128-1/05-440-1E-PDF.

Islam, N. (2015). *Water quality management in small to medium sized distribution networks: Optimizing chlorine disinfection strategies* (p. 207). Dissertation, University of British Columbia.

Islam, N., Farahat, A., Al-Zahrani, M., Rodriguez, M. J., & Sadiq, R. (2015). Contaminant intrusion in water distribution networks: Review and proposal of an integrated model for decision making. *Environmental Reviews, 23*(3), 337–352.

Leblanc, I. (2015). *Special advisor for housing, infrastructure and emergency issues*. Ottawa: Management at Assembly of First Nations, personal communication.

Jadas-Hécart, A., El Moher, A., Stitou, M., Bouillot, P., & Legube, B. (1992). The chlorine demand of a treated water. *Water Research, 26*(8), 1073–1084.

Kanitz, S., Franco, Y., Patrone, V., Caltabellotta, M., Raffo, E., Riggi, C., Timitilli, D., & Ravera, G. (1996). Association between drinking water disinfection and somatic parameters at birth. *Environmental Health Perspectives, 104*(5), 516–520.

Krasner, S. W., Weinberg, H. S., Richardson, S. D., Pastor, S. J., Chinn, R., Sclimenti, M. J., Onstad, G. D., & Thurston, A. D., Jr. (2006). Occurrence of a new generation of disinfection byproducts. *Environmental Science & Technology, 40*(23), 7175–7185.

Levallois, P., Gingras, S., Marcoux, S., Legay, C., Catto, C., Rodriguez, M. J., & Tardif, R. (2012). Maternal exposure to drinking water chlorination by-products and small-for-gestational age neonates. *Epidemiology, 23*(2), 267–276.

McBean, E., Zhu, Z., & Zeng, W. (2008). Systems analysis models for disinfection by-product formation in chlorinated drinking water in ontario. *Journal of Civil Engineering and Environmental Systems, 25*(2), 127–138.

Montgomery, J. M., & Consulting Engineers. (1985). *Water treatment principles and design*. Walnut Creek: Wiley.

Morris, R. D. (2007). *The blue death, disease, disaster and the water we drink*. New York: Harper Collins.

Murphy, H. M, Corston-Pine, E., & McBean, E. A. (2015a). *Insights and opportunities: Challenges of Canadian First Nations drinking water operators* (pp. 20–21). Water Canada (January–February).

Murphy, H. M., Corston-Pine, E., Post, Y., & McBean, E. A. (2015b). Insights and opportunities: challenges of Canadian First Nations drinking water operators. *The International Indigenous Policy Journal, 6*(3), 17.

Murphy, H. M., Bhatti, M. A., Harvey. R., & McBean, E. A. (2015c). Using decision trees to predict drinking water advisories and understand vulnerabilities in small water systems. *Journal of the American Water Works Association*, Accepted for publication August 2015.

Nagata, J., Valeggia, C., Smith, N., Barg, F., Guidera, M., & Bream, K. (2011). Criticism of chlorination: Determinants of drinking water beliefs and practices among the Tz'utujil Maya. *Revista Panamericana Salud Pública, 29*(1), 9–16.

Nieuwenhuijsen, M. J., Toledano, M. B., Eaton, N. E., Fawell, J., & Elliott, P. (2000). Chlorination disinfection byproducts in water and their association with adverse reproductive outcomes: A review. *Occupational & Environmental Medicine, 57*(2), 73–85.

Norton, C. D., & LeChevallier, M. W. (1997). Chloramination: Its effect on distribution system water quality. *Journal of the American Water Works Association, 89*(7), 66–77.

Piriou, P., Mackey, E. D., Suffet, I. H., & Bruchet, A. (2004). Chlorinous flavor perception in drinking water. *Water Science and Technology, 49*(9), 321–328.

Richardson, S. D. (2002). The role of GC-MS and LC-MS in the discovery of drinking water disinfection by-products. *Journal of Environmental Monitoring, 4*(1), 1–9.

Rodriguez, M. J., Milot, J., Sérodes, J.-B., & Pacaud, A. (2002). Estimation of bench-scale chlorine decay in drinking water using nth-order kinetic and neural network models. *Water Quality Research Journal of Canada, 37*(3), 613–635.

Rook, J. J. (1974). Formation of haloforms during chlorination of natural waters. *Water Treatment Examination, 23*(2), 234–243.

Rossman, L. A., Clark, R. M., & Grayman, W. M. (1994). Modeling chlorine residuals in drinking water distribution systems. *Journal of Environmental Engineering, 120*(4), 803–820.

Ruddick, J. A., Villeneuve, D. C., Chi, I., & Valli, V. E. (1983). A teratological assessment of four trihalomethanes in the rat. *Journal of Environmental Science and Health B, 18*(3), 333–349.

Savitz, D. A., Andrews, K. W., & Pastore, L. M. (1995). Drinking water and pregnancy outcome in Central North Carolina: Source, amount and trihalomethane levels. *Environmental Health Perspectives, 103*(6), 592–596.

Scheili, A. (2015). Variabilité spatio-temporelle de la qualité de l'eau potable des petits réseaux : Facteurs explicatifs et développement d'outils de gestion (p. 215). Dissertation, Université Laval.

Smith, H. V., Smith, A. L., Girdwood, R. W. A., & Carrington, E. G. (1990). The effect of free chlorine on the viability of *Cryptosporidium* sp. oocysts isolated from human faeces. In J. Badenoch (Ed.), *Cryptosporidium in water supplies: Report of the group of experts* (pp. 185–204). London: HMSO.

Sterling, C. R., Korich, D. G., Madore, M. S., & Sinclair, N. A. (1989, November 12–16). Chlorine and ozone inactivation of *Cryptosporidium* oocysts. In: *AWWA water quality technology conference and exposition proceedings*, Philadelphia.

Turgeon, S., Rodriguez, M. J., Thériault, M., & Levallois, P. (2004). Perception of drinking water in the Quebec City region (Canada): The influence of water quality and consumer location in the distribution system. *Journal of Environmental Management, 70*(4), 363–373.

US EPA. (1998, December 16). *National primary drinking water regulations: Disinfectants and disinfection by-products; Final rule*. Federal Register, vol 63, N°241.

US EPA. (1991). *Guidance manual for compliance with the filtration and disinfection requirements for public water systems using surface water sources*. Washington, DC: United States Environmental Protection Agency, EPA 570391001.

US EPA. (1999). *Microbial and disinfection by-product rules – Simultaneous compliance guidance manual*. Washington, DC: United States Environmental Protection Agency, EPA 815-R99-014.

US EPA. (2000). *The history of drinking water treatment*. Washington, DC: United States Environmental Protection Agency, EPA 816-F-00-006.

US EPA. (2015). http://water.epa.gov/lawsregs/rulesregs/sdwa/mdbp/tools.cfm. Accessed 16 Oct 2015.

Wikipedia. (2015). *Drinking water*. https://en.wikipedia.org/wiki/Drinking_water. Accessed 5 Oct 2015.

Wilczak, A., Hoover, L. L., & Lai, H. H. (2003). Effects of treatment changes on chloramines demand and decay. *Journal of the American Water Works Association, 95*(7), 94–106.

World Health Organization. (2005). *Bromate in drinking water: Background document for development of WHO guidelines for drinking-water Quality*. Geneva: World Health Organization, WHO/SDE/WSH/05.08/78.

Part VI
Case Studies

Chapter 20
Patchy Resources for the Governance of Canada's Resource Patches: How Hydraulic Fracturing Is Illuminating the Need to Improve Water Governance in Canada

Michele-Lee Moore, Karena Shaw, Heather Castleden, and Joanna Reid

Abstract Hydraulic fracturing emerged in the mid-2000s, with polarizing results across Canada. Provincial governments either embraced the technology for energy development as a key component of their economic strategy, or put moratoriums in place. These two extreme responses have emerged worldwide, and thus, much can be learned from the varying experiences across the Canadian landscape. However, everywhere that exploration and development has taken place, has been met with significant opposition by Indigenous and non-Indigenous communities, largely due to the concerns about the associated use of water in the hydraulic fracturing process, and the potential impacts to quality and quantity. Therefore, we argue that hydraulic fracturing has shone a spotlight on, although not caused, existing water governance concerns in Canada and the contestation that underpins many water decision-making

---

This chapter includes an edited version of some components from a 2015 report, *Building Capacity to Build Trust: Key Challenges for Water Governance in Relation to Hydraulic Fracturing* (Moore et al. 2015), developed as part of an initiative on water and hydraulic fracturing – see http://www.cwn-rce.ca/focus-areas/energy-and-resources/water-and-hydraulic-fracturing-report/.

M.-L. Moore (✉)
Department of Geography, and the Water, Innovation, and Global Governance Lab, University of Victoria, 3800 Finnerty Rd, V8W 2Y2 Victoria, BC, Canada
e-mail: mlmoore@uvic.ca

K. Shaw
School of Environmental Studies, University of Victoria, 3800 Finnerty Rd, V8W 2Y2 Victoria, BC, Canada
e-mail: shawk@uvic.ca

H. Castleden
Health, Environment, and Communities Research Lab, Department of Geography and Planning, Queen's University, 99 University Ave, Kingston, ON K7L 3N6, Canada
e-mail: heather.castleden@queensu.ca

J. Reid
Independent Scholar, 1631 Grant Street, V5L 2Y4 Vancouver, BC, Canada
e-mail: jiereid@gmail.com

S. Renzetti, D.P. Dupont (eds.), *Water Policy and Governance in Canada*, Global Issues in Water Policy 17, DOI 10.1007/978-3-319-42806-2_20

processes. This chapter will examine three main challenges. First, we illustrate how water governance approaches across the provinces and territories are patchy and fragmented, which leads to a perceived lack of transparency in the system. Second, we explain how capacity across the many actors involved in governing water or using water is not uniform and we reveal how this affects individual and collective abilities to make substantive changes that would improve the various water governance approaches across the country. On this matter, we add that the allocation of water for industrial use is often perceived as government favouring economic interests over all other values, which can lead to a perceived lack of accountability and conflicts of interest by those responsible for governing water use. Third, we describe how water governance has failed to adequately include Indigenous peoples vis-à-vis free, prior, and informed consent. Many Indigenous peoples are Treaty and Aboriginal rights holders with active land-based livelihood practices in regions affected by extractive activities. Thus, there are unique and significant impacts related to water use for hydraulic fracturing that must be resolved through Nation-to-Nation negotiations with Indigenous peoples. Each of these challenges will be explored in detail in the remainder of this chapter as a means to identify priority areas for change and governance innovation.

20.1 Introduction

Hydraulic fracturing ("fracking") is an extractive technique that involves injecting hydraulically pressurized liquid into shale rock in order to fracture it and release the natural gas inside (Rivard et al. 2014). Water use in this process is extensive and intensive: high volumes of water are typically combined with sand and chemicals for injection, and then the resulting contaminated wastewater is disposed of in holding pools or injected into aquifers. While hydraulic fracturing is not new, technological advancements in horizontal drilling and multi-stage fracturing have intensified both its profitability and thus, interest in development, but also the related and substantial water use. The growth in this industry has also brought an increase in public concerns about the impacts of hydraulic fracturing in general, and, more specifically, about the risks to water quality and quantity.

In recent years, research into the impacts of hydraulic fracturing has become clearer about the ecological and public health risks linked to surface and groundwater contamination; however, questions remain about the extent, probability, and best mitigation of such risks (Entrekin et al. 2011; Osborn et al. 2011; Rozell and Reaven 2012; Olmstead et al. 2013; Vengosh et al. 2014; Konschnik and Boling 2014; Elliot et al. 2016). Other concerns include multiple and cumulative impacts of industrial development in a watershed, such as habitat fragmentation that results from increased roads and traffic, air quality concerns, and the ecological impacts from hydraulic fracturing-associated seismic events (Rubinstein and Babaie Mahani 2015; Konschnik and Boling 2014; Garvie et al. 2014; Rivard et al. 2012).

In Canada, beyond apprehensions about specific land and water impacts, several Indigenous Nations have voiced significant concerns about the lack of meaningful consultation in their territories – a fiduciary responsibility of the federal government before such activities take place on their land. Many of these voices are coming from those who have faced hydraulic fracturing proposals, exploration, or full development on their traditional territories. For instance, in New Brunswick, proposals for exploration of shale gas resources in 2013 resulted in protests by the Elsipogtog First Nation who asserted that the province had failed to properly consult the Nation (Schwartz and Gollom 2013). In the fall of 2014, the premier imposed a moratorium on hydraulic fracturing and formed a commission that was responsible for determining when and if the conditions to lift the moratorium could be met (Bissett 2015). In British Columbia, the Fort Nelson First Nation filed an appeal with the BC Environmental Appeal Board, challenging the province's decision to allocate a water license to the Nexen Corporation for the purposes of hydraulic fracturing. The appeal was successful, and in 2015 the Board rejected the license (Environmental Appeal Board, Decision No. 2012-WAT-013(c)), finding that the license was "fundamentally flawed in both concept and operation" and that the company had failed to consult the Nation in good faith (Fort Nelson First Nation 2015).

Taken together, these and other political and legal conflicts illustrate the tensions surrounding the allocation and use of water for hydraulic fracturing in Canada. At the same time, these conflicts reveal systemic "patchwork" problems in municipal, provincial, federal, and Indigenous regimes for water governance.[1] In fact, we suggest that the rapid expansion of hydraulic fracturing of Canada's shale gas resources has not caused but rather *illuminated* the patchiness and inadequacy of many water governance arrangements in Canada.[2] Moreover, we argue that in this context of contestation and volatility around hydraulic fracturing, Canada is in dire need of creative, flexible, and socially responsive water governance arrangements. We define water governance as: (1) who decides who may use water and for what purposes; (2) what standards must be met during that use to protect ecological,

---

[1] While water governance is largely devolved to the provinces, there are jurisdictional challenges for water governance in Indigenous communities/territories that fall under the regulations of the federal Indian Act.

[2] This chapter involves the results of a study conducted in 2013/2014 that was supported by the Canadian Water Network (CWN). The findings are informed by the following methods: we analyzed the existing literature; ran a 3-round Delphi study of water and energy experts in Canada in which participants respond to each other's ideas and tried to build consensus on key themes; shared knowledge and findings with participants through "regional briefing notes"; and held a 2-day workshop in Victoria involving 25 people from 16 organizations in five regions, which was designed to establish priority challenges and paths forward with regards to water governance and hydraulic fracturing. While the Delphi participants had broad geographic representation, many workshop participants came from Eastern and Western Canada, and their critical insights shaped the more in-depth emphasis on these two regions in this chapter. In March 2015, we presented a draft report to CWN, other investigative teams, the CWN Research Management Board, and an ad-hoc research review committee, whose comments were then integrated or otherwise addressed (see *Building Capacity to Build Trust: Key Challenges for Water Governance in Relation to Hydraulic Fracturing* (Moore et al 2015)).

economic, social, and cultural values; and (3) how that decision process is undertaken.

Recently, significant and rapid declines in global markets for natural gas have slowed activity and led to the selling off of assets, including in provinces that had approved developments (International Energy Agency 2015; Apache 2014). Such a moment of relief in industry-driven pressure provides a critical opportunity in Canada to address existing water governance issues, and we focus this chapter specifically on three key challenges and priority areas for governance innovation. First, we illustrate how water governance approaches across the provinces and territories are patchy and fragmented, which leads to a perceived lack of transparency in the system. Second, we explain how capacity across the many actors involved in governing water or using water is not uniform and we reveal how this affects individual and collective abilities to make substantive changes that would improve the various water governance approaches across the country. On this matter, we add that the allocation of water for industrial use is often perceived as government favoring economic interests over all other values, which can lead to a perceived lack of accountability and conflicts of interest by those responsible for governing water use. Third and finally, we elucidate how water governance has failed to adequately include Indigenous peoples vis-à-vis free, prior, and informed consent. Many Indigenous peoples that are Treaty and Aboriginal rights holders have active land-based livelihood practices in regions affected by extractive activities. Thus, there are unique and significant impacts related to water use for hydraulic fracturing that must be resolved through Nation-to-Nation negotiations with Indigenous peoples. Each of these challenges will be explored in detail in the remainder of this chapter as a means to identify priority areas for change and governance innovation.

20.2 Challenge 1: "Patchy" Approaches to a Sustainable Water-Energy Future

Shale gas deposits exist in several parts of Canada, though the extent and accessibility of these plays vary significantly by region, as do patterns of development (see Fig. 20.1 Shale gas reserves in Canada).

This variation is based on a number of factors, but settlement patterns and historical land uses play a key role, as do the economic development strategies of each region. In British Columbia and Alberta, for example, which have histories of oil and gas development, provincial governments have encouraged widespread development of shale gas resources through hydraulic fracturing, even as the most comprehensive studies have argued that this type of development outpaces scientific knowledge and effective regulatory structures (e.g. CCA 2014; Rivard et al. 2012; Wiseman 2009). In other jurisdictions, including Quebec, Nova Scotia, and New Brunswick, provincial governments have identified the need for research and public consultation, enacting moratoriums on hydraulic fracturing (e.g. Mauro 2014).

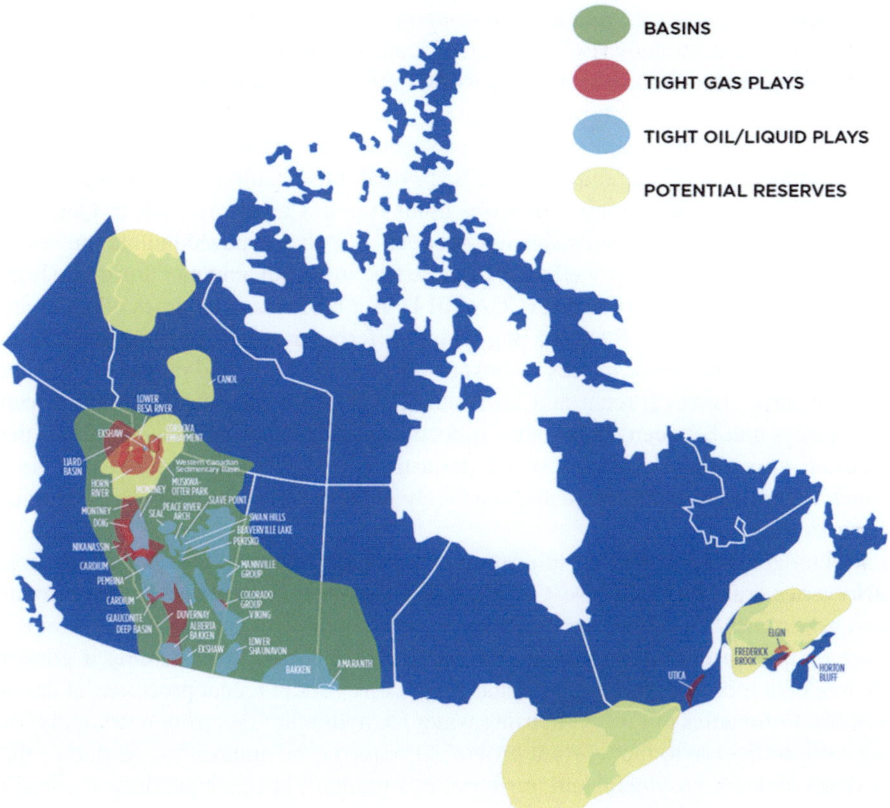

Fig. 20.1 Map of unconventional oil and gas plays in Canada (Canadian Water Network 2015, and for illustrative purposes only. Note that tight gas refers to gas found in low permeability sandstone formations, whereas shale gas typically refers to natural gas trapped between layers of shale)

These moratoriums delay development (but not necessarily exploration) and provide an opportunity for research and knowledge exchange across regulatory agencies, industry, Indigenous Nations, non-governmental organizations (NGOs), and the public. At the same time, these moratoriums do not constitute a governance regime within which water-related decision-making can occur on a long-term basis.

In Canada, each region has a vastly different history of resource extraction and energy development, which in turn creates variability in water use for energy. In Quebec, for example, initial producer interest led to a long process of research and public engagement and ultimately, a temporary moratorium on shale gas development in the St. Lawrence lowlands until further studies and consultation with the public and Indigenous nations can be completed (Rivard et al. 2014). But Alberta, the most significant oil and gas producer in the country, has extensive experience with hydraulic fracturing, a technique that has been used on over 180,000 oil and gas wells in the province since the 1950s (Natural Resources Canada 2015). Similarly, development in BC – which has a well-established industry in the Northeast of the province – has advanced quickly and remains a key part of the provincial government's economic strategy (BC Ministry of Energy and Mines 2012).

Existing evidence from around the world indicates that the amount of water used in different hydraulic fracturing operations varies widely, depending on the operating strategy of the company working on a specific shale play and the characteristics of the play itself (Chen and Carter 2016; Nicot and Scanlon 2012; Rivard et al. 2014). This trend in varying water uses holds true in Canada as well. In Quebec's Utica shale play, early results showed between 12,000 and 20,000 cubic metres of water needs to be used per well for hydraulic fracturing, an amount considered high relative to other plays in Canada (CCA 2014). But the highest volumes of water are used in the Horn River Basin in BC, where differing accounts have reported such volumes as 34,900 cubic metres (Johnson and Johnson 2012) and ranges of 25,000–75,000 cubic metres (Precht and Dempster 2015). With such differences between companies and between shale plays, it becomes difficult for anyone outside of the relevant government agencies or the industry actors to understand exactly how much water has been allocated and is actively being used for the purpose of hydraulic fracturing.

Water governance, and the allocation of water for hydraulic fracturing activities, also varies dramatically across Canada (see Table 20.1). In general though, provincial governments hold responsibility for water-related decision-making (and, in the case of the Northwest Territories, the territorial government is gaining a greater degree of authority over these governance processes under recent processes of devolution). Companies interested in using water for hydraulic fracturing must apply for a permit or license to the relevant provincial or territorial authorities. As part of the process, federal, provincial and territorial governments have a legal duty to consult and accommodate Indigenous peoples on any decisions that may affect Aboriginal or Treaty rights (Newman 2009).

Beyond these key similarities, provinces and territories have different types of entities and legislation that deal with hydraulic fracturing. In some provinces, such as Ontario, Quebec, New Brunswick, and Nova Scotia, a number of different agencies and authorities shape water-related hydraulic fracturing decisions. But in Alberta and BC, provincial governments have sought to streamline decision-making by creating "single window" regulators that deal with all issues related to oil and gas development, including the allocation of water licenses. The "single window" approach has been criticized due to the potential of the agency's conflicts of interest (Environmental Law Centre 2012). As one example, the BC Oil and Gas Commission's (OGC) funding model relies on industry fees, and thus, has been criticized for the limitations this poses to its independence and ability to hold industry accountable (Environmental Law Centre 2012). The Office of the BC Auditor General (2010) has also expressed concern that while the OGC is mandated to foster a healthy environment, no formal province-wide program exists to manage environmental impacts from the development it enables.

Over and above the variations across provinces, dissimilarities also exist within provinces, depending on the relationship among government agencies, specific companies, and communities – Indigenous and non-Indigenous – and the wide con-

Table 20.1 Provincial and territorial comparison of the general status of the hydraulic fracturing industry and associated governance frameworks (Moore et al. 2015)

Province	Status of industry	Overall governance framework
Alberta	Large and well-established oil and gas industry; 10,000 wells fracked since 2008.	Alberta Energy Regulator (AER) is responsible for all oil and gas development and accepts applications for water licences.
		Key legislation: *Water Act, Responsible Energy Development Act*
British Columbia	Fast-moving industry (between 2005 and 2009); 1560+ wells fracked by 2013	B.C. Oil and Gas Commission (OGC) is responsible for the development of petroleum resources and the management of industry impacts. The OGC has delegated authority to approve water licences.
		Key legislation: *Oil and Gas Activity Act, Water Act*, and newly established *Water Sustainability Act* (regulations still being finalized for new groundwater licensing).
New Brunswick	Long history of oil and gas development but shale gas industry is small; moratorium in place	Department of Energy and Mines has authority over natural gas; Ministry of Environment regulates water.
		Key legislation: *Clean Environment Act, Oil and Natural Gas Act, Clean Water Act.*
Nova Scotia	Moratorium on high volume hydraulic fracturing (except for testing and research)	Department of Energy holds authority for shale gas and Department of Environment receives application on potential impacts and water permits.
		Key legislation: *Water Act; Petroleum Resources Act*; *Environment Act;* and *Environmental Goals and Sustainable Prosperity Act.*
Northwest Territories	Companies exploring in the Canol shale oil play (Central Mackenzie Valley), but development is slow moving	Governance changing with current process of devolution under Bill C- 15, including the creation of a new eleven-person territorial board in Yellowknife. Department of Industry, Tourism, and Investment responsible for administration of onshore oil and gas, and the Department of Environment and Natural Resources regulates water.
		Key legislation: *Waters Act*, *Oil and Gas Operations Act*
Ontario	Starting in 2010, geological researchers identified potentially productive reserves; province currently lacks regulatory framework to govern hydraulic fracturing	Multiple agencies would be involved in regulation of shale gas development and water governance: the Ministry of Natural Resources, the Ontario Energy Board, and the Ministry of Environment.
		Key legislation: *Ontario Water Resources Act* and the Water Taking Regulation; *Ontario Clean Water Act*; *Environmental Protection Act; Oil, Gas, and Salt Resources Act*

(continued)

Table 20.1 (continued)

Province	Status of industry	Overall governance framework
Quebec	Quebec has a high potential shale gas landscape; the provincial government instituted a moratorium in 2011	Ministry of Environment governs surface and groundwater withdrawals, while the Ministry of Mines regulates mining, oil, and gas activities.
		Key legislation: *Environmental Quality Act*, *Water Act*, Water Policy legislation, *Mining Act*, and the moratorium bill.

Note – although shale deposits were initially explored in SK, reports on activity are not available. This table focuses on those areas where horizontal drilling and exploration has occurred, been reported and discussed widely, and where water has been allocated or at the source of conflict

tinuum of baseline knowledge about the watershed. This means that decisions about whether water will be allocated to hydraulic fracturing, and what type of consultation and/or accommodation may take place, is dependent on the specific regional water manager, the specific Nation, the specific company, along with the specifics of the hydrology and geology of the area. Sometimes there is even patchiness within a single region, or single Treaty area with regards to consultation and accommodations. The challenge is that the differences contribute to a lack of understanding and certainty about what will and what can happen, and a perceived sense that transparency in decision-making is lacking, since it is difficult to identify and understand who is responsible for these decisions in every watershed, or for addressing concerns once they are voiced (see Moore et al. 2015).

While this patchiness is a challenge, to some extent, such an approach to water governance in Canada makes at least some sense, given that there is no "one size fits all" approach, and each community and/or Indigenous Nation, resource patch, and company should be recognized for the different context, culture, and values involved in a particular watershed. But such a solution would require a clear governing strategy around a more "customized" approach, whereby each watershed and actor is treated in recognition of those differences but still with a coordinated, transparent, and accountable customized approach. Instead, the existing differences have emerged in an ad-hoc manner in response to a rapidly developing industry and continue to fall along political boundaries rather than natural or other innovative boundaries. Trying to understand the differences among patches, what that means for water use requirements, which company is operating and where, what their practices involve, and who is allocating water and enforcing standards for the protection of the environment and public health and safety across each of these operations is no simple task – and it is further complicated by the multiple, competing water uses in the same region (e.g. recreation, other industries, agriculture, commercial, etc.). Therefore, we emphasize that "patchiness" itself is not a problem; rather, uncoordinated, and opaque, as opposed to transparent, patchiness creates opportunities for problems to emerge.

20.3 Challenge 2: Capacity and Accountability Across the Existing Water Governance Systems in Canada

Shale gas development is complex and fast changing in ways that present unique governance challenges for provincial government regulators, who are understood to be largely responsible for water governance in relation to hydraulic fracturing. The water-related risks associated with shale gas development are different – and, some scholars argue, higher – than with conventional gas development (Konschnik and Boling 2014). For instance, the techniques frequently involve greater volumes of water, the use of which creates concerns about water stress (Freyman and Salmon 2013) and the associated seismic activity (Rubinstein and Babaie Mahani 2015). Additionally, different chemicals are used in the fracturing process and their use creates new concerns about risks of leaks, surface management of flowback water, and groundwater contamination the seismic activity the water use and fracturing process induces, and the volumes of water involved (Gregory et al. 2011; Llewellyn et al. 2015; Elliot et al. 2016; Chen and Carter 2016). Hydraulic fracturing is also, per well pad, more intense and expansive than conventional; after 2–3 years, production decreases and new sources of gas need to be exploited in order to maintain supply (Konschnik and Boling 2014). Moreover, technological innovation results in fast-changing extractive practices. Further complicating matters, hydraulic fracturing developments are often spread over large areas of land, through thousands of wells, and therefore, the possible pollution sources (e.g. storage tanks, borrow pits, and pipelines) are numerous (Konschnik and Boling 2014; Willow and Wylie 2014).

The complexity and dynamic nature of unconventional gas development raises significant capacity challenges for nearly every actor involved. With respect to industry, multiple actors may be involved in a single operation (e.g., investors, primary operators, sub-contractors, and consultants). Although these actors will have high levels of technical expertise relevant to their own role, their regional knowledge, their training in social, political, and cultural issues relevant to those regions, attitudes towards environmental protection and health and safety, and their experience with regards to implementing legal duties to consult with Indigenous nations may be lacking (Breiddal 2015; Moore et al. 2015).

For either Indigenous or non-Indigenous communities in regions where shale gas development has proceeded, awareness, training, and the capacity to respond to the demands associated with consultation may be limited. As just one example, in NE British Columbia, hydraulic fracturing activities have been approved by the OGC in Treaty 8 territories. The Fort Nelson First Nation is a signatory of Treaty 8 and the provincial government has a constitutional duty to consult with FNFN over impacts to treaty rights from the development. This duty is implemented through either the consultation process associated with an environmental impact assessment of a proposed development, or through the Oil and Gas Commission's consultation protocol agreements. In the "boom" period, particularly in 2008–2011, FNFN faced an average of 1000 referrals per year – although, these were not just for water licenses, but also for the development of well pads and all the related infrastructure necessary to support the development. Comment and consultation periods can involve time restrictions, and may range from 5 to 20 days (Breiddal 2015). Very few small to medium communities in Canada, characterized by Statistics Canada as less than 1,000 and up to 99,999 people – rural or urban, Indigenous or non-Indigenous, would be able to meet such intense demands, particularly when knowledge about a new industry is relatively limited.

Capacity challenges also exist within the provincial government agencies trying to ensure each actor is engaging in socially, ecologically, and culturally responsible practices with regards to water use. Specifically, the rapidly developing and changing nature of hydraulic fracturing technology raises questions about whether regulatory agencies have sufficient capacity to (1) understand the complexity of ecosystem functions in the watersheds where developments are proposed or occurring; (2) develop relationships with industry, community, and Indigenous nations based on trust and mutual respect; (3) fully understand the fast-changing operational practices of the industry and how best to regulate those practices; and (4) from there, enforce any regulations that do exist, all while meeting goals of fairness and efficiency, particularly when hydraulic fracturing may be just one of many uses for water in the watershed.

Simply adding capacity, however, will not equate to better governance without other changes to policy and agency function more broadly. Specifically, concerns exist about government partiality and "regulatory capture," meaning that communities are known to often perceive regulators' work as that which seeks to ensure regulations favour industry development over any other public interest or value. Other scholars have found that, in some jurisdictions, because politicians and regulators believe that new regulation-related costs might mean that producers move elsewhere, they have allowed industry to heavily influence legislative and regulatory agendas in natural gas producing regions (see Centner and O'Connell 2014 for analysis of this effect in the US; see also Stephenson and Shaw 2013; Allen 2012; Rahm 2011). The consequence of even the appearance of regulatory capture, let alone actual capture, is that governments are perceived as not holding industry accountable for their actions related to water use for hydraulic fracturing, and not upholding environmental protection responsibilities or sufficiently representing the interests and values of the public (Garvie and Shaw 2014).

We emphasize that many provincial initiatives have worked hard to address these concerns. For example, due in large part to the size and age of its industry, Alberta has developed extensive regulatory frameworks to address energy-related water governance, including more stringent regulations around well casings. Also, Alberta recently launched a pilot project to test a "play-based approach" to governing energy-water use, which requires full projects to be assessed, instead of the site-specific and well-by-well approaches that have traditionally been used (Alberta Energy Regulator 2015). In BC, the provincial government launched the NorthEast Water Tool, a new hydrological, GIS-based decision-support tool that proponents and the public may use and that aims to ensure science-informed governmental decision-making about water allocations. In Nova Scotia, the provincial government undertook extensive consultation in 2014 as part of the Nova Scotia Hydraulic Fracturing Review (i.e., the "Wheeler Report" (Government of Nova Scotia 2014)). The report was widely heralded for its fair consideration of Mi'kmaq-specific and public input. Across the board, however, we argue that the concerns about limited capacity and a lack of accountability and impartiality are still front and centre in spite of these initiatives. Thus, simply doing "more" may not be the answer; instead, we need to consider innovative alternative water governance approaches (see for instance, Moore et al. 2014) discussion on water policy innovations).

20.4 Challenge 3: Water Governance that Is Inclusive of Indigenous Peoples

The third significant water governance challenge is that of adequately including Indigenous Nations in decision-making, as they hold distinct Treaty and legal status as the original peoples of this land and are uniquely affected by all resource extraction industries on their treatied, unceded, and traditional lands. Some researchers have suggested that Indigenous peoples are "canaries in the coal mine" when it comes to the impacts of resource extraction and environmental degradation (Weaver 2010). Specifically, rural and remote Indigenous communities are particularly vulnerable to environmental impacts, including localized industrial and long-range transport contamination, because many of their members continue to harvest from the land and waters for the purposes of food security, cultural and spiritual revitalization, and social reproduction (Mascarenhas 2012; Turner 2014). In Canada, resource development projects that both fail to fully consult and accommodate Indigenous interests often result in opposition by Indigenous peoples and/or negative impacts on Treaty and Aboriginal rights – resulting in protracted legal suits and substantial delays.

Indigenous Nations in Canada hold a unique position as constitutionally protected rights holders with respect to both Aboriginal identity and Treaties. In any process to develop and implement new governance arrangements, attention to international declarations on Indigenous rights, constitutional and treaty rights, and evolving legal precedents, is needed (Prno and Slocombe 2012). Several scholars of Indigenous governance have noted that land-based practices are critical to maintaining the community-homeland relationship that is fundamental to cultural identity and self-determination (e.g. Alfred and Corntassel 2005; Corntassel 2008). The protection of these practices and connections over the long-term, scholars argue, requires an approach to governance broader than one based on rights alone, as a "rights-based" approach often focuses – too narrowly – on the ability to continue to procure food in a traditional way (Corntassel 2008; Schreiber 2006). Some are suggesting a "responsibility-based" approach may be the way forward, one that recognizes a relational approach to managing our relationships with each other, resources, and systems (Castleden et al. 2015).

The imbalance in authority – where Indigenous authority is generally under-recognized by federal and provincial governments – has been a major ongoing issue in governance related to land and resource governance in Canada (von der Porten and de Löe 2013). Yet, the federal government has a fiduciary responsibility and legal duty to consult and "where appropriate" accommodate Indigenous peoples whose traditionally territories comprise the proposed spaces of industrial development (Natcher 2001; Newman 2009). A number of recent court cases have set out the legal meanings of Aboriginal rights and title and provide substantial guidance regarding the responsibilities of governments in this realm (see CCA 2014; Castleden et al. 2016). Most recently, in June 2014, the Supreme Court of Canada passed down a unanimous decision that "significantly alters the legal landscape in Canada relating to land and resource entitlements and their governance" when it declared that consent was needed on territories recognized as "owned" by a Nation (Pinder 2014).[3] There has also been a dramatic shift in the way the federal government is approaching its relationship to Indigenous peoples in Canada. Where the former (Conservative) government took the approach that decisions they took would hold until the Supreme Court of Canada told them otherwise, the current (Liberal) government is taking the approach that such negotiations are between Nations, and that the 2007 United Nations Declaration on the Rights of Indigenous Peoples (UNDRIP) forms the basis of such conversations (Barrera 2015; Smith 2015).

Given these circumstances, the legal and constitutionally guaranteed rights of Indigenous peoples in Canada will inevitably shape water governance for hydraulic fracturing, though in ways that vary greatly by region and in the context of existing historic and modern treaties, land claim agreements, or ongoing claims of Aboriginal title in areas without treaties. In the context of hydraulic fracturing, research has explored instances where governance processes have failed to protect treaty rights but also, more broadly, to meaningfully include Indigenous Nations in decision-making related to water resources in their traditional territories (Garvie and Shaw 2014). In short, the lack of meaningful inclusion in decision-making limits genuine "government-to-government" relationships; the consequence is further discord over final decisions. When a Nation has not been included, and when the provincial or federal governing body does not indicate an understanding that Indigenous rights include more than food procurement for cultural or ceremonial purposes, it becomes difficult to understand how governments can claim that they have appropriately accommodated Treaty rights.

In some regions, hydraulic fracturing has already had an impact on the constitutionally guaranteed treaty rights of Indigenous peoples, a fact that has informed

---

[3] *Tsilhqot'in Nation v. British Columbia* 2014 SCC 44. In the Tsihlqot'in case, the court for the first time in Canada recognized Aboriginal title in a specific region of British Columbia and gave the concept meaning by defining it as "the right to control the land conferred by Aboriginal title means that governments and others seeking to use the land must obtain the consent of the Aboriginal title holders" (Pinder 2014). The full impacts of the Tsihlqot'in case remain uncertain and will be unfolding over years to come; however, the case suggests that the legal burden that Aboriginal title imposes on the Crown will necessitate the creation of new institutions of consultation and that respect the need to obtain *consent* prior to development taking place.

their responses to existing and proposed decision-making processes. Returning to the example of the Fort Nelson First Nation in northeast BC, the Nation has already seen extensive changes in the landscape as a result of shale gas development, changes that have an impact on well established Treaty practices (Garvie et al. 2014). The FNFN's concern over shale gas governance has often focused on water. For example, when the Nation appealed to the courts over the province's issuance of a water license in their territory, no government action was taken for over 6 months, during which time the licensee withdrew more than 180,000 cubic metres of water from a lake during a drought (Chapman 2013). While FNFN has filed a successful appeal of the license at the Environmental Review Board and the licensee has since been charged under the *Water Act* for breaching the license and diverting more water than authorized, it is really a case of "too little, too late"; the water has been used.

In other regions where shale gas development is proposed or nascent, such as on the East coast, concerns from Indigenous Nations regarding the uncertainty of the environmental impacts have driven opposition to hydraulic fracturing. In some jurisdictions Indigenous Nations have meaningful authority to help shape decision-making around hydraulic fracturing. For example, the Kwilmu'kw Maw-klusuaqn Negotiation Office (KMKNO), which is part of the Nova Scotia-based Mi'kmaq Rights Initiative, seeks to "address the historic and current imbalances in the relationship between Mi'kmaq and non-Mi'kmaq people in Nova Scotia" (Mi'kmaq Rights Initiative 2015).[4]

In this case, corporate and government entities in the province must go through a negotiation process with KMKNO and must receive approval from the Assembly of Nova Scotia Mi'kmaq Chiefs before proceeding with development. Among this Assembly there was (as noted earlier in the Chapter) no support for hydraulic fracturing, a position revealed during consultation as part of the Wheeler Report. The Review included a discussion paper that summarized the importance of addressing Aboriginal rights in Nova Scotia, noting that:

> The Mi'kmaq people possess robust treaty rights, as well as Aboriginal rights in Nova Scotia. These rights have considerable consequences for provincial deliberations over hydraulic fracturing, as the Province is constitutionally obliged to honour these rights (MacIntosh 2014, p 282).

Despite the legal requirements, public statements about the importance of government-to-government relationships, and the acknowledgement of Indigenous rights in documents such as the Wheeler Report, the actual processes of consultation (or any governance arrangement that is inclusive beyond consultation), or the principle of free, prior and informed consent (as per UNDRIP) remains a key issue in relation to water governance for hydraulic fracturing and elusive in practice (Garvie and Shaw 2014).

---

[4]The KMKNO seeks to "*undertake the necessary research, develop consensus positions on identified issues, and create public and community awareness in a manner that supports the ability of the Assembly to fully guide the negotiations and the implementation and exercise of constitutionally protected Mi'kmaq rights*" (Mi'kmaq Rights Initiative 2015).

20.5 Conclusions

The rapid expansion of the shale gas industry in Canada, and its consequent demands for water to facilitate hydraulic fracturing, has occurred amongst a patchwork of water governance arrangements. Although efforts have not been made to coordinate water governance approaches to hydraulic fracturing across Canada, efforts to streamline industrial land and water application processes within individual provincial jurisdictions have, in some cases, led to "single window" approaches. In turn, concerns have emerged about regulatory capture and an overemphasis on the economic benefits of energy development above all other social, ecological, and cultural values at the community level. In many cases, capacity within provincial government agencies is perceived as too limited and inexperienced with water use for hydraulic fracturing to govern using leading accountability and transparency practices, while simultaneously building knowledge about the short- and long-term social-ecological impacts of water use for hydraulic fracturing.

Existing conflicts over water use for hydraulic fracturing have exposed the limitations within and across Canada's water governance patchwork: in some places, water governance arrangements are failing to inspire adequate public trust that development can safely proceed in some places; in others, are failing to adequately protect Aboriginal and treaty rights, and public values. Despite the many differences in how water is governed in different jurisdictions, we have emphasized here three common challenges that must be addressed if water governance is to effectively manage water resources, and thus inspire public trust. First, water governance arrangements must be made more transparent and accountable. Second, their design and implementation must be adequately resourced, to ensure that regulators have the capacity to meet the challenges of regulating the complex, and dynamic industry before further developments proceed, or in the case that decisions are made to lift moratoria. Third, governance arrangements must be restructured to ensure that Indigenous peoples are adequately engaged in, and by, these improved processes and that their rights are protected. Although we have treated the three challenges described in this chapter as separate, they are deeply intertwined. Arguably, all of these governance challenges can and must be tackled together.

Several more problems will result if these challenges are not addressed. That is, if regulators are unable to stay ahead of this fast-moving global industry, policies will become reactionary and lacking in vision, the water cycle will not be governed in a coordinated way, and industrial impacts will be poorly understood and mitigated. A failure of water governance of this kind will have far-reaching impacts not only on Aboriginal rights and all Canadians, but on ecological resilience and public confidence in government. We have already observed the outcomes that follow from problematic and patchy water governance, including active resistance to industrial development. Improved water governance, then again, would provide a key focal point to ensure that decisions about industrial development are grounded in the best available evidence, are responsive to the rights of those affected, and ensure that a broad range of social, ecological, and cultural values are proactively protected.

We recognize that developing and testing new approaches to governance will not be easy – it will require the willingness of all key actors groups involved in, and affected by, water governance to participate, whether improved approaches are related to regulations, transparency mechanisms, or new organizational models and decision-making processes for collaboration. But such steps will be necessary to rebuild trust and to ensure that decision-making is competent and coordinated across Canada's resource patches.

References

Alberta Energy Regulator. (2015). *Play-based regulation pilot*. https://www.aer.ca/about-aer/spotlight-on/pbr-pilot-project. Accessed 11 Mar 2015.

Alfred, T., & Corntassel, J. (2005). Being indigenous: Resurgences against contemporary colonialism. *Government and Opposition, 40*(4), 597–614.

Allen, K. (2012). The big fracking deal: Marcellus Shale – Pennsylvania's untapped resource. *Villanova Environmental Law Journal, 23*(1), 57–87.

Apache. (2014). News release: Apache corporation announces the sale of Wheatstone and Kitimat LNG projects to Woodside Petroleum Limited. 15 December 2014. http://files.shareholder.com/downloads/APA/927117014x0x799519/C43BCC43-E162-4DFB-BD2D-74A998049B93/APA_News_2014_12_15_Financial.pdf. Accessed 27 Oct 2015.

Barrera, J. (2015, October 15). Trudeau: A Liberal government would repeal, amend all federal laws that fail to respect Indigenous rights. *The Aboriginal Peoples Television Network*. http://aptn.ca/news/2015/10/15/trudeau-a-liberal-government-would-repeal-amend-all-federal-laws-that-fail-to-respect-indigenous-rights/. Accessed 15 Mar 2016.

BC Office of the Auditor General. (2010). *Oil and gas site contamination risks: Improved oversight needed 2009/2010*, report number. http://www.prrd.bc.ca/board/agendas/2010/2010-13-8819986574/pages/documents/06-D-1aOGCNewsReleaseandAuditorGeneralReport.pdf. Accessed 15 Mar 2016.

Bissett, K. (2015). New Brunswick introduces fracking moratorium. *CBC News*, 18 December 2015. http://www.theglobeandmail.com/news/politics/new-brunswick-introduces-fracking-moratorium/article22139797/. Accessed 27 Oct 2015.

Breiddal, R. (2015). *When consultation becomes a checkbox, what's the fracking point?: Colonial constraints on social learning processes in Northeast BC and the Fort Nelson First Nation's new approach to resource governance*. Master's thesis, University of Victoria.

Canadian Water Network. (2015). *Water and hydraulic fracturing: Where knowledge can best support decisions in Canada*. Waterloo: Canadian Water Network.

Castleden, H., Cunsolo Willox, A., Harper, S., Martin, D., Hart, C., Stefanelli, R., Day, L., & Lauridsen, K. (2015). *Living with water: Examining methods and models for integrative Indigenous and Western Knowledge to inform – and transform – water research and management in Canada*. Report prepared for the Canadian Water Network.

Castleden, H., Cunsolo-Willox, A., Harper, S., Martin, D., & Hart, C. (2016). Reconciliation and relationality in water research and management in Canada: Implementing indigenous ontologies, epistemologies, and methodologies. In S. Renzetti & D. Dupont (Eds.), *Water policy and governance*. Cham: Springer.

Centner, T. J., & O'Connell, L. K. (2014). Unfinished business in the regulation of shale gas production in the United States. *Science of the Total Environment, 476–477*, 359–367.

Chapman, A. (2013). *A technical review of the various pieces of information provided to you by Nexen Inc.*, Related to water withdrawals from North Tsea Lake during the period of June to August, 2012. In: Chief Sharleen Gale in her own right and on behalf of the members of the

Fort Nelson First Nation v Assistant Regional Water Manager and Nexen Energy ulc. January, 2014. BC Environmental Appeal Board, Final Argument of the Appellant 2012-WAT-013, Victoria.

Chen, H., & Carter, K. E. (2016). Water usage for natural gas production through hydraulic fracturing in the United States from 2008 to 2014. *Journal of Environmental Management, 170*, 152–159.

Corntassel, J. (2008). Toward sustainable self-determination: Rethinking the contemporary indigenous-right discourse. *Alternatives: Global, Local, Political, 33*(1), 105–132.

Council of Canadian Academies. (2014). *Environmental impacts of Shale gas in Canada*. Ottawa: Council of Canadian Academies. http://www.scienceadvice.ca/en/assessments/completed/shale-gas.aspx

Elliot, E. G., Ettinger, A. S., Leaderer, B. P., Bracken, M. B., & Deziel, N. C. (2016). A systematic evaluation of chemicals in hydraulic-fracturing fluids and wastewater for reproductive and developmental toxicity. *Journal of Exposure Science and Environmental Epidemiology, 1*, 1–10.

Entrekin, S., Evans-White, M., Johnson, B., & Hagenbuch, E. (2011). Rapid expansion of natural gas development poses a threat to surface waters. *Frontiers in Ecology and the Environment, 9*(9), 503–511.

Environmental Law Centre. (2012). *Oil and gas reform (Summer)*. http://www.elc.uvic.ca/publications/index.htm#LawReform. Accessed 21 Mar 2015.

Fort Nelson First Nation. (2015). *Fort Nelson First Nation wins legal challenge stopping Nexen water license for fracking in BC*. Press release, 7 September 2015.

Freyman, M., & Salmon, R. (2013). *Hydraulic fracturing and water stress: Growing competitive pressures for water*. Boston: CERES.

Garvie, K. H., & Shaw, K. (2014). Oil and gas consultation and Shale gas development in British Columbia. *BC Studies (Winter), 184*, 73–79, 81–102, 183–184.

Garvie, K. H, Lowe, L., & Shaw, K. (2014). Shale gas development in Fort Nelson First Nation Territory: Potential regional impacts of the LNG Boom. *BC Studies (Winter), 184*, 45–51, 53–61, 63–72, 183–184.

Government of Nova Scotia. (2014, August 28). *Independent hydraulic fracturing review* (Also called the "Wheeler Review"). Nova Scotia: Halifax.

Gregory, K. B., Vidic, R. D., & Dzombak, D. A. (2011). Water management challenges associated with the production of shale gas by hydraulic fracturing. *Elements, 7*(3), 181–186.

International Energy Agency. (2015). *Medium-term gas market report: Market analysis and forecasts to 2020*. Paris: OECD/IEA.

Johnson, E. G., & Johnson, L. A. (2012). *Hydraulic fracture water usage in northeast British Columbia: Locations, volumes and trends*. In Geoscience Reports 2012, British Columbia Ministry of Energy and Mines, pp. 41–63.

Konschnik, K. E., & Boling, M. K. (2014). Shale gas development: A smart regulation framework. *Environmental Science & Technology, 48*(15), 8404–8416.

Llewellyn, G. T., Dorman, F., Westland, J. L., Yoxtheimer, D., Grieve, P., Sowers, T., Humston-Fulmer, E., & Brantley, S. L. (2015). Evaluating a groundwater supply contamination incident attributed to Marcellus Shale gas development. *Proceedings of the National Academy of Sciences, 112*(20), 6325–6330.

MacIntosh, C. (2014). *Hydraulic Fracturing and the aboriginal, treaty and statutory rights of the Mi'kmaq*. Nova Scotia Hydraulic Fracturing Review and Public Engagement Process. Cape Breton University and the Verschuren Centre. https://www.cbu.ca/hfstudy/resources/correspondence#.VG-QuIVj6XU. Accessed 21 Nov 2014.

Mascarenhas, M. (2012). *Where the waters divide: Neoliberalism, white privilege, and environmental racism in Canada*. Landham: Lexington Books.

Mauro, I. (2014). *Discussion Paper: The environmental impacts of hydraulic fracturing in Nova Scotia – A public participatory risk assessment*. Discussion paper prepared for Nova Scotia Hydraulic Fracturing Review and Public Engagement Process. Sydney: Verschuren Centre, Cape Breton University.

Mi'kmaq Rights Initiative. (2015). *KMKNO mission statement*. Millbrook: KMKNO. http://mikmaqrights.com/about-us/kmkno-mission-statement/. Accessed 20 Nov 2014.

Ministry of Energy and Mines. (2012). *British Columbia's natural gas strategy*. Victoria: Ministry of Energy and Mines. http://www.gov.bc.ca/ener/natural_gas_strategy.html. Accessed 29 Sept 2014.

Moore, M.-L., Shaw, K., Castleden, H., Breiddal, R., Kot, M., & Murray, M. (2015). *Building capacity to build trust: Key challenges for water governance in relation to hydraulic fracturing*. Ottawa: Canadian Water Network. http://www.cwn-rce.ca/assets/resources/pdf/Hydraulic-Fracturing-Research-Reports/Moore-et-al-2015-CWN-Report-Water-Governance-and-Hydraulic-Fracturing.pdf?u=keyword%3Dmoore

Moore, M.-L., von der Porten, S., Plummer, R., Brandes, O., & Baird, J. (2014). Water policy reform and innovation: A systematic review. *Environmental Science & Policy, 38*, 263–271.

Natcher, D. C. (2001). Land use research and the duty to consult: A misrepresentation of the aboriginal landscape. *Land Use Policy, 18*(2), 113–122.

Natural Resources Canada. (2015). *Alberta's Shale and tight resources*. Ottawa: Government of Canada. http://www.nrcan.gc.ca/energy/sources/shale-tight-resources/17679#a1

Newman, D. G. (2009). *The duty to consult: New relationships with Aboriginal peoples*. Saskatoon: Purich Publishing.

Nicot, J.-P., & Scanlon, B. R. (2012). Water use for Shale-gas production in Texas, U.S. *Environmental Science and Technology, 46*(6), 3580–3586.

Olmstead, S. M., Muehlenbachs, L. A., Shih, J. S., Chu, Z., & Krupnick, A. J. (2013). Shale gas development impacts on surface water quality in Pennsylvania. *Proceedings of the National Academy of Sciences, 110*(13), 4962–4967.

Osborn, S. G., Vengosh, A., Warner, N. R., & Jackson, R. B. (2011). Methane contamination of drinking water accompanying gas-well drilling and hydraulic fracturing. *Proceedings of the National Academy of Sciences, 108*(20), 8172–8176.

Precht, P. & Dempster, D. (2012). *Jurisdictional review of hydraulic fracturing regulation* (Final Report). Retrieved from: http://www.novascotia.ca/nse/pollutionprevention/docs/Consultation.Hydraulic.Fracturing-Jurisdictional.Review.pdf.

Pinder, M. (2014). *Tsilhqot'in Nation v. British Columbia 2014 SCC 44 – Case summary*. 27 June 2014. http://www.mandellpinder.com/tsilhqotin-nation-v-british- columbia-2014-scc-44-case--summary. Accessed 21 Nov 2014.

Prno, J., & Slocombe, D. S. (2012). Exploring the origins of 'social license to operate' in the mining sector: Perspectives from governance and sustainability theories. *Resources Policy, 37*, 346–357.

Rahm, D. (2011). Regulating hydraulic fracturing in shale gas plays: The case of Texas. *Energy Policy, 39*(5), 2974–2981.

Rivard, C., Lavoie, D., Lefebvre, R., Séjourne, S., Lamontagne, C., & Duchesne, M. (2014). An overview of Canadian shale gas production and environmental concerns. *International Journal of Coal Geology, 126*(1), 64–76.

Rozell, D. J., & Reaven, S. J. (2012). Water pollution risk associated with natural gas extraction from the Marcellus Shale. *Risk Analysis, 32*(8), 1382–1393.

Rubinstein, J. L., & Babaie Mahani, A. (2015). Myths and facts on wastewater injection, hydraulic fracturing, enhanced oil recovery, and induced seismicity. *Seismology Research Letters, 86*(4), 1060–1067. doi:10.1785/0120150067.

Schreiber, D. (2006). First Nations, consultation, and the rule of law: Salmon farming and colonialism in British Columbia. *American Indian Culture and Research Journal, 30*(4), 19–40.

Schwartz, D., & Gollom, M. (2013). N.B. Fracking protests and the fight for Aboriginal rights. *CBC News*. 19 October 2014. http://www.cbc.ca/news/canada/n-b-fracking-protests-and-the-fight-for-aboriginal-rights-1.2126515. Accessed 30 Sept 2014.

Smith, J. (2015, November 12). Canada will implement UN declaration on rights of indigenous peoples, Carolyn Bennett says. *The Toronto Star*. http://www.thestar.com/news/canada/2015/11/12/canada-will-implement-un-declaration-on-rights-of-indigenous-peoples-carolyn-bennett-says.html. Accessed 15 Mar 2016.

Stephenson, E., & Shaw, K. (2013). A dilemma of abundance: Governance challenges of reconciling Shale Gas development and climate change mitigation. *Sustainability, 5*, 2210–2232.
Turner, N. J. (2014). *Ancient pathways, ancestral knowledge: Ethnobotany and ecological wisdom of Indigenous peoples of Northwestern North America*. Montreal: McGill-Queen's University Press.
Vengosh, A., Jackson, R. B., Warner, N., Darrah, T. H., & Kondash, A. (2014). A critical review of the risks to water resources from unconventional Shale gas development and hydraulic fracturing in the United States. *Environmental Science and Technology, 48*(15), 8334–8348.
von der Porten, S., & de Löe, R. C. (2013). Collaborative approaches to governance for water and Indigenous peoples: A case study from British Columbia, Canada. *Geoforum, 50*(1), 149–160.
Weaver, J. (2010). *Notes from a miner's canary: Essays on the state of Native America*. Albuquerque: University of New Mexico Press.
Willow, A., & Wylie, S. (2014). Politics, ecology, and the new anthropology of energy: Exploring the emerging frontiers of hydraulic fracking. *Journal of Political Ecology, 21*, 222–236.
Wiseman, H. J. (2009). Untested waters: The rise of hydraulic fracturing in oil and gas production and the need to revisit regulation. *Fordham Environmental Law Review, 20*, 115.

Chapter 21
Agricultural and Water in Canada – Challenges and Reform for the 21 C

Marian Weber and Marius Cutlac

Abstract Agriculture is the dominant water use in Canada and the main contributor to non-point source pollution in agricultural watersheds contributing to toxic algal blooms in some of Canada's and the world's largest freshwater lakes. Water governance in Canada is fragmented, with water resources managed separately from land uses that contribute to water challenges. The performance of agri-environmental policies encouraging the adoption of beneficial management practices is also mixed. Efficient and effective farm level water management strategies will increase in importance over the next century as climate change and increasing demands for food put pressure on water quality and quantity. This chapter examines farm level decisions that affect water quality and quantity, and the factors that contribute to adoptability of beneficial practices. Decentralized and fragmented governance contributes to weak institutions for integrating water and agricultural land management resulting in poor monitoring and governance gaps at scales required to manage nutrient loads into major freshwater lakes as well emerging threats from unregulated pollutants. The potential for water quality trading to address risks from non-point source pollution is examined, along with opportunities for reform in Canada.

21.1 Introduction

Agriculture is Canada's most significant land use and largest consumer of water. Agricultural water management affects most of the Canadian population. Historically, water on agricultural land was managed through drainage and irrigation to encourage Canadian settlement (Fowke 1957). Today agriculture anchors rural communities and is fundamental to the Canadian identity (CCA 2013; AAFC 2013). However, impacts of agriculture on water and the environment are a growing

M. Weber (✉) • M. Cutlac
Alberta Innovates Technology Futures, 250 Karl Clark Road, T6N 1E4 Edmonton, AB, Canada
e-mail: marian.weber@albertainnovates.ca; cutlac@arc.ab.ca

S. Renzetti, D.P. Dupont (eds.), *Water Policy and Governance in Canada*,
Global Issues in Water Policy 17, DOI 10.1007/978-3-319-42806-2_21

public concern. Agriculture relies on clean water for irrigation, stock watering, and industrial cooling and cleaning. At the same time, agriculture has an impact on aquatic ecosystems through water consumption and the release of nutrients and other pollutants to water bodies. Agricultural non-point source pollution from diffuse sources and runoff is one of the leading contributors to water quality problems (CCA 2013).

Water stresses from agriculture are a challenge in nearly every province in Canada. While Canada's overall agri-environmental performance is good, Environment Canada's Water Quality Agri-Environmental Performance Index has declined over the last decade, largely due to increased application of nutrients (Environment Canada 2013). Iconic watersheds, including the St. Lawrence, Lake Erie and Lake Winnipeg are experiencing eutrophication and increased frequency of extensive algal blooms causing risks to human health and economic losses from fishery and beach closures. In many locations, high nitrate and phosphorous concentrations exceed drinking water standards (CCA 2013; Conference Board of Canada 2015). In some places, like the South Saskatchewan River, water demand frequently exceeds availability causing both quantity and quality challenges. Climate change is expected to exacerbate these challenges with reduced snow packs and snow cover, and more frequent and severe drought causing water shortages and further deteriorating water quality (Pomeroy et al. 2010; Westbrook et al. 2011). In addition, there are human health risks due to pathogens and toxins entering drinking water supplies from livestock, and emerging threats related to endocrine disruptors found in hormones and pesticides.

In the twenty-first century, global population pressures coupled with rising incomes and stresses from climate change will result in even more intensive use of water by agriculture (CCA 2013). Managing agricultural impacts on water will be critical for both the sustainability of aquatic ecosystems, as well as the sector. So while over the last century agricultural water management in Canada focused on irrigation and drainage to bring marginal lands into production, the challenge in this century is to identify practices that will allow farmers to intensify production sustainably (Corkal et al. 2007).

In this chapter we review risks to water from agricultural practices, drawing attention to key uncertainties and examining new approaches to restore degraded watersheds. Because of the diffuse nature of agricultural pollution the focus of farm water management has been on incenting beneficial management practices (BMPs) that reduce water use and minimize pollution. However, to date programs and incentives have been ad-hoc, with continued issues in basins with high agricultural land use. The poor results are partly due to failed policy and governance approaches to water management. We begin the chapter with a review of the current state of knowledge of the impacts of farm management practices on water. We then review water governance and policy in Canada, highlighting gaps that create risks for the public as well as policies that interact with farm level decisions. We examine factors that influence BMP adoption and evaluate the potential for Canada to develop more innovative approaches to address agricultural non-point source pollution. We summarize governance challenges for ensuring sustainable agricultural water management, and conclude with opportunities for reform.

21.2 Risks to Water Quantity and Quality from Agriculture

Agriculture affects water quantity and quality through water use and consumption, land management, and the management of waste as well as nutrient, chemical, and pharmaceutical inputs. Hydrology is the primary process through which land use affects water, therefore it is useful to distinguish between practices, such as irrigation and drainage, that intervene directly in the hydrologic cycle, and practices that intervene indirectly through changes in vegetation cover, soil moisture, and compaction (Elliott et al. 2014; Arnold et al. 2012). Agricultural management practices and their effects vary widely across the country due to variations in climate, vegetation, underlying soils and geology, and farm enterprise type. BMPs protect water quality by managing tillage and nutrient inputs, and by reducing runoff and nutrient losses. However science based evidence of BMP effectiveness is sparse and mixed. Below we summarize the impacts of agricultural management practices on water, highlighting key scientific uncertainties.

21.2.1 Irrigation

In 2011, the Canadian agricultural sector consumed approximately 1.5 billion m^3 of water for crop and animal production (Statistics Canada 2014). Irrigation demand is driven by soil moisture conditions, climate and weather patterns, and choice of crop and crop rotation. Irrigation is used primarily in Alberta and central British Columbia (BC) where annual precipitation is insufficient to meet crop evapotranspiration demands (Harker et al. 2004). Irrigation affects hydrologic processes such as deep percolation and runoff, and subsequent discharges into aquifers and streamflow (Rahbeh et al. 2013). Return flows from irrigation (approximately 30 %) contain excess nutrients, sediments, and trace metals from soil salinization (Nakamoto and Hassler 1992; Westbrook et al. 2011). In the South Saskatchewan River Basin annual gross diversion requirements sometimes exceed licensed allocation limits, a factor that will exacerbate water quality challenges under climate change (Bennett et al. 2014; Weber and Cutlac 2014). The majority of irrigation in Alberta is through sprinkler systems which have less impact on surface hydrology than flood irrigation. In BC the use of micro-irrigation optimizes use of water and significantly reduces surface and subsurface losses (Harker et al. 2004). Irrigation scheduling may increase irrigation efficiency, reducing water use by 8–25 % on average and as much as 50 % in dry years (Wang et al. 2015). However, higher efficiency irrigation systems may also reduce return flows, negatively impacting instream flows with combined effects on water quality not well understood (Harker et al. 2004).

21.2.2 Drainage

In Central and Eastern Canada, where precipitation is higher and soils are less permeable, drainage is necessary for earlier and more efficient spring seeding. Drainage is accomplished through pumping and ditches, and through tile drainage. Tile drains, underground channels that remove water from the sub-surface, are used extensively in Ontario and Quebec (Harker et al. 2004). Drainage accelerates discharge and flushes contaminants into receiving water bodies. However, controlled tile drainage uses water control structures at outlets to manage the water table beneath fields, allowing producers to optimize the level of water available for crops during the growing season and to manage the flow of water and nutrients leaving the field (Sunohara et al. 2014).

Wetland drainage is extensive. Since the 1800s, approximately 20 million ha (or 1/7) of Canada's total wetland area has been drained including more than 60 % in Southern Ontario and over half of the prairie potholes in Western Canada. Agriculture accounts for the majority of wetland loss (Environment Canada 1991; Rubec and Hanson 2009). Wetland degradation from agriculture also comes from location of farm infrastructure, the operation of farm equipment, livestock crossing, and intensive crop production in riparian areas. Wetlands are important for filtering nutrients and toxins, as well as slowing overland flows and reducing flood risk (Yang et al. 2010). Wetland drainage has contributed to changes in nutrient runoff and hydrologic connectivity with cumulative effects that are poorly understood (Pomeroy et al. 2010; Yang et al. 2010). In Southern Manitoba it is estimated that wetland drainage since 1968 has increased the area contributing runoff to Lake Winnipeg by 4518 km^2 and total phosphorous loads by 6 % (Yang et al. 2010).

21.2.3 Land Management

Land management practices which affect soil health and runoff include tillage, cultivar type and vegetation cover. The intensity of agricultural land use is increasing. Between 1971 and 2011 the area of cropland increased by 27 % and tame and seeded pasture increased by 34 %. At the same time the amount of summerfallow declined by 81 % and natural areas including pasture, woodland, and wetlands declined by 16 % (Statistics Canada 2014). Permanent and perennial cover crops reduce runoff as well as soil erosion. Riparian management, natural cover, grassed waterways and buffer strips are practices that increase water recharge and infiltration, reducing erosion and runoff near water bodies. In 2011, 54 % of farms maintained riparian buffer on waterways with the practice most common in Atlantic regions, and Quebec and Ontario (Statistics Canada 2013).

Climate, soils, and crop type influence tillage practice. Cereals, oilseeds, and bean crops are most amenable to conservation or no-till (the practice of leaving crop residue on the field after harvest), and the practice is widely adopted in the Prairies with adoption rates of 86 % (Statistics Canada 2014). The practice drops off significantly east of Saskatchewan as soil moisture increases. Conservation tillage and no-till slow runoff by trapping water and snow in residue, reducing evapotranspiration, and reducing soil compaction from machinery. However impacts on phosphorous are uncertain. Some studies show that decomposing organic matter from reduced-tillage can increase concentrations of soluble phosphorus in surface runoff (Harker et al. 2004). In the end, definitions of tillage systems may not be refined enough to validate the practices and understand biophysical processes, suggesting a need for better monitoring to understand water quality impacts from tillage (Lobb et al. 2007).

21.2.4 Livestock Management

Approximately 4 million beef cattle and 26 million pigs are raised in Canada each year (AAFC 2013). Over 50 % of manure is produced in the Prairies and the Great Lakes drainage regions of southern Ontario and the St. Lawrence (Statistics Canada 2014). Manure is used for fertilizing crops, but improper storage and application is a significant risk to water, both from introduction of nutrients as well as pathogens and endocrine disrupting hormones. Manure application is associated with high concentrations of nitrates in shallow ground water. Many farmers allow livestock to drink from surface water during at least part of the year, resulting in damage to riparian areas and increased contamination from manure (Statistics Canada 2014). Livestock grazing in riparian areas also contributes to increased runoff from soil compaction. Major effects of livestock grazing on stream and riparian ecosystems include changes in stream channel morphology, hydrology, riparian soil health, and instream and stream bank vegetation; these impacts manifest at local as well as watershed and regional scales (Belsky et al. 1999).

21.2.5 Nutrient and Pesticide Management

Fertilizer is necessary to optimize plant growth, but excessive nutrient application leads to increased phosphorous and nitrates in soil and is eventually carried through underground percolation or overland flows into rivers, lakes and estuaries. Rates of phosphorous application have increased over time due to an increase in the amount of farm land area treated with fertilizer. Surface runoff combined with soil erosion is the most important factor for nutrient contamination; however, atmospheric transport and groundwater leaching are regionally significant. Between 2001 and 2011, the total area in Canada treated with fertilizers increased by 4 %; similarly the area treated with herbicides, pesticides and fungicides increased by 3 %, 42 %, and 114 % respectively (Statistics Canada 2014). These trends are thought to be related to adaptation to climate change, the introduction of new genetically engineered crops that are tolerant to certain chemicals, and the adoption of other BMPs such as conservation tillage (Statistics Canada 2014). Crop rotation reduces the need for pesticides and is a form of pest control. In terms of nutrient management, soil testing and precision agriculture maximize the efficiency of application of manure and commercial fertilizers. Nutrient testing is done annually on about 20 % of farms while 13 % do not do nutrient testing (Statistics Canada 2014). Precision technologies that allow farmers to obtain irrigation scheduling data from local weather stations can also be used to reduce chemical and fertilizer applications (Nicol et al. 2010). The challenge remains to have the technology adopted by the majority of crop producers in order that the efficiencies gained become fully realized (Yang 2016).

21.2.6 *Tradeoffs and Uncertainties*

There is significant scientific uncertainty on the effectiveness of agricultural BMPs due to complex interactions between land management and nutrient transport, as well as uncertainties about the impacts of BMPs on crop demands for water, nutrients and pesticides. Most BMP studies focus on specific practices and combinations of practices that are not replicable or generalizable. Evidence is mixed on the effectiveness of BMPs with benefits dependent on local soils and climate conditions as well as how they interact with other management practices (CCA 2013). Little is known about the relationship between field level practice change and large scale cumulative effects (AARD 2014). Evaluating impacts at sub-basin and basin scales is a challenge because hydrologic and ecological processes involve many contributing factors that are hard to control for.

There are also field scale tradeoffs. For example, more efficient water management from conservation tillage can lead to more intensive land use as it allows for extended crop rotations and reduced summerfallow. Some BMPs may simply redistribute the movement of nutrients between surface and sub-surface pathways. For example, minimum tillage may reduce losses of water and some pollutants to surface drainage, but may also increase infiltration into the soil and leaching to groundwater (Harker et al. 2004). Predicting BMP effects requires understanding runoff and soil saturation, as well as predictions of nutrient reductions. Even where it is possible to understand the spatial distribution of runoff producing areas, dissolved phosphorous reductions from different BMPs are uncertain (Rao et al. 2009).

21.3 Agriculture and Water Policy

Water governance in Canada is highly decentralized with important implications for agricultural water management. A key characteristic of decentralized governance is delegation of responsibility and decentralization of authority leading to fragmentation and spillovers in decision making (Dunn et al. 2014). Canada is a federation with a national government as well as ten provincial and three territorial governments. The Canadian Constitution outlines the division of responsibilities and authorities between the Federal, Provincial and Territorial jurisdictions. The Federal government has responsibility for federal lands and international waters, as well as several sectors, including agriculture, that have an interest in or impact on water. Federal responsibility for water is fragmented across 19 different departments which have some degree of responsibility for water (Corkal et al. 2007). The departments that directly affect agriculture are Environment Canada, Health Canada, and Agriculture and Agri-Food Canada. These departments share responsibilities for health, environment and agriculture with the provinces. In areas of shared responsibility the typical role of the Federal government is to set standards, and leave implementation to the Provinces (Klimas and Weersink 2006). The management and allocation of land and water resources is the responsibility of the Provinces and Territories. Within Provinces and Territories much of the responsibility for water supply and treatment is delegated to municipalities and regional authorities often with insufficient budget and authority (Dunn et al. 2015). Operationally, the safeguarding of water resources is delivered by municipalities who are responsible for land use zoning.

Water has traditionally been managed as an input into different uses such as agriculture, or municipal uses. Regulation of water, in terms of quality and quantity is usually through permitting individual uses or projects. Water quantity and quality challenges have highlighted governance gaps at watershed scales (e.g., Dunn et al. 2014). Recently there has been a movement towards shared governance at a watershed scale through integrated water resource management, but integration of institutional mandates remains a challenge (Nowlan and Bakker 2007). Overlapping mandates and governance gaps have resulted in lack of policy harmonization with implications for data collection and management of cross-boundary issues (Dunn et al. 2014; Bakker and Cook 2011).

Source water protection is challenged by jurisdictional fragmentation between land and water authorities (Dunn et al. 2015). Health Canada sets guidelines for drinking water quality while responsibility for safe drinking water rests with the provinces, territories and municipalities. Standards are not binding and tend to be lower than in other jurisdictions (Dunn et al. 2015). Provincial responsibilities for water allocation and permitting of water treatment facilities and industrial point sources are largely within the purview of environment ministries, while agricultural and natural resource ministries and municipalities are responsible for land use. The results have been lethal. In May of 2000, 2300 individuals in Walkerton Ontario became sick and seven died when heavy rains following manure application on farmland flushed pathogens into community wells. The contamination occurred even though proper manure practices were being followed at the time. Since then, a multi-barrier approach has been adopted for source water protection, however, outbreaks of illness are still common (CCA 2013).

Regulation of nutrients and pesticides is primarily done through application standards. For example, the Ontario Water Resources Act and the Alberta Agricultural Operation Practices Act set manure management standards and practices for the application, handling and storage of nutrients and pesticides. However, they do not address land management practices. Provincial water and environmental legislation (including legislation triggering Environmental Impact Assessments for point sources such as intensive livestock operations) contain standards and regulations to protect water quality through set-backs, water crossing guidelines and nutrient and pesticide storage and handling procedures. Water quality guidelines have not been established for the majority of pesticides used in Canadian agriculture, nor are there guidelines or standards for emerging risks from endocrine disrupting substances in pharmaceuticals and pesticides. Bureaucratic fragmentation has allowed different levels of government to offload and delegate fiscal and regulatory responsibilities for water management resulting in reduced fiscal and administrative capacity for national monitoring to detect emerging risks from non-regulated sources (e.g., Bakker and Cook 2011).

Because of the diffuse nature of non-point sources the focus for on-farm water quality management has been on extension and incentives for voluntary adoption of BMPs. The Federal Government, through Agriculture and Agri-Food Canada (AAFC), influences farm practices through Growing Forward, a 5-year policy framework for the agricultural and agri-food sector. Renewed every 5 years Growing Forward is implemented through joint agreements between federal, provincial and territorial governments and is the foundation for jointly delivered agricultural programs and services including stewardship payments to improve the environmental performance of the agriculture and agri-food sector.

To access funds, producers are required to complete an Environmental Farm Plan (EFP) which identifies environmental risks, and develop action plans to mitigate these risks. Farm plans are confidential and farmers do not have to report baselines. Thus it is nearly impossible to evaluate the performance of the Growing Forward program (Vercammen 2011). Numerous studies highlight the importance of spatially targeting BMPs (e.g., Yang and Weersink 2004). For example, in the U.S. agri-environmental indices are used to allocate funds for the U.S. Conservation Reserve Program to target the most effective practices, locations, and issues. In Canada agri-environmental indices have been developed in Manitoba and Ontario, but they have been developed for different program objectives with little standardization making it difficult to make inter-program or inter-regional comparisons of investment efficiency (Hajkowicz et al. 2009).

The focus on voluntary incentives for BMP adoption in Canada mirrors the experience of other jurisdictions where the tradition is that regulated point sources pay agricultural non-point sources to reduce pollution (Shortle 2013). Perhaps the most significant factor favoring voluntary incentives is historic precedent and the assumption that regulation is a restriction on private landowner rights (Cortus et al. 2011). Farmers oppose regulation for water quality (Filson et al. 2009). However, taxes and command and control approaches, such as fertilizer standards and nitrate zoning limits, could be more effective (Tanaka and Wu 2004; Ribaudo et al. 2001; Worrall et al. 2009). Canadian experience in regulating agricultural land is limited. In 1999 Prince Edward Island used legislation to mandate BMPs in response to agricultural contamination of its main aquifer and drinking water supply. The legislation mandated buffer zones and crop rotation, but these policies were eventually abandoned in favor of stewardship payments due to the impossibility of monitoring and enforcement of field practices, as well as push back from producers (CCA 2013).

21.3.1 Integrated Water Resource Management

Integrated Water Resource Management (IWRM) brings together government and non-government decision makers and stakeholders with a role in water management to develop coordinated strategies to manage water and land resources (e.g., Roy et al. 2009; Morin and Cantin 2009). IWRM provides the coordination to address cross-sectoral challenges in water management and provides a means to address agricultural contributions. Roy et al. (2009) identified at least 119 non-government organizations and regional watershed authorities promoting water conservation in agriculture and involved in implementation of IWRM.

IWRM is based on partnership and collaborative governance. However, there are key differences across provinces in terms of how much support is received from different levels of government. In B.C., IWRM is driven by municipal interests coming together on shared issues and working through collaboration to achieve common objectives, primarily related to municipal development of land and green infrastructure (Roy et al. 2009). Alberta has Basin Wide Watershed Planning and Advisory Councils which receive technical and funding support from government for watershed planning and basin reporting. Sub-basin stewardship groups are also involved in implementation of IWRM but both Councils and stewardship groups lack funding for implementation. In Saskatchewan, IWRM is carried out by the Saskatchewan Watershed Authority with local Watershed Advisory Committees which consist of municipal government partners and community stakeholders. Manitoba Conservation Districts, Ontario Conservation Authorities, and in Quebec, Organisations de Bassin Versant, have a similar structure with combined municipal-provincial governance and funding. These organizations are co-funded by municipalities and their boards consist of elected municipal councillors who are tied to the agricultural community through rate payers and the local tax base. The committees receive technical support from Federal and Provincial staff. In some cases, the government authorities have also coordinated environmental group farm planning to leverage BMP funds from Growing Forward as well as from NGOs such as Ducks Unlimited who are also important funders of IWRM (Roy et al. 2009).

Challenges associated with implementing IWRM include integrating the mandates of different organizations and government departments, and developing appropriate financing strategies to address water issues at the right scale (e.g., Morin and Cantin 2009). All groups lack adequate funds to implement plans. Problems in the Great Lakes and Lake Winnipeg require management strategies to be developed at a large basin scale, while IWRM tends to target sub-basins which lack capacity as well as normative values for effective on the ground management of non-point source agricultural pollution (e.g., Cohen and Bakker 2010). The delegation of watershed planning to local and regional authorities assumes that local agencies are in the best position to deliver IWRM, but capacity varies. Local authorities are more likely to represent local interests at the expense of downstream basin wide interests, particularly when outcomes from non-point source controls are long term and uncertain and it is necessary to show demonstrable local benefits to keep rate-payers happy. One of the biggest hurdles for non-point sources management through IWRM is that specific costs are borne by landowners and municipalities while benefits are diffuse and difficult to attribute to individual actions. Thus IWRM, in its current form and practice throughout Canada, is unlikely to address large scale challenges such as eutrophication. There is an opportunity for the Federal government to improve IWRM by funding science and coordinating the development of indicators, particularly where there are cross boundary issues (e.g., Morin and Cantin 2009; Dunn et al. 2015) and tie Growing Forward payments to IWRM plans that address significant water quality challenges.

21.4 BMP Programs

21.4.1 Factors Affecting BMP Adoption

Adoption of BMPs is motivated by stewardship and environmental factors as well as on-farm benefits. In Canada the level of environmental farm planning and adoption of BMPs is mixed. In 2011 35 % of Canadian farms had EFPs. Of these, 95 % had at least partially implemented recommended practices. There is significant variation in EFP and BMP adoption rates across provinces, with farm planning and BMP adoption more likely in the East, particularly in the Atlantic Provinces and Quebec (Statistics Canada 2013), possibly because of more stringent regulation for nutrient and manure management than in the Prairies and more acute water quality issues.

Several studies have been carried out in Canada on factors affecting adoption of BMPs. Significant factors include: education, farm size, interaction with extension, enterprise type, debt level, farm diversification, age, and income. Biophysical characteristics such as type of land and elevation, and proximity to water bodies are significant (Kim et al. 2004). Social factors such as participation in environmental organizations and contact with extensionists also have a positive effect (Ghazalian et al. 2009; Kim et al. 2004; Paudel et al. 2008; Tamini et al. 2012; Nicol et al. 2010; Kutz et al. 2014; Hadrich 2012). Trust in government is an important factor in participation in government supported BMP programs (e.g., Kehrig 2002; Wang et al. 2015) as are attitudes and world views about the environment (Parker et al. 2007; Mitchell 2005). Barriers to BMP adoption include lack of awareness, lack of understanding of the feasibility and the costs of BMP adoption, and lack of funding (Amy et al. 2012; Ghazalian et al. 2009; Van Winkle and Hadrich 2011). Vercammen (2011) observes that while these results are not surprising, the influence of individual variables can be inconsistent from study to study and there is no theoretical framework for understanding which factors are most important for adoption. In a recent meta-analysis of BMP adoption Baumgart-Getz et al. (2012) identify 31 independent variables, many of which have only small effects individually. On this evidence it is hard to prioritize strategies to encourage adoption.

The greatest barrier to BMP adoption is cost (Statistics Canada 2013). Payments under Growing Forward are on a cost-share basis, but typically fall short of adoption costs for most BMPs. BMPs often have private benefits due to increased efficiency of inputs, as well as improved soil productivity and drought resilience. As technologies have improved and costs of water have increased there has been widespread adoption of BMPs which have private benefits. Nutrients are expensive and nutrient management BMPs can be profitable on average (e.g., Valentin et al. 2004). Similarly, zero till and conservation till are on average profitable (Sparling and Brethour 2007). On the other hand, BMPs to protect riparian areas are relatively high cost because they involve land use change and the reduction/removal of land from crop production (e.g. Jeffrey et al. 2014). For practices that do not have private benefits there has been little uptake even though in many cases the social benefits outweigh private costs and justify subsidization (Boxall et al. 2013; Cortus et al. 2011; Amy et al. 2012; Jeffrey et al. 2014).

21.4.2 Contract Design

The participation of producers in BMP programs depends on contract terms and their interactions with business risk management programs. Risks from BMP adoption include: production risk due to weather, pests and disease; price risk due to changes in prices for crops such as oilseeds; technological risk from not understanding the feasibility of practices; and institutional risk from program failure and regulatory change. Changing business risk management programs such as crop insurance and income support programs could increase adoption (Mitchell 2005). For example, Prince Edward Island has reduced insurance rates for crops that use nutrient management plans (CCA 2013).

The literature on BMP contract design is primarily concerned with producer preferences for contract length. However, there is more scope to understand the influence of business risk management on producer decisions to adopt BMPs. Major issues which must be addressed through contract design include high up-front costs, long time periods with no revenue, and high project risk due to long time horizons and uncertainty in realizing improvements in water quality (Goldstein et al. 2006). There are tradeoffs since fixed up-front payments result in moral hazard with over-compensation and underperformance. Long term agreements require premiums for loss of option value and may not compare favorable to short term agreements from a cost perspective. Lennox and Armsworth (2011) show how the choice between short and long contracts is affected by certainty about the future availability of alternatives. Understanding risk perspectives of producers is also important for understanding leakage and unanticipated consequences. For example reducing one input such as water might result in farmers applying higher rates of other inputs such as fertilizer and pesticides to reduce risk (Vercammen 2011). These interactions highlight the need for cross-compliance and a whole-farm approach in designing effective BMP programs.

The first-come-first-serve model for allocating Growing Forward funds has undoubtedly reduced program effectiveness, attracting producers and practices with low opportunity cost (Boxall et al. 2013). There is evidence of weak correlation between low cost practices and BMP effectiveness (Ribaudo et al. 2010; Boxall et al. 2013). Asymmetric information from land owners knowing more about their private costs and benefits than funding agencies results in adverse selection, with low cost low value contracts crowding out higher cost higher quality contracts (Wu and Babcock 1995).

Procurement (reverse) auctions for water management actions can be used to reveal costs and benefits of practices and to allocate funds more efficiently. Auctions are standard practice in other jurisdictions – for example they are used in the U.S. Conservation Reserve Program (Claassen et al. 2008). Properly designed auctions can increase producer participation and reduce the problem of adverse selection by targeting high benefit practices. Auctions include a number of design features such as the payment format, information about the budget, and the benefits of management practices that affect efficiency and participation rates. Unfortunately, there are few generalizable rules to guide the optimal design of auctions for BMPs (Latacz-Lohmann and Van der Hamsvoort 1998). There is evidence that over time producers learn about prices which reduces the competitiveness of auctions. Furthermore, indices to score and rank environmental benefits can lead to distributional consequences that must be considered in program design (Claassen et al. 2008; Rolfe et al. 2009). Nonetheless, overall the evidence seems to suggest auctions out-perform fixed price schemes.

Unlike other jurisdictions, Canada has been slow to use auctions to allocate budgets for procuring BMPs. There are notable exceptions. For example, Ducks Unlimited Canada and the Saskatchewan Watershed Authority used a reverse auction to procure wetland restoration in Saskatchewan's Assiniboine Watershed. In Manitoba Conservation Districts are using reverse auctions to allocate funds for IWRM. The auctions help to identify willing participants and practices, and the Conservation Districts have developed an Environmental Benefits Index to support water management objectives in Manitoba's IWRM plans. This information has been helpful for leveraging funds from Growing Forward, and other non-government agencies, which is a key to success when budgets are insufficient to fully implement programs.

21.4.3 *Water Quality Trading*

Limited budgets for the environment raise the difficult question of how to finance watershed restoration and who should pay for pollution reductions. Water quality trading is seen as a way to bring non-point source agricultural pollution under the regulatory umbrella (Shortle 2013; Shortle and Horan 2008). Water quality trading involves the trade of nutrient reductions between point and non-point sources and can be viewed as an innovative financing mechanism for non-point source reductions. The most common type of trading program involves an offset agreement between a regulated point source with specific emission limits – such as a waste water treatment plant – and other point or non-point sources. The U.S. has several such trading programs to meet requirements for permitted facilities under the National Pollution Discharge Elimination System. Over the last decade several multi-source trading programs have been established to meet Total Maximum Daily Load (TMDL) requirements under the U.S. Clean Water Act. This type of program involves trading between multiple point and non-point sources to collectively achieve a joint load limit or ambient target.

Water quality trading for agriculture poses unique challenges because of uncertainties in BMP effectiveness and difficulties monitoring non-point sources. However, there are increasing examples of water quality trading in agriculture as programs are being developed to meet TMDL requirements for large drainages such as Chesapeake Bay and the Gulf of Mexico. Two of the more significant programs in terms of number of participants and expected cost savings are the Pennsylvania Nutrient Credit Trading Program, which addresses the flow of nutrients from point and nonpoint sources in Pennsylvania to Chesapeake Bay, and the Ohio River Interstate Trading Program launched in March 2014. The Ohio River Basin covers 14 states and drains into the Mississippi River and eventually the Gulf of Mexico. The current program involves three states, Ohio, Indiana, and Kentucky and relies on all states operating under the same rules so that nutrient reduction credits can be traded between states. At full scale the project is expected to include eight states, with up to 230,000 farmers creating credits for 46 power plants and thousands of other industrial and municipal point sources (EPRI 2014).

Non-point to non-point trading programs are rare, though there are examples including the California Grassland Areas Program, which caps agricultural sources, as well as the Lake Taupo Nitrogen Trading Program in New Zealand, which has capped nitrogen loads for agriculture with landowners receiving allowances based on historical land uses (Selman et al. 2009; Shortle 2013). The California Grassland Areas Program operates like a point source program since it involves trading collected drainage water between irrigation districts (Shortle 2013). Both programs operate like cap and trade programs with the Lake Taupo program using public funds to purchase a 20 % permanent reduction in nitrogen. Both programs are notable for overcoming the political hurdle of imposing environmental regulations on the agricultural sector (Shortle 2013). Monitoring non-point sources in the Lake Taupo program is an ongoing challenge and the Lake Taupo Protection Trust, which administers the program, is currently exploring methods for farm self-monitoring and reporting through the use of Water Quality Management Plans. Water Quality Management Plans are used by the Environmental Protection Agency (EPA) for state and facility wide compliance. Their use for non-point sources from individual agricultural lands is evolving as a means of documenting compliance for individual operations.

One of the most important barriers to agricultural participation in water quality trading is onerous eligibility requirements and baselines. Eligibility requirements state which practices must be undertaken in order to create a credit. Baselines establish what practices are considered "additional" to business as usual and eligible for nutrient reduction offsets or credits. Programs can use current practice as the baseline or require a higher standard, such as an improved or best practice baseline. Controlled tile drainage is a potentially important practice for Ontario and Quebec, and thus the eligibility of this practice under EPA programs could be significant for any trading program addressing transboundary pollution in the Great Lakes. The US Department of Agriculture (USDA) and EPA have been examining whether the benefits of controlled tile drainage can be sold as water quality credits under EPA water quality trading programs (e.g., Skaggs and Youssef 2009). To date, EPA has not made a decision; however, the fact that controlled drainage is financially beneficial on average begs the question of whether this practice meets the test of financial additionality and requires additional incentives. Interestingly, controlled tile drainage may also be an effective way to develop a cap and trade program for agriculture, since the drainage outlets effectively become point sources.

The economic benefits of water quality trading arise from transferring high cost load reductions to low cost load reductions. This usually means rewarding "bad" actors who can make significant water quality improvements by undertaking relatively low cost BMPs and penalizing "good" actors who have already voluntarily adopted the low hanging fruit with further nutrient reductions only feasible through higher cost practices. This approach is often perceived as unpalatable by the public and seen as a way of rewarding polluters (Shortle 2013; O'Grady 2011). Furthermore, regulators are worried about paying for practices that are not financially additional since in theory the practice should be adopted without payment under business as usual. However, high baselines reduce participation and increase the cost of credits, ultimately reducing the abatement cost savings that can be achieved through trading between point and non-point sources (Shortle 2013).

21.5 Opportunities for Canada

In Canada there are no strong legislative barriers to water quality trading. Instead, barriers are related to lack of understanding of the specific form of pollution being addressed, concerns about hot spots, and developing the information systems and models as well as monitoring programs to support trading (Cantin 2006). The South Nation Conservation Authority Clean Water Program in Ontario is one of the first trading programs developed in North America. In this program point sources purchase offsets by funding BMPs in the agriculture sector. The program is interesting because the program is peer led – farmer to farmer – with the Conservation Authority acting as an intermediary between the buyers and the producers. South Nation bases nutrient reduction requirements for permitted entities on coefficients for BMPs derived from the literature and expert opinion. High trading ratios are then used to address uncertainties in the coefficients and ensure nutrient reductions. Agreements for nutrient reductions are between the point sources and the Conservation Authority which is liable for the nutrient reductions. From the farmer's perspective, the program is a BMP program and not a water quality trading program. In 2014, the Lake Simcoe Conservation Authority in Ontario launched an urban storm water Phosphorous Offset Program with a similar structure to offset new urban development. The program is administered by the Conservation Authority and will fund existing BMP programs run by the authority, as well as deliver specific protocols and mechanisms for procuring offsets for new development as demand emerges.

There are a number of opportunities to develop broader water quality trading programs to address some of Canada's urgent priorities around restoration of the Great Lakes and Lake Winnipeg Basins. In addition, there are opportunities to further develop smaller programs similar to the South Nation and Lake Simcoe programs where municipal waste treatment plants are facing the need to upgrade, or where there are local water quality challenges such as in the Bow River Basin in Alberta. Under the Canada-U.S. Great Lakes Water Quality Agreement new programs and approaches are required to reduce phosphorous loadings from urban, rural, industrial and agricultural sources in order to meet load reduction targets for phosphorous concentrations in Lake Erie (Environment Canada 2013; Goucher and Maas 2014).

Lake Erie is the third largest lake in North America. The Lake's water quality is very poor, with agricultural non-point sources blamed for eutrophication and a massive algal bloom which cut off water supply in Toledo Ohio for 3 days in 2014 (Goucher and Maas 2014; Carter 2014). Market based approaches have been recommended to address non-point source pollution from agriculture. With a surface area of 25,667 km^2 Lake Erie is bounded by Ontario to the North, Michigan to the West, and Ohio, Pennsylvania, and New York to the South and East. The inlet is the Detroit River which divides Canada and the U.S. and has a drainage of 1813 km^2. With several major cities including Detroit MI, Windsor ON, Toledo OH, Cleveland OH and Buffalo NY, it would be possible to develop a point-non-point trading program; the question is whether, similar to the Ohio

River interstate trading program, it would be possible to develop a program that could cross both international and state boundaries. At the least, Environment Canada, the Province of Ontario, and the EPA would have to harmonize rules for water quality protection and establish common protocols and indicators for BMPs. Increased education and trust building with farmers will likely be required; concerns include how reducing fertilizer might affect returns, especially when commodity prices are strong, and whether practices like no-till are effective or feasible for crops like corn (Carter 2014).

In June 2011, the Province of Manitoba passed the Save Lake Winnipeg Act included a number of initiatives and regulations to reduce the phosphorus load to Lake Winnipeg by 50 % to pre-1990 levels. The Canada-Manitoba Memorandum of Understanding "Respecting Lake Winnipeg and the Lake Winnipeg Basin" provides financial and technical support from both the Federal and Provincial Government to the Lake Winnipeg Basin Stewardship Fund, to reduce nutrient loads throughout the Lake Winnipeg Basin and its sub-watersheds. Similar to Lake Erie, Lake Winnipeg drains an enormous inter-jurisdictional basin covering approximately 1 million km^2, and encompassing parts of Alberta, Saskatchewan, Manitoba, Ontario, North and South Dakota and Montana. The basin contains 90 % of the Canadian Prairie's agricultural land and serves over 6 million people (Voora et al. 2009). The lake is one of the most eutrophic lakes in the world, and most loads come from non-point sources. Currently there are no efforts to address non-point sources through trading programs. In any case, it is not clear that demand from point sources would be sufficient to reduce phosphorous problems in the Lake. In order to seriously address the problem, given its scope, agriculture will have to face regulation making the basin a candidate for testing inter-jurisdiction trading between non-point sources.

Although there are no legal barriers to water quality trading in Canada, there are institutional ones. In 2006 the Policy Research Initiative engaged a group of experts to identify conditions that would have to be in place for a trading program to be successful (Cantin 2006). These included: the existence of a clearly documented problem, well-developed BMPs and the ability to quantify pollution reductions; an understanding of pollutant behavior and watershed dynamics for determining critical load and trading ratios; and a watershed that is well understood and well monitored (Cantin 2006). Canada would fail to meet most of these criteria. However, as the South Nation example illustrates they are overly stringent. Moreover, trading can provide a driver for filling in some of the gaps as market requirements emerge, particularly for coordinated science and monitoring and baseline data collection. Cantin (2006) also found a lack of flexibility on the part of regulators to relax regulatory standards for permitted sources. The expansion of programs in the U.S. and the urgent need to develop new programs for nutrient reductions in Canada may improve the appetite for risk taking.

21.6 Conclusions

Expansion and intensification of agriculture has remained possible, even with water constraints, due to better tillage practices and increased efficiencies in farm technologies and adoption of BMPs (Harker et al. 2004; CCA 2013). However, Canada's largest water bodies are showing stresses from agricultural impacts. The sustainability of both agriculture and water depend on addressing the impacts of agriculture on water quality. The need to act is urgent, however uncertainties, concerns about who should pay, and fragmented governance hinder action. Federal-Provincial and international agreements to address water quality in Lakes Erie and Winnipeg should encourage risk taking and experimentation. Building off a decade of BMP incentive programs, the time is ripe for coordinated effort between the Provinces and the Federal government to synthesize lessons learned, deepen investments in science and monitoring, and develop pilot projects in a coordinated fashion over large drainages so that initiatives can be knit together and rolled up into meaningful scale action.

Several opportunities to improve the effectiveness of agricultural water management have been identified. First, the Federal government can use its funding to link IWRM processes throughout the Provinces to provincial and national scale challenges related to eutrophication of major water bodies. Secondly, development of standardized monitoring and indicators across the provinces can help identify the contributions of BMPs from different sources to downstream water quality improvements, and can facilitate inter-jurisdictional program development. A combination of local, regional, and provincial water quality initiatives is necessary for managing drinking water as well as ensuring sustainable aquatic ecosystems.

Significant education and awareness building is required to engage the agricultural sector, in order to ensure that programs are designed to address norms, and that monitoring and enforcement of agreements is feasible. New monitoring strategies such as Water Quality Management Plans could be built off of Environmental Farm Plans and lessons can be learned from New Zealand and other jurisdictions that are testing these for farm level monitoring.

The Canadian decentralization experience can be compared to that of the United States where the Clean Water Act (1972) clearly outlines lines of authority and responsibility for water quality, and where decisions about water quality management are buffered from local interest through Federal control. In the US, Federal responsibility for water quality has led to standardization of monitoring and development of science and protocols for nutrient management and trading. The EPA provides a national regulatory backstop for water quality management with authority delegated to States, and USDA provides a national coordination role on private land efforts to meet water quality objectives. There is a need for the Federal Government in Canada to overcome bureaucratic fragmentation through multi-jurisdictional agreements with the Provinces and sufficient funding to implement programs at scale. Cross-ministry coordination is required and existing bodies such as the Canadian Council for Ministers of the Environment need to work more with their agricultural counterparts in developing joint agreements.

A tentative approach to agricultural water management will not reduce uncertainty, nor is it feasible to rely on other sectors to finance and address water challenges from agriculture. Canadian governments overall are risk averse, and Canada is a persistent laggard in innovative environmental management, often following the lead of other jurisdictions before developing its own programs. In the case of water management, programs in other jurisdictions provide a rich foundation of lessons and initiatives from which to build.

References

Agriculture and Agri-Food Canada (AAFC). (2013). *We grow a lot more than you may think*. Minister of Agriculture and Agri-Food. http://www.agr.gc.ca/resources/prod/doc/info/pdf/aafc-aac_brochure_mar2013_eng.pdf. Accessed 14 Jan 2016.

Alberta Agriculture and Rural Development (AARD). (2014). *Nutrient beneficial management practices evaluation project: Volume 2 – Field study*. Lethbridge: Alberta Agriculture and Rural Development.

Amy, S. P., Trenholm, R., & Haider, W. (2012). *Socio-economic and environmental assessment of beneficial management practices*. Report for the Sustainable Agriculture Management Branch, BC Ministry of Agriculture (AGRI).

Arnold, J. G., Moriasi, D. N., Gassman, P. W., et al. (2012). SWAT: Model use, calibration, and validation. *Transactions of the ASABE, 55*(4), 1491–1508.

Bakker, K., & Cook, C. (2011). Water governance in Canada: Innovation in the context of jurisdictional fragmentation. *International Journal of Water Resources Development, 27*(2), 275–289.

Baumgart-Getz, A., Prokopy, L. S., & Floress, K. (2012). Why farmers adopt best management practice in the United States: A meta-analysis of the adoption literature. *Journal of Environmental Management, 96*(1), 17–25.

Belsky, A. J., Matzke, A., & Uselman, S. (1999). Survey of livestock influences on stream and riparian ecosystems in the western United States. *Journal of Soil and Water Conservation, 54*(1), 419–431.

Bennett, D. R., Harms, T. E., & Entz, T. (2014). Net irrigation water requirements for major irrigated crops with variation in evaporative demand and precipitation in southern Alberta. *Canadian Water Resources Journal, 39*(1), 63–72.

Boxall, P. C., Perger, O., & Weber, M. (2013). Reverse auctions for agri-environmental improvements: Bid-selection rules and pricing for beneficial management practice adoption. *Canadian Public Policy, 39*(Supplement 2), S23–S36.

Cantin, B. (2006). *Can water quality trading help to address agricultural sources of pollution in Canada? Project Report* (Policy Research Initiative project: Sustainable development). Ottawa: Policy Research Initiative.

Carter, J. (2014, September 4). *Farming contributes to Lake Erie algal bloom*. The Western Producer. http://www.producer.com/2014/09/farming-contributes-to-lake-erie-algal-bloom/. Accessed 14 Jan 2016.

Claassen, R., Cattaneo, A., & Johansson, R. (2008). Cost-effective design of agri-environmental payment programs: U.S. experience in theory and practice. *Ecological Economics, 65*, 737–752.

Cohen, A., & Bakker, K. (2010). Groundwater governance: Explaining regulatory non-compliance. *International Journal of Water, 5*(3), 246–266.

Conference Board of Canada. (2015). *How Canada performs: A report card on Canada*. http://www.conferenceboard.ca/hcp/default.aspx. Accessed 14 Jan 2016.

Corkal, D. R, Inch, B., & Adkins, P. E. (2007). *The case of Canada – Institutions and water in the South Saskatchewan River Basin*. Working paper for Institutional Adaptations to Climate Change. http://www.parc.ca/mcri/pdfs/CaseofCanada-PFRA_FINAL_REV_0702 07.pdf. Accessed 14 Jan 2016.

Cortus, B. G., Jeffrey, S. R., Unterschultz, J. R., & Boxall, P. C. (2011). The economics of wetland drainage and retention in Saskatchewan. *Canadian Journal of Agricultural Economics, 59*(1), 109–126.

Council of Canadian Academies (CCA). (2013). *Water and agriculture in Canada: Towards sustainable management of water resources*. Ottawa: The Expert Panel on Sustainable Management of Water in the Agricultural Landscapes of Canada.

Dunn, G., Bakker, K., & Harris, L. (2014). Drinking water quality guidelines across Canadian Provinces and Territories: Jurisdictional variation in the context of decentralized water governance. *International Journal of Environmental Research and Public Health, 11*, 4634–4651.

Dunn, G., Harris, L., & Bakker, K. (2015). Microbial risk governance: Challenges and opportunities in fresh water management in Canada. *Canadian Water Resources Journal, 40*(3), 237–249.

Electric Power Research Institute (EPRI). (2014). *Ohio River Basin water quality trading project*. http://wqt.epri.com/pdf/3002001739_WQT-Program-Summary_2014-03.pdf. Accessed 14 Jan 2016.

Elliott, J., Deryng, D., Müller, C., Frieler, K., et al. (2014). Constraints and potentials of future irrigation water availability on agricultural production under climate change. *Proceedings of the National Academy of Sciences, 111*(9), 3239–3244.

Environment Canada. (1991). *The federal policy on wetland conservation*. Ottawa: Government of Canada.

Environment Canada. (2013). *Planning for a sustainable future: A federal sustainable development strategy for Canada 2013–2016*. Sustainable Development Office. https://www.ec.gc.ca/dd-sd/F93CD795-0035-4DAF-86D1-53099BD303F9/FSDS_v4_EN.pdf. Accessed 14 Jan 2016.

Filson, G., Sethuratnam, S., Adekunle, B., & Lamba, P. (2009). Beneficial management practice adoption in five Southern Ontario watersheds. *Journal of Sustainable Agriculture, 33*(2), 229–252.

Fowke, V. C. (1957). *The national policy and the wheat economy*. Toronto: University of Toronto Press.

Ghazalian, P. L., Larue, B., & West, G. E. (2009). Best management practices to enhance water quality: Who is adopting them? *Journal of Agricultural and Applied Economics, 3*, 663–682.

Goldstein, J. H., Naylor, R. L., Vitousek, P., et al. (2006). Business strategies for conservation on private lands: Koa forestry as a case study. *Proceedings of the National Academy of Sciences of the United States of America, 103*(26), 10140–10145.

Goucher, N., & Maas, T. (2014). *Clean, not green: Tackling algal blooms in the Great Lakes*. Toronto: Environmental Defence and Freshwater Future Canada.

Hadrich, J. C. (2012). *A survey of the awareness and use of BMPs on North Dakota beef operations* (Agribusiness and applied economics report no. 698). Fargo: North Dakota State University.

Hajkowicz, S., Collins, K., & Cattaneo, A. (2009). Review of agri-environment indexes and stewardship payments. *Environmental Management, 43*(2), 221–236.

Harker, B., Lebedin, J., Goss, M. J., Madramootoo, C., et al. (2004). Land-use practices and changes – Agriculture. In *Environment Canada. Threats to water availability in Canada* (NWRI scientific assessment report series no. 3 and ACSD science assessment series no. 1). Burlington: National Water Research Institute.

Jeffrey, S. R., Koeckhoven, S., Trautman, D., et al. (2014). Economics of riparian beneficial management practices for improved water quality: A representative farm analysis in the Canadian Prairie region. *Canadian Water Resources Journal, 39*(4), 449–461.

Kehrig, R. (2002). *Agricultural practices and water quality in Saskatchewan: The social ecology of resource management*. Master thesis, University of Saskatchewan.

Kim, S. A, Gillespie, J. M., & Paudel, K. P. (2004, February 18) *The effect of economic factors on the adoption of best management practices in beef cattle production*. Paper presented at Southern Agricultural Economics Association Annual meeting, Tulsa.

Klimas, M., & Weersink, A. (2006). The agricultural policy framework: Water effects. *Horizons, Policy Research Initiative, Government of Canada, 9*(1), 1525–1532.

Kutz, A. M., Clark, C. D., & Lambert, D. M. (2014, February 1–4). *Willingness to adopt best management practice bundles by beef cattle operations in an East Tennessee watershed*. Paper presented at Southern Agricultural Economics Association Annual meeting, Dallas.

Latacz-Lohmann, U., & Van der Hamsvoort, C. (1998). Auctions as a means of creating a market for public goods from agriculture. *Journal of Agricultural Economics, 49*(3), 334–345.

Lennox, G. D., & Armsworth, P. R. (2011). Suitability of short or long conservation contracts under ecological and socio-economic uncertainty. *Ecological Modelling, 222*(15), 2856–2866.

Lobb, D. A., Huffman, E., & Reicosky, D. C. (2007). Importance of information on tillage practices in the modelling of environmental processes and in the use of environmental indicators. *Journal of Environmental Management, 82*(3), 377–387.

Mitchell, P. D. (2005). Nutrient best management practice insurance and farmer perceptions of adaptation risk. *Journal of Agricultural and Applied Economics, 36*(3), 657–673.

Morin, A., & Cantin, B. (2009). *Strengthening integrated water resource management in Canada*. Ottawa: Policy Research Initiative.

Nakamoto, R. J., & Hassler, T. J. (1992). Selenium and other trace elements in bluegills from agricultural return flows in the San Joaquin Valley, California. *Environmental Contamination and Toxicology, 22*(1), 88–98.

Nicol, L., Bjornlund, H., & Klein, K. K. (2010). Private irrigators in Southern Alberta: A survey of their adoption of improved irrigation technologies and management practices. *Canadian Water Resources Journal, 35*(3), 339–350.

Nowlan, L., & Bakker, K. (2007). *Delegating water governance: Issues and challenges in the BC context*. Vancouver: Program on Water Governance, University of British Columbia.

O'Grady, D. (2011). Socio-political conditions for successful water quality trading in the South River Nation watershed, Ontario, Canada. *Journal of the American Water Resources Association, 47*, 39–51.

Parker, J. S., Moore, R., & Weaver, M. (2007). land tenure as a variable in community based watershed projects: Some lessons from the Sugar Creek Watershed, Wayne and Holmes Counties, Ohio. *Society and Natural Resources, 20*(9), 815–833.

Paudel, K. P., Gauthier, W. M., Westra, J. V., & Hall, L. M. (2008). Factors influencing and steps leading to the adoption of best management practices by Louisiana dairy farmers. *Journal of Agricultural and Applied Economics, 1*(April), 203–222.

Pomeroy, J. W., Fang, X., Westbrook, C., Minke, A., Guo, X., & Brown, T. (2010). *Prairie hydrological model study: Final report* (Center for hydrology report no. 7). Saskatoon: Centre for Hydrology, University of Saskatchewan.

Rahbeh, M., Chanasyk, D. S., & Patterson, S. J. (2013). Modeling the irrigation requirements for an experimental site in Northern Alberta, Canada. *Canadian Water Resources Journal, 38*(4), 280–295.

Rao, N. S., Easton, Z. M., Schneiderman, E. M., et al. (2009). Modeling watershed-scale effectiveness of agricultural best management practices to reduce phosphorus loading. *Journal of Environmental Management, 90*, 1385–1395.

Ribaudo, M. O., Heimlich, R., Claassen, R., & Peters, M. (2001). Least-cost management of nonpoint source pollution: Source reduction versus interception strategies for controlling nitrogen loss in the Mississippi Basin. *Ecological Economics, 37*(2), 183–197.

Ribaudo, M., Greene, C., Hansen, L., & Hellerstein, D. (2010). Ecosystem services from agriculture: Steps for expanding markets. *Ecological Economics, 69*, 2085–2092.

Rolfe, J., Windle, J., & McCosker, J. (2009). Testing and implementing the use of multiple bidding rounds in conservation auctions: A case study application. *Canadian Journal of Agricultural Economics, 57*(3), 287–303.

Roy, D., Oborne, B., & Venema, H. D. (2009). *Integrated Water Resources Management (IWRM) in Canada. Recommendations for agricultural sector participation*. Winnipeg: International Institute for Sustainable Development.

Rubec, C. D. A., & Hanson, A. R. (2009). Wetland mitigation and compensation: Canadian experience. *Wetlands Ecology and Management, 17*, 3–14.

Selman, M., Greenhalgh, S., Branosky, E. et al. (2009). Water quality trading programs: An international overview. *World Resources Institute*, Issue Brief 1, 1–15.

Shortle, J. (2013). Economics and environmental markets: Lessons from water-quality trading. *Agricultural and Resource Economics Review, 42*, 57–74.

Shortle, J. S., & Horan, R. D. (2008). The economics of water quality trading. *International Review of Environmental and Resource Economics, 2*(2), 101–133.

Skaggs, R., & Youssef, M. A. (2009). *Modern drainage water management to reduce drainage volumes and nitrogen losses*. International Commission on Irrigation and Drainage (ICID). ICID Newsletter (4). http://www.icid.org/drg_article_2009_4.pdf. Accessed 14 Jan 2016.

Sparling, B., & Brethour, C. (2007). *An economic evaluation of beneficial management practices for crop nutrients in Canadian agriculture*. Paper presented at the Canadian Agricultural Economics Society Annual Meeting, Portland, 29 July–1 August.

Statistics Canada. (2013). *Farm environmental management survey*. http://www.statcan.gc.ca/pub/21-023-x/21-023-x2013001-eng.pdf. Accessed 14 Jan 2016.

Statistics Canada. (2014). *Human activity and the environment*. Agriculture in Canada. Catalogue no 16-201-X. Environment, Energy and Transportation Statistics Division, Minister of Industry. http://www.statcan.gc.ca/pub/16-201-x/16-201-x2014000-eng.pdf. Accessed 14 Jan 2016.

Sunohara, M. D., Craiovan, E., Topp, E., et al. (2014). Comprehensive nitrogen budgets for controlled tile drainage fields in eastern Ontario, Canada. *Journal of Environmental Quality, 43*(2), 617–630.

Tamini, L. D., Larue, B., & West, G. (2012). Technical and environmental efficiencies and best management practices in agriculture. *Applied Economics, 44*(13), 1659–1672.

Tanaka, K., & Wu, J. (2004). Evaluating the effect of conservation policies on agricultural land use: A site-specific modeling approach. *Canadian Journal of Agricultural Economics, 52*(3), 217–235.

Valentin, L., Bernardo, D. J., & Kastens, T. L. (2004). Testing the empirical relationship between best management practice adoption and farm profitability. *Applied Economic Perspectives and Policy, 26*(4), 489–504.

Van Winkle, A., & Hadrich, J. (2011, July 24–26). *North Dakota beef cow producers: Identifying current management practices and factors that influence adoption rates of best management practices relating to surface water pollution*. Paper presented at the Agricultural and Applied Economics Association's AAEA and NAREA Joint Annual Meeting, Pittsburgh.

Vercammen, J. (2011). Agri-environmental regulations, policies, and programs. *Canadian Journal of Agricultural Economics, 59*, 1–18.

Voora, V., McCandless, M., Roy, D., Venema, H. D., Oborne, B., & Grosshans, R. (2009). *Water quality trading in the Lake Winnipeg Basin*. Winnipeg: International Institute for Sustainable Development.

Wang, J., Klein, K. K., Bjornlund, H., et al. (2015). Adoption of improved irrigation scheduling methods in Alberta: An empirical analysis. *Canadian Water Resources Journal, 40*(1), 47–61.

Weber, M., & Cutlac, M. (2014). Economic and environmental trade-offs from alternative water allocation policies in the South Saskatchewan River Basin. *Canadian Water Resources Journal, 39*(4), 409–420.

Westbrook, C. J., Bruner, N., Phillips, I., & Davies, J.-M. (2011). *Wetland drainage effects on Prairie water quality: Final report* (Centre for hydrology report no. 9). Saskatoon: Centre for Hydrology, University of Saskatchewan.

Worrall, F., Evans, M. G., Bonn, A., et al. (2009). Can carbon offsetting pay for upland ecological restoration? *Science of the Total Environment, 408*, 26–36.

Wu, J., & Babcock, B. A. (1995). Optimal design of a voluntary green payment program. *Journal of Agricultural and Resource Economics, 20*, 316–327.

Yang, W. (2016). Geomatics & water policy. In S. Renzetti & D. Dupont (Eds.), *Water policy and governance*. Cham: Springer.

Yang, W., & Weersink, A. (2004). Cost-effective targeting of riparian buffers. *Canadian Journal of Agricultural Economics, 52*(1), 17–34.

Yang, W., Wang, X., Liu, Y., et al. (2010). Simulated environmental effects of wetland restoration scenarios in a typical Canadian prairie watershed. *Wetlands Ecology Management, 18*, 269–279.

Part VII
Emerging Issues and Perspectives

Chapter 22
Shifting Perspectives in an Era of Complexity and Change: Incorporating Resilience into the Water Governance of Canadian Drainage Basins

Ryan Plummer, Julia Baird, Katrina Krievins, Jennifer Fresque-Baxter, Jack Imhof, and Simon J. Mitchell

Abstract Governance has emerged as a central issue in addressing contemporary and future water challenges. Many shortcomings of past approaches to water policy in Canada are revealed in this volume as they relate to conservation (Changing Currents: A Case Study in the Evolution of Water Law in Western Canada and Patchy resources for the governance of Canada's resource patches: How hydraulic fracturing is illuminating the need to improve water governance in Canada) and

R. Plummer (✉)
Environmental Sustainability Research Centre (ESRC), Brock University,
1812 Sir Isaac Brock Way, L2S 3A1 St. Catharines, ON, Canada
e-mail: rplummer@brocku.ca

J. Baird
Research Associate, Environmental Sustainability Research Centre (ESRC), Brock University, 1812 Sir Isaac Brock Way, L2S 3A1 St. Catharines, ON, Canada
e-mail: jbaird@brocku.ca

K. Krievins
Masters Student, Sustainability Science and Society Program, Brock University,
1812 Sir Isaac Brock Way, L2S 3A1 St. Catharines, ON, Canada
e-mail: katrina.krievins@brocku.ca

J. Fresque-Baxter
Department of Environment & Natural Resources, Government of the Northwest Territories,
X1A 2 L9 Yellowknife, NT, Canada
e-mail: jennifer_fresque-baxter@gov.nt.ca

J. Imhof
Director of Conservation Ecology, Trout Unlimited Canada, 6712 Fisher St SE,
T2H 2A7 Calgary, AB, Canada
e-mail: jimhof@tucanada.org

S.J. Mitchell
St John River Advisor, WWF, Waterlution and the Canadian Rivers Institute,
St. John River, NB, Canada
e-mail: smitchell@WWFCanada.org

S. Renzetti, D.P. Dupont (eds.), *Water Policy and Governance in Canada*,
Global Issues in Water Policy 17, DOI 10.1007/978-3-319-42806-2_22

health (Public Health at the Watershed Scale). A fundamental shift in the prevailing mindset of government control of the hydrological cycle for human use is necessary. Resilience offers a radical departure from dominant approaches of the past and conceptual developments inform the future of water governance in an era of complexity and change. Incorporating resilience thinking into the governance of drainage basins is critical in this context. Four cases from Canada are presented to illustrate how resilience is emerging in policy and practice. Taken together, resilience thinking and resilience practice, provide a fertile ground for re-envisioning water resources and their governance.

22.1 Introduction

Canada, home to the Great Lakes and an abundance of waterways, is perceived to be a water-rich nation; however, the reality is that the majority of this water is not renewable and much of the renewable supply flows to regions where populations are sparse, creating a myth of abundance (Sprague 2007). Drought, aging infrastructure and increased consumption have led to renewable water supply shortages in many municipalities (Bakker and Cook 2011; de Loë and Plummer 2010). Water quality concerns have also been evident, including tragic incidents of drinking water contamination as well as boil water advisories that exceeded 1000 at the time of writing (www.watertoday.ca). Issues pertaining to safe drinking water are pronounced in First Nations reservations, where 93 communities (excluding British Columbia) face drinking water advisories (Health Canada 2015). Climate change is altering past assumptions about water management (see Milly et al. 2008) and is anticipated to substantially impact water related issues across the globe (Bates et al. 2008) including in Canada (Lemmen and Warren 2004). Together, these myriad water stresses will manifest in drainage basins, often referred to as watersheds in Canada, and create long term challenges for sustainability (National Round Table on the Economy and Environment 2010).

Internationally, water issues are being recognized as a crisis of governance (Global Water Partnership Technical Advisory Committee 2000; Cooley et al. 2013; OECD 2011). Water governance, in the frequently cited definition by the Global Water Partnership refers to:

> the range of political, social, economic and administrative systems that are in place to develop and manage water resources, and the delivery of water services, at different levels of society (Rogers and Hall 2003, p 7).

Policy makers, practitioners and scholars alike argue that governance is similarly at the heart of Canada's water quality and quantity concerns (de Loë and Kreutzwizer 2007; Bakker and Cook 2011).

The water governance landscape in Canada is complicated, evolving, and laden with issues. Responsibility for water governance was, and to a considerable extent still is, the purview of governments in Canada. As Brandes et al. (2005, ii) observe, Canada's "…myriad public agencies share authority in 'a bewilderingly complex administrative galaxy' that fails to address the underlying problems" and "…is in need of sober reform". Perhaps not surprising, given such observations, is the shift away from governments acting alone in a top-down manner and using command and control approaches or tools (Simms and de Loë 2010). Precipitating this movement in Canada is recognition that: government-centric approaches and tools cannot adequately address water-related problems (Simms and de Loë 2010); the complexity of water management is increasing given its essential nature in many ecosystem functions (de Loë and Kreutzwiser 2007); and, an increased emphasis on integrated management and drainage basin management (Bakker and Cook 2011).

While governments and regulations remain an essential part of water governance, a rich dialogue has opened about the appropriateness of other models in Canada. For example, the *Water Act* modernization process in British Columbia gave impetus to envision transformative watershed governance (see Brandes et al. 2014). Simms and de Loë (2010) identified several water governance challenges in Canada as described by practitioners, including leadership and commitment; legitimacy; actors, roles and relationships; knowledge; integrating institutions; resources and capacity; accountability; learning; adaptation; and, evaluation.

Adding to the discussion of appropriate water governance models is the acknowledgement that we are in an era of complexity, uncertainty and change, prompting reconsideration of the assumptions about water and society. Folke (2003, p 2033), accordingly, argued for a transition to thinking for resilience and stressed that it requires:

> …a shift from trying to control and allocate freshwater flows in an optimal manner for various human uses to recognition of the necessity to actively manage the essential role of freshwater in dynamic landscapes faced with uncertainty and surprise.

Over a decade later, Schoeman et al. (2014, p 378) documented emergence of such a water paradigm.

This chapter unpacks the concept of resilience and explores the body of scholarship in relation to water governance. While conceptually rich, there is much less experience with resilience practice. Evidence that resilience is informing water governance of drainage basins in Canada is explored through four cases. In line with the work of Clancy (2014), drainage basins focus the exploration because they represent interconnected systems of nature and humans that have long been considered when making decisions about freshwater and its management in Canada. Insights from scholarship, as well as these emerging experiences, provide a growing basis from which to consider resilience in Canadian water governance.

22.2 Resilience Thinking, Water Systems, and Their Governance

Resilience is a concept that has evolved over time (see Folke 2006; Plummer 2010; Krievins et al. 2014). With acknowledgement of its multiple meanings, the term here is used in the sense of social-ecological resilience. Social-ecological resilience is underpinned by two complimentary perspectives: complex adaptive systems, which takes an evolving view of nature characterized by uncertainty, self-organization and non-linearity (Holling and Gunderson 2002; Holling et al. 2002) and social-ecological systems where ecological and social systems are inextricably linked and delineations between them are considered artificial (Berkes and Folke 1998; Berkes et al. 2003).

Social-ecological resilience emphasizes the ability to navigate complexity, uncertainty and change. It is understood as:

> (1) the amount of disturbance a system can absorb and still remain within the same state or domain of attraction, (2) the degree to which the system is capable of self-organization (versus lack of organization, or organization forced by external factors), and (3) the degree to which the system can build and increase the capacity for learning and adaptation (Folke 2006, p 259–260).

Resilience thinking (Walker and Salt 2006; Folke et al. 2010) provides a theoretical framework for understanding social-ecological systems and integrating the core elements of resilience, adaptability and transformability. Folke et al. (2010) explain in this framework:

> Resilience is the tendency of a SES [social-ecological system] subject to change to remain within a stability domain, continually changing and adapting yet remaining within critical thresholds. Adaptability is a part of resilience. Adaptability is the capacity of a SES to adjust its responses to changing external drivers and internal processes and thereby allow for development within the current stability domain, along the current trajectory. Transformability is the capacity to create new stability domains for development, a new stability landscape, and cross thresholds into a new development trajectory.

Insights into building resilience are developing as resilience research accumulates. Reflecting upon experiences conveyed in their seminal work on navigating social-ecological systems, Folke et al. (2003, p 354–355) identified four critical factors for building resilience and adaptive capacity:

> …learning to live with change and uncertainty; nurturing diversity for reorganization and renewal; combining different types of knowledge for learning; and creating opportunity for self-organization toward social-ecological sustainability.

More recently, efforts have been made to identify universal principles that indicate resilience or may enhance it. Biggs et al. (2012) identified seven principles for enhancing resilience of ecosystem services. Three of the principles (diversity and redundancy, connectivity, and slow variables and feedbacks) relate to system properties to be managed, and the remaining four principles (understanding social-ecological systems as complex adaptive systems, learning and experimentation, participation, and polycentricity) are considered key attributes of the governance system. Frequent co-occurrence and high interdependencies of these principles are observed which requires future research to enhance understanding and make them operationally applicable.

Freshwater in relation to resilience was identified as an unexplored area by Folke in 2003, but attention to it has increased since that time (e.g., Folke 2003; Galaz 2007; Milman and Short 2008; Rockström et al. 2014; Plummer et al. 2014, 2016a). While research focusing specifically on resilience and water governance is more limited, a few salient works have explored the connection. Folke's (2003) foundational effort outlines the resilience perspective and contrasts difference in world-views by comparing examples of modern aquaculture with two examples of adaptive co-management in Sweden, an approach to catchment-based stewardship. In so doing, he offered insights into social dimensions of managing freshwater, social features of resilience, and multi-level catchment governance. Galaz (2007, p 7) examined integrated water resources management (IWRM) as a strategic approach in the international policy arena and understanding of freshwater resources in terms of resilience and their governance. His findings revealed:

> …a substantial lag between advances in research on what constitutes resilient interconnected freshwater resources and their governance, and what is being promoted by policy makers at several policy scales, from the international to the national arena.

Most recently, Plummer et al. (2014) conducted a Delphi study of global experts in aquatic systems governance and resilience to reconcile diverse terminology and develop consensus on essential key attributes. These attributes related to specified resilience (i.e., resilience to specific disturbances) and general resilience (i.e., general capacity to respond to disturbances of all kinds including surprises) (Table 22.1). While the study by Plummer et al. (2014, p 15) consolidated these attributes, it recognized that:

> ensuring the core aspects of resilience are held, while infusing the spirit of the concept into the important dialogue on governance and aquatic systems, is a challenge for researchers, decision-makers, and citizens moving forward.

Practitioners and policy makers will immediately be confronted by the need to tradeoff specified and general attributes (Plummer et al. 2014). Moreover, the strengths of resilience attributes will need to be considered in light of their costs. For example, polycentric governance can also display disadvantages such as difficulties in making collective decisions, increased transaction costs, and loss of democratic accountability (Huitema et al. 2009).

Finally, it is important to acknowledge the critical discourse emerging about resilience. Criticism of resilience thinking has largely come from scholars in the

Table 22.1 Resilience attributes essential for governance

Specified resilience attributes	General resilience attributes
Adaptive planning	Institutional flexibility
Polycentric network structure and the presence of boundary organizations	Decentralized system
Diverse actor participation	Inclusive participation and building a shared understanding
Authority/leadership for efficient, adaptive responses	Strong, not individually concentrated, leadership
Equity and transparency	Wide range of ecosystem services included in planning
Capacity to self-organize	
Social memory to maintain knowledge base	
Precautionary risk assessment and reduction strategies	

Adapted from Plummer et al. (2014)

social sciences (e.g., Davidson 2010; Cote and Nightingale 2012; Hornborg 2013; Brown 2014; Olsson et al. 2015). As Brown (2014, p 107) summarizes:

> the application of resilience concepts to social and ecological systems and dilemmas has been roundly critiqued for undertheorizing social dimensions, and human geographers in particular have been an important critical voice in highlighting the omission of social, political and cultural dynamics from different resilience literatures.

22.3 Signs of Resilience Thinking in Canadian Drainage Basin Governance

While the concept of resilience has grown in importance and gained widespread uptake in terms of research and policy, a gap is evident with the "…demonstrated capacity to govern for resilience in practice…" (Wilkinson 2012, p 319).

Difficulties in applying the idea of resilience have been noted. For example, the need for definitions and metrics to facilitate making resilience operational is necessary (Kerner and Thomas 2014). Fragmentation by the variety of disciplines and strands within resilience research itself has led to limited coherence of important factors, in particular context, to build resilience and the ways they may be made operational (Biggs et al. 2015). Walker and Salt (2012) argue, in light of such challenges, that applying resilience thinking in practice is a logical next step.

To the best of our knowledge, the manner in which resilience thinking is informing water governance practice in Canada is not well understood. However, we contend that evidence is emerging throughout Canada that resilience ideas are shaping conversations about water governance as well as being put into practice. The following examples highlight signs of resilience thinking in Canadian drainage basin governance and illustrate how resilience concepts may be put into practice.

22.3.1 St. John River Basin

The St. John River Basin is an international transboundary basin (situated in the state of Maine in the USA, and the provinces of New Brunswick and Quebec in Canada) that is over 55,000 km^2 in area and has a main stem of 678 km that flows from the north Maine woods to the Bay of Fundy at Saint John, New Brunswick. It is home to over half a million people, has a diversity of flora and fauna and a number of rare, threatened and endangered species. It has some of the oldest settlements in Canada and a strong and rich natural and cultural history that dates back some 8500 years to the arrival of the Wolastoquiyik, known today as the Maliseet. The home river of the Maliseet is the St. John, or Wolastoq, "the good and bountiful river".

World Wildlife Fund (WWF) (Canada) has been working in the region for 3 years, with much of their effort centered on the health of the river and surrounding communities, habitats and species. WWF's recently released Watershed Reports (www.watershedreports.wwf.ca) is a key tool for building a common understanding of freshwater health and threats nationally, regionally and locally. On the St. John River, this tool has helped to refine the understanding of river health; and, when combined with the Social Ecological Inventory (SEI) and social network analysis through a joint research project with Brock University, a greater understanding of who is doing what and where in support of a healthy river has emerged (see Plummer et al. 2016b). With over 160 actors identified in the study as contributing to river health, the next obvious question is what is the nature of the relationship among these actors?

Understanding who is funding these efforts, who is contributing to regulation and policy, who is responsible for enforcement, who is acting on the ground to ensure healthy waters, and how they are connected, builds new knowledge about the watershed and presents opportunities to bridge important gaps between actors within and among sectors and across this transboundary basin. In so doing, a number of the specified and general resilience attributes for governance (e.g., equity and transparency, diverse actor participation, etc.) are flourishing. Moreover, application of the SEI demonstrates a technique for identifying key actors and engaging them at the drainage basin scale. In drawing social and ecological connections among scales (e.g., specific sites, sub-basins) it informs decision-making in the system and contributes to resilience.

As WWF's efforts on the St. John River / Wolastoq mature, resilience is gaining prominence. Key aspects of resilience thinking, including diverse actor participation and interconnectedness (in terms of interests and scales), inclusion and consideration of different knowledges, and a focus on learning and adaptation are being integrated into municipal climate vulnerability assessments and the development of adaptation plans. Healthy waters are understood to be complex systems in which social, economic, ecological, traditional and spiritual dimensions are constantly interacting and influencing one another. While resilience thinking is not a panacea and does not offer a prescription of how to make hard choices/tradeoffs between investments in specified and general resilience, it helps to understand how interacting systems of people and nature operate within the river basin, assists in navigating different values and interests, and avails management approaches with capacity to deal with complexity, uncertainty and unexpected changes. Ultimately, resilience thinking provides a conceptual framework for devising new ways to mitigate impacts and seize opportunities resulting from rapid change, and can catalyze the transition towards a future where humans and nature thrive, and are better prepared to engage in continuous learning and adaptation, in support of healthy freshwaters.

22.3.2 *Water Stewardship in Canada's Northwest Territories*

In Canada's Northwest Territories (NWT), water is viewed as a fundamental human right, which was recognized in March 2007 by the 15th Assembly of the Northwest Territories in a unanimous motion (Government of the Northwest Territories (GNWT) and Aboriginal Affairs and Northern Development Canada (AANDC) 2010). Water is critical for social, cultural, economic and spiritual purposes, and supports Northern livelihoods (including subsistence and commercial harvesting, travel, and recreation).

Protecting aquatic ecosystems – and the people, plants, birds, fish, wildlife and processes that rely on these – is paramount for NWT water governance. Northerners have expressed concerns about impacts to water from a wide range of drivers, including climate change, industrial development and activities in upstream jurisdictions. These concerns pointed to a need for a coordinated and collaborative approach to NWT water stewardship.

Released in 2010, *Northern Voices, Northern Waters: The NWT Water Stewardship Strategy*[1] (the 'Strategy'), was a collaborative endeavour between Aboriginal, municipal, territorial and federal governments, regulatory boards, environmental organizations, industry, academic partners and the public (GNWT and AANDC 2010). The vision of the Strategy, as the guiding policy document for water management in the NWT, is that "the waters of the Northwest Territories will remain clean, abundant and productive for all time" (GNWT and AANDC 2010).

The development of the Strategy shifted water-related work from a focus on government responsibilities and agency-specific mandates towards a broader collective mandate jointly shared by NWT Water Partners. Through its more coordinated approach, the Strategy has created collaborative opportunities for water stewardship, meaning partners can better address their broader interests, needs, priorities and responsibilities through working together. As partners learn and work together, improvements to collaboration and coordination continue to progress.

In 2011, an associated Action Plan identified key priorities, deliverables, timelines and lead agencies for stewardship activities to ensure accountability and progress towards the Strategy's vision and goals (GNWT and AANDC 2011). Water partners are currently implementing activities that support priorities under the Strategy, including (but not limited to): negotiation of bilateral water management agreements with neighbouring Mackenzie River Basin jurisdictions; community-based aquatic ecosystem health research and monitoring; source water protection planning; education and outreach; youth engagement; capacity-building and training; water regulatory initiatives; traditional knowledge activities; on-the-land programs; and, information management (GNWT and AANDC 2011).

[1] For more information on the Strategy, partners involved, and past and current activities, please visit: nwtwaterstewardship.ca.

The Strategy promotes resilience of NWT waters through a number of attributes, including[2]:

Collaboration and Integration Core to the success of Strategy development and implementation has been collaboration of multiple, diverse partners in tackling complex issues. The Strategy recognizes stewardship as a collective responsibility, and promotes people with different mandates, experiences and knowledge working together to make sound decisions for protection of aquatic ecosystems.

Ecosystem-Based Approach The Strategy adopts a holistic approach, which considers the multiple nested scales at which stewardship, management and decision-making are enacted (from the local to Basin-wide), and the interconnectedness between all parts of the ecosystem, including people as part of that system.

Adaptability Information gathered through research and monitoring supports better understanding of aquatic ecosystems to make appropriate decisions. The Strategy allows sufficient flexibility to adapt management to social and ecological changes, and to adjust practices as new information is gathered.

Multiple Knowledge Systems The Strategy is built on the importance of using multiple knowledge systems to make decisions. Drawing on the strengths of traditional and local knowledge and western science together can provide a more holistic understanding of the environment, change, and strategies to address change.

The attributes above align with recognized attributes of resilience for aquatic system governance (Plummer et al. 2014). In 2015, implementation of the Strategy underwent an independent evaluation, and a new Action Plan is being collaboratively developed to set priorities for 2016–2020. The new Action Plan will continue to reflect the above attributes in efforts to promote resilient ecosystems, people, communities and livelihoods.

22.3.3 The Cowichan Watershed

Resilience concepts are inherently compatible with watershed governance: notions of inclusive participation, building a shared understanding, ensuring social memory (redundancy), capacity for self-organization, and considering ecosystem services in planning (Biggs et al. 2012; Plummer et al. 2014) all resonate with multi-stakeholder governance at the watershed scale. But, these attributes may not be considered in resilience terms by those engaged in governance. Researchers from Brock University and the University of Victoria facilitated a resilience workshop research study to assess the potential for learning about resilience to enhance resilience at a watershed

[2]Note: This is a not an exhaustive list of attributes in the Strategy that promote resilience, but are rather selected examples.

scale. One of the watersheds of focus was the Cowichan Watershed on Vancouver Island in British Columbia.

The Cowichan Water Board (CWB), established in 2010, has been active in leading water management efforts in the Cowichan Watershed. They experience a range of issues related to both water quantity and quality, including management of a weir and associated water supply and demand issues and water pollution from several sources including agriculture, industry and sewage (Hunter et al. 2014). Several members of the CWB and other interested individuals from the community took part in a 2-day workshop that introduced resilience concepts and illustrated how these concepts could be applied in practice. The workshop offered opportunities to build resilience among participants in terms of watershed governance by working through exercises:

1. Building a shared understanding of what was valued about the watershed;
2. Building a shared understanding of what threatens the watershed (considering scales above and below the focal scale of the watershed);
3. Identifying opportunities to build specified resilience (i.e., resilience to specific, known disturbances) and working through an example; and,
4. Identifying attributes of general resilience (i.e., resilience to unknown, unexpected disturbances or attributes that confer resilience to multiple specific disturbances) held by the CWB and attributes that could be strengthened.

The workshop provided an introduction to a much larger discussion and process of incorporating resilience thinking into their governance practice. It offered an opportunity to approach governance from a different perspective – considering scales, known and unknown disturbances, and adaptive capacity – and engage in decision-making based on this perspective. To assess the impacts of the workshop immediately and over time, a questionnaire was administered. The focus of the questionnaire was on learning, as learning is a key aspect of resilience and encouraging learning is considered to enhance resilience (Biggs et al. 2012). A framework that measured three learning types (Baird et al. 2014) was used: cognitive (learning new facts or restructuring knowledge); normative (changing viewpoints, norms, and/or values); and, relational (improved understanding of others' mindsets, developing relationships, developing trust). Some evidence of all three types of learning was measured immediately after the workshop, and persisted 6 months after the workshop was held (Baird et al. 2016). Some participants indicated that efforts have been made to incorporate resilience practice into governance of the watershed, and two thirds of respondents stated that they increased the network of those they communicate with and sharing information about resilience.

Does learning about resilience enhance resilience in watershed governance? The results of the study provide positive indications that it does. While learning is critical for resilience, it is insufficient on its own to ensure a resilient system. However, the discussions during the workshop indicate that other aspects of resilience, including building a shared understanding and including ecosystem services in planning, may have been enhanced as well and this bodes well for the CWB and the Cowichan Watershed.

22.3.4 Stream Rehabilitation, From Form to Function: Trout Unlimited Canada Training Program

Examples of degraded streams and stream corridors are prevalent throughout Canada as a result of historic and contemporary uses of, and alterations to, these systems and the watersheds in which they are situated (Fisheries and Oceans Canada 2006). Further degradation as a result of continued exploitation and manipulation of streams and their corridors, coupled with the effects of changing temperature and precipitation patterns associated with climate change, is likely and the consequences are uncertain (Imhof and FitzGibbon 2014).

Rehabilitation, the "reparation of ecosystem processes, productivity, and services rendered" (Clewell and Aronson 2013, p 203), can serve as a critical part of restoring and enhancing watershed resilience by rebuilding the functional characteristics of a stream within its watershed so that it is capable of dealing with higher variability while maintaining its health. Increasingly, communities and local organizations are playing an important and expanding role in achieving the goal of enhanced watershed resilience through rehabilitation (Imhof and FitzGibbon 2014). To support volunteer groups and communities in this endeavour, a consortium of conservation organizations and individuals developed the *Stream Rehabilitation, From Form to Function* training program (formerly the *Aquatic Renewal Stream Restoration Training Program*). Led by Trout Unlimited Canada (TUC), a national not-for-profit organization, the program's series of six workshops provide volunteers and young professionals with a basic level of training on the development and implementation of stream and watershed rehabilitation projects and programs. More specifically, the program helps groups and individuals understand that simply placing structures in a stream will not restore resilience. Rather, understanding the causes of dysfunction in the stream, and the watershed more broadly, and strategically applying the appropriate techniques and approaches to restore the functions of the system will aid in enhancing watershed resilience.

One of the primary reasons for the development of the training program was to foster social memory to maintain a knowledge base around stream and watershed rehabilitation. Engaging and mentoring the next generation of watershed stewards with the capacity to lead rehabilitation projects in their local watersheds is a central aim of the program. Building this community of practice sets the stage for watershed stewardship with local leadership that is efficient and well suited to responding adaptively to local issues as they arise.

The approach to rehabilitation taught in TUC's training program differs from more traditional approaches in that streams and watersheds are treated as complex, dynamic social-ecological systems for which consideration must be given to uncertainty, surprise, feedbacks, and temporal and spatial scale. In taking this novel approach, the program emphasizes several resilience principles identified by Plummer et al. (2014) including adaptive planning, diverse actor participation, inclusive participation and building a shared understanding, and including a wide range of ecosystem services in planning. Rather than simply discussing these principles, the program goes one step further and illustrates how they can be put into practice in the context of stream rehabilitation.

22.4 Concluding Reflections

Effectively navigating the landscape of water governance in Canada is essential and doing so will be no easy task given the complexity, uncertainty and change that characterize water resources. Positioned within a different world view, resilience thinking provides a distinct way for understanding water systems, and correspondingly, offers alternative approaches to drainage basin governance than have been used in the past.

Resilience and water governance is an emerging area of scholarship. Unpacking the concept of resilience and its relation to water governance reveals a fertile conceptual ground. Inroads into frameworks for resilience of governance in aquatic systems, and more broadly for environmental governance, are being made by some scholars (e.g., Walker and Salt 2006; Biggs et al. 2012; Plummer et al. 2014). Continued efforts in this direction may enhance understanding across varied settings.

Incorporating resilience into the unfolding dialogue about water governance in Canada is essential. Despite this imperative "it still remains a huge challenge to bring about a convergence between resilience and adapting institutions" (Boyd and Folke 2012, p 277). One aspect of this challenge is how to incorporate resilience thinking into existing institutions and dominant modes of governance. In the United States, for example, Benson and Garmestani (2011) examined the question of integrating resilience thinking into existing governing federal agencies. Some of the several challenges they found to integration include existing laws and regulations which ignored ecological complexity, the tendency to treat social and ecological systems in an uncoupled fashion, and the absence of enforceable standards regarding resilience. A more encompassing aspect is to incorporate principles for building resilience (Biggs et al. 2012, 2015) when addressing the water governance challenges, such as those identified in Canada by Simms and de Loë (2010).

Translating resilience thinking into resilience practice is an identified gap (Wilkinson 2012) and necessary next step (Walker and Salt 2012). To some extent, this will be facilitated or constrained by the extent of encouragement and support from existing agencies as well as the degree of incorporation at various scales within the water governance landscape. The four cases in this chapter demonstrate aspects of resilience being incorporated into drainage basin governance in Canada. Collating and communicating experiences from the field are essential. A primer on incorporating resilience into drainage basin management (Krievins et al. 2015) is intended to begin a discussion in Canada about resilience at the drainage basin scale, with those who work within it, and the broader water policy community.

References

Baird, J., Plummer, R., Haug, C., et al. (2014). Learning effects of interactive decision-making processes for climate change adaptation. *Global Environmental Change, 27*, 51–63.

Baird, J., Plummer, R., Moore, M.-L., et al. (2016). Introducing resilience practice to watershed groups: What are the learning effects? *Society & Natural Resources, 29*, 1214–1229.

Bakker, K., & Cook, C. (2011). Water governance in Canada: Innovation and fragmentation. *International Journal of Water Resources, 27*(2), 275–289.

Bates, B. C., Kundzewicz, Z. W., & Wu, S. et al. (Eds.). (2008). *Climate change and water*. Technical Paper of the Intergovernmental Panel on Climate Change (IPCC). Geneva: IPCC Secretariat.

Benson, M. H., & Garmestani, A. S. (2011). Can we manage for resilience? The integration of resilience thinking into natural resource management in the United States. *Environmental Management, 48*, 392–399.

Berkes, F., & Folke, C. (1998). Linking social and ecological systems for resilience and sustainability. In F. Berkes & C. Folke (Eds.), *Linking social and ecological systems: Management practices and social mechanisms for building resilience* (pp. 1–25). New York: Cambridge University Press.

Berkes, F., Colding, J., & Folke, C. (Eds.). (2003). *Navigating social-ecological systems: Building resilience for complexity and change*. Cambridge: Cambridge University Press.

Biggs, R., Schlüter, M., Biggs, D., et al. (2012). Toward principles for enhancing the resilience of ecosystem services. *Annual Review of Environment and Resources, 37*, 421–448.

Biggs, R., Schlüter, M., Schoon, M. L., et al. (2015). *Principles for building resilience: Sustaining ecosystem services in social-ecological systems*. New York: Cambridge University Press.

Boyd, E., & Folke, C. (2012). Conclusions: Adapting institutions and resilience. In E. Boyd & C. Folke (Eds.), *Adapting institutions: Governance, complexity and social-ecological resilience* (pp. 264–280). Cambridge: Cambridge University Press.

Brandes, O. M., Ferguson, K., M'Gonigle, M., et al. (2005). *At a watershed: Ecological governance and sustainable water management in Canada*. Victoria: POLIS Project on Ecological Governance.

Brandes, O. M., O'Riordan, J., O'Riordan, T., et al. (2014). *A blueprint for watershed governance in British Columbia*. Victoria: POLIS Project on Ecological Governance.

Brown, K. (2014). Global environmental change: A social turn for resilience? *Progress in Human Geography, 38*(1), 107–117.

Clancy, P. (2014). *Freshwater politics in Canada*. North York: University of Toronto Press.

Clewell, A. F., & Aronson, J. (2013). *Ecological restoration: Principles, values, and structure of an emerging profession* (2nd ed.). Washington, DC: Island Press.

Cooley, H., Ajami, N., Ha, M. L., et al. (2013). *Global water governance in the 21st century*. Oakland: Pacific Institute.

Cote, M., & Nightingale, A. J. (2012). Resilience thinking meets social theory: Situating social change in socio-ecological systems (SES) research. *Progress in Human Geography, 36*(4), 475–489.

Davidson, D. J. (2010). The application of the concept of resilience to social systems: Some sources of optimism and nagging doubts. *Society & Natural Resources, 23*, 1135–1149.

de Loë, R., & Kreutzwizer, R. (2007). Challenging the status quo: The evolution of water governance in Canada. In K. Bakker (Ed.), *Eau Canada: The future of Canada's water* (pp. 85–104). Vancouver: UBC Press.

de Loë, R., & Plummer, R. (2010). Climate change, adaptive capacity, and governance for drinking water in Canada. In D. Armitage & R. Plummer (Eds.), *Adaptive capacity and environmental governance* (pp. 157–178). Heidelberg: Springer.

Fisheries and Oceans Canada. (2006). *Ecological restoration of degraded aquatic habitats: A watershed approach*. Available via Fisheries and Oceans Canada. http://www.dfo-mpo.gc.ca/library/321286.pdf. Accessed 18 Sept 2015.

Folke, C. (2003). Freshwater for resilience: A shift in thinking. *Philosophical Transactions of the Royal Society of London B Biological Sciences, 358*(1440), 2027–2036.

Folke, C. (2006). Resilience: The emergence of a perspective for social–ecological systems analyses. *Global Environmental Change, 16*(3), 253–267.

Folke, C., Colding, J., & Berkes, F. (2003). Synthesis: Building resilience and adaptive capacity in social-ecological systems. In F. Berkes, J. Colding, & C. Folke (Eds.), *Navigating social-ecological systems: Building resilience for complexity and change* (pp. 352–387). New York: Cambridge University Press.

Folke, C., Carpenter, S. R., Walker, B., et al. (2010). Resilience thinking: Integrating resilience, adaptability and transformability. *Ecology and Society, 15*(4), 20.

Galaz, V. (2007). Water governance, resilience and global environmental change – A reassessment of integrated water resources management (IWRM). *Water Science and Technology, 56*(4), 1–9.

Global Water Partnership Technical Advisory Committee. (2000). *Integrated water resources management* (TAC background papers, 4). Stockholm: Global Water Partnership.

Government of the Northwest Territories and Aboriginal Affairs and Northern Development Canada. (2010). *Northern voices, northern waters: NWT Water Stewardship Strategy*. Yellowknife: Government of the Northwest Territories.

Government of the Northwest Territories and Aboriginal Affairs and Northern Development Canada. (2011). *NWT water stewardship: A plan for action, 2011–2015*. Yellowknife: Government of the Northwest Territories.

Health Canada. (2015). *Drinking water and wastewater*. Available via Health Canada. http://www.hc-sc.gc.ca/fniah-spnia/promotion/public-publique/water-eau-eng.php. Accessed 18 Sept 2015.

Holling, C. S., & Gunderson, L. H. (2002). Resilience and adaptive cycles. In L. H. Gunderson & C. S. Holling (Eds.), *Panarchy: Understanding transformations in human and natural systems* (pp. 25–62). Washington, DC: Island Press.

Holling, C. S., Gunderson, L. H., & Ludwig, D. (2002). In quest of a theory of adaptive change. In L. H. Gunderson & C. S. Holling (Eds.), *Panarchy: Understanding transformations in human and natural systems* (pp. 3–22). Washington, DC: Island Press.

Hornborg, A. (2013). Revelations of resilience: From the ideological disarmament of disaster to the revolutionary implications of (p)anarchy. *Resilience, 1*(2), 116–129.

Huitema, D., Mostert, E., Egas, W., et al. (2009). Adaptive water governance: Assessing the institutional prescriptions of adaptive (co)management form a governance perspective and defining a research agenda. *Ecology and Society, 14*(1), 26.

Hunter, R., Brandes, O., & Moore, M. -L. et al. (2014). *The Cowichan watershed board: An evolution of collaborative watershed governance*. Available via POLIS Project on Ecological Governance. http://poliswaterproject.org/publication/761. Accessed 18 Sept 2015.

Imhof, J. G., FitzGibbon, J. E. (2014, June 24–26). *Restoring streams and their watersheds by building the capacity of local communities: A watershed and stream rehabilitation training program.* Presentation at the Adaptation in the Great Lakes Region conference, Ann Arbor, Michigan.

Kerner, D. A., & Thomas, J. S. (2014). Resilience attributes of social-ecological systems: Framing metrics for management. *Resources, 3*, 672–702.

Krievins, K., Plummer, R., & Baird, J. (2014). *Resilience: An annotated bibliography*. Available via Environmental Sustainability Research Centre, https://dr.library.brocku.ca/handle/10464/5239. Accessed 18 Sept 2015.

Krievins, K., Baird, J., Plummer, R., et al. (2015). *Resilience in a watershed governance context: A primer*. St. Catharines: Environmental Sustainability Research Centre.

Lemmen, D. S., & Warren, F. J. (2004). *Climate change impacts and adaptation: A Canadian perspective*. Ottawa: Natural Resources Canada.

Milly, P. C. D., Betancourt, J., Falkenmark, M., et al. (2008). Stationarity is dead: Whither water management. *Science, 319*, 73–574.

Milman, A., & Short, A. (2008). Incorporating resilience into sustainability indicators: An example for the urban water sector. *Global Environmental Change, 18*(4), 758–767.

National Round Table on the Economy and Environment. (2010). *Changing currents: Water sustainability and the future of Canada's natural resource sectors*. Ottawa: NRTEE.

OECD. (2011). *Water governance in OECD countries: A multi-level approach*. Paris: OECD Publishing. doi:10.1787/9789264119284-en

Olsson, L., Jerneck, A., Thoren, H., et al. (2015). Why resilience is unappealing to social science: Theoretical and empirical investigations of the scientific use of resilience. *Science Advances, 1*(4), e1400217.

Plummer, R. (2010). Social-ecological resilience and environmental education: Synopsis, application, implications. *Environmental Education Research, 16*(5–6), 493–509.

Plummer, R., Baird, J., Moore, M.-L., et al. (2014). Governance of aquatic systems: What characterizes resilience? *International Journal of Water Governance, 2*(4), 1–18. doi:10.7564/14-IJWG51.

Plummer, R., Baird, J., Bullock, R., et al. (2016a). *Probing the relationship between ecosystem perceptions and approaches to environmental governance: An exploratory content analysis of seven water dilemmas*. Resilience: International Policies, Practices and Discourses. Online. http://www.tandfonline.com/doi/full/10.1080/21693293.2016.1202903#abstract.

Plummer, R., Baird, J., Krievins, K., et al. (2016b). Improving river health: Initiating collaboration in a transboundary river basin. *International Journal of River Basin Management, 14*, 119–132.

Rockström, J., Falkenmark, M., Allan, T., et al. (2014). The unfolding water drama in the Anthropocene: Towards a resilience based perspective on water for global sustainability. *Ecohydrology, 7*(5), 1249–1261.

Rogers, P., & Hall, A. W. (2003). *Effective water governance* (TEC background papers, 7). Stockholm: Global Water Partnership.

Schoeman, J., Allan, C., & Finlayson, C. M. (2014). A new paradigm for water? A comparative review of integrated, adaptive and ecosystem-based water management in the Anthropocene. *International Journal of Water Resources, 30*(3), 377–390.

Simms, G., & de Loë, R. C. (2010). *Challenges for water governance in Canada: A discussion paper* (Governance for source water protection in Canada report no. 2). Waterloo: Water Policy and Governance Group.

Sprague, J. (2007). Great wet north? Canada's myth of water abundance. In K. Bakker (Ed.), *Eau Canada: The future of Canada's water* (pp. 23–36). Vancouver: UBC Press.

Walker, B., & Salt, D. (2006). *Resilience thinking: Sustaining ecosystems and people in a changing world*. Washington, DC: Island Press.

Walker, B., & Salt, D. (2012). *Resilience practice: Building capacity to absorb disturbance and maintain function*. Washington, DC: Island Press.

Wilkinson, C. (2012). Urban resilience: What does it mean in planning practice. *Planned Theory & Practice, 13*(2), 299–333.

Chapter 23
Geomatics and Water Policy

Wanhong Yang

Abstract Geomatics including remote sensing and geographic information system (GIS) is the geospatial technology for gathering, management, analysis, and dissemination of spatially referenced information. Water policy is the governance framework including legislation/regulation, standard, planning and management of water resources. Water policy involves complex physical, biological, economic, social, and political processes, and manifests at location, subbasin, watershed, regional, national, and international scales. Water policy has an inherently spatial dimension which offers opportunities for Geomatics support. The applications of Geomatics for supporting water management and policy can be classified into three categories: Firstly, Geomatics supports acquisition, storage, management, visualization, and distribution of water and related datasets. Secondly, Geomatics supports spatial analysis of water resources including GIS and water modelling integration. Lastly, Geomatics supports decision making in water policy including the development of water related spatial decision support systems. While Geomatics has a great potential to support water management and policy in Canada, several strategic issues needs to be addressed. A Geomatics framework needs to be developed for water data gathering, analyzing and visualizing. Technology standards need to be developed for integrating Geomatics with water models. Particularly, Geomatics visualization technologies needs to be further developed to improve accessibility of water information for non-technical users.

Keywords Remote sensing • GIS • Water policy • Water models • Decision support systems

W. Yang (✉)
Department of Geography, University of Guelph, 50 Stone Rd E,
N1G 2W1 Guelph, ON, Canada
e-mail: wayang@uoguelph.ca

S. Renzetti, D.P. Dupont (eds.), *Water Policy and Governance in Canada*,
Global Issues in Water Policy 17, DOI 10.1007/978-3-319-42806-2_23

23.1 Introduction

Geomatics including remote sensing and geographic information system (GIS) is the geospatial technology for gathering, management, analysis, and dissemination of spatially referenced information (Natural Resources Canada 2013). Water policy is the governance framework including legislation/regulation, standard, planning and management of water resources (Dinar 1998). Water policy involves complex physical, biological, economic, social, and political processes, and manifests at location, subbasin, watershed, regional, national, and international scales (Wilson et al. 2000). Water policy has an inherently spatial dimension which offers opportunities for Geomatics support. As an innovative enabling technology, Geomatics has had widespread applications in many fields including water policy. Geomatics applications in water resources range from data gathering, georeferencing, to analyzing complex relationships between water conditions and related drainage area characteristics (Maidment 2002). In this chapter, the discussions on Geomatics applications for supporting water policy will be classified into three interconnected categories based on increasing levels of complexity: Firstly, Geomatics supports acquisition, storage, management, visualization, and distribution of water and related datasets. Secondly, Geomatics supports spatial analysis of water resources. Lastly, Geomatics supports decision making in water policy. In each category, both the current status and further development of Geomatics applications for supporting water policy will be discussed. After that, the strategies for improving Geomatics support for water policy will be proposed.

23.2 Data Support

Water policy needs to be supported by extensive water datasets and also water related datasets (Wilson et al. 2000). While not a complete list, water datasets include surface water (water flow and levels in streams and lakes), groundwater (water levels in wells), water quality (nutrients, pesticides, bacteria and other constituents in streams, lakes, springs, wells, and other sites), and water use data. Water related datasets include climate, landcover/landuse, soil, topography, land management, and ecological datasets. Water related datasets represent those causing factors for the status of water resources such as water quantity and quality.

Geomatics provides support for generating or acquiring water and related datasets (Wilson et al. 2000). GIS functions can be used to digitize or derive water features such as wetlands, ponds, reservoirs, lakes, and streams, and calculate dimensions such as area, length, and volume. For example, Ducks Unlimited Canada is using GIS technology to develop a wetland inventory for selected areas in Canada (http://www.ducks.ca/what-we-do/cwi/), which contributes to supporting water management and policy related to wetlands. Remote sensing has been widely used to conduct landcover/landuse mapping, identify land management practices, and estimate soil moisture. For example, Agriculture and Agri-Food Canada has initiated annual crop mapping based on remote sensing imageries since 2009 (http://open.canada.ca/data/en/dataset/ba2645d5-4458-414d-b196-6303ac06c1c9) and in conjunction with other datasets, crop mapping can be used to examine the impacts of crop production on water resources. LiDAR (Light Detection And Ranging) technology has also been increasingly used to generate Digital Elevation Model (DEM) data for landscapes, which provides high-resolution terrain data. Particularly, remote sensing has been applied to detect the visible change of water bodies. With field data validation, remote sensing has the ability to estimate major water quality variables such as chlorophyll-a, total suspended solids, turbidity, total nitrogen and phosphorus, and surface water temperature (Chen et al. 2007). For example, remote sensing has been applied to assess water temperature in rivers and streams in eastern and western Oregon (Torgersen et al. 2001), derive data on algal and mineral turbidity in Lake Erie (Binding et al. 2012), and conduct water quality monitoring in support of the EU Water Framework Directive (Chen et al. 2004). This technology advancement has the potential to be applied to monitor water bodies in a timely manner.

Geomatics supports the storage and management of water and related datasets (Wilson et al. 2000). GIS sets up the spatial framework for storing various datasets in layer structure, which provides the basis for conducting spatial analysis of different features. For example, GIS stores climate and water monitoring stations as points, streams as lines, and ponds and reservoirs as polygons to establish spatial relationships among these features. GIS data model for water resources such as Arc Hydro has been developed specifically to characterize water and related features including hydrography, soil, landuse, climate, stream flow, and water quality (Maidment 2002). GIS also supports management of various datasets including data transfer, interpolation and other processing. For example, GIS can support transferring flow data at the outlet of one watershed to the outlet of a nearby watershed without flow data based on the similarities of climate and landscape characteristics and comparison of drainage areas (Yang et al. 2010). GIS functions can be used to support the interpolation of climate point data into continuous areal data such as gridded climate data. For example, US National Oceanic & Atmospheric Administration (NOAA) publishes a series of gridded global climate data including precipitation and soil moisture (http://www.esrl.noaa.gov/psd/data/gridded/). GIS functions can be also applied to resample or interpolate data layers with different resolutions into the same resolution, such as resampling DEM data to fit into the resolution of landuse data or vice versa. These data management functions facilitate development of consistent water and related datasets at various spatial scales.

In addition to map display, GIS has functions to visualize water and related data in charts and tables (Wilson et al. 2000). These functions can be customized to serve water management and policy needs. For example, for lake management, maps can be used to display distribution of algae bloom and water temperature, charts can be used to show variation of water level, and tables can be used to provide details on water quality parameter values. The advancement of GIS technology has also been increasingly utilized to display and distribute water and related datasets to a wide range of users. Applications of internet-based mapping technology allow users to visualize and download water and related datasets (Verma et al. 2012). For example, U.S. Geological Survey (USGS)'s National Water Census Portal (http://cida.usgs.gov/nwc/) provides map based estimates of water budget components for local watersheds, water withdrawal data for counties, tools to calculate statistics of daily streamflow records, modeled daily streamflow at ungauged stations, and access to records of aquatic biology observations. A user survey in the Great Lakes region of the United States also showed increasing use of WebGIS for data visualization and distribution (Rousseau and Lawrence 2013). The Grand River Conservation Authority (GRCA) in southern Ontario of Canada developed a WebGIS portal for public users to view and download various watershed datasets (http://maps.grandriver.ca/). These Geomatics technology developments greatly facilitated the dissimilation of water and related datasets to stakeholders and other users.

While Geomatics has been providing data support for water policy including acquisition, storage, management, visualization, and distribution of water and related datasets, its potential can be significantly improved. Firstly, various data gaps such as limited spatial and temporal coverage and incomplete data variables exist for water and related datasets (Dworak et al. 2005). For example, a significant amount of watersheds are ungauged and for those locations with water data, the time series could be short. Some of the soil data maybe based on soil survey a long time ago and some key soil variables for water resources assessment such as initial soil nutrient concentration may have missing data. In some areas landcover and landuse data are outdated, at coarse resolution or with limited time series. Data on cropland management practices such as tillage and nutrient application are not gathered or incomplete. While providing ongoing support on data acquisition to address these gaps, Geomatics has the potential to be applied to develop strategic planning for acquisition of water and related datasets. For example, the network of climate and water monitoring stations/locations may be not sufficient for supporting water policy. Geomatics can support the spatial analysis of monitoring network and identify locations for strengthening or adding monitoring efforts. Secondly, water and related datasets are typically originated from and resided in various sources or organizations. These datasets have different scales, resolutions, and qualities. Water management and policy agencies may use different processing approaches in various applications, which cause inconsistencies in data analysis. In addition to support raw data acquisition, Geomatics can be applied to develop processed or enhanced datasets to more effectively support water policy. For example, Ontario Ministry of Natural Resources and Forestry developed the Ontario Flow Assessment Tool (OFAT) based on climate, DEM, stream network and gauged flow data. The tool can be used to delineate watersheds, characterize watersheds, estimate watershed flow (mean annual, flood and drought condition flow), and view flow monitoring gauge statistics (http://www.ontario.ca/page/watershed-flow-assessment-tool). This tool is very useful for supporting water management and policy. Thirdly, water policy needs to be supported by timely data. For example, data on landcover/landuse, land management, and water body status need to be regularly updated to establish relationships between drainage area and receiving water body characteristics.

Remote sensing technologies need to be further utilized to provide timely monitoring of landscape and water bodies (Chen et al. 2004). Finally, GIS, particularly WebGIS technologies need to be developed to improve accessibility of water and related data to stakeholders and the general public. The development of public participation GIS can support the engagement of stakeholders and the general public in water policy through viewing, interpreting water and related data and also providing feedback (Pease and Murray 2014). While all these efforts may face various constraints, Geomatics has the potential to overcome institutional barriers to improve the effectiveness of data support for water policy.

Geomatics has the advantage of providing strong data support to water policy. However, Geomatics users also need to understand limitations of these datasets and related implications. Firstly, Geomatics data quality may cause uncertainty on water data analysis (Heywood et al. 2012; Devillers et al. 2010). For example, spatial positioning of water and related datasets may have errors due to instrument limitation or human error. Interpolation of spatial data such as precipitation from points to areas may have errors due to limited original data points and landscape complexities. Secondly, scaling issues in Geomatics data may cause challenges on water data analysis (Beven 1995; Zhang et al. 2009). For example, scaling up observed soil erosion rates at specific points to watershed level may be constrained by representativeness of the data points. Using DEM data at fine and coarse resolutions may derive different slope and slope length data for the same area and affect related runoff and soil erosion rate estimation. Inevitably Geomatics users will encounter various data limitations. It is important to understand the implications of these data limitations on water resources analysis and keep errors in the acceptable range.

23.3 Analysis Support

Geomatics, particularly GIS can provide various levels of analysis support for water policy (Maidment 2002). GIS has functions to conduct basic spatial analysis of water and related features. GIS can provide a spatial framework for water models such as water allocation models. GIS can also provide interface support on setting up water models including modelling parameterization. Furthermore, GIS can support integration of various water and related models.

GIS spatial analysis functions can be utilized to gain insights on landscape conditions related to water. For example, in a study area with given climate conditions, location-specific soil erosion risk is related to topography, landcover/landuse, and soil characteristics. Typically topography and landcover/landuse data are in raster format and soil data are in vector format. GIS rasterization function can be used to convert soil vector data into raster data format. If necessary, GIS resampling functions can be used to interpolate the three raster layers into consistent data layers with the same cell size. After that, reclassifying function can be used to assign standardized scores on erosion risk to each layer, and raster calculation functions can be used to combine the three layers together into one layer with overall erosion risk based on weights for each factor (Ormsby et al. 2010). The final erosion risk GIS layer can be used as a reference for water management and policy.

GIS can provide a spatial framework for water models, such as water allocation models. For example, an optimization model was developed and applied to a rural area in South Cyprusto to examine landuse and water allocation planning. The decision problem was to maximize economic benefit, while minimizing water extraction and transportation cost under ecological constraints (Fotakis and Sidiropoulos 2014). In this study, GIS provided georeferencing for water supply from wells and also water demand from nearby land. The spatial framework supported the optimization model to identify optimal water allocation to land blocks in the study area. GIS also supports spatially explicit modelling of water bodies. In a river basin study in Kentucky, USA, GIS provided a spatial framework for a neural network based water quality model, which included setting up the stream network and also incorporating landuse factors (Anmala et al. 2015). In an application of a three-dimension hydrodynamic-ecological model to examine the phytoplankton spring bloom in Lake Erie, GIS was applied to support the grid-based parameterization of the model, which included characterizing lake bathymetry based on 2-km horizontal grids and 40 vertical layers of variable thickness (Bocaniov et al. 2014). In these applications, water models can utilize the spatial identification and relationship for data preparation and analysis and also result display.

In recent decades, GIS interface development has made a significant contribution to the advancement and application of hydrologic and hydraulic models. GIS framework specifically for water resources analysis such as Arc Hydro has been developed. In addition to a geographic data model for water and related thematic layers including streams, drainage areas, channels, surface terrain, and rainfall response, Arc Hydro further develops a set of tools to support hydrologic analysis and modelling (Maidment 2002). For watershed hydrologic models, GIS interface utilizes the DEM data to delineate a watershed into subbasins and estimate related parameters such as slope and slope length. GIS interface supports the preparation of model parameters from landcover/landuse, land management, and soil data. The climate data can be interpolated to subbasins to characterize the driving forces such as temperature and precipitation. GIS interface also supports model calibration and validation through examining and adjusting model parameters. Finally, GIS supports the display of modelling results (George and Leon 2008). While various options of GIS and hydrologic model interactions such as loose coupling, tight coupling, and integration exist, most of the hydrologic and hydraulic models have developed GIS interfaces (Table 23.1).

GIS further supports the integration of different models for water policy. For example, a farm economic model estimates the economic costs of agricultural beneficial management practices (BMPs) at field and farm scales. A watershed hydrologic model simulates water quantity and quality effects of agricultural BMPs at subbasin level. GIS provides support to interpolate modelling parameters and also outputs into the same spatial unit such as field/farm or subbasin to examine cost effectiveness of agricultural BMPs for protecting water resources (Yang et al. 2007). The integration can be further developed to include other dimensions. For example,

Table 23.1 Examples of key water models with GIS interface

Water Model	GIS	Model domain	Reference
HEC-RAS	ArcGIS	Flood modelling	USACE (2011)
SWAT	ArcGIS, MapWindow GIS	Watershed hydrology, water quality	Arnold et al. (2012), George and Leon (2008)
AnnAGNPS	ArcGIS	Watershed hydrology, water quality	Xiao (2003)
WEPP	ArcGIS	Watershed hydrology, water erosion	Flanagan et al. (2013)
APEX	ArcGIS	Field and small watershed hydrology, water quality	Gassman et al. 2010
MODFLOW	ArcGIS	Groundwater modelling	Jones et al. (2015)
SWMM	ArcView, ArcGIS	Stormwater hydrology and hydraulics	Heineman (2001), Shamsi and Smith (2005)
MIKE SHE	ArcGIS	Watershed hydrology, water quality	DHI (2012), Jaber and Shukla (2012)

Source: Adapted from Martin et al. 2005

the integration of farm economic, watershed hydrologic, and landscape ecological models was developed to identify riparian buffers in a southern Ontario watershed in Canada to minimize economic costs while achieving water quality benefit target only, wildlife habit improvement target only, or both (Yang et al. 2014). In model integration, GIS provides the support for spatial referencing, model parameterization, and harmonization of spatial units for economic, environmental, and ecological models. With GIS support, integrated water models have been increasingly developed in academia and also for supporting water policy (Table 23.2).

In order to meet the demand for water policy, Geomatics/GIS support for data analysis and modelling needs to be further developed. Firstly, GIS has the potential to further support water analysis/model transfer and scale-up. In most cases, GIS-based water modelling is conducted for specific areas and for academic purposes. For water management and policy, modelling efforts in some cases were only conducted for selected areas due to modelling complexity and/or data and resource constraints. For example, the Watershed Evaluation of BMPs (WEBs) program was established in Canada in 2004, with the mission to assess the environmental and economic performance of agricultural BMPs at a watershed scale. As a collaborative initiative between Agriculture and Agri-Food Canada and Ducks Unlimited Canada, the WEBs program selected nine representative micro-watersheds across the country to conduct field experiments on BMPs, monitoring for water quantity and quality and modelling for assessing BMP effects (http://www.agr.gc.ca/eng/?id=1297085765144). This program was running for 9 years from 2004 to 2013 and made an important contribution to developing the science base for quantifying BMP effects at field and watershed scales (Yang et al. 2007). However, the program was ended in 2013 without a following-up initiative to transfer the knowledge to similar watersheds and scale up the watershed analysis to regional scales.

Table 23.2 Examples of integrated water modelling with GIS support

Modelling name	Study area	Focus	Reference
Integrated hydrologic-agronomic-economic model	Syr Darya River Basin, Central Asia	Evaluating economic and environmental consequences of various irrigation policy choices	Cai et al. (2003)
Integrated economic-hydrologic model	Rio Grande Basin, southwestern United States and northern Mexico	Evaluating tradeoffs among efficiency, equity, and sustainability for the water pricing program	Ward and Pulido-Velázquez (2008)
Integrated watershed-farm economic model	Rock River watershed, Vermont, USA	Targeting critical sources areas for agricultural BMPs based on economic and water quality criteria	Ghebremichael et al. (2013)
Integrated economic valuation and catchment model	George Catchment, Tasmania, Australia	Evaluating economic, ecological and hydrologic effects of landuse change and stream protection	Kragt et al. (2011)
Integrated surface-groundwater and economic model	Campo de Dalias coastal plain in southeastern Spain	Evaluating opportunity costs of water allocation to urban and agricultural water users	Pulido-Velazquez et al. (2008)
Integrated hydrodynamic and economic model	Flood-prone areas in the Netherlands	Assessing direct and indirect flood damage	Jonkman et al. (2008)
Integrated hydrological, biological, and economic model	Grande Ronde River Basin, USA	Evaluating economic costs and temperature targets and fishery benefits of riparian restoration scenarios	Watanabe et al. (2006)
Integrated economic-hydrologic modelling	Illinois, USA	Identifying land retirement based on economic and environmental tradeoffs	Yang et al. (2003)
Integrated economic-hydrologic-carbon model	Ontario, Canada	Targeting conservation tillage to achieve water quality and carbon benefits	Yang et al. (2005)
Integrated economic-hydrologic-ecological model	Ontario, Canada	Targeting riparian buffer to achieve water quality and habitat benefits	Yang et al. (2014)

To expand spatial coverage of modelling efforts, one approach is to develop water modelling for representative areas and then transfer the modelling approach and/or results to those areas with similar climate and landscape characteristics. Another approach is to scale up water modelling to regional scales. In recent years, several initiatives have been developed to conduct regional modelling to support water policy. Since 2004, U.S. Department of Agriculture has led the "Conservation Effects Assessment Project (CEAP)" to quantify the environmental effects of conservation practices and programs and develop the science base for managing the agricultural landscape for environmental quality (http://www.nrcs.usda.gov/wps/

portal/nrcs/main/national/technical/nra/ceap/). In the CEAP, a cascaded modelling approach was developed to scale up field-level conservation effects to regional scale. In the approach, the field-scale model APEX (Agricultural Policy Environmental Extender) characterized hydrologic processes and estimated water quantity and quality effects of agricultural conservation practices at field scale. The field-scale results were fed into the watershed model SWAT (Soil and Water Assessment Tool) and with the HUMUS (Hydrologic Unit Model of the United States) framework, watershed-level conservation effects were scaled up to the regional level. Up to now, this GIS-based modelling system has been applied to 12 regions in the United States including the Upper Mississippi River Basin, the Great Lakes Region, and the Chesapeake Bay Region (Johnson et al. 2015). These modelling efforts have provided important information for supporting agricultural conservation programs and water policy in the United States. There also existed several GIS-based regional modelling efforts with coarse spatial resolution, with which the areas in the magnitudes of a few hundred square kilometers were defined as the basic modelling units or subbasins. For example, the SWAT was applied to the continent of Africa to examine fresh water availability under climate change scenarios (Faramarzi et al. 2013). Recently the SWAT was applied to the province of Alberta in Canada to estimate water resource supply in over 2000 subbasins (Faramarzi et al. 2015). These GIS based modelling initiatives demonstrated the utility of large-scale landscape analysis on supporting water policy.

Secondly, uncertainties associated with water modelling have been an ongoing concern for water managers and decision makers (Refsgaard et al. 2007). Due to data, model structure, and other related limitations, it is reasonable to expect that uncertainties of water modelling will persist. While Geomatics can support various efforts on reducing uncertainties of water modelling such as improving data quality and modelling structure, GIS can play an important role on communicating uncertainties to users (Kubíčeka and Šašinkaa 2011). Visualization of modelling or analysis uncertainties at both spatial and temporal dimensions can facilitate informed water policy.

Thirdly, typically water modelling is complex. It is necessary to communicate water modelling results to the stakeholder and decision makers in a user friendly manner (Yang 2011). Furthermore, there is an important need to allow decision makers and water managers to operate water models. This leads to the development of decision support systems for water policy.

23.4 Decision Support

Water policy needs support from readily available analytical information. However, there are various gaps between water science and policy, particularly for providing timely and user friendly decision support (Quevauviller et al. 2005). These gaps exist for several reasons. Firstly, water research is typically for answering scientific questions, not necessarily for addressing specific water policy questions. Secondly, water science, analysis and model maybe too complex for decision makers and water managers to operate and use. These gaps lead to the development of water decision support systems (DSS) in recent decades. Typically these DSS are supported by Geomatics include remote sensing and GIS.

Water DSS are developed based on a set of analytical approaches or models to address specific water issues. For example, an integrated Riverine Environmental Flow Decision Support System (REFDSS) was developed based on hydrodynamic and ecological models and MapWindow Open Source GIS platform. The REFDSS was applied to the Upper Delaware River, USA to estimate local hydrologic conditions (e.g., depth, velocity, shear stress, etc.) at a fine pixel-scale (1 m^2) and habitat suitability for a variety of taxa. The results can be used for evaluating competing flow scenarios and other variables on instream habitat (Maloney et al. 2015). Another example is the irrigation management DSS developed for the San Diego County Water Authority in California, USA. This ArcGIS based DSS was developed based on computational methods for water supply, demand, and operation in the irrigation district considering municipal, industrial, and environmental/ecological interests. The DSS can be applied to maximize water savings from on-farm conservation measures and delivery system efficiency improvements through integrated management of large-scale irrigation systems (Triana and Labadie 2012).

Majority of water DSS have been developed based on desktops with two key features. The first feature is scenario analysis. In most cases, it is very challenging to conduct scenario analysis in modelling with a manual approach. The modelers may need to re-setup the analysis procedure or modelling system, which may include step-by-step data preparation and model parameterization. This could be a time consuming task if the modelling system is complex in terms of number of model parameters and size of study area. Water DSS have various interface functions to automate the model re-setup and parameterization processes. With DSS, users have the option to define a scenario based on spatial locations and tasks, and then user selection can be passed onto the DSS to run the models. For example, an irrigation management DSS allows users to define a scenario based on water conservation measures and locations and the DSS has the capacity to directly parameterize the scenario for running irrigation models to examine water allocation based supply and demand relationship (Triana and Labadie 2012). The second feature of DSS is to display analytical results based on specific tasks. Water modelling results could be extensive and post-processing is necessary to distill output data into useful information. DSS have customized visualization functions which allow users to examine modelling results in a user friendly manner. For example, users have the option to compare the results of multiple scenarios and water and related effects (Triana and Labadie 2012).

In recent years, water DSS have evolved from desktop systems to WebGIS platforms (Table 23.3). For example, a WebGIS based water right DSS was developed for Yakima River Basin, Washington, USA. This DSS provided water right mapping which included the water right document ID, priority date of the water right, the quantity of water for irrigation, and the size of the parcel linked to the water right. This spatial referenced water right DSS provided important utility to users in

Table 23.3 Examples of water DSS with GIS support

Water DSS name	Study area	Focus	Reference
Integrated riverine environmental flow DSS	The Upper Delaware River, USA	Evaluating competing flow scenarios and other variables on instream habitat.	Maloney et al. (2015)
Integrated irrigation management DSS	San Diego County, California	Maximizing water savings from efficiency improvements and conservation measures	Triana and Labadie (2012)
Integrated flood management DSS	The Sinclair Dam in Georgia, USA	Evaluating flood management strategies under dam break	Qi and Altinakar (2011)
Stormwater management DSS	Oregon, USA	Evaluating runoff mitigation strategies for rural stormwater management	Wing and Godwin (2011)
Integrated GIS-based water quality management DSS	The Upper Litani Basin (ULB) river system in Lebanon	Evaluating wastewater and sewage management plans for improving surface water quality	Assaf and Saadeh (2008)
WebGIS based water rights DSS	Yakima River Basin, Washington State, USA	Mapping water rights to facilitate public participation	Pease and Murray (2014)
Web-based DSS for watershed management	Jinjiang Basin, China	Evaluating the soil and water conservation effects of a vegetation filter strip	Zhang et al. (2015)
Integrated Web-based water resources management DSS	Daegu city, Republic of Korea	Optimizing allocation of water resources among different regions within a city	Zeng et al. (2012)
WebGIS based integrated flood assessment DSS	Navi Mumbai, India	Integrating multiple models for coastal urban flood simulation	Kulkarni et al. (2014)

comparing to traditional means, with which most of the information were stored in documents. Further development of the DSS to link water rights data with infrastructure, hydrologic, and other environmental data has the potential to improve decision making in environmental management (Pease and Murray 2014). WebGIS has also been developed to support complex modelling based DSS. Based on current technology, the Internet based DSS are developed based on both WebGIS Server and Web server. The WebGIS server manages the communications between GIS databases and modelling. A user interface is developed to manage user inputs from Web browser into the Web server and then communicate to WebGIS server for modelling execution, and in turn, communicate modelling results from WebGIS server to Web server, and then to the Web browser. The data communication needs to be supported by WebGIS mapping functions (Kulkarni et al. 2014). For example, a WebGIS based integrated water management DSS was developed for water allocation in the Daegu city in the Republic of Korea. In the DSS, water supply was simulated by hydrologic model HEC-HMS, water demand was estimated by an artificial neural network (ANN) based model and a genetic algorithm (GA) was used to optimize water supply and demand for efficient allocation of water resources among different regions within a city. This WebGIS based DSS addressed complex water resources management problems in urban areas (Zeng et al. 2012). In another application, WebGIS technology was utilized to develop a watershed management DSS for reducing agricultural non-point source pollution in Jinjiang Basin in southwestern China. The core component of the DSS was the SWAT model for evaluating water quantity and quality effects of various agricultural BMP scenarios. In order to make the SWAT operational, a series of interface functions including map operation tools, interactive BMP scenario interface, BMP identification interface, and result display interface were developed. The DSS was supported by spatial databases and map tile files. Interestingly, the DSS was deployed over a cloud computing platform which supported quasi-real-time assessment of agricultural BMPs and decision making (Zhang et al. 2015).

Geomatics based water DSS have had significant advancements in recent years. However, Geomatics decision support for water policy faces several challenges, which indicate the importance of further development of Geomatics technologies in this direction. Firstly, majority of DSS developments are originated from academic purposes. These research efforts have made important contributions to developing the methodology and showcasing DSS performance. However, demands from water policy may be not well incorporated into the DSS. There is a necessity to reverse the DSS development process to start with practical user needs and tasks. Secondly, user friendliness of DSS needs to be further developed. The user friendliness is a relative concept with the goal of meeting user needs. In this process, one aspect is to further develop analysis and visualization functions to enhance DSS functionalities. The other aspect is to train users to gain technical skills to master DSS. The integration of the two aspects has the potential to maximize the utility of DSS for water policy. Thirdly, DSS need to be developed to engage stakeholders and further, the general public. In some settings water users have extensive conflicts such as upstream and downstream users, and urban and rural users. DSS have the potential to communicate these conflicts and various tradeoffs in decision making to the stakeholders and the general public. In recent years, public participation GIS has been gradually developed to engage users into various decision-making issues such as urban and regional planning (Bugs et al. 2010; Brown 2012). There is a necessity to further develop public participation GIS to support decision making in water policy. Finally, maintenance and further development of DSS need to be supported by continuing development of water and related data and analysis.

23.5 Discussions and Conclusions

Water issues are complex, which have multiple purpose/objectives, scales and time steps, and involve multiple sectors and jurisdictions. Geomatics technologies including remote sensing and GIS have provided important support for water management and policy and further development of Geomatics support has significant potential. The demand from water management and policy calls for the development of water information infrastructure empowered by Geomatics technologies.

It is necessary to conduct strategic development of water information infrastructure, which includes water data, analysis, and DSS. Water policy stakeholders involve different government departments, industries, non-government organizations, conservation authorities, and the general public. Inevitably duplicate and inconsistent efforts exist for water information development. On the other hand, resource constraints will be an ongoing concern. It is necessary to develop collaborative partnerships to overcome institutional barriers for further development of the water information infrastructure. These strategic developments have the potential to lead to improvements in cost effectiveness in developing water data, analysis, and DSS, which provide support for water regulation, standard, programming, planning, and management.

Secondly, it is important to develop an operational framework for water information infrastructure. In most cases, water data maybe collected and analyzed internally in specific institutions. For example, an agriculture ministry and an environment ministry may maintain their own data and analysis capacities. There is a necessity to develop a unified framework for water data acquisition, analysis, and DSS, while addressing specific needs of various institutions. For example, the water information infrastructure can be developed to support drought and irrigation analysis, and also flood and mitigation analysis. The information infrastructure can be further used for supporting water quality management. For example, in Australia, a common set of modelling systems/platforms has been developed and promoted for water resources analysis (http://ewater.org.au/). The unified framework has the advantage of developing data, analysis and DSS based on both spatial and temporal coverages, and incorporating maintenance and further improvement or upgrade. The unified framework also supports interoperability, transfer and scale-up efforts of data, analysis, and DSS.

Finally, it is necessary to develop water information infrastructure to support extensive engagement of stakeholders and users, particularly the general public. In water management and policy, the traditional paradigm is a top-down approach with which water agencies provide water data, analysis results, standard, and others to stakeholders and the general public. Geomatics technologies have the capacity to support the paradigm change, which fosters a down-top process to involve stakeholders and users in water management and policy. In this process, public participation GIS shows promise to support not only the top-down approach to disseminate data, analysis and decision making to stakeholders and users but also the down-top approach to set up channels for stakeholders and users to provide feedback on water management and policy. With Geomatics support, the dialogue between water policy authorities and stakeholders and the general public will have the potential to improve the development and implementation of water management and policy.

References

Anmala, J., Meier, O., Meier, A., & Grubbs, S. (2015). GIS and artificial neural network–based water quality model for a stream network in the Upper Green River Basin, Kentucky, USA. *Journal of Environmental Engineering, 141*(5), 04014082.

Arnold, J. G., Moriasi, D. N., Gassman, P. W., Abbaspour, K. C., White, M. J., Srinivasan, R., Santhi, C., Harmel, R. D., van Griensven, A., Van Liew, M. W., Kannan, N., & Jha, M. K. (2012). SWAT: Model use, calibration, and validation. *Transactions of the ASABE, 55*(4), 1491–1508.

Assaf, H., & Saadeh, M. (2008). Assessing water quality management options in the Upper Litani Basin, Lebanon, using an integrated GIS-based decision support system. *Environmental Modelling & Software, 23*, 1327–1337.

Beven, K. (1995). Linking parameters across scales: Subgrid parameterizations and scale dependent hydrological models. *Hydrological Processes, 9*, 507–525.

Binding, C. E., Greenberg, T. A., & Bukata, R. P. (2012). An analysis of MODIS-derived algal and mineral turbidity in Lake Erie. *Journal of Great Lakes Research, 38*(1), 107–116.

Bocaniov, S. A., Smith, R. E. H., Spillman, C. M., Hipsey, M. R., & Leon, L. F. (2014). The nearshore shunt and the decline of the phytoplankton spring bloom in the Laurentian Great Lakes: Insights from a three-dimensional lake model. *Hydrobiologia, 731*, 151–172.

Brown, G. (2012). Public Participation GIS (PPGIS) for regional and environmental planning: Reflections on a decade of empirical research. *URISA Journal, 24*(2), 7–18.

Bugs, G., Granell, C., Fonts, O., Huerta, J., & Painho, M. (2010). An assessment of Public Participation GIS and Web 2.0 technologies in urban planning practice in Canela, Brazil. *Cities, 27*(3), 172–181.

Cai, X., McKinney, D., & Lasdon, L. (2003). Integrated hydrologic-agronomic-economic model for river basin management. *Journal of Water Resources Planning and Management, 129*(1), 4–17.

Chen, Q., Zhang, Y., Ekroos, A., & Hallikainen, M. (2004). The role of remote sensing technology in the EU water framework directive (WFD). *Environmental Science & Policy, 7*, 267–276.

Chen, Q., Zhang, Y., & Hallikainen, M. (2007). Water quality monitoring using remote sensing in support of the EU water framework directive (WFD): A case study in the Gulf of Finland. *Environmental Monitoring and Assessment, 124*, 157–166.

Devillers, R., Stein, A., Bédard, Y., Chrisman, N., Fisher, P., & Shi, W. (2010). Thirty years of research on spatial data quality: Achievements, failures, and opportunities. *Transactions in GIS, 14*, 387–400. doi:10.1111/j.1467-9671.2010.01212.x.

DHI. (2012). *MIKE SHE user manual*. Hørsholm: Danish Hydrologic Institute. http://dssplanning.dhigroup.com/links/MIKE_SHE_Printed_V1.pdf.

Dinar, A. (1998). Water policy reforms: Information needs and implementation obstacles. *Water Policy, 1*(4), 367–382.

Dworak, T., Gonzalez, C., Laaser, C., & Interwies, E. (2005). The need for new monitoring tools to implement the WFD. *Environmental Science & Policy, 8*, 301–306.

Faramarzi, M., Abbaspour, K. C., Vaghefi, S. A., Farzaneh, M. R., Zehnder, A. J. B., Srinivasan, R., & Yang, H. (2013). Modeling impacts of climate change on freshwater availability in Africa. *Journal of Hydrology, 480*, 85–101.

Faramarzi, M., Srinivasan, R., Iravani, M., Bladon, K. D., Abbaspour, K. C., Zehnder, A. J. B., & Goss, G. G. (2015). Setting up a hydrological model of Alberta: Data discrimination analyses prior to calibration. *Environmental Modelling & Software, 74*, 48–65.

Flanagan, D. C., Frankenberger, J. R., Cochrane, T. A., Renschler, C. S., & Elliot, W. J. (2013). Geospatial application of the Water Erosion Prediction Project (WEPP) model. *Transactions of the ASABE, 56*(2), 591–601.

Fotakis, D., & Sidiropoulos, E. (2014). Combined land-use and water allocation planning. *Annals of Operations Research, 219*, 169–185.

Gassman, P. W., Williams, J. R., Wang, X., Saleh, A., Osei, E., Hauck, L. M., Izaurralde, R. C., & Flowers, J. D. (2010). The agricultural policy/environmental eXtender (APEX) model: An emerging tool for landscape and watershed environmental analyses. *Transactions of the ASABE, 53*(3), 711–740.

George, C., & Leon, L. F. (2008). WaterBase: SWAT in an open source GIS. *The Open Hydrology Journal, 2*, 1–6.

Ghebremichael, L. T., Veith, T. L., & Hamlett, J. M. (2013). Integrated watershed- and farm-scale modeling framework for targeting critical source areas while maintaining farm economic viability. *Journal of Environmental Management, 114*, 381–394.

Heineman, M. C. (2001). An Arc View GIS tool for SWMM. *Journal of Water Management Modeling* R207–R211. doi:10.14796/JWMM .R207-11

Heywood, I., Cornelius, S., & Carver, S. (2012). *An introduction to geographical information systems*. Harlow: Pearson Education Limited.

Jaber, F. H. & Shukla, S. (2012). MIKE SHE: Model use, calibration, and validation. *Transactions of the ASABE, 55*(4), 1479–1489.

Johnson, M.-V. V., Norfleet, M. L., Atwood, J. D., Behrman, K. D., Kiniry, J. R., Arnold, J. G., White, M. J., & Williams, J. (2015). The Conservation Effects Assessment Project (CEAP): A national scale natural resources and conservation needs assessment and decision support tool. *IOP Conference Series: Earth and Environmental Science, 25*, 012012.

Jones, D., Jones, N., Greer, J., & Nelson, J. (2015). A cloud based MODFLOW service for aquifer management decision support. *Computers & Geosciences, 78*, 81–87.

Jonkman, S. N., Bočkarjova, M., Kok, M., & Bernardini, P. (2008). Integrated hydrodynamic and economic modelling of flood damage in the Netherlands. *Ecological Economics, 66*(1), 77–90.

Kragt, M. E., Newham, L. T. H., Bennett, J., & Jakeman, A. L. (2011). An integrated approach to linking economic valuation and catchment modelling. *Environmental Modelling & Software, 26*, 92–102.

Kubíčeka, P., & Šašinkaa, Č. (2011). Hematic uncertainty visualization usability – Comparison of basic methods. *Annals of GIS, 17*(4), 253–263.

Kulkarni, A. T., Mohanty, J., Eldho, T. I., Rao, E. P., & Mohan, B. K. (2014). A web GIS based integrated flood assessment modeling tool for coastal urban watersheds. *Computers & Geosciences, 64*, 7–14.

Maidment, D. R. (2002). *Arc Hydro: GIS for water resources*. Redlands: ESRI Press.

Maloney, K. O., Talbert, C. B., Cole, J. C., Galbraith, H. S., Blakeslee, C. J., Hanson, L., & Holmquist-Johnson, C. L. (2015). An integrated Riverine Environmental Flow Decision Support System (REFDSS) to evaluate the ecological effects of alternative flow scenarios on river ecosystems. *Fundamental and Applied Limnology, 186*(1–2), 171–192.

Martin, P. H., LeBoeuf, E. J., Dobbins, J. P., Daniel, E. B., & Abkowitz, M. D. (2005). Interfacing GIS with water resource models: A state- of-the-art review. *Journal of the American Water Resources Association, 41*(6), 1471–1487.

Natural Resources Canada. (2013). *Geomatics*. http://www.nrcan.gc.ca/earth-sciences/geomatics/10776

Ormsby, T., Napoleon, E., Burke, R., Groessl, C., & Bowden, L. (2010). *Getting to know ArcGIS desktop*. Redlands: ESRI Press.

Pease, M., & Murray, J. (2014). Making water resource decisions more "Informationally" efficient: Development of a geospatial water rights decision support system for Kittitas County, Washington. *International Journal of Geospatial and Environmental Research, 1*(2), 1–12.

Pulido-Velazquez, M., Andreu, J., Sahuquillo, A., & Pulido-Velazquez, D. (2008). Hydroeconomic river basin modelling: The application of a holistic surface–groundwater model to assess opportunity costs of water use in Spain. *Ecological Economics, 66*(1), 51–66.

Qi, H., & Altinakar, M. S. (2011). A GIS-based decision support system for integrated flood management under uncertainty with two dimensional numerical simulations. *Environmental Modelling & Software, 26*, 817–821.

Quevauviller, P., Balabanis, P., Fragakis, C., Weydert, M., Oliver, M., Kaschl, A., Arnold, G., Kroll, A., Galbiati, L., Zaldivar, J. M., & Bidoglio, G. (2005). Science-policy integration needs in support of the implementation of the EU Water Framework Directive. *Environmental Science & Policy, 8*, 203–211.

Refsgaard, J. C., van der Sluijs, J. P., Højberg, A. L., & Vanrolleghem, P. A. (2007). Uncertainty in the environmental modelling processes – A framework and guidance. *Environmental Modelling & Software, 22*, 1543–1556.

Rousseau, K., & Lawrence, P. L. (2013). The application of GIS in watershed planning: The case of the western Lake Erie Basin. In J. D. Gatrell & R. R. Jensen (Eds.), *Geospatial tools for urban water resources* (Geotechnologies and the environment, Vol. 7, pp. 145–159). Dordrecht: Springer.

Shamsi, U. M., Smith, P. (2005). ArcGIS and SWMM integration. *Journal of Water Management Modeling*, 296–307. doi: 10.14796/JWMM. R223–15

Torgersen, C. E., Faux, R. N., McIntosh, B. A., Poage, N. J., & Norton, D. J. (2001). Airborne thermal remote sensing for water temperature assessment in rivers and streams. *Remote Sensing of the Environment, 76*, 386–398.

Triana, E., & Labadie, J. W. (2012). GIS-based decision support system for improved operations and efficiency conservation in large-scale irrigation systems. *Journal of Irrigation and Drainage Engineering, 138*(10), 857–867.

US Army Corps of Engineers (USACE). (2011). HEC-GeoRAS – GIS tools for support of HEC-RAS using ArcGIS: User's manual.

Verma, S., Verma, R. K., Singh, A., & Naik, N. S. (2012). Web-based GIS and desktop open source GIS software: An emerging innovative approach for water resources management. In D. C. Wyld et al. (Eds.), *Advances in computer science, engineering & applications, AISC 167* (pp. 1061–1074). Berlin/Heidelberg: Springer.

Ward, F. A., & Pulido-Velázquez, M. (2008). Efficiency, equity, and sustainability in a water quantity–quality optimization model in the Rio Grande basin. *Ecological Economics, 66*(1), 23–37.

Watanabe, M., Adams, R. M., & Wu, J. (2006). Economics of environmental management in a spatially heterogeneous River Basin. *American Journal of Agricultural Economics, 88*(3), 617–631.

Wilson, J. P., Mitasova, H., & Wright, D. J. (2000). Water resource applications of geographic information systems. *URISA Journal, 12*(2), 61–79.

Wing, M. G. & Godwin, D. (2011). SWAMP GIS: A spatial decision support system for predicting and treating stormwater runoff. *Journal of Spatial Hydrology, 11*(2), 21–32.

Xiao, H. (2003). *An integrated GIS-AnnAGNPS modeling interface for non-point source pollution assessment*. Twenty-third annual ESRI International User Conference proceedings: GIS serving our world. ESRI, Redlands, California.

Yang, W. (2011). Developing open access in conservation research. *Journal of Soil and Water Conservation, 66*(1), 6A–8A.

Yang, W., Khanna, M., Farnsworth, R., & Onal, H. (2003). Integrating economic, environmental and GIS modeling to determine cost effective land retirement in multiple watersheds. *Ecological Economics, 46*, 249–267.

Yang, W., Sheng, C., & Voroney, P. (2005). Spatial targeting of conservation tillage to improve water quality and carbon retention benefits. *Canadian Journal of Agricultural Economics, 53*, 477–500.

Yang, W., Rousseau, A. N., & Boxall, P. (2007). An integrated economic-hydrologic modeling framework for the watershed evaluation of beneficial management practices. *Journal of Soil and Water Conservation, 62*(6), 423–432.

Yang, W., Wang, X., Liu, Y. B., Gabor, S., Boychuk, L., & Badiou, P. (2010). Simulated environmental effects of wetland restoration scenarios in a typical Canadian prairie watershed. *Wetlands Ecology and Management, 18*(3), 269–279.

Yang, W., Liu, W., Liu, Y. B., Corry, R. C., & Kreutzwiser, R. D. (2014). Cost-effective targeting of riparian buffers to achieve water quality and wildlife habitat benefits. *International Journal of River Basin Management, 12*(1), 43–55.

Zeng, Y., Cai, Y., Jia, P., & Jee, H. (2012). Development of a web-based decision support system for supporting integrated water resources management in Daegu city, South Korea. *Expert Systems with Applications, 39*, 10091–10102.

Zhang, J. X., Wu, J. Q., Chang, K., Elliot, W. J., & Dun, S. (2009). Effects of DEM source and resolution on WEPP hydrologic and erosion simulation: A case study of two forest watersheds in northern Idaho. *Transactions of the ASABE, 52*(2), 447–457.

Zhang, D., Chen, X., & Yao, H. (2015). Development of a prototype web-based decision support system for watershed management. *Water, 7*, 780–793.

Index

S. Renzetti, D.P. Dupont (eds.), *Water Policy and Governance in Canada*, Global Issues in Water Policy 17, DOI 10.1007/978-3-319-42806-2

U

W

Y

MIX
Papier aus verantwortungsvollen Quellen
Paper from responsible sources
FSC® C105338

If you have any concerns about our products,
you can contact us on
ProductSafety@springernature.com

In case Publisher is established outside the EU,
the EU authorized representative is:
Springer Nature Customer Service Center GmbH
Europaplatz 3, 69115 Heidelberg, Germany

Printed by Libri Plureos GmbH
in Hamburg, Germany